# THE BIOLOGICAL FOUNDATIONS OF MUSIC

ANNALS OF THE NEW YORK ACADEMY OF SCIENCES

*Volume 930*

# THE BIOLOGICAL FOUNDATIONS OF MUSIC

*Edited by Robert J. Zatorre and Isabelle Peretz*

*The New York Academy of Sciences*
*New York, New York*
*2001*

**Library of Congress Cataloging-in-Publication Data**

The biological foundations of music / edited by Robert J. Zatorre and Isabelle Peretz.
p. cm. — (Annals of the New York Academy of Sciences, ISSN 0077-8923 ; v. 930)
Includes bibliographical references and indexs.
ISBN 1-57331-306-8 (cloth : alk. paper) — ISBN 1-57331-307-6 (pbk. : alk. paper)
1. Music—Physiological aspects—Congresses. 2.Music—Psychological aspects —Congresses. 3. Musical perception—Congresses. I. Zatorre, Robert J. II. Peretz, Isabelle. III. Series.

Q11.N5 vol. 930
[ML3820]
500 s—dc21
[781'.11] 2001030421

GYAT/PCP

*Printed in the United States of America*
**ISBN 1-57331-306-8** (cloth)
**ISBN 1-57331-307-6** (paper)
**ISSN 0077-8923**

ANNALS OF THE NEW YORK ACADEMY OF SCIENCES

*Volume 930*
*June 2001*

# THE BIOLOGICAL FOUNDATIONS OF MUSIC

*Editors*
ROBERT J. ZATORRE AND ISABELLE PERETZ

This volume is the result of a conference entitled **The Biological Foundations of Music** held by the New York Academy of Sciences on May 20–22, 2000 at The Rockefeller University, New York, New York.

CONTENTS

## Part IV. The Musical Brain



## Part V. Musical Expertise/Brain Plasticity



## Part VI. Music and Related Cognitive Abilities



## Part VII. Poster Papers

**Financial assistance was received from:**

***Major Funders***

- **THE CHARLES A. DANA FOUNDATION**
- **THE DAVID AND LUCILE PACKARD FOUNDATION**

***Supporters***

- **FONDAZIONE PIERFRANCO E LUISA MARIANI, PEDIATRIC NEUROLOGY, MILAN ITALY**
- **INTERNATIONAL FOUNDATION FOR MUSIC RESEARCH**
- **NATIONAL ASSOCIATION OF MUSIC MERCHANTS, INC., INTERNATIONAL MUSIC PRODUCTS ASSOCIATION**
- **NATIONAL INSTITUTES OF HEALTH**
  **NATIONAL INSTITUTE OF CHILD HEALTH AND HUMAN DEVELOPMENT (NICHD)**
  **NATIONAL INSTITUTE OF NEUROLOGICAL DISORDERS AND STROKE (NINDS)**
- **NATIONAL SCIENCE FOUNDATION**

# Preface

ROBERT J. ZATORRE[a] AND ISABELLE PERETZ[b]

[a]*Montreal Neurological Institute, McGill University, Montreal, Quebec H3A 2B4, Canada*

[b]*Department of Psychology, University of Montreal, Montreal, Quebec, H3C 3J7, Canada*

Over the past decade, we have witnessed an explosion in research activities on music perception and performance and their correlates in the human brain. This sudden increase in scientific work on music has been motivated in part by the idea that music offers a unique opportunity to better understand the organization of the human brain. Like language, music exists in all human societies. Like language, music is a complex, rule-governed activity that seems specific to humans. Similarly, music appears to be associated with a specific brain architecture. Sensitivity to musical structure develops early in life, without conscious effort, in a large majority of the population, but only a minority of individuals become proficient musicians through explicit study. This particularity in the distribution of acquired skills confers on music a privileged role in the study of brain plasticity.

Musical performance and appreciation is generally considered an exquisite product of human culture and, as such, is often assumed to be a cultural artifact. Yet, the ubiquity of music, its developmental features, and its neurobiological substrates suggest that it may be appropriate to consider it akin to functions that result from the action of evolutionary forces—that is, as having biological foundations. In this context, it is remarkable to note that the study of music as a major brain function has been relatively neglected. Neuropsychological questions related to musical abilities have been of occasional interest to neurologists and psychologists since the nineteenth century (e.g., see Bouillaud[1] for pioneering work), but systematic, sustained investigations have been rare until recently.

The objective of this conference, The Biological Foundations of Music, from which the present collection of papers is derived, was to demonstrate the potential dynamism and richness of this emerging discipline of the neuroscience of music cognition. The conference was held in May 2000 in New York City and was initiated and sponsored by the New York Academy of Sciences. Major scientists from six countries, working in a variety of interrelated disciplines and pursuing sustained investigations in the field of music research, were invited to present their work. Most responded enthusiastically and addressed issues crucial to a better understanding of the neural substrates underlying musical functions. Their presentations were grouped under six major categories, selected because they are interdisciplinary and are the object of intensive current research activities and continued theoretical development.

The first part, The Origins of Music, sets the stage for viewing music as a biological entity (Sandra Trehub, Carolyn Drake, Ian Cross, and David Huron). The Musical Mind presents the functional architecture of music cognition (Stephen McAdams and Carol Krumhansl). The Neurons of Music describes the neural hardware of the brain that allows all musical processes to take place (Mark Tramo, Timothy Griffiths, and Catherine Liégeois-Chauvel). The section entitled The Musical Brain is the core of the symposium. It features a wide sample of work covering diverse musical functions, neurological abnormalities, and technologies: brain lesions (John Brust, Isabelle Peretz, and Séverine Samson); brain imaging with fMRI and PET (Robert Zatorre and Lawrence Parsons); electrophysiological recording (Mireille Besson and Mari Tervaniemi); and convergence across methods (Andrea Halpern). The Musical Expertise/Brain Plasticity section shows effects of early versus late skill acquisition on brain plasticity (Alvaro Pascual-Leone, Gottfried Schlaug, and Christo Pantev). The part entitled Music and Related Cognitive Abilities includes the influence of musical training on nonmusical cognitive skills (Glenn Schellenberg) and the relationship between music and language (Fred Lerdahl).

In addition to these plenary sessions, two roundtables were organized on the emerging topics of music and emotion, and of the role played by music in other cognitive functions. Finally, a call for poster papers turned out to be quite successful, giving the opportunity to 42 presenters, a selection of whose work is presented here.

The contributions to the present proceedings reveal how much progress has been made in the field over the last decade. Besides providing research overviews, many papers also delineate significant avenues for future research. We hope that this recent productivity, coupled with an increase of interest in the biological foundations of music in the scientific and general community, provides the spark for building a solid and stimulating neuroscience of music cognition.

We take this opportunity to express our gratitude to Rashid Shaikh, Director of Science and Technology Meetings at the New York Academy of Sciences, who, with tact and efficacy, helped organize the conference; to Harvey Weiss for suggesting the topic of the meeting to the Academy; and to John Brust for endorsing it. We also wish to thank the various foundations and agencies (see the list of financial contributors) that generously provided financial support for the meeting. We are grateful to the Editorial Department of the Academy, and particularly to Steven Bohall, Richard Stiefel, and the late Cook Kimball, who shepherded this volume through the press, and to Stephanie J. Bludau who oversaw production.

## REFERENCE

1. BOUILLAUD, J. 1865. Sur la faculté du langage articulé. Bulletin de l'Académie de Médecine **30:** 752–768.

# Musical Predispositions in Infancy

SANDRA E. TREHUB

*Department of Psychology, University of Toronto at Mississauga, Mississauga, Ontario, Canada L5L 1C6*

**ABSTRACT: Some scholars consider music to exemplify the classic criteria for a complex human adaptation, including universality, orderly development, and special-purpose cortical processes. The present account focuses on processing predispositions for music. The early appearance of receptive musical skills, well before they have obvious utility, is consistent with their proposed status as predispositions. Infants' processing of musical or music-like patterns is much like that of adults. In the early months of life, infants engage in relational processing of pitch and temporal patterns. They recognize a melody when its pitch level is shifted upward or downward, provided the relations between tones are preserved. They also recognize a tone sequence when the tempo is altered so long as the relative durations remain unchanged. Melodic contour seems to be the most salient feature of melodies for infant listeners. However, infants can detect interval changes when the component tones are related by small-integer frequency ratios. They also show enhanced processing for scales with unequal steps and for metric rhythms. Mothers sing regularly to infants, doing so in a distinctive manner marked by high pitch, slow tempo, and emotional expressiveness. The pitch and tempo of mothers' songs are unusually stable over extended periods. Infant listeners prefer the maternal singing style to the usual style of singing, and they are more attentive to maternal singing than to maternal speech. Maternal singing also has a moderating effect on infant arousal. The implications of these findings for the origins of music are discussed.**

**KEYWORDS: Musical predispositions; Adaptation; Melodic contour; Intervals; Scale structure; Rhythm; Maternal music**

## MUSICAL PREDISPOSITIONS IN INFANCY

The notion of musical predispositions implies a biological basis for music. For contributors to a forum on the biological foundations of music, consideration of predispositions is hardly surprising. It is worth remembering, however, that the larger scientific community remains skeptical about links between music and biology. Steven Pinker, who popularized the notion of a *language instinct*,[1] dismisses music as "cultural cheesecake"—a confection that tickles our sensitive spots. "As far as biological cause and effect are concerned, music is useless. It shows no signs of design for attaining a goal such as long life, grandchildren, or accurate perception and prediction of the world. Compared with language, vision, social reasoning, and physical

Address for correspondence: Sandra E. Trehub, Department of Psychology, University of Toronto at Mississauga, 3359 Mississauga Road North, Mississauga, Ontario, Canada L5L 1C6. Voice: 905-828-5415; fax: 905-569-4326.

sandra.trehub@utoronto.ca

know-how, music could vanish from our species and the rest of our lifestyle would be virtually unchanged." [2]

Amusing words, perhaps, but muddled thinking. For one thing, the *competence–performance* distinction, which is central to theories of language, is accorded no role in music. Pinker emphasizes the capacity of *all* normal children to speak and understand language without explicit instruction, in contrast to the musical incapacity of many adults, particularly those who cannot carry a tune. He notes as well that the majority of adults cannot play an instrument; for the small minority who can, considerable training and practice are required. These apparent deficits in explicit musical knowledge are treated as deficits in implicit knowledge. So how does music still manage to tickle the appropriate spots?

Pinker also contrasts the wide variations in musical sophistication with the presumed uniformity in language ability. Such pronouncements confound language and music in general—the underlying knowledge that makes it possible to produce and understand language or music (i.e., competence)—with language or musical expression (i.e., performance). Instead of comparing musical performance with everyday speaking and listening, why not compare judgments of "grammaticality" for simple musical and speech sequences? Finally, Pinker's vision of life without music betrays his ethnocentrism—music as material for concerts, dance halls, and movie soundtracks rather than something entwined in the fabric of life. It ignores the historical and cross-cultural importance of music in ritual ceremonies,[3–6] work,[7,8] and childcare.[9]

Geoffrey Miller,[10] by contrast, considers music to exemplify many of the classic criteria for a complex human adaptation: (1) no culture in any period of recorded history has been without music—in other words, universality; (2) an orderly developmental schedule;[11] (3) high ability, with almost every normal adult capable of carrying a tune and appreciating music (unlike Pinker's acquaintances); (4) specialized memory that makes it possible to recognize thousands of melodies; (5) involvement of special-purpose cortical areas;[12,13] (6) analogues in the signals of other species, such as songbirds, gibbons, and whales,[10] raising the possibility of convergent evolution; and (7) the propensity of music to evoke strong emotions, which implies receptive as well as productive adaptations.

If music is a complex biological adaptation rather than a by-product of other evolutionary processes, it must have conferred survival or reproductive benefits through the cumulative effect, over generations, of natural selection or sexual selection.[10,15,16] The costs of music production—energy expenditure—must be balanced by the benefits. Because no clear survival benefits are evident, Miller argues for the evolution of music on the basis of its reproductive benefits—music as sexual courtship.[10] He contends that, whereas music is typically performed *in* groups, it is not *for* groups. He notes, for example, that group tribal dancing (i.e., vigorous movement to music) would have provided a convenient means for young women to assess the strength, endurance, and motor coordination of potential mates. Ancestral life in relatively stable social groups would have allowed women to scrutinize the contenders over an extended period. Miller acknowledges, however, that perceptual and cognitive preferences would be equally effective for purposes of sexual selection.

The operation of sexual (or natural) selection would result in the replication of genes related to receptive and productive musical ability. Although motor limitations might constrain the expression of productive abilities in early life, they would not interfere with receptive abilities. In this regard, it is important to emphasize that musical abilities with a biological basis may not be evident at birth. Moreover, such abilities may be influenced by exposure. Learning can be guided by innate preferences, as is the case for many songbird species.[17] Thus, musical skills that are evident in infancy, well before they have obvious utility, can be considered predispositions.

## PROPOSED MUSICAL PREDISPOSITIONS

A common means of assessing the music perception skills of 6- to 10-month-old infants is to present them with a repeating melody or tonal pattern from a laterally displaced loudspeaker. In this "game," infants must turn to the loudspeaker when they hear a change in the repeating pattern. When they respond correctly (i.e., turning within 4 seconds of a change in sound), they are rewarded by a colorful mechanical toy (e.g., dancing bear, acrobatic monkey). Failure to turn to a change or turning at other times has no consequences. The visual rewards are sufficient to motivate attentive listening for 10–15 minutes, or 20–30 test trials. Aside from the bells and whistles (e.g., bears, monkeys) that facilitate communication with nonverbal participants, the task is similar to same–different procedures used with adults. Infants respond (by turning) when they consider the comparison pattern different from the standard pattern (i.e., response of *different*), and they do not respond if they fail to notice any change (i.e., response of *same*). Evidence of detecting the change is provided by significantly more responses (turns) to changed patterns than to unchanged patterns. Infants' performance also provides information about the features that they encoded from the original (repeating) melody. For example, if infants failed to notice the tempo of the original melody, they would be unable to detect changes in tempo.

### *Relational Processing of Pitch and Duration*

Infants' resolution of frequency,[18,19] timing,[20,21] and timbre[22,23] is finer than that required for musical purposes. Moreover, perceptual grouping principles that are relevant to music[24,25] are operative in infancy. For example, infants group isochronous tone sequences on the basis of similarities in pitch, loudness, and timbre.[26,27]

Infants go well beyond such minimal criteria for music perception. Like adults, they recognize a melody when the pitch level is shifted upward or downward.[28,29] The critical issue is whether the relations between tones are preserved. An analogous situation prevails with respect to timing. Infants recognize a tone sequence when the tempo is altered, so long as the relative durations remain unchanged.[30] In other words, they focus on relational rather than absolute attributes, making an otherwise unmanageable feat—remembering the exact pitches and durations of melodies—potentially manageable.

Infants retain absolute pitch information in some circumstances. When the standard and comparison melodies are presented in the same key, or with the same starting pitch, infants notice some pitch changes that might otherwise go unnoticed.[31,32] As the task gets more difficult, however, they shift their focus from absolute to relational cues. For example, infants can discriminate a semitone change in the context of an augmented triad (C-E-G♯-E-C to C-E-G-E-C) when the standard and comparison melodies are presented at the same pitch level.[32] They are unable to do so when the patterns are presented at different pitch levels.[33] In the latter situation, infants focus on the melodic contour, or overall pitch configuration, of the melody rather than on its component pitches.

Melodic contour is an especially durable feature of melodies. Remarkably, infants can identify changes in contour when the standard and comparison melodies are separated by a full 15 seconds[28] or when distractor tones are inserted between standard and comparison sequences.[34] In fact, pitch contour seems to be the most salient musical feature for infant listeners.[35,36] It may also be the most salient feature of mothers' speech to prelinguistic infants.[37–40] Evidence from adult listeners is consistent with contour processing as fundamental and relatively impervious to musical experience. During contour-processing tasks, for example, the amplitude and latency of event-related potentials do not differ for musicians and nonmusicians; during interval-processing tasks, however, experience-dependent differences are apparent.[41]

### *Interval Processing*

For adults, the extraction of pitch information is not limited to melodic contour, except in the context of novel melodies.[42,43] When the melodies are reasonably simple and they conform to familiar musical conventions, adults typically remember more detailed pitch information than would otherwise be the case.[44–46] For highly familiar melodies, of course, adults retain the precise pitch distances between tones (i.e., intervals), but not the absolute pitches.[47,48] When singing their favorite pop song, however, adults deviate by about 2 semitones in pitch[49] and 8% in tempo from the canonical recording.[50]

In some circumstances, infants also extract rich pitch information. Interestingly, those circumstances are much like the ones that promote enhanced pitch processing for adults. For example, infants detect interval changes in brief melodies that conform to cultural conventions but fail to do so for melodies that violate those conventions.[33,51,52,53] I am not arguing that six-month-old infants have acquired conventions relating to musical structure. On the contrary, there is evidence that infants lack such implicit knowledge.[46,54,55] The apparent influence of musical conventions is the consequence of universal or near-universal features.[35,56]

In studies of infant melodic perception, the presence of a prominent seven-semitone interval, or *perfect fifth*, has been associated with success in interval discrimination, and its absence has been associated with failure.[33,31,51–53] The perfect fifth is not alone, however, in enhancing infants' perception of pitch relations[57] (see FIG. 1). Rather, infants, children, and adults more readily detect interval changes in the context of small-integer ratios—the octave (2:1), perfect fifth (3:2), and *perfect fourth* (4:3)—compared to large-integer ratios, such as the *tritone* (45:32 ratio).[57–59] In other words, it is likely that inherent differences in ease of processing resulted in consonant intervals assuming a prominent role in most musical systems.[35,60,61] The historical record is consistent with the inherent inequality of intervals, providing numerous references to the beauty of octaves, perfect fifths, and perfect fourths[62,63] and the ugliness of the tritone—*the devil in music* in medieval times.[64]

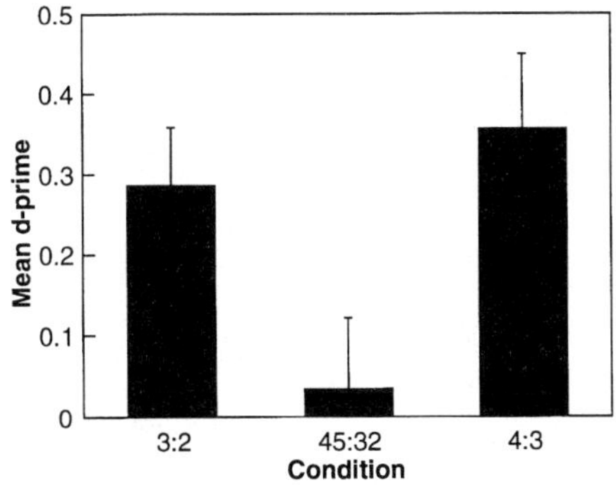


**FIGURE 1.** Infant discrimination scores as a function of the frequency ratios of tones in melodic (sequential) intervals. (Reproduced with permission from Schellenberg and Trehub.[57])

Consonant and dissonant intervals not only promote ease of processing; they also influence infant attention and affect. For example, infants are more attentive and exhibit more positive affect while listening to consonant music than to music with many dissonant intervals.[65–67] In principle, these preferences could arise from the predominance of consonant intervals in the infant's environment (e.g., speech, music, environmental sounds). Comparable preferences are demonstrable, however, in newborns whose deaf parents communicate by means of sign language.[68]

### *Processing of Scale Structure*

Despite the diversity of scales across cultures, a number of similarities are evident. One of these is the typical division of the octave into five to seven discrete pitches, a design feature that probably arises from cognitive constraints.[69] Another is the ubiquitous perfect fifth, which is associated with enhanced processing in listeners of all ages.[57,59,70] Yet another similarity is the presence of unequal scale steps[71]—for example, one- and two-semitone steps in diatonic scales—a feature that is thought to confer processing advantages.[72,73]

In a test of this hypothesis,[55] nine-month-old infants and adults listened to one of three ascending–descending scales: (1) the major scale, which is highly familiar to adults and potentially familiar to infants; (2) an artificial analogue of the major scale that was created by dividing the octave into 11 equal units, and selecting a seven-tone subset (featuring 1- and 2-unit steps); and (3) an artificial equal-step scale that was created by dividing the octave into 7 equal steps[74,75] (see FIG. 2). Infants and adults were required to detect a mistuned tone (0.75- and 0.5-semitone change for infants and adults, respectively). Although adults' performance was highly accurate on the familiar major scale, they performed equally poorly on the two unfamiliar scales. By contrast, infants showed a general facilitation for scales with unequal steps, performing equivalently on the major and artificial unequal-step scales, and significantly more poorly on the equal-step scale (see FIG. 3). These findings provide three important pieces of information. First, unequal-step scales, which are universal or near-universal, facilitate pitch processing in infancy. Second, infants' performance on the major scale cannot be attributed to familiarity. Thus, other instances of

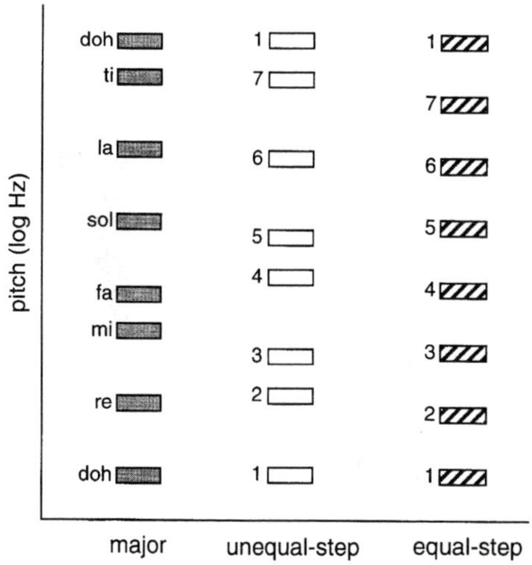


**FIGURE 2.** Schematic depiction of the major, artificial unequal-step, and equal-step scales. (Reproduced with permission from Trehub, Schellenberg, and Kamenetsky.[55])

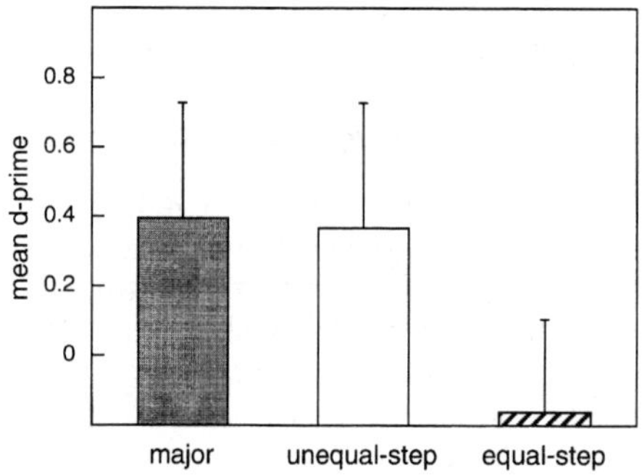


**FIGURE 3.** Infant discrimination scores on the ascending–descending major scale, artificial, unequal-step scale, and equal-step scale. (Reproduced with permission from Trehub, Schellenberg, and Kamenetsky.[55])

enhanced processing—small-integer ratios, for example—are likely to be unrelated to exposure. Third, culture-specific experience, such as adults' long-term exposure to the major scale, can override initial processing biases.

There are other examples of culture-specific exposure attenuating or eliminating initial processing biases. For example, pitch processing in infancy is enhanced when the standard and comparison patterns are presented in related keys—specifically, keys standing in a 3:2 ratio—rather than unrelated keys; for adults, such enhancement is limited to conventionally structured melodies.[51] Similarly, the conventionality of a melody is unrelated to infants' ability to profit from melodic redundancy, but it is a major determinant of adults' performance in such circumstances.[46]

### *Rhythm Processing*

Infants can differentiate tone sequences with identical pitches but contrasting rhythmic arrangements.[30,76,77] As noted, they also impose rhythmic groupings on isochronous tone sequences,[26,27] raising the possibility that they might be adult-like in other respects. Their detection of pitch and duration changes is best for melodies whose tone durations are related by small-integer ratios, specifically 2:1.[78] Moreover, there are indications that adults' preferred rhythmic arrangements of melodies generate the best performance on the part of infants.

When adults listen to music, they are presumed to generate an internal clock that reflects the organization of accented notes.[79] This dynamic mode of attending leads to more accurate reproduction of metric sequences compared to nonmetric sequences.[80] When nine-month-old infants listen to metric and nonmetric rhythmic patterns, their ability to detect rhythmic changes in one note is evident only in the context of metric patterns.[81] Thus, the available evidence on rhythm perception in infancy is consistent with processing predispositions for metric rhythmic patterns and for tone durations related by small-integer ratios.

### *Hemispheric Specialization*

Well before infants have any understanding of language, they show a right-ear (i.e., left hemisphere) advantage for speech and a left-ear (i.e., right hemisphere) advantage for music.[82,83] Moreover, eight-month-old infants exhibit a left-ear superiority for contour processing and a right-ear superiority for intervalic processing.[84] Such laterality effects parallel those reported for adults.[85,86]

## MATERNAL MUSIC

The relevance of these perceptual processing predispositions to the daily lives of infants is unclear. For one thing, the musical patterns in most laboratory studies (e.g., sequences of pure tones) differ dramatically from infants' usual musical input. Every day, indeed many times each day, infants hear distinctive performances of maternal songs, a situation that is common to all cultures and historical periods.[87–89]

Until very recently, research on maternal vocal behavior focused exclusively on speech to infants—*motherese*, or infant-directed speech—and its effects on infant listeners.[90,91] When mothers sing to their infants, they typically use a special-purpose repertoire consisting of lullabies and play songs.[9,88] To a naïve observer, unfamiliar lullabies, including those from foreign musical cultures, sound much like familiar lullabies. For example, when adults listen to pairs of unfamiliar lullabies and nonlullabies that are matched on culture of origin and tempo, they readily identify the lullabies.[92] Structural simplicity or repetitiveness seems to be the principal cue that guides identification.[93]

Mothers' style of singing to infants differs from their usual style by virtue of a higher pitch level, slower tempo, and more emotive voice quality.[94–96] Mothers also vary their performance in subtle ways for infant or preschool listeners.[97] Specifically, their pitch level is approximately one-semitone higher and their articulation of words more slurred for infant than for preschool audiences. Features of this caregiv-

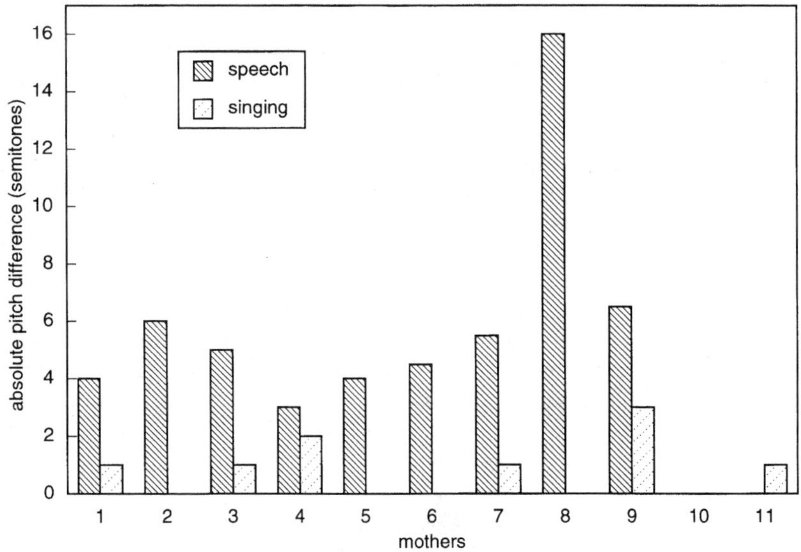


**FIGURE 4.** Pitch differences, in semitones, of repetitions of songs and verbal phrases by 11 mothers. Note that mothers' repetitions were separated by one week or more. (From Bergeson and Trehub. In press.[100])

ing style are evident in fathers' singing to infants[96,98] and in preschoolers' singing to their infant siblings.[99]

Although researchers have commented on parallels between speech and song for infants,[9,36,88] they have been slow to delineate differences. In general, women's singing tends to be higher-pitched than their speech, but speech to infants is higher pitched than singing to infants.[100] This situation results from the large increase in pitch level (three to four semitones) for infant-directed speech[101] coupled with the small increase (one semitone) for infant-directed singing.[94,96]

There is considerable interindividual variation in maternal singing, but surprising intraindividual stability. When mothers sing the same song to their infant on two occasions separated by a week or more, their pitch level differs by less than a semitone (see FIG. 4), and their tempo (in beats per minute) differs by only 3% (see FIG. 5).[100,102] Such pitch and tempo variations are smaller than those reported for adults' repeated renditions of pop or folk songs.[49,50,103] Mothers' repetition of identical verbal phrases across the same period shows considerably greater variability in pitch and tempo (see FIGS. 4 and 5). If the act of singing affects mothers' mood or state, that might account, in part, for the stability of pitch and tempo over extended periods—a possible consequence of state-dependent memory. Frequent singing to infants may also implicate motor memory. In any event, mothers' performances of songs seem to become ritualized, which may facilitate their use as communicative signals to prelinguistic infants.

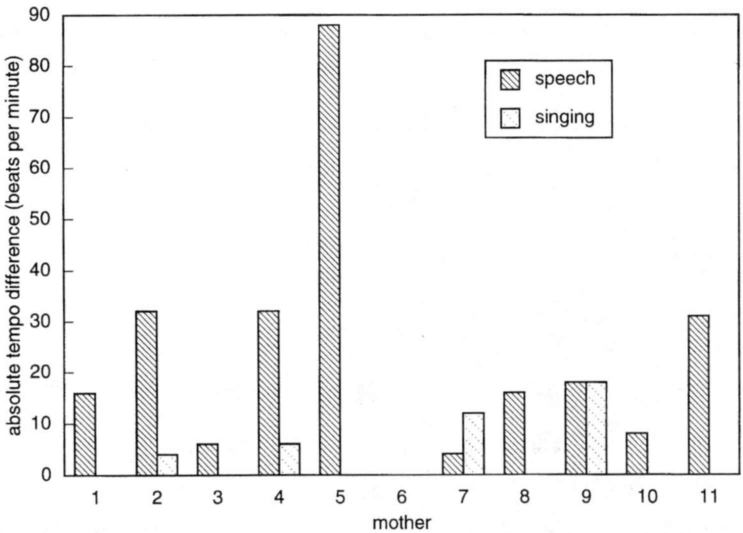


**FIGURE 5.** Tempo differences, in beats per minute, of repetitions of songs and verbal phrases by 11 mothers. Note that mothers' repetitions were separated by one week or more. (Reproduced with permission from Bergeson and Trehub. In press.[100])

### *Responsiveness to Maternal Music*

How do infants respond to these maternal singing rituals? In the newborn period and beyond, infants listen significantly longer to women's singing (audio recordings) in the maternal style than to these singers' usual informal style.[104,105] They also exhibit greater attention to higher- over lower-pitched versions of the same song.[106] These preferences indicate the potency of singing, especially in view of the unfamiliar singers and the absence of expressive gestures and movements that typically accompany maternal singing.

When six-month-old infants view videotaped performances of their own mothers (recorded while singing to their infant), which provide access to visual as well as acoustic features, they show more sustained attention to mothers' singing episodes than to their speaking episodes (see FIG. 6).[107,108] Infants tend to be mesmerized by these sung performances, remaining glued to the monitor for extended periods. Mothers' speaking is also engaging, but not nearly as engaging as her singing.

Although there has been much conjecture about maternal speech and singing as modulators of infant arousal,[9,90,109] the measurement focus has been on attention rather than arousal. In a recent investigation, six-month-old infants furnished saliva samples before and after mothers sang to them for 10 minutes.[110] Changes in salivary cortisol from pre- to posttest levels were highly but inversely correlated with pretest levels (see FIG. 7). Specifically, infants with initially lower levels of salivary cortisol showed cortisol increases, but those with initially higher levels showed decreases. The arousal-modulating effect of maternal singing was also reflected in reduced variance from pre- to posttest levels. These findings are impressive, given the narrow range of initial cortisol levels and the absence of stressful circumstances.

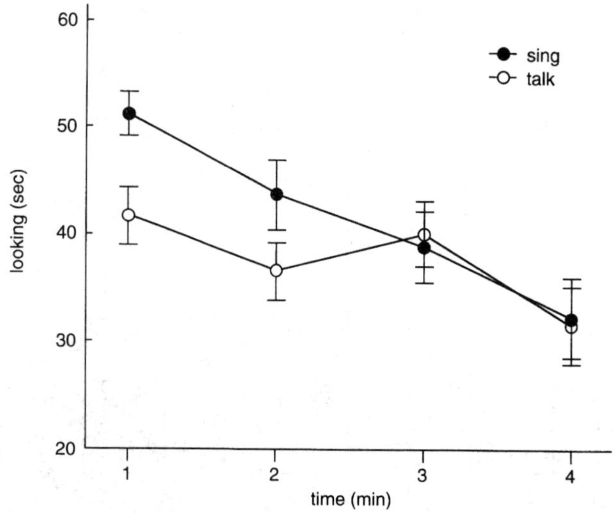


FIGURE 6. Infant looking time, minute by minute, during episodes of mothers' speech and singing. (Reproduced with permission from Bergeson and Trehub.[108])

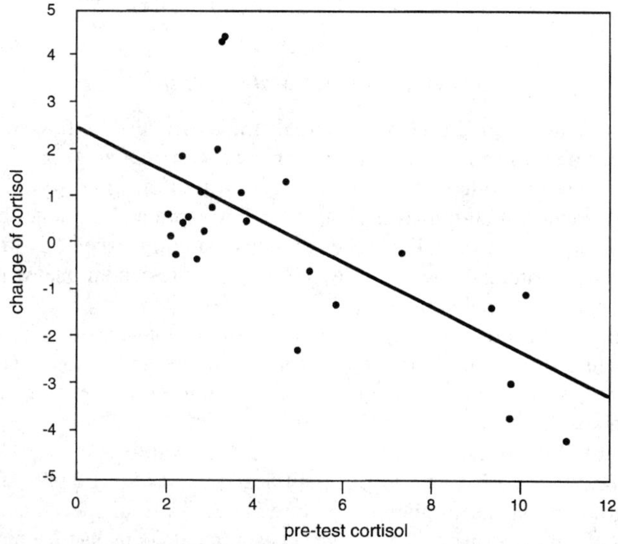


FIGURE 7. Changes in infant salivary cortisol following maternal singing. Data are plotted as a function of pretest levels. (Reproduced with permission from Shenfield and Trehub.[110])

### *Biological Significance of Singing to Infants*

Mothers' propensity to sing to infants and the impact of their singing on infant attention and arousal raise the possibility that maternal singing could have enhanced infant survival in difficult ancestral conditions. Children's extended period of helplessness would have created intense selection pressures for parental commitment,[111] including pressures for infant displays to sustain such commitment. It is likely that singing to infants promotes reciprocal emotional ties,[98] just as singing in other circumstances reduces the psychological distance between singer and listener.[112,113] Presumably, mother's growing attachment to infants would lead them to generate increasingly expressive performances. To balance the enormous physical and psychological costs of parenting, infant recipients of such largesse would have to advertise their worth. Falling asleep to lullabies or entering trance-like states to performances of other songs might qualify in this respect. In general, favorable consequences of maternal singing on infant arousal, whether through cry reduction, sleep induction, or positive affect, would contribute to infant well-being while promoting the continuation of such maternal behavior. The healthy and contented offspring of singing mothers would be more likely to pass on their genes than would the offspring of nonsinging mothers.

## MUSICAL COURTSHIP

An alternative, or supplementary, means of gene transmission could operate on mate selection through female musical courtship. According to this scenario, young adult males succumb to the expressive singing of young adult females, a situation for which there are nonhuman parallels. In some avian species, for example, females sing to attract mates and lead duets with their male partner.[114,115] There is speculation, moreover, that such duetting fosters long-term pair bonding.[116]

What about Miller's proposal of music as male sexual courtship? [10] In an unusual test of this hypothesis, Miller compared the recorded output of prominent jazz, rock, and classical musicians. The data confirmed Miller's predictions. Males produced ten times as much music as females, their output reaching its zenith near the age of peak mating effort. From this unconventional data set, Miller concluded that music continues to function as a male courtship display to attract females. I could offer more parsimonious interpretations, but they are beyond the scope of the present paper. In the spirit of Miller's real-world analysis, however, I propose an alternative test—assessing the musical skills of ordinary males and females (i.e., nonmusicians), perhaps those without musical training. My prediction is that more females than males will be capable of tuneful, expressive singing.

## CODA

It is clear that infants do not begin life with a blank musical slate. Instead, they are predisposed to attend to the melodic contour and rhythmic patterning of sound sequences, whether music or speech. They are tuned to consonant patterns, melodic as well as harmonic, and to metric rhythms. Mothers cater to infants' musical inclinations by singing regularly in the course of caregiving. They also adapt their usual singing style in ways that are congenial to infant listeners. Such maternal singing has the features of a unique vocal signature.

Perhaps it is not surprising that infants are predisposed to attend to and appreciate the species-typical vocalizations of their primary caregiver. Infants' predisposition to attend to particular aspects of musical structure is surprising only if music is viewed in a narrow sense, as fully developed musical systems of particular cultures. Viewed broadly, however, music embraces what all musical systems have in common. In that sense, infants begin life as musical beings, being responsive to the musical primitives or universals that are the foundation of all styles of music.

## ACKNOWLEDGMENTS

Preparation of this paper was assisted by grants from the Natural Sciences and Engineering Research Council and the Social Sciences and Humanities Research Council of Canada.

## REFERENCES

1. PINKER, S. 1994. The Language Instinct: How the Mind Creates Language. William Morrow. New York.
2. PINKER, S. 1997. How the Mind Works: 528. Norton. New York.
3. BASSO, E. 1985. A Musical View of the Universe: Kalapalo Rhythmic and Ritual Performance. University of Pennsylvania Press. Philadelphia.
4. DISSANAYAKE, E. 1992. Homo Aestheticus: Where Art Comes From and Why. Free Press. New York.
5. MERRIAM, A.P. 1964. The Anthropology of Music. Northwestern University Press. Evanston, IL.
6. SARNO, L. 1993. Song from the Forest: My Life among the Ba-Benjelle Pygmies. Houghton Mifflin. Boston.
7. DENG, F.M. 1973. The Dinka and Their Songs. Oxford University Press. London.
8. FIRTH, R. 1961. Elements of Social Organization. Watts. London.
9. TREHUB, S.E. & L.J. TRAINOR. 1998. Singing to infants: lullabies and play songs. *In* Advances in Infancy Research. C. Rovee-Collier, L. Lipsitt & H. Hayne, Eds.: 43–77. Ablex. Stamford, CT.
10. MILLER, G. 2000. Evolution of human music through sexual selection. *In* The Origins of Music. N.L. Wallin, B. Merker & S. Brown, Eds.: 329–360. MIT Press. Cambridge, MA.
11. DOWLING, W.J. 1999. The development of music perception and cognition. *In* The Psychology of Music. D. Deutsch, Ed.: 603–625. Academic Press. San Diego, CA.
12. PATEL, A.D. & I. PERETZ. 1997. Is music autonomous from language? A neuropsychological appraisal. *In* Perception and Cognition of Music. I. Deliege & J. Sloboda, Eds.: 191–215. Psychology Press. Hove, UK.
13. ZATORRE, R.J. 1994. Neural mechanisms underlying melodic perception and memory for pitch. J. Neurosci. **14:** 1908–1919.
14. MARLER, P. 2000. Origins of music and speech: insights from animals. *In* The Origins of Music. N.L. Wallin, B. Merker & S. Brown, Eds.: 31–48. MIT Press. Cambridge, MA.
15. DAWKINS, R. 1996. Climbing Mount Improbable. Penguin Books. London.
16. WILLIAMS, G.C. 1966. The Selfish Gene. Adaptation and Natural Selection. Princeton University Press. Princeton, NJ.
17. MARLER, P. & S. PETERS. 1977. Selective vocal learning in a sparrow. Science **198:** 519–521.
18. OLSHO, L.W., C. SCHOON, R. SAKAI, *et al.* 1982. Preliminary data on frequency discrimination in infancy. J. Acoust. Soc. Am. **71:** 509–511.
19. WERNER, L.A. 1992. Interpreting developmental psychoacoustics. *In* Developmental Psychoacoustics. L.A. Werner & E.W. Rubel, Eds.: 47–88. American Psychological Association. Washington, DC.

20. TREHUB, S.E., B.A. SCHNEIDER & J.L. HENDERSON. 1995. Gap detection in infants, children, and adults. J. Acoust. Soc. Am. **98:** 2532–2541.
21. Werner, L.A., G.C. Marean, C. Halpin, *et al.* 1992. Infant auditory temporal acuity: gap detection. Child Dev. **63:** 260–272.
22. CLARKSON, M.G. & R.K. CLIFTON. 1995. Infants' pitch perception: inharmonic tonal complexes. J. Acoust. Soc. Am. **98:** 1372–1379.
23. TREHUB, S.E., M. ENDMAN & L.A. THORPE. 1990. Infants' perception of timbre: classification of complex tones by spectral structure. J. Exp. Child Psychol. **49:** 300–313.
24. BREGMAN, A.S. 1990. Auditory Scene Analysis. MIT Press. Cambridge, MA.
25. DEUTSCH, D. 1999. Grouping mechanisms in music. *In* The Psychology of Music. D. Deutsch, Ed.: 299–348. Academic Press. San Diego, CA.
26. THORPE, L.A. & S.E. TREHUB. 1989. Duration illusion and auditory grouping in infancy. Dev. Psychol. **25:** 122–127.
27. THORPE, L.A., S.E. TREHUB, B.A. MORRONGIELLO & D. BULL. 1988. Perceptual grouping by infants and preschool children. Dev. Psychol. **24:** 484–491.
28. CHANG, H.W. & S.E. TREHUB. 1977. Auditory processing of relational information by young infants. J. Exp. Child Psychol. **24:** 324–331.
29. TREHUB, S.E., L.A. THORPE & B.A. MORRONGIELLO. 1987. Organizational processes in infants' perception of auditory patterns. Child Dev. **58:** 741–749.
30. TREHUB, S.E. & L.A. THORPE. 1989. Infants' perception of rhythm. Categorization of auditory sequences by temporal structure. Can. J. Psychol. **43:** 217–229.
31. LYNCH, M.P., R.E. EILERS, D.K. OLLER & R.C. URBANO. 1990. Innateness, experience, and music perception. Psychol. Sci. **1:** 272–276.
32. TREHUB, S.E., A.J. COHEN, L.A. THORPE & B.A. MORRONGIELLO. 1986. Development of the perception of musical relations: semitone and diatonic structure. J. Exp. Psychol. Hum. Percept. Perform. **12:** 295–301.
33. COHEN, A.J., L.A.THORPE & S.E. TREHUB. 1987. Infants' perception of musical relations in short transposed tone sequences. Can. J. Psychol. **41:** 33–47.
34. TREHUB, S.E., D. BULL & L.A. THORPE. 1984. Infants' perception of melodies: the role of melodic contour. Child Dev. **55:** 821–830.
35. TREHUB, S., E. SCHELLENBERG & D. HILL. 1997. The origins of music perception and cognition: a developmental perspective. *In* Perception and Cognition of Music. I. Deliege & J. Sloboda, Eds.: 103–128. Psychology Press. East Sussex, UK.
36. TREHUB, S.E., L.J. TRAINOR & A.M. UNYK. 1993. Music and speech processing in the first year of life. *In* Advances in Child Development and Behavior. Vol. **24:** 1–35. Academic Press. New York.
37. FERNALD, A. 1991. Prosody in speech to children: prelinguistic and linguistic functions. Ann. Child Dev. **8:** 43–80.
38. FERNALD, A. & P.K. KUHL. 1987. Acoustic determinants of infant preference for motherese. Infant Behav. & Dev. **10:** 279–293.
39. PAPOUŠEK, M. 1992. Early ontogeny of vocal communication in parent-infant interactions. *In* Nonverbal Vocal Communication: Comparative and Developmental Approaches. H. Papoušek, V. Jürgens & M. Papoušek, Eds.: 230–261. Cambridge University Press. Cambridge.
40. LEWIS, M.M. 1951. Infant Speech. Routledge & Kegan Paul. London.
41. TRAINOR, L.J., R.N. DESJARDINS & C. ROCKEL. 1999. A comparison of contour and interval processing in musicians and nonmusicians using event-related potentials. Aust. J. Psychol. **51:** 147–153.
42. BARTLETT, J.C. & W.J. DOWLING. 1980. Recognition of transposed melodies: a key-distance effect in developmental perspective. J. Exp. Psychol. Hum. Percept. Perform. **6:** 501–515.
43. DOWLING, W.J. 1978. Scale and contour: two components of a theory of memory for melodies. Psychol. Rev. **85:** 341–354.
44. CUDDY, L.L., A.J. COHEN & D.J.K. MEWHORT. 1981. Perception of structure in short melodic sequences. J. Exp. Psychol. Hum. Percept. Perform. **7:** 869–883.
45. KRUMHANSL, C.L., J.J. BHARUCHA & E.J. KESSLER. 1982. Perceived harmonic structure of chords in three related keys. J. Exp. Psychol. Hum. Percept. Perform. **8:** 24-36.

46. SCHELLENBERG, E.G. & S.E. TREHUB. 1999. Culture-general and culture-specific factors in the discrimination of melodies. J. Exp. Child Psychol. **74:** 107–127.
47. ATTNEAVE, F. & R.K. OLSON. 1971. Pitch as a medium: a new approach to psychophysical scaling. Am. J. Psychol. **84:** 147–166.
48. DOWLING, W.J. & D. FUJITANI. 1971. Contour, interval, and pitch recognition in memory for melodies. J. Acoust. Soc. Am. **49:** 524–531.
49. LEVITIN, D.J. 1994. Absolute memory for musical pitch: evidence from the production of learned melodies. Percept. Psychophys. **56:** 414–423.
50. LEVITIN, D.J. & P.R. COOK. 1996. Memory for musical tempo: additional evidence that auditory memory is absolute. Percept. Psychophys. **58:** 927–935.
51. TRAINOR, L.J. & S.E. TREHUB. 1993. Musical context effects in infants and adults: key distance. J. Exp. Psychol. Hum. Percept. Perform. **19:** 615–626.
52. TRAINOR, L.J. & S.E. TREHUB. 1993. What mediates infants' and adults' superior processing of the major over the augmented triad? Music Percept. **11:** 185–196.
53. TREHUB, S.E., L.A. THORPE & L.J. TRAINOR. 1990. Infants' perception of *good* and *bad* melodies. Psychomusicology **9:** 5–15.
54. TRAINOR, L.J. & S.E. TREHUB. 1992. A comparison of infants' and adults' sensitivity to Western musical structure. J. Exp. Psychol. Hum. Percept. Perform. **18:** 394–402.
55. TREHUB, S.E., E.G. SCHELLENBERG & S.B. KAMENETSKY. 1999. Infants' and adults' perception of scale structure. J. Exp. Psychol. Hum. Percept. Perform. **25:** 965–975.
56. TREHUB, S.E. & L.J. TRAINOR. 1993. Listening strategies in infancy: the roots of music and language development. *In* Thinking in Sound: The Cognitive Psychology of Human Audition. S. McAdams & E. Bigand, Eds.: 278–327. Oxford University Press. London.
57. SCHELLENBERG, E.G. & S.E. TREHUB. 1996. Natural musical intervals: evidence from infant listeners. Psychol. Sci. **7:** 272–277.
58. SCHELLENBERG, E.G. & S.E. TREHUB. 1994. Frequency ratios and the discrimination of pure tone sequences. Percept. Psychophys. **56:** 472–478.
59. SCHELLENBERG, E.G. & S.E. TREHUB. 1996. Children's discrimination of melodic intervals. Dev. Psychol. **32:** 1039–1050.
60. MEYER, L.B. 1956. Emotion and Meaning in Music. University of Chicago Press. Chicago.
61. SACHS, C. 1943. The Rise of Music in the Ancient World: East and West. Norton. New York.
62. BOWER, C. 1980. Boethius, Anicius, Manlius Severinus. *In* The New Grove Dictionary of Music and Musicians. S. Sadie, Ed.Vol. 2: 844. Macmillan. London.
63. HILL, C.C. 1986. Consonance and dissonance. *In* The New Harvard Dictionary of Music. D.M. Randel, Ed.: 197–199. Belknap Press. Cambridge, MA.
64. KENNEDY, M. 1994. Song. The Oxford Dictionary of Music, 2nd ed. Oxford University Press. Oxford.
65. TRAINOR, L.J. & B.M. HEINMILLER. 1998. Infants prefer to listen to consonance over dissonance. Infant Behav. & Dev. **21:** 77–88.
66. ZENTNER, M.R. & J. KAGAN. 1996. Perception of music by infants. Nature **383:** 29.
67. ZENTNER, M.R. & J. KAGAN. 1998. Infants' perception of consonance and dissonance in music. Infant Behav. & Dev. **21:** 483–492.
68. MASATAKA, N. 1999. Personal communication, June.
69. BURNS, E.M. 1999. Intervals, scales, and tuning. *In* The Psychology of Music. D. Deutch, Ed.: 215–264. Academic Press. San Diego, CA.
70. SCHELLENBERG, E.G. & S.E. TREHUB. 1994. Frequency ratios and the perception of tone patterns. Psychon. Bull. Rev. **1:** 191–201.
71. SLOBODA, J.A. 1985. The Musical Mind: The Cognitive Psychology of Music. Clarendon Press. Oxford.
72. BUTLER, D. 1989. Describing the perception of tonality in music: a critique of the tonal hierarchy theory and proposal for a theory of intervallic rivalry. Music Percept. **6:** 219–242.
73. SHEPARD, R.N. 1982. Geometrical approximations to the structure of musical pitch. Psychol. Rev. **89:** 305–333.

74. JORDAN, D.S. & R.N. SHEPARD. 1987. Tonal schemas: evidence obtained by probing distorted scales. Percept. Psychophys. **41:** 489–504.
75. SHEPARD, R.N. & D.S. JORDAN. 1984. Auditory illusions demonstrating that tones are assimilated to an internalized scale. Science **226:** 1333–1334.
76. CHANG, H.W. & S.E. TREHUB. 1977. Infants' perception of temporal grouping in auditory patterns. Child Dev. **48:** 1666–1670.
77. DEMANY, L., B. MCKENZIE & E. VURPILLOT. 1977. Rhythm perception in early infancy. Nature **266:** 718–719.
78. TREHUB, S.E., D.S. HILL & S.B. KAMENETSKY. 1997. Infants' perception of melodies with "good" or "bad" rhythms. Presented at the biennial meeting of the Society for Research in Child Development. Washington, DC. April.
79. POVEL, D.-J. 1981. Internal representation of simple temporal rhythms. J. Exp. Psychol. Hum. Percept. Perform. **7:** 3–18.
80. POVEL, D.-J. & P. ESSENS. 1985. Perception of temporal patterns. Music Percept. **2:** 411–440.
81. BERGESON, T.R. 2000. Infants' perception of rhythmic patterns. Presented at the Society for Music Perception and Cognition. Toronto, ONT. November.
82. BEST, C.T., H. HOFFMAN & B.B. GLANVILLE. 1982. Development of infant ear asymmetries in speech and music. Percept. Psychophys. **31:** 75–85.
83. BERTONCINI, J., J. MORAIS, R. BIJELJAC-BABIC, *et al.* 1989. Dichotic perception and laterality in neonates. Brain Lang. **37:** 591–605.
84. BALABAN, M.T., L.M. ANDERSON & A.B. WISNIEWSKI. 1998. Lateral asymmetries in infant melody perception. Dev. Psychol. **34:** 39–48.
85. PERETZ, J. 1987. Shifting ear-asymmetry in melody comparison through transposition. Cortex **23:** 317–323.
86. PERETZ, J. & M. BABAÏ. 1992. The role of contour and intervals in the recognition of melody parts: evidence from cerebral asymmetries in musicians. Neuropsychology **30:** 277–292.
87. BRAKELEY, T.C. 1950. Lullaby. *In* Standard Dictionary of Folklore, Mythology, and Legend. M. Leach & J. Fried, Eds.: 653–654. Funk & Wagnalls. New York.
88. TREHUB, S.E. & E.G. SCHELLENBERG. 1995. Music: its relevance to infants. Ann. Child Dev. **11:** 1–24.
89. TUCKER, N. 1984. Lullabies. Hist. Today **34:** 40–46.
90. FERNALD, A. 1991. Prosody in speech to children: prelinguistic and linguistic functions. Ann. Child Dev. **8:** 43–80.
91. FERNALD, A. 1992. Meaningful melodies in mothers' speech to infants. *In* Nonverbal Vocal Communication: Comparative and Developmental Aspects. H. Papoušek, V. Jurgens & M. Papoušek, Eds.: 262–282. Cambridge University Press. Cambridge, UK.
92. TREHUB, S.E., A.M. UNYK & L.J. TRAINOR. 1993. Adults identify infant-directed music across cultures. Infant Behav. & Dev. **16:** 193–211.
93. UNYK, A.M., S.E. TREHUB, L.J. TRAINOR & E.G. SCHELLENBERG. 1992. Lullabies and simplicity: a cross-cultural perspective. Psychol. Mus. **20:** 15–28.
94. TRAINOR, L.J., E.D. CLARK, A. HUNTLEY & B. ADAMS. 1997. The acoustic basis of infant preferences for infant-directed singing. Infant Behav. & Dev. **20:** 383–396.
95. TREHUB, S.E., A.M. UNYK & L.J. TRAINOR. 1993. Maternal singing in cross-cultural perspective. Infant Behav. & Dev. **16:** 285–295.
96. TREHUB, S.E., A.M. UNYK, S.B. KAMENETSKY, *et al.* 1997. Mothers' and fathers' singing to infants. Dev. Psychol. **33:** 500–507.
97. BERGESON, T.R. & S.E. TREHUB. 1999. Mother's singing to infants and preschool children. Infant Behav. & Dev. **22:** 51–64.
98. TREHUB, S.E., D.S. HILL & S.B. KAMENETSKY. 1997. Parents' sung performances for infants. Can. J. Exp. Psychol. **51:** 385–396.
99. TREHUB, S.E., A.M. UNYK & J.L. HENDERSON. 1994. Children's songs to infant siblings: parallels with speech. J. Child Lang. **21:** 735–744.
100. BERGESON, T.R. & S.E. TREHUB. Speech and song to infants: Do mothers have a unique vocal signature? Psychol. Sci. In press.

101. FERNALD A. & T. SIMON. 1984. Expanded intonation contours in mothers' speech to newborns. Dev. Psychol. **20:** 104–113.
102. TREHUB, S.E & T.R. BERGESON. 1999. Mothers' speech and song for infants [abstract]. J. Acoust. Soc. Am. **106:** 2235.
103. HALPERN, A.R. 1989. Memory for the absolute pitch of familiar songs. Mem. Cognit. **17:** 572–581.
104. MASATAKA, N. 1999. Preference for infant-directed singing in 2-day-old hearing infants of deaf parents. Dev. Psychol. **35:** 1001–1005.
105. TRAINOR, L.J. 1996. Infant preferences for infant-directed versus non-infant-directed play songs and lullabies. Infant Behav. & Dev. **19:** 83–92.
106. TRAINOR, L.J. & C.A. ZACHARIAS. 1998. Infants prefer higher-pitched singing. Infant Behav. & Dev. **21:** 799–806.
107. TREHUB, S.E. & T. NAKATA. 2000. Musical emotions: a perspective from development. Presented at the biennial meeting of the International Conference on Music Perception and Cognition, Keele, UK. August.
108. NAKATA, T. & S.E. TREHUB. 2000. Maternal speech and singing to infants. Presented at the Society for Music Perception and Cognition. Toronto, ON. November.
109. TRAINOR, L.J., C.M. AUSTIN & R.N. DESJARDINS. Is infant-directed speech prosody a result of the vocal expression of emotion? Psychol. Sci. **11:** 188–195.
110. SHENFIELD, T. & S.E. TREHUB. 2000. Infants' response to maternal singing. Poster presented at the biennial meeting of the International Conference on Infant Studies. Brighton, UK. July.
111. DALY, M. & M. WILSON. 1995. Discriminative parental solicitude and the relevance of evolutionary models to the analysis of motivational systems. *In* The Cognitive Neurosciences. M. S. Gazzaniga, Ed.: 1269–1286. MIT Press. Cambridge, MA.
112. LOMAX, A. 1968. Folk Song Style and Culture. American Association for the Advancement of Science. Washington, DC.
113. PANTALEONI, H. 1985. On the Nature of Music. Welkin Books. Oneonta, NY.
114. LEVIN, R.N. 1996. Song behaviour and reproductive strategies in a duetting wren, *Thryothorus nigricapillus* I: Playback experiments. Anim. Behav. **52:** 1107–1117.
115. SLATER, J.B. 2000. Birdsong repertoires: their origins and use. *In* The Origins of Music. N.L. Wallin, B. Merker & S. Brown, Eds.: 43–49. MIT Press. Cambridge, MA.
116. HOOKER T. & B.I. HOOKER. 1969. Duetting. *In* Bird Vocalizations. R.A. Hinde, Ed.: 185-205. Cambridge University Press. Cambridge.

# The Quest for Universals in Temporal Processing in Music

CAROLYN DRAKE AND DAISY BERTRAND

*Laboratoire de Psychologie Expérimentale, CNRS UMR 8581, Université René Descartes, Institut de Psychologie, Centre Universitaire de Boulogne, 92774 Boulogne-Billancourt Cedex, France*

*One would ultimately hope to specify these cognitive principles or "universals" that underlie all musical listening, regardless of musical style or acculturation. To what extent is it learned, and to what extent is it due to an innate musical capacity or general cognitive capacity?*

LEHRDAL & JACKENDOFF[1]

**ABSTRACT: Music perception and performance rely heavily on temporal processing: for instance, each event must be situated in time in relation to surrounding events, and events must be grouped together in order to o vercome memory constraints. The temporal structure of music varies considerably from one culture to another, and so it has often been supposed that the specific implementation of perceptual and cognitive temporal processes will differ as a function of an individual's cultural exposure and experience. In this paper we examine the alternative position that some temporal processes may be universal, in the sense that they function in a similar manner irrespective of an individual's cultural exposure and experience. We first review rhythm perception and production studies carried out with adult musicians, adult nonmusicians, children, and infants in order to identify temporal processes that appear to function in a similar fashion irrespective of age, acculturation, and musical training. This review leads to the identification of five temporal processes that we submit as candidates for the status of "temporal universals." For each process, we select the simplest and most representative experimental paradigm that has been used to date. This leads to a research proposal for future intercultural studies that could test the universal nature of these processes.**

**KEYWORDS: Temporal processing; Rhythm perception; Segmentation; Grouping; Regularity; Duration ratios**

Address for correspondence: Dr. Carolyn Drake, Laboratoire de Psychologie Expérimentale, CNRS UMR 8581, Université René Descartes, Institut de Psychologie, Centre Universitaire de Boulogne,71, avenue Edouard Vaillant, 92774 Boulogne-Billancourt Cedex, France. Voice: +33-1-5520-5930; fax: +33-1-5520-5854.

http://lpe.psycho.univ-paris5.fr

drake@idf.ext.jussieu.fr

## TOWARDS INTERCULTURAL RESEARCH IN TEMPORAL PROCESSING

We plead guilty to the charge of cultural egocentricity. In previous research, we have proceeded as if the musical environment the world over is identical to that in France, Belgium, England, and the USA (the countries in which we have carried out our experiments). The musical environment in these countries is relatively homogenous in the sense that it is dominated by a Western tonal tradition (although we acknowledge significant differences even between these environments). From a rhythmic point of view, both our "classical" and "popular" musics are dominated by relatively simple rhythmic structures, organized around a regular beat, with binary or ternary multiplications and subdivisions of this beat.[1,2] By focusing on the auditory world around us, we have identified some of the psychological rhythmic processes that appear to function in our particular instance when listening to or producing "our" type of music. Following the results of this developmental prospective, we have even made the claim that some of these processes function the same way whatever the musical environment in which a listener is immersed.[3] Despite this previous highly focused position, we are fully aware that our assumptions concerning the generalizability of these processes to other cultures are only that—assumptions. This paper is an attempt to make amends for our previous cultural egocentricity by suggesting how our psychological developmental prospective could be adapted to an intercultural one, into which existing knowledge from the field of ethnomusicology and future cross-cultural studies could be incorporated.

Our previous research has focused on how people hear musical and nonmusical rhythms in order to identify the underlying psychological processes that make the perception of music such a fulfilling and satisfying activity. Music is typical of all forms of sustained activity over time, in that it can be successfully perceived or performed only if the individual events from which it is composed are perceptually integrated into larger units spread over time. Indeed, music has been defined as the art of organizing events in time. As such, it provides an ideal opportunity to investigate the perceptual and cognitive temporal processes that make such activity possible.

In the past we have adopted the experimental principle of comparing performances on both perceptual and motor rhythmic tasks by people varying in levels of rhythmic skill, be it by age (as children get older there is a gradual increase in their exposure to, and experience with, the rhythmic structures around them) or musical training (the musical training common to our culture is particularly characterized by the development of explicit knowledge about musical structure). Such an experimental principle allows us to tease apart the processes that appear to be "innate" or "hardwired" (functioning at birth, determined by genes, independent of environmental influence and experience) and those that develop with maturation, acculturation (learning by immersion in the auditory world around us), or explicit training. The principle has been that if young infants, children, and nonmusician and musician adults display similar functioning modes on a particular task, then we conclude that this process may be "innate" or at least "functional" at an early age. Alternatively, if differences are observed between these populations, we then conclude that this type of functioning is acquired, either through acculturation or explicit learning.

We have, in the past avoided the word *innate* due to strong theoretical and philosophical connotations, referring rather to processes that may be universal—that is, that function in the same way in everyone. However, as has been pointed out to us, this wording is ambiguous. If we claim that a process is universal, we must demonstrate that it occurs the world over, irrespective of the cultural environment in which the individual lives and grew up. The present paper proposes such a research project. Whereas we have the know-how about both the psychological processes involved and appropriate paradigms to demonstrate the functioning of these processes, we are quite ignorant of the enormously exciting field of ethnomusicology, and our attempts at contact have so far been limited. Such a project cannot be accomplished on our own, or even with just one or two colleagues, but requires numerous sites the world over with researchers collaborating within a network.

## TEMPORAL PROCESSING IS LIMITED BY MEMORY SPACE AND PROCESSING TIME

If a computer programmer wanted to make a computer reproduce a musical rhythm, the program would probably record the precise duration (in milliseconds or computer clicks) of each interval in the sequence, put them in a lookup table, and then recall these values to produce the rhythm. The result would be that the reproduced rhythmic structure would be identical to that of the model. Is this a good model for the functioning of the human perceptual system? Probably not. Whereas human beings are able to reproduce musical rhythms so that they sound satisfactorily similar to the model, previous research suggests that our perceptual system does not function in the same way as the computer. The main difference in the way computers and humans function concerns memory limitations: computer memory is usually not a problem, while human memory is severely limited. As psychologists, our task is thus to describe the way in which our perceptual system analyses incoming temporal information and how it overcomes the main enemies: memory space and processing time.

Sound events usually do not occur in isolation, but rather are surrounded by other events, with each sound embedded within a sequence. Each event takes its existence from its relation with these surrounding events, rather than from its own specific characteristics. The task of situating each event in relation to surrounding events is quite simple when the sequence is short in duration and when it contains a limited number of events (i.e., three or four intervals). However, as the sequence becomes longer we very quickly run into problems of memory space and processing time. Imagine the number of intervals that would need to be stored and accessed when listening to a Beethoven sonata!

The idea is that all events would be stored in a memory buffer lasting several seconds (psychological present, probably corresponding to echoic or working memory), allowing the system to extract all relevant information (interval duration). One can imagine a temporal window gliding gradually through time, with new events arriving at one end and old events disappearing at the other due to decay. Thus only events occurring within a span of a few seconds would be accessible for processing at any one time, and only events occurring within this limited time window can be situated in relation to each other by the coding of relevant relational information. However, when the sequence becomes more complicated (longer and/or more events), the number of events that must be processed quickly goes beyond the limits of this buffer. Consequently, simple concatenation models that propose the maintenance in memory of the characteristics of each event are not able to account for the perception of long sequences because of problems of memory overload.

In this paper, we present a set of five temporal processes that overcome these constraints, at least partially. These processes have been selected because previous research in the field of rhythmic perception suggests that they function in a similar fashion in all the populations examined so far. They are therefore candidates for the status of "temporal universal." This is not an exhaustive list, and other candidates could certainly be added as the project advances. As mentioned above, this list emerges from previous developmental and comparative studies. In each case we select key paradigms that we have used successfully in the past to demonstrate the lack of difference between populations differing in age and musical experience. An essential characteristic of developmental research is that the tasks used must be conceptually simple; must not require reading, writing, or any other specific knowledge such as musical notation; must be short (usually not more than 20 minutes of experimenting time); and must be technically transportable (into schools). As such, the paradigms we have used for children should be easily adaptable for intercultural research, which faces similar constraints. Taken together, this set of experimental paradigms could be used as a basis for answering the question of the true "universal" nature of these psychological processes.

## CANDIDATE 1: SEGMENTATION AND GROUPING

**We tend to group into perceptual units events that have similar physical characteristics or that occur close in time.**

One way to overcome processing limitations and to allow events to be processed together is to group the events into small perceptual units. These units result from a comparison process that compares incoming events with events that are already present in memory. If a new event is similar to those that are already present, it will be assimilated. If the new event differs too much (by its acoustical and/or temporal characteristics), the sequence will be segmented. This segmentation leads to the closure of one unit and the opening of the next. Elements grouped together will be processed together within a single perceptual unit and thus can be situated in relation to each other.

Several studies support these ideas. Listeners usually segment sequences as a function of the surface characteristics (timbre, pitch, intensity, event duration, pauses, etc.), following the principles laid down by the Gestalt psychologists: a change in any sound parameter leads to the perception of a break in the sequence and thus to the creation of groups separated by the changes (see Ref. 5 for a summary). For instance, the occurrence of a longer temporal gap or a major change in pitch leads to the segmentation of the sequence at that point, with the termination of one perceptual unit and the beginning of the next.

### *Paradigm 1: Online Segmentation*

Many variations on a simple segmentation paradigm have been used. Usually participants listen to a musical excerpt and are asked to indicate, by pressing a button or drawing a line on a musical score, whenever they hear a "break" in the sequence, so that events that belong together go together. The sequences are constructed in such a way as to establish whether each type of segmentation is used, as well as to indicate the relative importance of each segmentation principle.

### *Arguments in Favor of Universal Status*

*Comparisons across musical skill levels.* Grouping appears unaffected by musical training, as similar segmentation principles are observed for adult musicians and nonmusicians, although musicians are more systematic in their responses.[6–8]

*Infants.* Grouping also appears to function from an early age. For instance, studies using a gap detection paradigm suggest that six- to eight-month-old infants segment in a similar fashion to adults, at least for timbre and pitch, although a much greater pitch change was necessary for infants compared with adults.[9,10]

*Comparisons across ages.* Findings are even less clear in children. Among the few previous studies on rhythmic development, we often find suggestions that this sort of grouping process is functioning in relatively young (five- to seven-year-old) children. However, these studies employ extremely indirect methodologies, and the conclusions are tentative.[10–15] With this criticism in mind, a major research project has been undertaken to examine the principles governing segmentation in 4- to 12-year-old musician and nonmusician children.[16] The children's task was to listen to short musical sequences composed of nine tones and to indicate, during a second listening, the point at which they would break the sequence into two (group boundary). Each sequence contained two changes in either pitch, intensity, tone duration, or pause duration. The question was whether or not their perceived boundaries were induced by the physical changes in the sequence, as observed previously in adults. The results provide a clear answer to this question: segmentations were well above chance level for all four segmentation principles, even for the youngest children (four years). On average, more than three-quarters of the segmentations corresponded to the physical change; this effect was equally strong for the four indices (pitch, intensity, tone duration, and pause duration). There was a slight improvement with age and musical experience, probably due to improved task-related skills.

*Comparisons across cultures.* The on-line segmentation paradigm can easily be adapted for intercultural research, as the stimuli are non-culture-specific, and the task does not require any specific learned skill. If these basic processes of segmentation and grouping are unaffected by experience with a particular type of sequence, then the same types of segmentation should be observed in all cultures.

## CANDIDATE 2: PREDISPOSITION TOWARDS REGULARITY

**Processing is better for regular than irregular sequences. We tend to hear as regular sequences that are not really regular.**

As mentioned above, rather than coding the precise duration of each interval, our perceptual system compares each newly arriving interval with preceding ones. If the new interval is similar in duration to preceding intervals (within an acceptable temporal window, the "tolerance window"), it will be categorized as "same;" if it is significantly longer or shorter than the preceding intervals (beyond the tolerance window), it will be categorized as "different." There may be an additional coding of "longer" or "shorter." Thus, our system may code two or three categories of durations (same/different, or same/longer/shorter); but note that this is a relative, rather than absolute, coding system.

One consequence of this type of processing is that if a sequence is irregular (each interval has a different duration) but all the intervals remain within the tolerance window, then we will perceive this sequence as the succession of "same" intervals and so perceive a regular sequence. Such a tolerance in our perceptual system is quite understandable when we examine the temporal microstructure of performed music: local lengthenings and shortenings of more than 10% are quite common[17,18] and are not necessarily picked out by listeners as irregularities.[19]

### *Paradigm 2: Tempo Discrimination for Regular and Irregular Sequences*

This predisposition towards regularity has been investigated using a tempo discrimination paradigm. Listeners hear two sequences that differ slightly in tempo, and they must say which is the fastest. The degree of regularity of the sequences is varied from completely regular (isochronous) to extremely irregular (based on the standard deviation of the interval durations).[20] Results show that up to a certain degree of irregularity, irregular sequences are processed as well as the regular sequences: they are "assimilated" towards regularity. However, above a certain degree of irregularity, discrimination performance drops considerably. This cut-off point corresponds to the tolerance window mentioned above.

### *Arguments in Favor of Universal Status*

*Comparisons across musical skill levels.* The tolerance window appears to function in a similar fashion for both musicians and nonmusicians, with slightly irregular sequences being assimilated to regular sequences, although temporal thresholds are lower in musicians than nonmusicians.[20]

*Comparison across ages.* To our knowledge, nothing is known about children's ability to process irregular sequences. We do know, however, that from the age of four years, children are able to detect a small change in tempo of an isochronous sequence.[21] Another way of demonstrating the importance of temporal regularity in processing sequences involves rhythm reproduction tasks. For instance, when five- and seven-year-old children reproduced both regular and irregular rhythms, performance was much better for the regular ones.[22]

*Infants.* We also know little about infants' ability to process irregular sequences. We do know, however, that the capacity to detect a small change in tempo of an isochronous sequence is already functional at two months in infants.[23] They are able to habituate to a particular tempo, and there is a reaction to novelty if the tempo changes.

*Comparison across cultures.* This tempo discrimination task for regular and irregular sequences is easily adaptable to people from other cultures. If this process is universal, we should observe the same low tempo discrimination thresholds for regular and slightly irregular sequences, with considerably higher thresholds for very irregular sequences.

## CANDIDATE 3: ACTIVE SEARCH FOR REGULARITY

**We spontaneously search for temporal regularities and organize events around this perceived regularity.**

Coding events in terms of temporal regularity is thus an economical processing principle, and it has implications. If an incoming sequence can be coded in such a fashion, the needed processing resources are reduced, thus making it easier to process such a sequence. Indeed, we can say that the perceptual system exploits this predisposition, by actively "looking for" temporal regularities in all types of sequences. We therefore suggest that when listening to a piece of music, we are predisposed to finding a regular pulse, that which is emphasised by tapping our foot in time with the music (*tactus* in musical terms). Once this underlying pulse has been identified, it is used as an organizational framework around which other events are situated.

### *Paradigm 3: Synchronization with Musical Sequences*

A simple way of investigating this process is to ask people to listen to a musical sequence, and then to tap in time with the music in a regular fashion at the rate that they think "goes best" with the music. Synchronization is considered to be accomplished if successive taps coincide with tones within the music (at a particular hierarchical level) within a 10% window. This is quite a strict criteria when you take into consideration the considerable temporal variations observed in performed music. In order to accomplish this task, listeners must abstract an underlying regularity, even if the performance variations tend to mask "pure" temporal regularity.

### *Arguments in Favor of Universal Status*

*Comparison across musical skill levels.* Synchronization success rates for both musicians and nonmusicians are very high (more than 90%) when they are asked to tap in synchrony with music, even if the music contains many temporal and other performance microvariations.[24]

*Comparison across ages.* All 20 four year olds (as well as the older children aged 6 to 10 years) examined in our study were able to successfully synchronize with the beginning of Ravel's *Boléro*.[21]

*Infants.* Infants are able to adapt their spontaneous sucking rate to the rate of an auditory sequence,[25] at least under certain circumstances.

*Comparison across cultures.* In order to adapt this paradigm to intercultural research, musical excerpts must be selected to include both music that respects the temporal structure with which the participants are familiar (their own musical idiom) and music that does not.

## CANDIDATE 4: TEMPORAL ZONE OF OPTIMAL PROCESSING

**We process information best if it arrives at an intermediate rate.**

A fourth processing principle concerns the rate of temporal sequences. People spontaneously "listen for" important events occurring at equally spaced moments in time, and the rate at which they "search for" important information is specific to each individual.[26,27] Thus, the search for temporal regularities described above occurs at a particular rate. A zone of optimal processing has been demonstrated with numerous paradigms and types of sequences. The results are concordant: sensitivity to change is highest if events occur about every 600 ms, with a range stretching between about 300 and 800 ms interonset interval (IOI).

### *Paradigm 4: Irregularity Detection in Complex Sequences*

How can we demonstrate that people focus spontaneously on events occurring at intermediate rates? A first method is simply to compare tempo discrimination thresholds of different rates,[20] as mentioned above. However, a more ecologically valid paradigm is currently being developed that would be ideal for the present requirements. When participants listen to complex sequences composed of two cooccurring isochronous subsequences, they should focus on the subsequence closest in rate to the optimal processing zone (300–800 ms IOI). We therefore introduce a temporal irregularity into one of the subsequences, which listeners should be able to detect only if they are focusing on that particular subsequence. The overall rate of the complex sequence is varied from very fast to very slow. Our results indicate that, as expected, detection was better for the slowest subsequence when the complex sequence was fast, and better for the fastest subsequence when the complex sequence was slow.[28] This promising paradigm has been applied only to nonmusician adults.

### *Arguments in Favor of Universal Status*

*Comparison across musical skill levels.* Although musicians demonstrate a wider range of optimal tempi than nonmusicians, this optimal zone is still centered around the same value of 600 ms IOI.[20]

*Comparison across ages.* Children between the ages of 4 and 10 years also demonstrate the same zone of optimal tempi, although the range increases with age.[21]

*Infants.* Two-month-old infants demonstrate a reaction to novelty only for sequences at 600 ms IOI, demonstrating the same optimal zone from a very young age.[23]

*Comparison across cultures.* This paradigm can be easily adapted to people of other cultures.

## CANDIDATE 5: PREDISPOSITION TOWARD SIMPLE DURATION RATIOS

**We tend to hear a time interval as twice as long or short as previous intervals.**

A fifth principle concerns the perceptual status of the "longer" and "shorter" durations: longer time intervals tend to be perceived and produced twice as long as the "same" intervals, and shorter intervals tend to be perceived and produced twice as short as the same intervals.[29–31] The implication is that a categorization process is involved here, with clear-cut passages from one category to another.[30,32] Such an organization principle results in the dominance of binary, rather than ternary or more complex ratios between the three categories of intervals.

### *Paradigm 5: Rhythm Reproduction*

One consequence of this process is that people are better able to reproduce rhythms containing only 1:2 ratios than rhythms containing 1:3 ratios[33] or even more complicated ratios.[34] Also, when people reproduce complex musical rhythms, the duration of some intervals undergoes a distortion towards one of the categories—that is, the rhythm is simplified so that the produced intervals respect a 1:2 ratio.[29]

### *Arguments in Favor of Universal Status*

*Comparison across musical skill levels.* The same pattern of perceptual and motor distortions are observed in musicians and nonmusicians.[34]

*Comparison across ages.* From the age of 5 years, children are able to reproduce short rhythms based on binary (1:2), but not ternary (1:3) ratios.[33] When 5- and 7-year-old children reproduce complex rhythms, their reproductions undergo simplification towards simple ratios.[22] Their incorrect reproductions contain only two durations, in a 2:1 ratio.[22]

*Infants.* To our knowledge, no data exists concerning the functioning of this duration categorization process in infants.

*Comparison across cultures.* A rhythm reproduction task can be adapted for intercultural research. Participants can be asked to reproduce rhythms varying in the number of different durations they contain. Reproductions should be best when they contain only two different durations in a ratio of 1:2. Ratios of 1:3 should be harder, and more complex ratios (which are perceived as irregular—see Candidate 2) even harder. The reproductions of complex rhythms should demonstrate simplification toward simple ratios.

## CONCLUSION

We have thus proposed a series of five experimental paradigms, designed and tested to demonstrate the functioning of five temporal processes. If our hypotheses are correct concerning the universal nature of these processes, then whatever the culture or origin of the people being tested, we should observe the same results. This list of potential candidates for the status of temporal universals is probably far from complete, but it provides a starting point. Further suggestions are welcome.

## REFERENCES

1. Lerdahl, F. & R. Jackendoff. 1983. A generative theory of tonal music. MIT Press. Cambridge, MA.
2. Cooper, G. & L.G. Meyer. 1960. The rhythmic structure of music. University of Chicago Press. Chicago.
3. Yeston, M. 1976. The stratification of musical rhythm. Yale University Press. New Haven, CT.
4. Drake, C. 1998. Psychological processes involved in the temporal organization of complex auditory sequences: universal and acquired processes. Music Percept. **16:** 11–26.
5. Handel, S. 1989. Listening. An introduction to the perception of auditory events. MIT Press. Cambridge, MA.

6. DELIÈGE, I. 1987. Grouping conditions in listening to music: an approach to Lerdahl and Jackendoff's grouping preference rules. Music Percept. **4:** 325–360.
7. FITZGIBBONS, P.J., A. POLLATSEK & I.B. THOMAS. 1974. Detection of temporal gaps within and between perceptual tonal groups. Percept. Psychophys. **16:** 522–528.
8. PERETZ, I. & J. MORAIS. 1989. Music and modularity. Contemp. Music Rev. **4:** 279–293.
9. KRUMHANSL, C.L. & P.W. JUSCZYK. 1990. Infants perception of phrase structure in music. Psychol. Sci. **1:** 70–73.
10. THORPE, L.A. & S.E. TREHUB. 1989. Duration illusion and auditory grouping in infancy. Dev. Psychol. **25:** 122–127.
11. DRAKE, C. 1993. Influence of age and experience on timing and intensity variations in the reproduction of short musical rhythms. Psychol. Belg. **33:** 217–228.
12. DRAKE, C. 1993. Perceptual and performed accents in musical sequences. Bull. Psychon. Soc. **31:** 107–110.
13. DRAKE, C., J. DOWLING & C. PALMER. 1991. Accent structures in the reproduction of simple tunes by children and adult pianists. Music Percept. **8:** 313–332.
14. GÉRARD, C. & C. DRAKE. 1990. The inability of young children to reproduce intensity differences in musical rhythms. Percept. Psychophys. **48:** 91–101.
15. BAMBERGER, J. 1980. Cognitive structuring in the apprehension and description of simple rhythms. Arch. Psychol. **48:** 177–199.
16. BERTRAND, D. 1999. Groupement rythmique et representation mentale de mélodies chez l'enfant. PhD. Thesis, Liège University, Belgium.
17. PENEL, A. & C. DRAKE. 1997. Perceptual and cognitive sources of timing variations in music performance: a psychological segmentation model. Psychol. Res. **61:** 12–32.
18. REPP, B.H. 1992. Diversity and commonality in music performance: an analysis of timing microstructure in Schumann's Träumerei. J. Acoust. Soc. Am. **92:** 2546–2568.
19. REPP, B.H. 1992. Probing the cognitive representaion of musical time: structural constraints on the perception of timing perturbations. Cognition **44:** 241–281.
20. DRAKE, C. & M.C. BOTTE. 1993. Tempo sensitivity in auditory sequences: evidence for a multiple-look model. Percept. Psychophys. **54:** 277–286.
21. DRAKE, C., M.R. JONES & C. BARUCH. 2000. The development of rhythmic attending in auditory sequences: attunement, reference period, focal attending. Cognition **77:** 251–288.
22. DRAKE, C. & C. GÉRARD. 1989. A psychological pulse train: how young children use this cognitive framework to structure simple rhythms. Psychol. Res. **51:** 16–22.
23. BARUCH, C. & C. DRAKE. 1997. Tempo discrimination in infants. Infant Behav. Dev. **20:** 573–577.
24. DRAKE, C., A. PENEL & E. BIGAND. 2000. Tapping in time with mechanically and expressively performed music. Music Percept. **18**(1): 1–24.
25. POUTHAS, V. 1995. The development of the perception of time and temporal regulation of action in infants and children. *In* Musical Beginnings: The Origins and Development of Musical Competence. I. Deliège & J.A. Sloboda, Eds.: 115–141. Oxford University Press. New York.
26. JONES, M.R. 1976. Time, our last dimension: toward a new theory of perception, attention, and memory. Psychol. Rev. **83:** 323–355.
27. JONES, M.R. & M. BOLTZ. 1989. Dynamic attending and responses to time. Psychol. Rev. **96:** 459–491.
28. BROCHARD, R. 1997. Role of attention in the perceptual organisation of complex auditory sequences. Ph.D. thesis, University of Paris.
29. FRAISSE, P. 1956. Les structures rythmiques. Publications Universitaires de Louvain. Louvain.
30. CLARKE, E.F. 1987. Categorical rhythm perception: an ecological perspective. *In* Action and Perception in Rhythm and Musica. A. Gabrielsson, Ed.: 19–34. Royal Swedish Academy of Music. Stockholm.
31. PARNCUTT, R. 1994. A perceptual model of pulse salience and metrical accent in musical rhythms. Music Percept. **11:** 409–64.
32. SCHULZE, H.H. 1989. Categorical perception of rhythmic patterns. Psychol. Res. **51:** 10–15.

33. DRAKE, C. 1993. Reproduction of musical rhythms by children, adult musicians and adult nonmusicians. Percept. Psychophys. **53:** 25–33.
34. STERNBERG, S., R.L. KNOLL & P. ZUKOFSKY. 1982. Timing by skilled musicians. *In* The Psychology of Music. D. Deutsch, Ed.: 182–237. Academic Press. New York.

# Music, Cognition, Culture, and Evolution

IAN CROSS

*Faculty of Music, University of Cambridge, Cambridge CB3 9DP, United Kingdom*

**ABSTRACT: We seem able to define the biological foundations for our musicality within a clear and unitary framework, yet music itself does not appear so clearly definable. Music is different things and does different things in different cultures; the bundles of elements and functions that are music for any given culture may overlap minimally with those of another culture, even for those cultures where "music" constitutes a discrete and identifiable category of human activity in its own right. The dynamics of culture, of music as cultural praxis, are neither necessarily reducible, nor easily relatable, to the dynamics of our biologies. Yet music appears to be a universal human competence. Recent evolutionary theory, however, affords a means for exploring things biological and cultural within a framework in which they are at least commensurable. The adoption of this perspective shifts the focus of the search for the foundations of music away from the mature and particular expression of music within a specific culture or situation and on to the human capacity for musicality. This paper will survey recent research that examines that capacity and its evolutionary origins in the light of a definition of music that embraces music's multifariousness. It will be suggested that music, like speech, is a product of both our biologies and our social interactions; that music is a necessary and integral dimension of human development; and that music may have played a central role in the evolution of the modern human mind.**

**KEYWORDS: Music; Cognition; Culture; Evolution**

We can express our understanding of biology within a framework that enables us to relate it, if not reduce it, to our understanding of the world in physical and material terms. Biological and physical understandings of the world are *commensurable*, in at least one of the senses that Lakoff[1] (p. 322) proposes. An understanding of ourselves as biological beings appears to be an understanding of "natural kinds."[a] But

Address for correspondence: Ian Cross, Faculty of Music, University of Cambridge, West Road, Cambridge CB3 9DP, UK. Voice: +44 (0)1223 335185.
ic108@cus.cam.ac.uk

[a]This is not to endorse the idea that there are "natural kinds," that science provides an account of the essences of things in the world. The term "natural kind" is used here simply as a concise way of referring to the objects of scientific discourse. As Rose[45] (p. 42) points out, even a concept as seemingly "natural" and unambiguous as a protein is susceptible to multiple and differing levels of definition that are dependent on "the purposes for which we need to make the definition." There is an incontestably societal dimension to the makeup of what is taken to constitute science at any time. The notion, however, that scientific procedures and understandings are simply varieties of social practice definable by their particular vocabularies[46] or by their poverty and abstraction[47] is insufficient to account for the instrumentality of those procedures and for the commensurability of the understandings that they afford.

is music a natural kind, comprehensible within the generalized framework that is science?

Many argue that music is not a natural kind. Indeed, following a conventional dictionary definition of music—"The art of combining sounds of voices or instruments to achieve beauty of form and expression of emotion"—it would be difficult to do so. The consensual view from within the humanities appears to be that music is cultural rather than natural; music is viewed as constituted of practices, concepts, and perceptions that are grounded in particular social interactions and constructions. Molino[2] (p. 169), in questioning the status of music as a natural kind, proposes that "Nothing guarantees that all the forms of human music contain a nucleus of common properties that would be invariant since the origination of music."

As Geertz[3] (p. 5) put it, in promoting a semiotic and interpretive approach to culture, "man is an animal suspended in webs of significance he himself has spun," and within Geertz's humanocentric web of culture there is little room for the "natural." For Treitler[4] (p. 203), "Meaning in music is a function of the engagement of codes or orders by the note-complexes of which the musical event is comprised," and musical phenomena are thus "intelligible only in the light of an interpretation which intuits the purpose or intention that they embody." Tomlinson, inciting musicologists to embrace Geertz's concept of culture, makes explicit the idea that scientific generalization is incompatible with musicological method; he asserts[5] (p. 352) that the essence of cultural—and hence musicological—explanation is "not to codify abstract regularities but to make thick description possible, not to generalize across cases but to generalize within them." Indeed, Abbate[6] has suggested (p. xv) that "There is nothing immanent in a musical work (beyond the material reality of its written and sonic traces) and our perceptions of forms, configurations, meanings, gestures and symbols are always mediated by verbal formulas, as on a broader scale by ideology and culture." And Garnett[7] proposes that "there is...no extra-cultural locus from which to observe music, nor extra-cultural meaning to observe." Music is seen as the expression of discrete, contingent, socially conditioned factors in respect of which a generalizable—and hence scientific—account is neither relevant nor possible.

Such an approach to understanding music appears justified in view of the heterogeneity of forms that music can take. What "nucleus of common properties" other, perhaps, than the very concept of the musical work[8] underlies such diverse products of Western culture (or, as Slobin[9] has put it, musical microcultures) as (i) a performance of a Machaut motet; (ii) the autograph score of a Beethoven string quartet; (iii) the concept of Robert Ashley's "I am sitting in a room..."; (iv) a live broadcast of Brian Ferneyhough's *Transit*; (v) the grooves in the vinyl that constitute the recording of the Holy Modal Rounders' track "Half a mind to have a mind;" (vi) and the samples that make up part of the mastering materials of a Dr. Dre CD? Indeed this last example undermines the very concept of the musical work itself. It might be suggested that "the art of combining sounds of voices or instruments..." provides such a common property (though the issue of "the achievement of beauty..." seems moot). But if we look beyond what Tomlinson[5] refers to as a "presentist"[b] view of our own culture, that common property evanesces.

[b]Tomlinson[5] (p. 358): "The presentist view of art works as transcendent entities fully comprehensible without reference to the conditions of their creation sacrifices Geertz's expansion of human discourse for a solipsistic and ultimately narcissistic aestheticism."

Indeed, when we look to other cultures, the notion of music per se is called into question. For example, Malm[10] (p. 5) questions whether or not the sound of the bull-roarer used in some Arnhem Land aboriginal ceremonies constitutes music, as "its sounds in secret rituals are not considered as independent sonic events but rather are thought to be the sounds of the supernatural itself." Moreover, a culture may be constituted such that it does not distinguish a discrete category of practices that map onto those that would comprise music from a Western perspective. Gourlay[11] notes that some cultures employ terms far more inclusive than the Western notion of music; for the Igbo of Nigeria, *nkwa* denotes "singing, playing instruments and dancing." Thus the anthropologist approaching Igbo culture with a view to examining its "music" is confronted by a dilemma; as Gourlay puts it (p. 35), "By forcing the Igbo concept into the Procrustean bed of Western conceptualization, she is in fact surrendering to the dominance of Western ideas—or at least to the dominance of the English language. How different things would be if the Igbo tongue had achieved the same 'universality' as English! We should then have been seeking for universals in *nkwa*, and regarding the whole process of Western 'serious' music as an aberration because it excluded dance." If the very concept of music is so variable and inextricable from its cultural context, how can we expect to seek, far less find, its biological foundations?

Perhaps some space for the natural in conceptions of music might be found through recourse to cognitive anthropology rather than to anthropology per se, using the notion of mind to connect culture with biology. As D'Andrade [12] notes, the tacit notion of culture even within the anthropological consensus is that it is *in the mind*. He suggests (p. 146) that "Since the 1950's most anthropologists have defined culture as a purely *mental* phenomenon." Hence if culture is a mental phenomenon, "the structures that exist in the physical world as objects or events...are all thought to be...more or less a reflection of these mental cultural structures." Given the success of cognitive science in instating mind in the material, it seems reasonable to expect that notions of mind might furnish terms that could connect the discourses of anthropology and of the natural sciences.

However, cognitive anthropology, or from another perspective, cultural psychology,[13] offers little comfort. For there is little consensus about the depth to which culture permeates the grain of our experience. The question of whether we should conceive of some psychic unity upon which culture forms an overlay, or a psychic diversity that is principally constructed by culture, remains unresolved. D'Andrade,[12] for example, leaves space for the notion of a primitive and perceptual psychic unity when he suggests (p. 217) that "culture seems to have its greatest effect...on semantic memory and complex reasoning." He undermines that notion by suggesting (p. 184) that "In general, naive perception can be influenced by cultural schemas." A practical musical example might uphold the second suggestion by showing that culture indeed appears to determine the grain of our unreflective perceptions.

One factor that appears to apply to almost all the world's musics is that there is a level of temporal organization that is regular and periodic, sometimes called the tactus. It is taken to correspond to the regular points in the music where one would tap one's foot or clap along. In listening to a piece in a familiar idiom, all listeners are capable of tapping along without thinking (even if it is frowned upon in the concert hall). Even when encountering previously unheard music from an unknown cul-

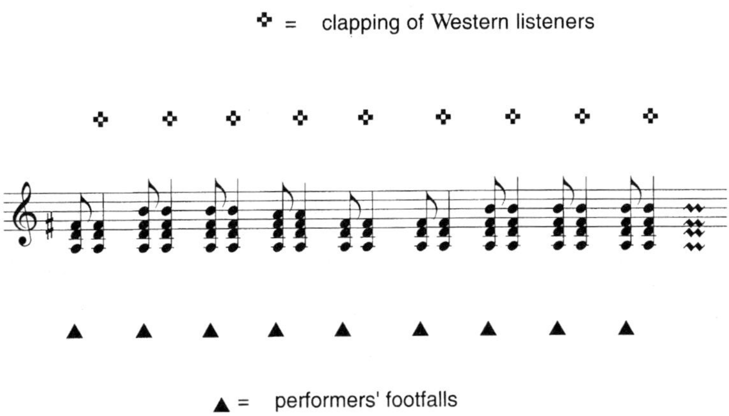


**FIGURE 1.** Differently perceived tactus in a piece from Northern Potosí in Bolivia.

ture, a listener can still "keep a beat." Most Western listeners confronted by this particular piece from Northern Potosí in Bolivia (recorded by my friend and colleague Henry Stobart, an ethnomusicologist specializing in Andean music) are likely to clap or tap along with the regularly spaced longer and louder notes (FIG. 1).

A recording of this piece can be heard on the Internet by accessing http://www.mus.cam.ac.uk/~cross/BJE/ and selecting Example III(a). At the end of the recording something occurs that gives the game away; for the first time the regular footfalls that accompany—that are part of—the performance can be heard. And they coincide not with the longer and louder notes, but with the short, sometimes almost inaudible notes that alternate with the long notes. For the people of the culture from which this music comes, these appear to be the appropriate places to tap the feet—that is, the point at which the tactus occurs. For Western listeners, while one can *learn* to clap or tap at those points in the piece, it does not appear to *feel*—and I speak from long experience—as though one is tapping on the beat. It always feels as though one is tapping on an offbeat.

While certain features of this music's organization can be accounted for by invoking the operation of perceptual processes underpinning the experience of time that appear genuinely universal, the fact that longer, more intense, notes do not mark out the tactus cannot be explained in this way. In fact, it seems likely that prosodic features of the language of Northern Potosí, Quechua, relate to the way in which tactus is organized, projected, and experienced in that culture's music.

That even such an apparently unreflective act as regularly tapping the foot in time to a piece of music is so susceptible to cultural differentiation appears to suggest either that tapping one's foot in time to music has a semantic component (if D'Andrade's proposal that culture impacts cognition primarily at the conceptual level is accepted) or that human cognitive capacities are so grounded in culture that any elementary commonalities are overridden, and that minds are only susceptible to explanation in terms specific to the particular cultures in which those specific minds are rooted—a return to the position of Geertz. In other words, culture is in the bones, and science has no place in its understanding.

We appear to have reached an impasse; it seems that music is cultural, variable and particular, and not susceptible to explanation in general and scientific terms. Yet there are those who argue that music is, nevertheless, a human universal. Blacking[14] (p. 224) states that "every known human society has what trained musicologists would recognize as 'music,'" and Merriam[15] (p. 227) bluntly asserts that music "is a universal behavior." How can these claims be squared with music's cultural particularity?

As a first step we must enquire what Blacking and Merriam mean by music in this universal manifestation, for neither is music just sound. Indeed, the musical example given above is not "the music;" it was a recording of the sound of a musical activity in a particular cultural context. To experience the music, you might have to undergo what has been called[16] the "Total Turing Test" of lifelong immersion in the culture; at the least you would have to get the feel of the instrument and of the movements involved. The music would involve embodied action as much as disembodied sound. Even in our own culture it is only in the last hundred years, with music becoming an increasingly commodified aural consumable, that the self-evident ties between musical sound and human movement have been rendered obscure. For John Blacking, the claim that music is a human universal explicitly involves acknowledgment of the embodied nature of music, the *indivisibility* of movement and sound in characterizing music across times and cultures; he claims[14] (p. 241) that "'music' as a human capability is a cognitive, and hence affective, activity of the body." For the greater number of cultures in the world, and for the greater part of the historical existence of our culture, music appears to have involved and to involve *movement* just as much as sound.

Music in its universal guise, however, involves not only sound and movement, but also multiplicity of reference and meaning; for Blacking and Merriam music is intrinsically polysemic. For example, music can function as a means of communication with the dead for the Kaluli of Papua New Guinea, binding birds, souls, places, and people at a time of transformation; or music can function in the restructuring of social relations, as in the *domba* initiation of the Venda. In each of these two very different ceremonies, music is central, its meaning rarely if ever explicit but its fugitive significances essential. Blacking[14] notes (p. 237) "the 'same' sound patterns...can...have different meanings within the same society because of different social contexts." As Merriam[15] (p. 221) suggests, a defining characteristic of the musical utterance is its property of being "unrepudiable in form but repudiable as to content." In other words, music has the capacity to *lack* consensual reference; it can be *about* something, but its *aboutness* can vary from context to context and even within context.

Not only can music's significance vary according to social context, but also the significances of a singular musical activity can vary from individual to individual. We know this from our experience of music in our own cultures, and we can see it in others. For example, in the *gisalo* ceremonies of the Kaluli, some performers seek to dominate and direct, while the performances of others appear to emerge from performer–audience interaction;[17] and in any particular performance some participants weep while others do not.[18]

Finally, music in general has a further, peculiarly negative, feature; it appears to have no immediate and evident efficacy. Music neither ploughs, sows, weaves, nor feeds; in itself, if it can be considered to exist outside its context of use, it does not seem to be capable of being a material cause of anything other than a transient hedonic encounter. It is inefficacious.

From these considerations of the "universal" characteristics of music we can return to the original question of whether music can be construed as a natural kind. We now appear to have a basis for proposing an operational definition of music that might afford the commonalities that would allow an instatement of the natural in the musical. It seems that a generalizable definition of music would refer to music's two roots in sound and movement, to music's heterogeneity of meaning, to its grounding in social interaction yet personalized significance, and to its inefficacy. Putting these four premises together yields the following operational definition:

> Musics can be defined as those temporally patterned human activities, individual and social, that involve the production and perception of sound and have no evident and immediate efficacy or fixed consensual reference.

When applied to the mature expressions of music in particular cultures, this definition does no more than provide a conceptual umbrella for otherwise potentially heterogeneous ethnographies; in that context it is descriptive rather than explanatory. It constitutes a general proposition that applies to the universal presence of something like "music" in all human cultures. However, in claiming that music is universal, Blacking goes further than this; he suggests that music is ubiquitous not only across human societies but across all members of those societies. As he states[14] (p. 236), "the almost universal distribution of musical competence in African societies suggested that musical ability [is] a general characteristic of the human species rather than a rare talent." This suggestion squares with recent research into the precursors of musical ability in a Western context, notwithstanding those such as Barrow[19] who assert (p. 194) that "musical ability is...limited in its distribution." While it is self-evidently true that the production of music in contemporary Western society is in the hands of a specialized class of performers and composers (and lawyers), musical ability cannot be defined solely in terms of productive competence; (almost) every member even of our own, highly specialized, society is capable of listening to and hence of *understanding* music. Indeed, recent research[20] can be interpreted as suggesting that musical productive abilities in a Western context, rather than being rare capacities that are evidence of some inborn "talent," are better explained in terms of the effects of motivation and of practice. In other words, music is not just universal across cultures; it appears that everyone has the capacity to be musical, though this capacity is likely to be realized to different degrees and in different ways in different cultural and social environments.

It is within this broader notion of music as a universal human attribute that the operational definition of music given above might be informative, as applied to the human *capacity* for musicality; if borne in mind in the exploration of the propensities for, and the functionalities of, music for infants and children, it might yield an understanding of the commonalities that appear necessary in order to relate music to our biologies. As Sandra Trehub notes in this volume, investigations of infant and childhood musical capacities do appear to reveal cross-cultural invariants.

Sandra Trehub and her collaborators have shown that even young infants possess the capacities to perceive significant structural and affective features of musical sounds.[21] The real-world context in which such capacities are most evident is the typically affect-laden interaction of the infant with a caregiver (and it is notable that, irrespective of culture, even adults appear sensitive to features of infant-directed song[22]). Other researchers such as the Papouseks and, more recently, Colwyn Trevarthen and his collaborators have focused closely on the musicality of such interactions.

The Papouseks have been particularly concerned with infant vocal capacities and interactions. Hanus Papousek[23] (p. 43) has noted that "musical elements participate in the process of communicative development very early," suggesting that "they pave the way to linguistic capacities earlier than phonetic elements." He sees (pp. 46–47) infant and early childhood musical behaviors as forms of play, involving higher level integrative processes that act to nurture "exploratory competence" (a notion that seems to rely on the idea of musical signification as transposable); these exploratory competences entail the participation of emotions and constitute precursors of artistic or scientific competencies.

Mechthild Papousek focuses on the musicality of infant–caregiver interactions, stressing the indivisibility in these of music and movement and the fact that they appear to involve patterns of infant *and* caregiver behavior that are singularly invariant across cultures. She notes[24] (p. 100) that "parents' multimodal stimulation is tailored to infants' early competence for perceiving information through different senses as coordinated wholes," and that "regular synchronization of vocal and kinaesthetic patterns provides the infant with multimodal sensory information including tactile, kinaesthetic and visual information."

The work of Trevarthen and his colleagues has centered on these temporal characteristics of infant–caregiver interaction. Trevarthen[25] states that, from birth, central to our neuronal anatomy is a "body-imaging core system" that comes to act so as to integrate attention, learning, and self-regulating physiology with actions of expression and execution; this he terms the *intrinsic motive formation* (IMF). In operation, the IMF incorporates periodic timing mechanisms that give rise to a "hierarchy of motor rhythms;" these, governing movement and binding affect in rhythmic time, he calls the *intrinsic motive pulse*. For Trevarthen (p. 160) "musicality...is the aurally appreciated expression of the IMF with the intrinsic motive pulse as its agent." From these premises, Trevarthen develops a conceptual framework to explore the expression and development of communication—of intersubjectivity, in his terms—through empirical observations and analyses of infant–caregiver interaction.

The rhythmicity of caregiver–infant interaction, in terms of the capacity of the infant to follow and respond in kind to temporal regularities in vocalization and in movement, and in time to initiate temporally regular sets of vocalizations and movements, is seen here as central to the development of human significative and communicative capacities; its embodied nature enables the sharing of patterned time with others and facilitates harmonicism of affective state and interaction. For Trevarthen, that rhythmicity is also a manifestation of a fundamental musical competence. As he frames it (p. 194), "musicality is part of a natural drive in human socio-cultural learning which begins in infancy."

There is, thus, an increasing amount of evidence that musicality is in our birthright; the capacity for music is an integral component of the infant mind. However, the notion of music as innate that emerges from the research just cited does not sit easily with current general theories of the infant mind. While these theories are increasingly nativist in suggesting that the infant mind, rather than being domain general, is endowed with either modular or domain-specific competences, they tend to account for the existence of these competences on the basis of their adaptive value in evolution, and most theorists see no adaptive role for human musicality in evolution. The present consensus[26] suggests that the infant mind is primed for the rapid emergence of competence in (at the least) interpreting social relations, physical and mechanical interactions, and the behavior of biological systems. This view is supported by a substantial quantity of empirical research, in particular by recent work[27] that supports the cross-cultural generality of some of these domains.

There have been some suggestions that musicality might constitute one of these "native domains." Gelman and Brenneman[28] propose that a domain-specific competence in music is evidenced by the results of Trehub and others, suggesting that sensitivities to harmonicism of tonal relations and to melodic contour formations constitute evidence of a music-specific competence. However, their conclusions are rather undermined by their focus on perceptual capacities of which the existence might equally well be accounted for by their utility in other cognitive domains; the tendency to link sounds that are perceived as harmonically related, and to differentiate between sequences of sounds that differ in their contour, appears more likely to derive from a general capacity for auditory scene analysis[29] than to testify to the existence of an early and specifically musical competence. In other words, the suite of perceptual capacities that Gelman and Brenneman identify as making up the domain of musical competence might be epiphenomenal; each capacity might be more securely considered as being proper to other domains that are more self-evidently and immediately functional than is music.

Indeed when music has been viewed from an evolutionary perspective, it has often been viewed as contingent, at best exaptive, a view most clearly exemplified by Steven Pinker[30] and endorsed by others such as Barrow[19] and Sperber.[31] For Pinker, music is, famously "auditory cheesecake;" while music in his view is bound to the domains of language, auditory scene analysis, habitat selection, emotion, and motor control, it does no more than exploit the capacities that have evolved to subserve each of these areas. Music is thus "exaptive," an evolutionary by-product of the emergence of other capacities that have direct adaptive value. Barrow[19] similarly suggests that human musicality has had no role in our survival as a species, suggesting that it derives from an "optimal instinctive sensitivity for certain sound patterns" that itself arose because it proved adaptive. Sperber[31] goes furthest in condemning music as an evolutionary "parasite," though he explicitly disavows serious intent in formulating that view. Nevertheless, he does suggest that music is a human activity that arose to exploit parasitically the operation of a cognitive capacity to "process complex sound patterns discriminable by pitch variation and rhythm" that was originally functional in primitive human communication but that fell into disuse with the emergence of the modern vocal tract and the finer shades of differentiation in sound pattern that it afforded. For Pinker, Sperber, and Barrow, music exists simply because of the pleasure that it affords; its basis is purely hedonic, and, as Pinker puts it, "compared with language, vision, social reasoning, and physical know-how, music could vanish from our species and the rest of our lifestyle would be virtually unchanged."

These three views appear to constitute the beginnings of a consensus that would relegate music to the status of evolutionary footnote and would seem to vitiate the idea that its biological foundations deserve any attention. However, all three theories suffer from an attribute that disqualifies their conclusions from serious consideration, that of ethnocentricity. Theirs is a "culture-lite" view of music. They take no account in their conclusions of the *indivisibility* of movement and sound in music,[c] focusing on only one dimension of music as defined above, that of music's inefficacy in any domain other than the individually hedonistic. Despite lip service paid to the notion that music might take other forms in other cultures, music appears in these theories largely as disembodied sound oriented towards individual hedonism, a notion quite untenable before the advent of recording technology. Indeed, over the last hundred years, recording technology together with the reification of intellectual property and the globalization of its law has sanctioned the subsumption of music into the capitalist economy as a tradable and consumable commodity. It might well be that Pinker's view of music is an accurate reflection of what music is now for some within Western culture, but that culture-specific "music" is scarcely representative of the complex and embodied set of activities and interpretations that are evident in most non-Western "musics." To put it another way, what music is for some at present is not what music is for others, was for our predecessors, or could be for our children.

Even when music has been viewed as adaptive in human evolution, the problem of ethnocentricity can remain. Miller,[32] in promoting the notion that "Machiavellian intelligence" played and still plays a significant role as an agent in processes of sexual selection, suggests that musicality constitutes a marker for possession of such intelligence; musical performance constitutes a display of protean behaviors and functions so as to advertise to prospective mates the possession of the "protean" capacity to be "unpredictable," a capacity that he suggests is of value in social interaction. The putative link between music and sex certainly motivates many adolescents in our society to engage in "musical" behaviors, but most will realize only too quickly that mere presence onstage is no guarantee of successful subsequent sexual interaction. If it could be demonstrated from a comprehensive cross-cultural survey that music's primary function is as a vehicle for the display of "protean behaviors," it would be reasonable to infer that this was music's *raison d'être*. However, the available evidence does not sustain this view; music is and has been employed for many different ends by different societies, and in most the role of music in courtship is positively subsidiary to its value in activities of healing, praying, mourning, or instructing. Miller's view of music seems as bound to the peculiarities of current Western practice as does that of Pinker.

For Pinker, Barrow, Sperber, and Miller, the effects of music are at the level of the individual, whether in terms of affording hedonic experience or exhibiting protean attributes. It is notable that much of the research into infant musicality (particularly that of Trevarthen) suggests a different locus for music's functionality (if any), that of human interaction. Several recent theories of music as adaptive in human evolution have located its functionality at the level of the group, including the writings of

[c]Although Pinker explicitly addresses the link between music and movement he treats this simply in terms of music "tapping in" to systems of motor control.

Kogan and of Brown. Kogan[33] (p. 197) notes that current evolutionary theory suggests that "natural selection operates not only within groups but also between them." He follows McNeill's[35] notion of "muscular bonding" in proposing that the communal experience of affect elicited by moving together rhythmically in music and dance could have enhanced cooperative survival strategies for early humans, for example, in hunting or in intergroup conflict. This efficacy of rhythmic synchronicity in promoting group identity can be related back to the "time-sharing" capacities exhibited in infant–caregiver interaction, though here it seems limited to its impact on affect, ignoring any broader functionality.

For Brown,[34] the adaptive features of music for the group go beyond those on which Kogan relies. Brown adduces the notion of music as reinforcing "groupishness," which he defines as a "suite of traits that favor the formation of coalitions, promote cooperative behavior towards group members and create the potential for hostility towards those outside the group." Music supports these traits through the opportunities that it offers for the formation and maintenance of group identity, for the conduct of collective thinking (as in the transmission of group history and planning for action), for group synchronization—the sharing of time—between members of a group, and for group catharsis, the collective expression and experience of emotion. Ultimately, Brown sees music as having become instantiated in human cultures through its role as "ritual's reward system;" music, for him, is a type of "modulatory system acting at the group level to convey the reinforcement value of these activities...for survival." If Brown is correct in his portrayal of music's role in promoting "groupishness," music is likely to have been a major contributor to what Smith and Szathmáry[36] hold to have constituted one of the major transitions in evolution: the very emergence of human culture.

Dissanayake[37] sees the mature expression of music in human culture as intimately linked to the characteristics of mother–infant interaction. She views music (p. 390) as "multimodal or multimedia activity of temporally patterned movements" that has "the capacity to coordinate the emotions of participants and thus promote conjoinment." She suggests that features of the musicality of mother–infant interaction might lay the foundations for a "grammar of the emotions" that can be expressed in mature musical (and other artistic) activities. For the developing child, the musical characteristics of mother-infant interaction are of critical importance in the acquisition of capacities for "social regulation and emotional bonding;" these characteristics also provide the elements in the "musical play" of later childhood that will equip the adult with the predisposition and capacity to engage in the structured interactions of ceremony and ritual as well as in specifically musical behaviors. However, other significant and functional roles have been proposed for music in individual development and in the development of capacities for social interaction; music can be both a consequence-free means of exploring social interaction and a "playspace" for rehearsing processes that may be necessary to achieve cognitive flexibility.[38]

Music is consequence free in that it is not *directly* functional; it is nonefficacious. It is specifically suited to testing out aspects of social interaction by virtue of both its nonefficaciousness and its polysemic nature, its multiple potential meanings. For each child in a group ostensibly involved in a cooperative musical activity, that musical activity can mean something different, yet the singularity of the collective musical activity is not threatened by the existence of multiple simultaneous and potentially conflicting meanings. Music provides for a child a medium for the gestation of a capacity for social interaction, a risk-free space for the exploration of social behavior that can sustain otherwise potentially risky action and transaction.

Just as one can posit a role for music in the socialization of the child, one can also postulate a role for music in the development of the child's individual cognitive capacities that is quite distinct from its efficacy in the child's acculturation. Again, this role is motivated by the intrinsically polysemic nature of music, the fact that its significances can modulate from situation to situation and can even be simultaneously multiple. If music is *about* anything, it exhibits a deictic intentionality, a "transposable aboutness." It is conceivable that music's "transposable aboutness" is exploited in infancy and childhood as a means of forming connections and interrelations between different domains of infant and childhood competence such as the social, biological, and mechanical. To give a crude example; the arc of a ball thrown through the air, the prosodic contour of a comforting utterance, the trajectory of a swallow as it hawks an insect, the pendular ballistics of a limb swung in purposive movement, might, for a child, each underlie the significances of a single musical phrase or protomusical behavior on different occasions. Indeed, these heterogeneous incidents may be bound together simultaneously in the significance of that phrase or behavior, the music thus exhibiting what I have called elsewhere a "floating intentionality." The floating intentionality of the music can provide for the child a space within which she can explore the possible bindings of these multidomain representations. Hence one and the same musical activity might, at one and the same time, be about the trajectory of a body in space, the dynamic emergence or signification of an affective state, the achievement of a goal and the unfolding of an embodied perspective. All these "aboutnesses" exist not in respect of objects but events, ongoing structures in time, and music or protomusical behaviors afford the opportunity to explore the cross-domain mappings that the representation of temporal sequences of object-states as events makes available.

From this perspective one can advance a second definition of music, one that rests on the idea that what "music" is for any given culture may vary immensely but will derive from the same general human propensities:

> Musics are cultural particularizations of the human capacity to form multiply-intentional representations through integrating information across different functional domains of temporally extended or sequenced human experience and behavior, generally expressed in sound.

In this view, music, or protomusical behaviors, subserve a *metaphorical* domain or perhaps more appropriately, underpin a metaphorical stance, acting to create and to maintain the cognitive flexibility that marks off humans from all other species. It could be that the emergence of protomusical behaviors and their cultural realization as music (and, for the matter, dance) might themselves have been crucial in precipitating the emergence of the cognitive flexibility that marks the appearance of *Homo sapiens*. For if Smith and Szathmáry[36] are correct in maintaining that human culture constitutes one of the major transitions in evolution, and Mithen[39] is correct in claiming that the appearance of *Homo sapiens*, ourselves, is marked by the emergence of a flexible cross-domain cognitive capacity, then music is uniquely fitted to have played a significant role in facilitating the acquisition and maintenance of the skill of being a member of a culture—of interacting socially with others—as well as providing a vehicle for integrating our domain-specific competences so as to endow us with the multipurpose and adaptive cognitive capacities that make us human.

Of course, what music, or more appropriately protomusic, is for infants and children and what it might have been in evolution, is not necessarily what music is for a mature culture or society. Culture shapes and particularizes protomusical behaviors and propensities into specific forms for specific functions, and those, as noted at the outset of this paper, can be so divergent that they do not appear to be mutually reducible—they do not appear to exhibit a "nucleus of common properties" (after Molino[2]). However, it is noteworthy that for most cultures, music—and here one might almost prefer to use the Igbo term *nkwa*, as it seems to capture the interlinking of sound and action that characterizes music for most cultures—the functions that music fulfils, the contexts in which it appears most efficacious, often lie in the realm of ritual and of psychic healing. That is, music often functions in individual and group encounters with the numinous and in the modulation of affective state.

The affective functionality of music can certainly be referred to its embodiment in action and to the contexts within which protomusical activities occur for infants and children,[37] as well as to a broad range of different circumstances in specific cultures.[40] It seems, however, that music's ubiquity and efficacy in encounters with the numinous are best accounted for by reference to protomusic's polysemy, its "floating intentionality." This property of protomusical activities may facilitate the mature use of music in those cultural contexts that deal with what Sperber has called "relevant mysteries." Sperber applies this term to situations where beliefs or mental representations arise that are contradictory but are each separately related to (and hence relevant in respect of) other mental representations and beliefs. When simultaneously foregrounded by actions or circumstances, these contradictory beliefs then become "mysteries" and "achieve relevance because of their paradoxical character—that is, because of the rich background of everyday empirical knowledge from which they systematically depart" (Sperber,[31] p. 72). Within the framework of Sperber's theory, religious ideas are distinguished from everyday beliefs by their paradoxicality and their relevance, by their broad applicability and their ambiguity; and the view of music's functionality outlined above would suggest that music is also distinguished by just such a broad applicability and ambiguity. By virtue of these attributes music may thus be particularly appropriate as a means of amplifying, exemplifying, or reinforcing in the course of ongoing experience just these attributes of belief that are interpretable as religious; music's indeterminacy may suit it for use as a means of pursuing and perhaps even parsing the numinous.

The factors that endow music with its efficacy for individual cognitive development and socialization in infancy and childhood cannot by themselves determine the multiplicitous forms and functions that music takes and that music serves in mature cultural contexts. The meaning of a musical activity for a mature individual will necessarily depend at any given moment on that person's own history and narratives, and on the situational significances that culture's "shared system of meanings" confer on that activity. The polysemic potential that characterizes protomusical activity is likely to underpin the social functionality of music and to contribute to, but not determine, music's meaning. The functionalities and functions of music or protomusical behaviors for the individual, whether in their own cognitive development or in their socialization, must be set in the context of the functionalities and functions of music as a cultural phenomenon. Music, like language, cannot be wholly private; it is a property of communities, not individuals. These different levels at which music may be efficacious must be integrated in any understanding of its foundations.

Music's very existence is best evidenced in interaction. If music is of importance in human development, evolution, and life, then an attempt to render commensurable our understanding of music as interaction with our understanding of music's biological foundations is crucial in coming to terms with what Henry Plotkin[41] (p. 222) calls "the most complicated thing in the universe—the collective of human brains and their psychological processes that make up human culture."

To return to the beginning; what are the implications for this view of music as something more than patterned sound, for an understanding of its biological foundations? We would expect its neurophysiology to be complex, reaching beyond the auditory pathways to the limbic system and to centers of motor behavior. We would expect that the cultural context of music—the forces that shape music for any given culture—should condition its neurophysiological correlates, and we might expect music and language to share many, but not all, neurophysiological correlates.[42] While music and language might meet somewhere near poetry, music can never attain the unambiguous referentiality of language (which Deacon[43] holds to be language's primary defining characteristic), nor language the absolute ambiguity of music.

At the limit, however, while music may be in our biologies, our culture is in our music. If the roots of human musicality are to be found in infancy, particularly in infant–caregiver interaction, its potency might be tied to the support provided by society for those interactions. In an intriguing study Maya Gratier[44] investigated the coherence of interactions between caregivers and infants in three different contexts: French mothers in France, Southern Indian mothers in South India, and Southern Indian mothers who were recent immigrants to France. She found a difference between the coherence of the immigrant mother-infant interactions and those of the other two culturally embedded groups; interaction between the immigrant mothers and their infants was significantly less coherent than was interaction in the other two groups. She suggests that the cultural dislocation of the immigrant mothers had impacted directly on their capacities to interact "musically" with their infants. Something as individual and putatively innate as the capacity of a mother to interact coherently in time with her child seems to be dependent on the mother's rootedness in her cultural environment. In other words, if music is in our birthright, its inheritance appears to be a fragile gift that rests on the humaneness and sympathy of the culture that surrounds us.

## REFERENCES

1. LAKOFF, G. 1987. Women, Fire and Dangerous Things. University of Chicago Press. Chicago.
2. MOLINO, J. 2000. Towards an evolutionary theory of music and language. *In* The Origins of Music. N. Wallin, B. Merker & S. Brown, Eds.: 165–176. MIT Press. Cambridge, MA.
3. GEERTZ, C. 1973. The Interpretation of Cultures: Selected Essays. Basic Books. New York.
4. TREITLER, L. 1980. History, criticism and Beethoven's Ninth Symphony. Nineteenth Century Music **3:** 193–210.
5. TOMLINSON, G. 1984. The web of culture. Nineteenth Century Music **7:** 350–362.
6. ABBATE, C. 1991. Unsung Voices: Opera and Musical Narrative in the Nineteenth Century. Princeton University Press. Oxford.

7. GARNETT, L. 1998. Musical meaning revisited: thoughts on an 'epic' critical musicology. Critical Musicology. http://www.leeds.ac.uk/music/Info/CMJ/Articles/1998/01/01.html.
8. GOEHR, L. 1992. The Imaginary Museum of Musical Works. Clarendon Press. Oxford.
9. SLOBIN. M. 1993. Subcultural Sounds: Micromusics of the West. Weslyan University Press. Hanover.
10. MALM, M. 1977. Music Cultures of the Pacific, the Near East and Asia. Prentice-Hall. Englewood Cliffs, NJ.
11. GOURLAY, K.A. 1984. The non-universality of music and the universality of non-music. The World of Music **26:** 25–36.
12. D'ANDRADE, R. 1996. The Development of Cognitive Anthropology. Cambridge University Press. Cambridge, UK.
13. COLE, M. 1996. Cultural Psychology. Belknap Press of Harvard University Press. London.
14. BLACKING, J. 1995. Music, Culture and Experience. University of Chicago Press. London.
15. MERRIAM, A.P. 1964. The Anthropology of Music. Northwestern University Press. Chicago.
16. HARNAD, S. 1992. There is only one mind/body problem. Int. J. Psychol. **27:** 521.
17. SCHEIFFLIN, E.L. 1993. Performance and the cultural construction of reality: a New Guinea example. *In* Creativity/Anthropology. R. Rosaldo, S. Lavie & K. Narayan, Eds.: 270–295. Cornell University Press. London.
18. FELD, S. 1982. Sound and Sentiment: Birds, Weeping, Poetics and Song in Kaluli Expression. University of Pennsylvania Press. Philadelphia.
19. BARROW, J.D. 1995. The Artful Universe. Clarendon Press. Oxford.
20. SLOBODA, J. A., J.W. DAVIDSON, M.J.A. HOWE & D.G. MOORE. 1996. The role of practice in the development of performing musicians. Br. J. Psychol. **87:** 287–309.
21. TREHUB, S.E., G. SCHELLENBERG & D. HILL. 1997. The origins of music perception and cognition: a developmental perspective. *In* Perception and Cognition of Music. I. Deliège & J. Sloboda, Eds.: 103–128. The Psychology Press. Hove.
22. TREHUB, S.E., A.M. UNYK & L.J. TRAINOR. 1993. Adults identify infant-directed music across cultures. Infant Behav. & Dev. **16:** 193–211.
23. PAPOUSEK, H. 1996. Musicality in infancy research: biological and cultural origins of early musicality. *In* Musical Beginnings. I. Deliège & J. Sloboda, Eds.: 37–55. Oxford University Press. Oxford.
24. PAPOUSEK, M. 1996. Intuitive parenting: a hidden source of musical stimulation in infancy. *In* Musical Beginnings. I. Deliège & J. Sloboda, Eds.: 88–112. Oxford University Press. Oxford.
25. TREVARTHEN, C. 1999. Musicality and the intrinsic motive pulse: evidence from human psychobiology and infant communication. Musicae Sci. Special Issue: 155–215.
26. SPELKE, E. 1999. Infant cognition. *In* The MIT Encyclopedia of Cognitive Sciences. R.A. Wilson & F.C. Keil, Eds.: 402–404. MIT Press. Cambridge, MA.
27. MEDIN, D. & S. ATRAN. 1999. Folk Biologies. MIT Press. Cambridge, MA.
28. GELMAN, R. & K. BRENNEMAN. 1994. First principles can support both universal and culture-specific learning about number and music. *In* Mapping the Mind: Domain Specificity in Cognition and Culture. L.A. Hirschfeld & S.A. Gelman, Eds.: 396–390. Cambridge University Press. Cambridge, UK.
29. BREGMAN, A. 1990. Auditory Scene Analysis. MIT Press. Cambridge, MA.
30. PINKER, S. 1997. How the Mind Works. Allen Lane. London.
31. SPERBER, D. 1996. Explaining Culture. Blackwell. Oxford.
32. MILLER, G.F. 1997. Protean primates: the evolution of adaptive unpredictability in competition and courtship. *In* Machiavellian Intelligence II: Extensions and Evaluations. A. Whiten & R.W. Byrne, Eds.: 312–340. Cambridge University Press. Cambridge, UK.
33. KOGAN, N. 1997. Reflections on aesthetics and evolution. Crit. Rev. **11:** 193–210.
34. BROWN, S. Evolutionary Models of Music: From Sexual Selection to Group Selection. Forthcoming.
35. MCNEILL, W.H. 1995. Keeping Together in Time. Harvard University Press. London.
36. SMITH, J.M. & E. SZATHMÁRY. 1995. The Major Transitions in Evolution. Oxford University Press. Oxford.

37. DISSANAYAKE, E. 2000. Antecedents of the temporal arts in early mother–infant interactions. *In* The Origins of Music. N. Wallin, B. Merker & S. Brown, Eds.: 389–407. MIT Press. Cambridge, MA.
38. CROSS, I. 1999. Is music the most important thing we ever did? Music, development and evolution. *In* Music, Mind and Science. Suk Won Yi, Ed.: 10–39. Seoul National University Press. Seoul.
39. MITHEN, S. 1996. The Prehistory of the Mind. Thames & Hudson. London.
40. SLOBODA, J.A. & P. JUSLIN, Eds. 2001. Music and Emotion: Theory and Research. Oxford University Press. Oxford.
41. PLOTKIN, H. 1997. Evolution in Mind. Allen Lane. London.
42. PERETZ, I., S. BELLEVILLE & S. FONTAINE. 1997. Dissociation between music and language following cerebral hemorrhage: another instance of amusia without aphasia. Can. J. Exp. Psychol. **51:** 354–368.
43. DEACON. T. 1996. The Symbolic Species: The Co-evolution of Language and the Human Brain. Allen Lane. London.
44. GRATIER, M. 1999. Expressions of belonging: the effect of acculturation on the rhythm and harmony of mother-infant interaction. Musicae Sci. Special Issue: 93–122.
45. ROSE, S. 1996. Lifelines: Biology, Freedom, Determinism. Allen Lane. London.
46. RORTY, R. 2000. Being that can be understood is language. London Rev. Books **126:** 23–25.
47. FEYERABEND, P. 1981. Problems of Empiricism. Vol. 2. Cambridge University Press. Cambridge, UK.

# Is Music an Evolutionary Adaptation?

DAVID HURON

*School of Music, Ohio State University, Columbus, Ohio 43210, USA*

**ABSTRACT: In contemplating the function and origin of music, a number of scholars have considered whether music might be an evolutionary adaptation. This article reviews the basic arguments related to evolutionary claims for music. Although evolutionary theories about music remain wholly speculative, musical behaviors satisfy a number of basic conditions, which suggests that there is indeed merit in pursuing possible evolutionary accounts.**

**KEYWORDS: Evolutionary theories of music; Music industry; Evolutionary origin of language; Music and social bonding; Oxytocin; Mood regulation**

Addressing the question of music's origins has a long history and is patently speculative. Although several cultures have provided colorful stories describing how people acquired the capacity for music, most contemporary scholars have focused on possible psychological, social, and cultural beginnings. In this article, I propose to offer a social account of music's origins that is explicitly linked to one of the most successful theories yet devised: the theory of evolution by natural selection.

Evolution is often thought of in purely physiological rather than psychological terms.[1–5] It is not simply that evolution has shaped immune systems, digestive tracts, and knee caps. Evolution has also shaped our attitudes, dispositions, emotions, perceptions, and cognitive functions. Some of our deepest convictions can be traced to plausible evolutionary origins: we love life, we fear death, and we nurture our children because these dispositions better ensure the propagation of our genes.

The theory of evolution by natural selection is a distal theory. It is not a theory that explains specific behaviors, such as why you chose to cook ravioli for dinner last night, or why you parked in a particular parking spot this morning. Similarly, if music is an evolutionary adaptation, this will not allow us to account for particular musical acts, such as Mendelssohn's writing of the *Scottish Symphony.* Evolution proceeds by selecting traits that are adaptive to an organism's environment. For example, evolution did not "originate" or "create" the phenomenon of altruism. Instead, given a certain environment, natural selection favored individuals who exhibited certain altruistic traits. Evolution does not dictate our behavior: it selects which behaviors are likely to be passed on to subsequent generations—and it selects only those behaviors that have a genetic component. So in discussing possible evolutionary origins for musical behaviors, the question is not, What caused people to make music? but rather, How might music-making behaviors have escaped the hatchet of natural selection? or more precisely, What advantage is conferred on those individuals who exhibit musical behaviors over those who do not?

Address for correspondence: David Huron, Ph.D., School of Music, Ohio State University, 1866 College Road, Columbus, OH 43210. Voice: 614-688-4753; fax: 614-292-1102.
huron.1@osu.edu

## DOES MUSIC HAVE SURVIVAL VALUE?

Many knowledgeable people have concluded that music has no survival value. Indeed, a number of esthetic philosophers have argued that an essential, defining characteristic of the arts is that they serve no practical function.[6] Accordingly, any music that is created for biological (or even economic) reasons cannot be considered art. Even among evolutionary psychologists such as Steven Pinker, it has been common to suppose that music is not adaptive.[7]

Many linguists—Pinker included—believe that language is likely an evolutionary adaptation.[8] However, the evidence in support of language as an adaptation is not notably stronger than comparable evidence for music. Where pertinent evidence is available, music exhibits the essential properties of all adaptations. My goal in this article is not to attempt to prove that music is adaptive. (Like others, I am not at all convinced that music has evolutionary origins.) Rather, my goal here is to convince you that the question of music's origins remains open and warrants further investigation (see also several articles in Wallin, Merker, and Brown[9]).[10]

Of course there are a number of dangers attending evolutionary speculation.[11,12] Popper (1935/1959) pointed out that no scientist has yet formulated the theory of evolution by natural selection in such a way that a set of observations could, in principle, be used to falsify it.[13] Gould and Lewontin[14] have noted that evolutionary reasoning is plagued by *post hoc* reasoning. Evolutionary theory has been used to defend all sorts of nefarious ideologies, from racism to sexism. Philosophers note that evolutionary arguments often lead to the naturalist fallacy where what is is confused with what ought to be. In the case of music, there is an undistinguished history of polemical writing where certain kinds of music have been condemned for being "unnatural." Finally, by focusing on biological issues, one can leave the false impression that the effects of culture on music are minimal.

If music is an evolutionary adaptation, then it is likely to have a complex genesis. Any musical adaptation is likely to be built on several other adaptations that might be described as premusical or protomusical. Moreover, the nebulous rubric *music* may represent several adaptations, and these adaptations may involve complex coevolutionary patterns with culture (see Ref. 15). In biological matters, things are rarely straightforward.

Given these possible dangers, why bother attempting to formulate an evolutionary theory of music? Isn't it premature? First, as noted above, my goal here is not to convince you that music is adaptive; my goal is only to convince you that this is a worthwhile question. Understanding the possible origins of music might help inform us about some of the reasons we tend to respond in certain ways. Second, in the spirit of Popper, I will aim to tell an evolutionary story that is able to generate testable hypotheses. Like other evolutionary accounts, my own theory will draw on existing knowledge, and so be *post hoc* in character. As long as this account remains *post hoc*, Gould and Lewontin's criticisms raise justified and paramount difficulties. However, it is my hope that the theory can be developed to the point where testable hypotheses might be derived.

Before entertaining some possible evolutionary views of music's origins, let us first consider two pertinent complicating points of views. One view is that music is a form of nonadaptive pleasure seeking. A second view is that music is an evolutionary vestige.

## NAPS THEORY OF MUSIC

Most pleasurable activities, such as eating and sex, have clear links to survival. Such activities ultimately stimulate brain mechanisms that are specifically evolved to reward and encourage adaptive behaviors. Note that once brain mechanisms are in place that permit the experience of pleasure, it may be possible to stimulate those mechanisms in ways that do not confer a survival advantage. We can call these behaviors nonadaptive pleasure seeking (NAPS). An example of NAPS behavior is found in the human taste for sugars and fats. In premodern times, sugars and fats were rare in human diets, but highly nutritious in the amounts available. There are good reasons why human tastes would evolve to reward the ingestion of foods with high fat or sugar content. However, centuries of human ingenuity have succeeded in generating a modern diet that contains unnaturally high levels of fats and sugars—levels so high as to cause health problems such as diabetes and heart disease. Although such tastes originally conferred an increased chance of survival, in the modern environment, these behaviors have become less adaptive.

Another example of NAPS behavior is found in drug use, such as heroin or cocaine. These drugs can directly activate the brain's pleasure centers, simply by injecting or imbibing a substance. Although the channel for pleasure exists for good evolutionary reasons, it may be possible to exploit the channel without any concomitant survival-enhancing result.

As in the case of drugs, it is possible that musical behaviors are forms of nonadaptive pleasure seeking. That is, music itself may not enhance human survival; music may merely exploit one or more existing pleasure channels that evolved to reinforce some other adaptive behavior(s). We might call this view the "NAPS theory of music."

One way to determine whether some pleasure-seeking behavior is adaptive or nonadaptive is to consider how long the behavior has been around. In the long span of evolutionary history, nonadaptive pleasure-seeking behaviors tend to be short-lived. For example, heroin users tend to neglect their health and are known to have high mortality rates. Furthermore, heroin users make poor parents; they tend to neglect their offspring. Poor health and neglect of offspring are infallible ways of reducing the probability that one's genes will be present in a future gene pool. After many generations, natural selection will tend to militate against heroin use. Those individuals who are not disposed (for whatever reason) to use heroin, are much more likely to procreate and so pass along their aversion to the use of such drugs, provided that the aversive behavior is somehow linked to a gene or genes.

The use of alcohol already suggests how NAPS behaviors can transform a gene pool. Although no gene has been identified, either for alcohol susceptibility or for alcohol tolerance, the responses of different human populations to alcohol show a suggestive pattern. Large quantities of alcohol became possible only with the advent of agriculture. Modern European and Asian descendents of early agrarian cultures (such as originated in Mesopotamia) manage to deal with alcohol better than descendents of traditional hunter-gatherer societies, such as indigenous peoples in the Americas and in the arctic regions of Europe. Of course there are certain to be nongenetic factors influencing alcohol tolerance and abuse. However, alcohol researchers suspect that genetic factors are at work. Those people who have descended from traditional agricultural societies have a clear statistical advantage in dealing with the nonadaptive consequences of alcohol, and this would be expected if alcohol had been prevalent in these societies for thousands of years.

If music itself has no survival value (and merely exploits an existing pleasure channel), then any disposition towards musical behaviors would tend to worsen one's survival. Spending inordinate amounts of resources (such as time and money) on music would be expected to place music lovers at an evolutionary disadvantage. In other words, if the NAPS theory of music is true, then we would expect music appreciation to be correlated with marginal existence: as in the case of alcohol, music lovers would be disproportionally more likely to end up on "skid row."

In addition, if music is nonadaptive, then it must be the case that music is historically recent; otherwise music lovers would have become extinct some time ago. As we will see, the archaeological evidence indicates that music is very old—much older than agriculture—and this great antiquity is inconsistent with music originating as a nonadaptive pleasure-seeking behavior. In short, there is little evidence that musical behaviors have been selected against. All of this suggests that there is little support for the NAPS theory of music.

## MUSIC AS AN EVOLUTIONARY VESTIGE

Another possibility is that, although music at one time did indeed confer some survival value, it is now merely vestigial. Like the human appendix, at one time this "organ" may have contributed directly to human survival, but now it is largely irrelevant —a piece of evolutionary litter. If this view is true, then we would have to ask What advantage did music once confer? and How have things changed so that music is no longer adaptive?

## MEASURING THE ADAPTIVE VALUE OF MUSIC

The adaptive value of some function is often evident in the individual survival costs arising from that function. For example, the larynx of newborn infants is anatomically arranged so that breathing and swallowing can happen at the same time. When the larynx enlarges, our physiological capacity for speech is purchased at the price of the danger of choking. In effect, one measure of the evolutionary advantage of speech is the mortality rate due to choking.

Similarly, an estimate of the evolutionary advantage conferred by music is to measure the amount of time people spend in musical behaviors. In the Atlas mountains of Morocco, full-time Jujuka mountain musicians are supported by the local villagers. That is, there is an entire caste of people whose principal productive activity is music making. A ready index of the importance of music in such a society may be the ratio of the number of musicians to the number of farmers and herders.

## SOME EVOLUTIONARY THEORIES OF MUSIC

Of the various proposals concerning a possible evolutionary origin for music, eight broad theories can be identified:

*Mate selection.* In the same way that some animals find colorful or ostentatious mates attractive, music making may have arisen as a courtship behavior. For example, the ability to sing well might imply that the individual is in good health.

*Social cohesion.* Music might create or maintain social cohesion. It may contribute to group solidarity, promote altruism, and so increase the effectiveness of collective actions such as defending against a predator or attacking a rival clan.

*Group effort.* More narrowly, music might contribute to the coordination of group work, such as pulling a heavy object.

*Perceptual development.* Listening to music might provide a sort of "exercise" for hearing. Music might somehow teach people to be more perceptive.

*Motor skill development.* Singing and other music-making activities might provide (or have provided) opportunities for refining motor skills. For example, singing might have been a necessary precursor to the development of speech.

*Conflict reduction.* In comparison with speech, music might reduce interpersonal conflict. Campfire talk may well lead to arguments and possible fights. Campfire singing might provide a safer social activity.

*Safe time passing.* In the same way that sleep can keep an animal out of harm's way, music might provide a benign form of time passing. Evolutionary biologists have noted, for example, that the amount of sleep an animal requires is proportional to the effectiveness of food gathering. Efficient hunters (such as lions) spend a great deal of time sleeping, whereas inefficient feeders (such as grazing animals) sleep relatively little. Sleep is thought to help keep an animal out of trouble. A lion is more apt to injure itself if it is engaged in unnecessary activities. As early humans became more effective at gathering food, music might have arisen as a harmless pastime. (Note, for example, that humans sleep more than other primates.)

*Transgenerational communication.* Given the ubiquity of folk ballads and epics, music might have originated as a useful mnemonic conveyance for useful information. Music might have provided a comparatively good channel of communication over long periods of time.

### *Sexual Selection*

Before continuing, we should take a moment to discuss a variant of the mate selection theory. Charles Darwin identified a form of natural selection known as sexual selection. The classic example of sexual selection is the peacock's tail. The function of the tail is not to promote the survival of the peacock; rather, the function is to promote the survival of the peacock's genes. Sexual selection arises once a particular genetic preference is established by the opposite sex—in this case, the preference of the peahen for flashy tails. Even if one peahen is not particularly impressed by Las Vegas–style tails, it remains to the female's benefit to mate with the most colorful male if her offspring are more likely to be desired by other females who are fond of colorful tails.

Darwin himself suggested that music might have arisen due to sexual selection in mating calls.[16] Like the peacock's tail, the preferences of hominid women could create an escalating competition for ever more elaborate and beautiful melodies. Miller[17] has suggested that sexual selection accounts for why musical interests appear to peak in adolescence. However, the members of the all-male Vienna Philharmonic notwithstanding, there is nothing to indicate that one sex is more musical than the other, and so there is no evidence of the dimorphism commonly symptomatic of sexual selection. Women may be impressed by men who serenade them outside their balcony windows, but unlike female songbirds, female humans are perfectly capable of serenading men. Currently, there is no other known example of sexual selection that does not exhibit high sexual dimorphism—although one should note that biology is full of surprises.

## TYPES OF EVIDENCE

In presenting a case for the evolutionary origins of music, we can consider four types of evidence:

*Genetic evidence.* The best evidence of an evolutionary origin would be the identification of genes whose expression leads to the behavior in question. Unfortunately, it is rare for scientists to be able to link particular behaviors to specific genes. Although behavior-linked genes have been discovered in other animals (such as fruit flies), no behavior-linked gene has yet been conclusively established in humans. As in so many other areas, music has attracted a kind of folklore related to heritability. In some cultures, it is common for people to assume or believe that musical talent is partly inherited. More recently, work at the University of California, San Francisco by Baharloo *et al.*[18] appears to suggest a genetic component for absolute pitch.

*Neurological evidence.* The existence of specialized brain structures is neither a sufficient nor a necessary condition for music to be an evolutionary adaptation. Nevertheless, if stable anatomical brain structures exist for music, then this is consistent with music arising from innate development rather than being due solely to a generalized learning.

*Ethological evidence.* Are musical behaviors consistent with survival and the propagation of genes? In order for music to be an evolutionary adaptation, music-related behaviors must somehow increase the likelihood that the musical person's genes will be propagated.

*Archaeological evidence.* Since complex evolutionary adaptations arise over many thousands of generations, we must ask how widespread music is in biological history? If music originated in the past few thousand years, then it is highly unlikely to be an evolutionary adaptation. Evolution does not work that fast.

As noted, there is currently no evidence that links music to any gene. Let's consider the other areas of evidence in more detail.

## ARCHAEOLOGICAL EVIDENCE

Let's begin by considering some of the archaeological facts. The archaeological record shows a continuous record of music making in human settlements. Wherever you find evidence of human settlement, you find evidence of musical activities.

In 1995, paleontologist Ivan Turk discovered a bone flute while excavating an ancient burial mound in Divje Babe, Slovenia.[19,20] This flute has been determined to be between 43,000 and 82,000 years old, using electron spin dating. The instrument was fashioned from the femur of the now-extinct European bear.

Of course, finding this flute does not mean we have found the earliest musical instrument; this is just the earliest found instrument. As musical instruments go, flutes are rather complicated devices. If we look at contemporary hunter-gatherer societies, the most common instruments are rattles, shakers, and drums. For example, prior to the arrival of Europeans, by far the most common instruments in native American cultures were rattles and drums. The same pattern of preferred instruments is evident in African and Polynesian cultures. If we assume that rattles and drums typically predated the use of flutes, then the ancient music makers of Slovenia might well have been creating instrumental music somewhat earlier than 100,000 years ago.

What sort of music making, however, might have existed prior to the fashioning of musical instruments? It is not unreasonable to assume that singing preceded the making of musical instruments by some length of time. If we suppose that singing predated instrument making by 50% of the intervening time, then music making might have existed 150,000 years ago—roughly twice the age of the older estimate for the Divje Babe flute. Even this figure might be a conservative estimate, and the actual origin of music might be twice as old, say around 250,000 years ago.

On the other hand, the Divje Babe flute might truly be an early specimen, and singing might have developed about the same time. Using the most recent estimate for the Divje Babe flute would therefore place the origins of music making about 50,000 years ago.

In summary, the archaeological record implies that music making likely originated between 50,000 years ago and a quarter of a million years ago. Although Wurlitzer organs, American Bandstand, and MTV are relatively recent phenomena, music making, in general, is really quite old. The evidence pointing to the great antiquity of music satisfies the most basic requirement for any evolutionary argument. Evolution proceeds at a very slow pace, so nearly all adaptations must be extremely old. Music making satisfies this condition.

Incidentally, the antiquity of music raises problems for those who would wish to use evolutionary arguments in esthetic debates by claiming that one music is "more natural" than another. Whatever the origins of music, the vast majority of people have long ceased to live in Pleistocene conditions. In deciding whether Twisted Sister is better or worse than the Grateful Dead, appealing to the sounds of neolithic caves is unlikely to help.

## ANTHROPOLOGICAL EVIDENCE

Turning to contemporary anthropology, we can ask, What does the plethora of existing human cultures tell us about music? Without taking time to review the evidence, there is one overwhelming conclusion from the modern anthropological record. There is no human culture known in modern times that did not, or does not, engage in recognizably musical activities.

Not only is music making very old, it is ubiquitous; it is found wherever humans are found.[21] Moreover, I neglected earlier to mention one important fact about the bone flute at Divje Babe: the flute was found in a Neanderthal burial site. The Divje Babe flute isn't even a human artifact. In short, it may be that music making is not just ubiquitous among *Homo sapiens*; music making may possibly be characteristic of the entire genus *Homo*.

The evidence pointing to the ubiquity of music satisfies another important basic requirement for any evolutionary argument. Relatively few adaptations are not found throughout the entire population of the affected species. For example, if eyelashes confer an evolutionary advantage, then just about everyone should have eyelashes. There are some exceptions to this principle, some of which are very important. For example, humans divide into female and male versions, so there are some genes that are not shared by everyone. Another more subtle example is the gene that codes for sickle-cell—a gene that protects against malaria, but that can also cause anemia. In general, however, adaptive genes are typically ubiquitous throughout a gene pool.

## ETHOLOGICAL EVIDENCE

When studying a particular animal, ethologists often begin by making an inventory of observed behaviors. What does the animal do, and how often does it do it? Activities that require a great deal of time and large expenditures of energy are understandably considered important. Ethologists assume that behaviors are likely to be optimized. Even behaviors that seem unimportant (such as infant play or sleeping) often have a serious or critical purpose.

Primates, for example, spend an extraordinary amount of time grooming each other. Ethologists feel obliged to formulate theories that account for the various proportions of resources dedicated by an animal to different activities.

Let's apply the ethological approach to the behaviors we call musical. For the purposes of illustration, we will consider two case descriptions. The first case is that of the Mekranoti Indians of the Brazilian Amazon, and the second case is that of contemporary society in the United States.

### *The Mekranoti Indians*

The Mekranoti Indians are hunter-gatherers who live in the Amazon rain forest of Brazil. In Mekranoti culture, singing plays a prominent role in daily life. For several months of the year, every morning and evening the women lay banana leaves on the ground where they sit and sing for between one and two hours. The men sing early every morning, starting typically around 4:30, but sometimes as early as 1:30 A.M. The men sing for roughly two hours each day, and often they will also sing for a half hour or so before sunset.

When singing, the Mekranoti men hold their arms in a sort of cradling position and swing their arms vigorously. The men endeavor to sing in their deepest bass voices and heavily accent the first beats of a pervasive quadruple meter with glottal stops that make their stomachs convulse in rhythm. Anthropologist Dennis Werner[22] describes their singing as a "masculine roar." When gathering in the middle of the night, the men are obviously sleepy, and some men will linger in their lean-tos well after the singing has started. These malingerers are often taunted with shouted insults.

Werner reports that "Hounding the men still in their lean-tos [is] one of the favorite diversions of the singers. 'Get out of bed! The Kreen Akrore Indians have already attacked and you're still sleeping,' they [shout] as loudly as they [can].... Sometimes the harassment [is] personal as the singers [yell] out insults at specific men who rarely [show] up."[23]

What is extraordinary about the Mekranoti singing is the amount of time involved—roughly two hours per day. (Remember, this is a subsistence hunter-gatherer society.) For the evolutionary ethologist, the important question arising from the Mekranoti Indians is why music making would attract so much of the tribe's resources. We will return to this question later.

### *Modern United States of America*

By way of comparison, consider now the prevalence of music in a modern industrialized society like the United States. For the ethologist looking at modern human behaviors, a crude though ready index of the amount of resources we dedicate to a particular activity can be found by measuring economic activity.

There is a widespread misconception that the foremost export sector in the US economy is "high technology." In fact, the preeminent export sector in the US economy is entertainment. Of the various component areas—films, sports, television, toys, and games—it is music that ranks foremost.

How big is the music industry? The music industry is bigger than the pharmaceutical industry. People spend more money on music than on prescription drugs. We purchase recordings, go to concerts, buy sheet music, take our children to music lessons, listen to commercial radio, watch film accompanied by music, and encounter Muzak in the local shopping mall. The most active concert venues in the world are freeways: a major preoccupation for millions of drivers is listening to music.

Of course financial measures are crude indicators of behavioral significance. The ethological point is simple. In both a hunter-gatherer society and a modern industrial society, we find humans dedicating a notable proportion of the available resources to music making and listening. Music may not be more important than sex; but it is arguably more expensive, and it is certainly more time consuming.

In order to put these behaviors in perspective, suppose you were a Martian anthropologist visiting Earth. There are many aspects of human behavior that would have recognizable value. You would see people engaged in growing and preparing food, in raising and educating children, people involved in transportation, health, and governance; but even if Martian anthropologists had ears, I suspect they would be stumped by music.

If you are still not convinced that music attracts a peculiarly excessive proportion of human resources, consider another comparison. Think of how important food is to human well-being, of how tasty and enjoyable food is and can be. Now how many universities have departments of cuisine or nutrition, or departments of food sciences, or even departments of home economics? Now consider how many universities have departments of music. Why would music figure more prominently than food? To a visiting tourist from Mars, music sticks out; it is a remarkable and bizarre activity that earthlings do.

Of course, we must be careful in drawing any conclusions about adaptations based on observations of modern behaviors. If music making is an adaptive behavior, then it must have arisen long ago in the environment of evolutionary adaptedness—namely, the Pleistocene period, when the vast majority of human evolution occurred.

## ETHOLOGY AND EVOLUTION

Just because an animal spends a lot of time on certain activities does not mean that the activity represents an evolutionary adaptation. Ethologists must connect the behavior to an explicit evolutionary account. That is, there must exist a plausible explanation of how the behavior would be adaptive.

Before considering such a theory for music, let us examine a nonmusical example —an example that has a richer theoretical literature about its origins. Specifically, we will consider some of the evolutionary arguments that have been advanced to account for the origins of language.

## ON THE EVOLUTIONARY ORIGIN OF LANGUAGE

As in the case of music, views concerning the origins of language are necessarily speculative. Nevertheless, we can learn a great deal by considering some of the theories that have been advanced concerning its origin. Until recently, the principal view of language was that it facilitated complex collaborative activities such as coordinating actions during hunting. This account seems unlikely, first, because talking is a bad idea when tracking prey, and second, because men as a group display inferior language skills compared with women.

A number of anthropological psychologists have suggested that language (and even music) evolved as surrogates for social bonding.

### *The Grooming and Gossip Theory of Language Origins*

The most empirically grounded of the recent theories of language origins is what might be called the "grooming and gossip hypothesis." Its principal advocate is Robin Dunbar.[24] The theory proposes the following logic.

Animals often live in groups for mutual protection against predators. In general, larger groups are more effective in detecting and warding off predators than smaller groups, but there are costs associated with maintaining a large group. One cost is that feeding must be much more intensive in a given area and so a larger group must travel greater distances in search of food. A second cost is that as group size increases, threats are more likely to arise from internal conflict within the group rather than from external predators. That is, there is a point where group size effectively minimizes predation, but at the cost of threats from members of the group itself. Nowhere is this more evident than in primates. As a consequence of internal threats, animals within the group begin to form alliances with one another. These alliances reduce the likelihood of conflict due to the threat of group retaliation.

In primates, the principal means by which alliances are formed and bonds maintained is through grooming. Grooming accounts for between 10% and 20% of an individual's daytime activities.

There is good evidence to suggest that the principal purpose of grooming is to form alliances between individuals. First, grooming partners are much more likely to come to the defense of one another when threatened by another member of the group. Even more important evidence comes from relating the amount of time spent grooming to the size of the group. Different primate species have different typical group sizes. Gorillas, macaques, chimpanzees, and bonobos, among other groups, tend to form groups that have different average sizes. Primatologists have measured the different amounts of time each species engages in grooming.[25]

A major discovery has been that there is a consistent relationship between group size and the amount of time spent grooming. As the group size increases, the average grooming time also increases. This is an unusual finding: there is no reason to suppose that animals in larger groups tend to get dirtier than animals in smaller groups, so the increase in grooming is unlikely to be related to cleanliness. Primatologists widely agree that the increase in grooming time for larger groups arises from the need to form more extensive networks of alliances. In a large group, an individual fares better by having a wider circle of friends, and the way to build primate friendships is through mutual grooming.

In the case of humans, the common "group size" has been estimated at roughly 150 people. This is approximately the size of most rural villages in the world. This means that human groups are especially large when compared with other primates. As Dunbar has pointed out, "If modern humans tried to use grooming as the sole means of reinforcing their social bonds, as other primates do, then the equation for monkeys and apes suggests we would have to devote around 40% of our day in mutual mauling."[26]

Dunbar has suggested that language evolved as an alternative to physical grooming. In effect, physical grooming was replaced by "vocal grooming," whose purpose remains the formation and maintenance of friendships or alliances. Such vocal grooming has a distinct advantage over physical grooming: we can talk to several people simultaneously. This increases the number of people we can bond with at the same time.

Note, however, that even language has significant limitations for multiple concurrent social interaction. Dunbar[27] has noted that "there appears to be a decisive upper limit of about four on the number of individuals who can be involved in a conversation." When a fifth or sixth person joins a conversation there is a marked tendency for the group to subdivide into two or more concurrent conversations. It is only in hierarchical situations (such as in a formal lecture) where a single conversation can be maintained in a larger group.

All of this suggests that language is most useful in close interpersonal interactions, such as grooming, gossiping, courting, and conspiring. Note, however, that there are other activities that are of value to members of a social group that involve the entire group (or at least large segments) rather than groups of twos or threes. Chief among these group activities is defense. When under threat, uniform group action is indeed a mighty force, much more powerful than smaller groups of twos and threes.

## MUSIC AND SOCIAL BONDING

At this point, we might speculate how music might fit into this account. Let's assume, for the moment, that the hypothesis that language evolved as a surrogate for physical grooming is true, and that language thereby allowed humans to live in larger groups with their attendant complex social relations. We could certainly conceive of a similar function for music. In some ways, music provides several advantages over language. Singing is much louder than speaking, so singing may facilitate group interactions involving more than the four individuals posited as the upper limit for conversation.

This view of the possible origins for music was essentially proposed by Juan Roederer in 1984: "... the role of music in superstitious or sexual rites, religion, ideological proselytism, and military arousal clearly demonstrates the value of music as a means of establishing behavioral coherency in masses of people. In the distant past this could indeed have had an important survival value, as an increasingly complex human environment demanded coherent, collective actions on the part of groups of human society."[28]

In light of later work by primatologists such as Dunbar, there appears to be merit in Roederer's hypothesis. Music might have originated as an adaptation for social bonding—more particularly, as a way of synchronizing the mood of many individuals in a larger group. That is, music helps to prepare the group to act in unison.

Perhaps a helpful image is to imagine the cackling of geese prior to their taking off. How is it that individual geese manage to synchronize their actions so that the entire flock takes flight more or less simultaneously? For anyone who has watched geese take off, there is a clear increase in the volume of cackling: more and more geese start honking. The general hubbub of honking geese is apt to raise the arousal levels of all geese in the vicinity. This increased arousal (which includes increased heart rate) would prepare the geese for a significant collective expenditure of energy.

## MUSIC AND SOCIAL BONDING—FURTHER EVIDENCE

It is this theory of music and social bonding that I believe holds the greatest promise as a plausible evolutionary origin for music. For the remainder of this article, I would like to review further phenomena that provide support for this hypothesis. The evidence is going to come from the following five sources: various mental disorders imply a strong link between sociability and musicality; child development implies a social role for music; brain structures related to music are linked to social and interpersonal functions; the most popular musical works imply social functioning; music modifies hormone production in groups of people.

### *Complementary Disorders: Williams Syndrome and Asperger Autism*

Consider two mental disorders: Williams syndrome and Asperger-type autism. The principal feature of Williams syndrome is mental retardation. Williams syndrome is unique in that sufferers display three additional characteristics. One characteristic is high verbal abilities. Individuals suffering from Williams syndrome take a great interest in words. Their speech is fluent and peppered with a remarkably sophisticated vocabulary. In fact, when first encountering someone with Williams syndrome, the language fluency tends to mask the mental handicap.

In addition to high verbal abilities, Williams syndrome individuals also exhibit high sociability. They are gregarious and sociable. Coupled with the high verbal abilities, this makes Williams syndrome children a delight to work with. Finally, Williams syndrome children exhibit high musicality.

Daniel Levitin and Ursula Bellugi[29] have described the musical activities of Williams syndrome children at a summer camp in New York State. The children are remarkable. The entire camp is alive with music. Although the children have been shown to have no greater musical aptitude than normals, they clearly relish both the musical activities and the social environment in which their musical enthusiasms can flower.

Now consider the case of Asperger-type autism. Autism is characterized by a strong aversion to social interaction. Although most autism is associated with reduced mental functioning, mental retardation is not always evident. There are autistic individuals with normal and above-average intelligence as well. Autism is related to an emotional deficit—notably the failure to develop the so-called secondary or social emotions, including shame, pride, guilt, love, and empathy. For normal children, these secondary emotions typically appear by the age of about four.

Temple Grandin is a high functioning Asperger-type autistic who has become well known through her writings about her own condition. Concerning love, Grandin talks about her confusion in high school when reading Shakespeare's *Romeo and Juliet*. "I never figured out what it was all about," said Grandin. In a trip through the Rocky Mountains with Oliver Sacks, Grandin remarked, "The mountains are pretty, ... but they don't give me a special feeling, the feeling you seem to enjoy." "You get such joy out of the sunset," she said. "I wish I did, too. I know it's beautiful, but I don't 'get' it."[30] Grandin's experience of music is similar. Although Grandin has perfect pitch and what she describes as a tenacious and accurate auditory memory, she finds music leaves her cold. She finds the sounds "pretty," but in general, she just doesn't get it.[31] All the fuss about music leaves her mystified.

Grandin's own explanation is that not all of the "emotional circuits" are connected. Sacks interprets the phenomenon as follows: "An autistic person can have violent passions, intensely charged fixations and fascinations, or, like Temple [Grandin], an almost overwhelming tenderness and concern in certain areas. In autism, it is not affect in general that is faulty but affect in relation to complex human experiences, social ones predominantly, but perhaps allied ones—esthetic, poetic, symbolic, etc. No one, indeed, brings this out more clearly than Temple herself....She feels that there is something mechanical about her mind, and she often compares it to a computer.... She feels that there are usually genetic determinants in autism; she suspects that her own father, who was remote, pedantic, and socially inept, had Asperger's—or, at least, autistic traits—and that such traits occur with significant frequency in the parents and grandparents of autistic children."[32]

The contrast between Asperger-type autism and Williams syndrome is striking. On the one hand we have a group of people whose symptoms include high sociability linked with high musicality. On the other hand we have a group of people whose symptoms include low sociability linked with low musicality. Together, these mental conditions are consistent with a relationship between sociability and musicality—and this link is the principal assumption of a group-oriented evolutionary account.

## MUSIC AND SOCIAL FUNCTION

Consider the following question: What is the most successful piece of music in modern history? Of course the answer to this question depends on how we define success—and this is far from clear, as esthetic philosophers have shown. Nevertheless, I want to use a straightforward criterion: let us assume that the most successful musical work is the one that is most performed and most heard. Using this criterion, you might be surprised by the answer. The most successful musical work was composed by Mildred and Patti Hill in 1893, and revised in the 1930s.[33] The piece in question is, of course, "Happy Birthday." "Happy Birthday" has been translated into innumerable languages and is performed on the order of a million times a day. It remained under copyright protection until the middle of the twentieth century. For many people, the singing of "Happy Birthday" is the only time they sing in public. For other people, the singing of "Happy Birthday" constitutes the only time they sing.

In some ways, "Happy Birthday" is the quintessential feminist work. Its composers remain unknown and uncelebrated; the work was created by the collaboration of two women rather than as an egotistic expression of one man. It is a thoroughly domestic work: "Happy Birthday" is performed in the kitchen or lunch room rather than in the concert hall. No other musical work has evoked so much spontaneous music making. The work is domestic, amateur, and relationally oriented. Despite its extraordinary success, it remains undervalued as a musical creation.

"Happy Birthday" plays a role in our evolutionary story because I suspect that for the vast majority of human history, music making was of this ilk. In Western culture, it is surely the camp songs sung by Girl Scouts or the songs sung by British soccer fans that come closest to what might be imagined in Pleistocene *Homo sapiens*. In all these cases, the music serves an obvious social role and in helping to define a sense of identity and common purpose.

In light of our evolutionary hypothesis, let us return and reconsider the singing of the Mekranoti Indians. Recall some of the characteristic features—especially the singing done by the men: the men's singing is done late at night and in the early morning, and their singing is associated with a high degree of machismo. Like most native societies, the greatest danger facing the Mekranoti Indians is the possibility of being attacked by another human group. The best strategic time to attack is in the very early morning while people are asleep. Recall the insult shouted at men who continued to sleep in their lean-tos: "Get out of bed! The Kreen Akrore Indians have already attacked and you're still sleeping." The implication is obvious. It appears that the nightly singing by the men constitutes a defensive vigil. The singing maintains arousal levels and keeps the men awake.

Of course music making is also associated with stirring a war party. North American Indians famously sang and danced prior to initiating an attack on another tribe. One might suppose that engaging in an activity that publicly announces a hostile intention would be counterproductive. War dances might possible warn an enemy of an impending attack. However, the music making seems to serve a more important role: that of raising arousal and synchronizing individual moods to serve the larger goal of the group.

## SOCIAL BONDING AND HORMONES

Apart from arousing individuals, music can also pacify. Experimental work by Fukui has shown that listening to music can reduce testosterone levels. Fukui himself was quick to point out the possible social and evolutionary significance of this finding. In human social groups, lower levels of testosterone are likely to result in less aggression, less conflict, less sexual confrontation or sexual competition, and consequently more group cohesiveness.

A problem with Fukui's experiment is that he did not manipulate the type of music heard by his listeners. Listeners simply listened to their favorite music. Depending on his sample of listeners, we might expect whole genres of music were not represented. We might suppose, for example, that heavy metal, hard rock, or thrash music might well have increased rather than decreased testosterone levels. Further research is necessary to document the specific hormonal changes associated with different types of musical experiences. However, Fukui's work at least shows that music can have marked effects on hormone levels—specifically, hormones that relate especially strongly to sociability.

## OXYTOCIN AND THE BIOLOGY OF SOCIAL BONDING

An important question to ask is how precisely music might bring about social bonding. Neurophysiologist Walter Freeman[34] has proposed a pertinent theory related to the hormone oxytocin.

Oxytocin is most commonly associated with the "let-down" response in new mothers—that is, the response that enables the flow of breast milk following childbirth. The presence of oxytocin also has dramatic effects on the brain. For example, when a ewe gives birth to a lamb, the olfactory bulb in the ewe's brain is bathed in oxytocin. Following the birth of the new lamb, a ewe will imprint on the smell of the new lamb, but will subsequently fail to recognize the smell of her former offspring. The result is that the ewe will suckle only the newborn lamb.

Neurophysiological research has shown that oxytocin acts as a sort of "eraser" that wipes away previous memories and simultaneously facilitates the storage of new memories. When linked with significant life events, oxytocin is the cement that binds new memories. The amnesic properties of oxytocin are evident in all kinds of learning episodes. However, their strongest effects occur during major limbic activations, such as those resulting from trauma or from ecstasy. Pavlov discovered this phenomenon when serious spring flooding affected his lab and nearly drowned his caged dogs. Following their rescue it was discovered that the dogs had to be retrained from scratch.[35]

In his book, *Societies of Brains*, Freeman chronicles a number of circumstances where oxytocin release occurs and the effects of these releases on neural organization. As we have noted, oxytocin releases are associated with trauma and ecstasy. In addition to childbirth, oxytocin is released in males and females following sexual orgasm. Freeman also suggests that oxytocin is released during trance and while listening to music.

In many cases, the presence of oxytocin is correlated with human and animal bonding circumstances. For example, in the case of sexual orgasm, oxytocin may significantly facilitate pair bonding in the same way that oxytocin following childbirth facilitates mother–child bonding. Freeman's suggestion that music causes oxytocin to be released has important repercussions for instances of peer-group bonding and social identity. If Freeman is correct, there would be good neurophysiological reasons for lovers to enjoy music while courting, for union members to sing while on the picket line, for religious groups to engage in collective music making, for colleges to promote alma mater songs, and for warriors to sing and dance prior to fighting.

## MOOD REGULATION

Thayer and his colleagues have carried out a number of studies concerning how people regulate their moods. One study attempted to determine what people do to try to get out of a bad mood. Of 29 categories of activities, the foremost activity was calling or talking to a friend. The second most frequently reported activity was trying to think positive thoughts—to give oneself a sort of "pep talk." The third most frequently reported activity—ahead of a wide variety of behaviors—was listening to music. Forty-seven percent of respondents reported that they used music to temper or eliminate a bad mood.[36]

Thayer *et al.*[37] carried out a similar study to determine what people do to raise their alertness or energy level. Listening to music was reported by 41% of respondents, following activities such as sleeping, taking a shower, getting some fresh air, and drinking coffee. Finally, in a third study investigating what people do to reduce nervousness, tension, or anxiety, listening to music ranked third at 53%, following after only calling or talking to someone, and trying to calm down by thinking about a situation.

There are two points to highlight from these studies. The first is that the foremost category of behavior for mood regulation is being with or conversing with a friend. That is to say, our first tendency is to seek mood regulation through social interaction. Moods are contagious, and we rely to some extent on each other to modulate, reinforce, or temper our moods. Although we know that moods are highly influenced by the individual's physiological state—notably through food, exercise, or rest—behaviors such as eating, exercise, and rest are less frequently used for mood regulation than music.

The second point to highlight is the obvious point that music appears to figure prominently as a method for mood regulation. Although in contemporary society music tends to be experienced in a personalized or individualized listening context, we already know that this context is historically unprecedented. Most music making in hunter-gatherer societies occurs in a social or group context. Until the invention of the phonograph, the vast majority of music in Western culture was also experienced in social or group contexts. In short, music is not out of place in the list of socialized behaviors used for mood regulation.

## CONCLUSION

By way of conclusion, first let me reiterate that I do not think the evidence in support of music as an evolutionary adaptation is strong. The purpose of this article has been to show that there are no obvious or fatal impediments that rule out a possible evolutionary origin.

We might summarize the basic evidence as follows:

- Complex evolutionary adaptations arise only over many millennia. Accordingly, in order for a behavior to be adaptive, it must be very old. As we have seen, music making does indeed conform to the criterion of great antiquity.
- Behavioral specializations are often expected to be associated with specific anatomical or functional brain structures. Lesions and other neurological assaults can leave an individual with impaired musical functioning. There are double-dissociations between various amusias and virtually every other kind of functional mental loss. This does not prove that music is not acquired by general learning, but the neurological evidence is at least consistent with the possibility that there are specialized music-related brain structures.
- In order for a behavior to be adaptive, the behavior itself must enhance the propagation of the individual's genes. As we have seen, musical behaviors are consistent with mood modification and group mood synchronization—and these synchronous states are at times clearly associated with situations where group efforts are adaptive—such as in the case of defense against other human groups. In addition, high musical involvement is not associated with dereliction or poor survival (such as the case for alcohol); this raises problems for the view that music is a form of nonadaptive pleasure seeking.

The evidence we have for mood regulation and synchronization is suggestive:

- We have noted contrasting disorders in Williams syndrome and Asperger-type autism. In one case, we see a group of individuals who are highly sociable and also highly musical. In the other case, we see a group of individuals who display extremely low sociability and also low musical understanding or affinity.
- Although we did not review this literature, the emergence of the secondary or socialized emotions in child development is strongly associated with musical empathy, understanding, and sophistication. The pertinent research on child development implies a social role for music.
- We noted that the most popular musical works often imply some sort of social function. "Happy Birthday" is only one example. Group identity is often expressed through, for example, folk songs, Girl Scouts' camp songs, sports, and war dances.
- Although we did not review the literature, it is also known that the emergence of musical tastes relates to postpubescent socializing and group identity.
- Finally, we discussed how music modifies hormone production in groups of people.

As noted at the beginning of this essay, there is a long history of abuse of genetic claims serving ulterior and often nefarious motives. Even if we assume that musicality has some adaptive function, the repercussions for modern music making and modern musical enjoyment are likely to be minimal. Music is now deeply embedded in a cultural/historical context where human musical memories span centuries, and the fashion cycle is a significant engine of change. Music is now part of a Lamarckian system where acquired characteristics are transmitted in Dawkinsean "meme pool" rather than in Mendelian "gene pool."[38] Like language, the details of musical culture and tastes are largely a product of enculturation.

Nevertheless, it remains worthwhile to attempt to understand where music comes from and why it has achieved such a ubiquitous presence in human lives. Evolutionary theorizing about music may well remain in the realm of *Just-So* stories. There is always the possibility, however, of a testable hypothesis emerging, and if so, we'll all wait with interest to see the results.

## ACKNOWLEDGMENTS

I would like to extend my thanks to Dr. Kristin Precoda for drawing my attention to the work of Werner regarding the Mekranoti Indians, and to Dr. David Wessel for drawing my attention to Freeman's work concerning oxytocin. This lecture was originally presented at the Department of Music, University of California, Santa Barbara, March 6, 1998. A shortened version was presented at the Society for Music Perception and Cognition Conference, Evanston, Illinois, August 16, 1999.

## REFERENCES

1. Barkow, J., L. Cosmides & J. Tooby, Eds. 1992. The Adapted Mind: Evolutionary Psychology and the Generation of Culture. Oxford University Press. Oxford.
2. Tooby, J. & L. Cosmides. 1992. The psychological foundations of culture. *In* The Adapted Mind: Evolutionary Psychology and the Generation of Culture. J. Barkow, L. Cosmides & J. Tooby, Eds.: 19–136. Oxford University Press. Oxford.
3. Baron-Cohen, S., Ed. 1997. The Maladapted Mind: Classic Readings in Evolutionary Psychopathology. Psychology Press. East Sussex, UK.
4. Shepard, R.N. 1992. The perceptual organization of colors: an adaptation to regularities of the terrestrial world? *In* The Adapted Mind: Evolutionary Psychology and the Generation of Culture. J. Barkow, L. Cosmides & J. Tooby, Eds.: 495–532. Oxford University Press. Oxford.
5. Wright, R. 1994. The Moral Animal: The New Science of Evolutionary Psychology. Vintage Books. New York.
6. Dissanayake, E. 1988. What Is Art For? University of Washington Press. Seattle.
7. Pinker, S. 1997. How the Mind Works. W.W. Norton. New York.
8. Pinker, S. 1994. The Language Instinct. Morrow. New York.
9. Wallin, N.L., B. Merker & S. Brown. 2000. The Origins of Music. MIT Press. Cambridge, MA.
10. Williams, L. 1980. The Dancing Chimpanzee: A Study of the Origin of Music in Relation to the Vocalising and Rhythmic Action of Apes. Allison & Busby. London. Revised edition.
11. Lewontin, R.C. 1991. Biology as Ideology; The Doctrine of DNA. House of Anansi Press. Concord, Ontario.
12. Symons, D. 1992. On the use and misuse of Darwinism in the study of human behavior. *In* The Adapted Mind: Evolutionary Psychology and the Generation of Culture. J. Barkow, L. Cosmides & J. Tooby, Eds.: 137–159. Oxford University Press. Oxford.
13. Popper, K. 1935/1959. Logik der Forschung. Vienna, 1935. Translated as The Logic of Scientific Discovery. Basic Books. New York.
14. Gould, S.J. & R.C. Lewontin. 1979. The spandrels of San Marco and the Panglossian program: a critique of the adaptationist programme. Proc. R. Soc. Lond. **250:** 281–288.
15. Durham, W. 1991. Coevolution: Genes, Culture, and Human Diversity. Stanford University Press. Stanford, CA.
16. Darwin, C. 1872. The Expression of Emotion in Man and Animals. Murray. London.

17. MILLER, G. 2000. Evolution of human music through sexual selection. *In* The Origins of Music. N.L. Wallin, B. Merker & S.Brown, Eds.: 329–360. MIT Press. Cambridge, MA.
18. BAHARLOO, S., P.A. JOHNSTON, S.K. SERVICE *et al.* 1998. Absolute pitch: an approach for identification of genetic and nongenetic components. Am. J. Hum. Genet. **62:** 224–231.
19. Anon. 1997. Neanderthal notes: did ancient humans play modern scales? Sci. Am. **277:** 28–30.
20. TURK, I., Ed. 1997. Mousterian "Bone Flute" and Other Finds from Divje Babe I Cave Site in Slovenia. Zalozba ZRC. Ljubljana.
21. BROWN, D. 1991. Human Universals. McGraw-Hill. New York.
22. WERNER, D. 1984. Amazon Journey; An Anthropologist's Year among Brazil's Mekranoti Indians. Simon and Schuster. New York.
23. WERNER, D. 1984. Amazon Journey; An Anthropologist's Year among Brazil's Mekranoti Indians. 245–247. Simon and Schuster. New York.
24. DUNBAR, R. 1997. Grooming, Gossip and the Evolution of Language. Faber & Faber. New York.
25. AIELLO, L. & R.I.M. DUNBAR. 1993. Neocortex size, group size and the evolution of language. Curr. Anthropol. **34:** 184–193.
26. DUNBAR, R. 1997. Grooming, Gossip and the Evolution of Language. p. 78. Faber & Faber. New York.
27. DUNBAR, R. 1997. Grooming, Gossip and the Evolution of Language. p. 121. Faber & Faber. New York.
28. ROEDERER, J. 1984. The search for a survival value of music. Music Percept. **1:** 350–356.
29. LEVITIN, D.L. & U. BELLUGI. 1997. Musical abilities in individuals with Williams Syndrome. Paper presented at the 1997 Society for Music Perception and Cognition. Massachusetts Institute of Technology. Cambridge, MA.
30. SACKS, O. 1996. An Anthropologist on Mars. p. 124. Random House. New York.
31. SACKS, O. 1996. An Anthropologist on Mars. p. 122. Random House. New York.
32. SACKS, O. 1996. An Anthropologist on Mars. p. 123. Random House. New York.
33. FULD, J.J. 1995. The Book of World-famous Music; Classical, Popular and Folk. Dover. 4th edition. New York.
34. FREEMAN, W.J. 1995. Societies of Brains: A Study in the Neuroscience of Love and Hate. Lawrence Erlbaum Associates. Hillsdale, NJ.
35. PAVLOV, I.P. 1955. Selected Works. Translated by S. Belsky, edited by J. Gibbons. Foreign Languages Publishing House. Moscow.
36. THAYER, R.E. 1996. The Origins of Everyday Moods. Oxford University Press. New York.
37. THAYER, R.E., J.R. NEWMAN & T.M. MCLAIN. 1994. The self-regulation of mood: strategies for changing a bad mood, raising energy, and reducing tension. J. Pers. Soc. Psychol: **67:** 910–925.
38. DAWKINS, R. 1976. The Selfish Gene. Oxford University Press. Oxford.

# Similarity, Invariance, and Musical Variation

STEPHEN McADAMS AND DANIEL MATZKIN

*IRCAM-CNRS, F-75004 Paris, France*

**ABSTRACT: Perceptual similarity underlies a number of important psychological properties of musical materials, including perceptual invariance under transformation, categorization, recognition, and the sense of familiarity. Mental processes involved in the perception of musical similarity may be an integral part of the functional logic of music composition and thus underly important aspects of musical experience. How much and in what ways can musical materials be varied and still be considered as perceptually related or as belonging to the same category? The notions of musical material, musical variation, perceptual similarity and invariance, and form-bearing dimensions are considered in this light. Recent work on similarity perception has demonstrated that the transformation space for a given musical material is limited by several factors ranging from degree of match of the values of auditory attributes of the events composing the sequences to their relations of various levels of abstraction and to the degree that the transformation respects the grammar of the musical system within which the material was composed. These notions and results are considered in the light of future directions of research, particularly concerning the role of similarity and invariance in the understanding of musical form during listening.**

**KEYWORDS: Similarity in music; Invariance in music; Musical variation; Perception of music**

## INTRODUCTION

Perceptual similarity underlies a number of important psychological properties of musical materials, including perceptual invariance under transformation, categorization, recognition, and the sense of familiarity. Mental processes involved in the perception of musical similarity may be an integral part of the functional logic of music composition and may thus underly important aspects of musical experience. We are interested in the perception of musical materials and of their musical transformations—that is, "themes" and "variations," but in a much larger sense than is usually attached to this specific form in Western tonal music.

A longer-term goal of this approach is to understand the role of the perception of similarity and invariance, as well as that of the change produced by variations of the original materials in the experience of musical form. We hesitate to say "perception" of musical form since most musical forms are too long to be "perceived" as such. By understanding the mental processing underlying similarity perception, it may be possible to explain in part how listeners form hierarchical and associative relations

Address for correspondence: Dr. Stephen McAdams, IRCAM-CNRS, 1 place Igor Stravinsky, F-75004 Paris, France. Voice: +331-4478-4838; fax: +331-4478-1540.
mail: smc@ircam.fr

among features of a musical piece belonging to a given musical system and thereby explain in some musical situations how they manage or fail to establish these relations. How much and in what ways can musical materials be varied and still be considered as perceptually related or as belonging to the same category? Can there be a repertoire of possible transformational rules that the cognitive system can decipher, allowing listeners to perceive a relation between a variation and a theme and perhaps even the nature of the transformation process itself?

We will first of all consider in a general way the notions of musical material, musical variation, similarity, and invariance, and then what these all suggest for the development of the notion of form-bearing dimensions in music as originally proposed by McAdams.[1] Subsequently, preliminary experiments on musical similarity that we have conducted[2,3] will be summarized. In this work, pitch and duration variations are studied for both tonal/metric and nontonal/nonmetric musical systems. Finally, a number of theoretical issues will be developed in the hopes of stimulating and orienting further work in this area, notably concerning (1) the nature of the representation of musical materials and transformation processes, (2) modularity and dimensional interactions within the realm of music cognition, (3) parallelism and associative structures in music, and (4) the notion of musical development and musical process.

## MUSICAL MATERIALS

We take the notion of musical material in a very broad way. Materials may be more conventionally considered as simple figures or themes such as the opening five-note figure of Anton Webern's *Sechs Stücke für großes Orchester*, op. 6 (1909) (only slightly more complex than Beethoven's famous ta-ta-ta-tum, although not as explicitly developed) or slightly more lengthy themes such as a Bach fugue subject. They may also be more fully developed musical ideas such as the thematic materials used by Roger Reynolds in pieces such as *Archipelago* for chamber orchestra and computer-generated sound (1982–1983) or *The Behavior of Mirrors* for guitar (1986). He calls these materials *core elements*.[4] They may vary in duration anywhere from several seconds to several tens of seconds, and any given piece will have several of them, each designed to have distinctive characteristics of pitch materials, durational proportions, textures, and gestural movements. Musical materials can also be complex textures such as the dense blocks of sound used emblematically by Krzysztof Penderecki in *Threnody for the Victims of Hiroshima* for 52 strings (1960), the architectonic gestures often used by Iannis Xenakis in his orchestral works such as *Pithoprakta* (1956), and the rich micromelodic tapestries in early orchestral and choral works by György Ligeti such as *Atmosphères* (1961) and *Lux æterna* (1966). Finally, we should not leave out the realm of electroacoustic music in which spectral and temporal sound structures can be imagined from scratch and become the basis for compositional development or in which recorded sounds can become musical materials in their own right as the composer delves into the inner structure of the sound itself. Notable examples of the latter class are the deep church bell tone and the boy soprano's voice that are the two concrete sound sources from which the computer-generated piece *Mortuos Plango, Vivos Voco* (1980) was composed by Jonathan Harvey or the interpolations between unlikely partners such as a horse's laugh and a baby's cry in *Vox 5* (1986) by Trevor Wishart.

## MUSICAL VARIATION

Given the variety of types of musical materials, there is obviously an infinity of ways that even simple ones can be varied in order to generate new material to fill the needs of musical discourse, all the while maintaining (most often) some kind of perceptual link to the original material so that the listener not only recognizes the relation (implicitly or explicitly), but also senses the kind of variation that has been applied. This trajectory of change is one of the perceptual components of musical development, and our intuition is that it contributes to the sense of larger-scale movement and cohesion in a work. In the simplest (operational) terms, musical variation or transformation may be taken to be simply the change along one or more musical dimensions of a given musical material.

In most musics, the principal dimensions of materials targeted for transformation are pitch and duration, although space and timbre have also become the focus of musical structuring in the 20th century. A complete catalogue of transformation devices is, of course, impossible, so only a few clearly definable types will be mentioned below. The simplest kind of (non)variation is the repetition of the same material at a different point in time. In this case there is absolute identity in a physical sense, and the perception of similarity and the recognition of the return of the material depend only on the listener and the intervening material. Another version would keep the pitch and rhythmic values constant and change the timbre through orchestration. A blatantly clear example of this approach is the piece *Boléro* by Maurice Ravel (1928), in which two themes are alternated AABB four times and then AB at the end. However, each time, the A and B melodies and their accompaniment are orchestrated differently. The similarity and sense of repetition are strong; and it is the pattern of change in orchestration, dynamics, and articulation that creates the expressive trajectory of the work. A more complex approach is to change the timbre of each note, a true Klangfarbenmelodie, as proposed by Schoenberg,[5] and then have different timbral compositions for each new return of the theme. The danger in this case, as will be discussed below, is that the melody may become fragmented perceptually if the timbre change is too drastic, bringing about the assignment of different subsets of notes to different auditory streams and thus compromising the psychological coherence of the melody as a whole.[6]

Transposition of pitch and change of tempo are the simplest kinds of transformations and testify to the fact that the representation of relations between events (exact intervals or number of scale steps related to a given scale) are prominent in musical memory. The perception of similarity of transposed melodies is indicative of the influence of internalized musical systems on recognition and similarity perception. An exactly transposed melody (one that maintains the same pitch intervals between successive notes) will in many cases be perceived as a change in key. A melody that is transposed, keeping the same number of scale steps between successive notes but with pitches that remain within the current key, is often perceived as more similar, as suggested by recognition studies, although this ability depends on musical training, musical context, and task instructions.[7] At any rate, what seems to be the important aspect of the perceptual representation underlying melodic and rhythmic similarity here is the relations among notes—that is, the relative distance in intervals or scale steps. Slight interval changes generally preserve the contour (pattern of ups and downs), a component of melody representation that has been shown to be particularly cogent for the similarity of unfamiliar melodies as measured by false recognitions.[8]

Some studies have shown that the perception of timbral intervals is also possible.[9,10] This work was based on the notion of timbre space in which the perceptual relations among sounds equated for pitch, duration, and loudness are represented mathematically as a multidimensional space. Timbral intervals are defined as oriented vectors in the space, and timbral contours could be defined by the relative patterns of up and down along the various dimensions. A transformation of a timbre melody such as transposition would simply be the translation of the timbre vectors in the space, keeping their lengths and orientations constant. Although the precise reproduction of timbral intervals is problematic with instrumental music, such musical structures are quite possible with synthesized sounds.

A variant on the simple translation operation is the inversion operation, in which the sign of the intervals and contour directions are changed. This technique is used in mirror canon writing and is also quite prominent in the 20th century in serial music.

Another class of transformations involves temporal rearrangement of the musical material. For example, playing a melody backward is called retrogradation and is often found in music of the common practice, romantic, and contemporary periods of Western music. Mirror and cancrizans canons used this technique, and it also became a prominent operation in serial music. A particularly intriguing compositional problem-solving task involves creating a melody that can be played forwards and backwards (the cancrizans or "crab" canon). Often the material that is retrograded may not reproduce the exact interval structure for reasons related to the musical syntax. In such cases, certain prominent intervals or the inverted melodic contour may preserve the sense of similiarity. Take, for example, the opening five-note figure of Anton Webern's *Sechs Stücke*, op. 6 mentioned above, in which the flute plays C♯4-E4-F4-G4 followed by D5 on the muted trumpet. Subsequently the sustained trumpet tone becomes the first note of the retrograded pattern D5 (trumpet), B♭4-E♭4-D♭4 (flute), B3 (French horn). Although the intervals are slightly different, the pitch contour (small skip, two steps, large skip) is inverted, as is the timbral contour, moving from the low-register flute to a bright, muted trumpet in one direction and from the flute to a lower, darker French horn in the other. The mirror image is quite convincing perceptually. Various other techniques of recombination have been used as well, often involving the choice of particularly salient figures from the theme and their subsequent development—a practice quite prominent in Beethoven piano sonatas, for example.

All of the preceding examples are based on rearrangements of symbolically notated events. However, in electroacoustic music, where the material is a continuous sound waveform, the operations need not be limited to predefined events. Granular synthesis, for example, takes bits of recorded sound from different parts of the original sample, applies a smoothing envelope to them and then recombines them in ingenious ways to create rich sound textures.[11] The larger the grains, the more the original sound quality "comes through" in the recombination. Roger Reynolds has used a technique involving segmentation and rearrangement in several pieces, applied either to sound samples (as in *Archipelago*, 1982–1983) or to note lists representing symbolic events (as in *Variation* for piano, 1988).[4] He has developed two classes of algorithms (SPLITZ and SPIRLZ) that transform the original samples within a space of variation constrained by the variables of the algorithm. The first one segments the sound sequence into equal-sized units and then plays the odd ones in one order and the even ones in reverse order. The spacing and timing of the segmented unfolding can be controlled. This gives a kind of alternation between the front and back ends of the sound sequence that move toward the center and then back to the extremities. The technique is obvious when applied to musical gestures with a clear trajectory, but can give stunning results in other cases even though the listener may not be able to follow explicitly the algorithm's logic. As for granular synthesis, the transformed sequences may have a more or less strong similarity to the original material depending on the size and ordering of the segments. When applied within a piece to several different themes, the transfomation device itself may in some cases take on an identity across materials that is stronger than the relations between the original and transformed materials. The perception of these kinds of materials is currently under study in our laboratory in collaboration with the composer.

Staying within the realm of electroacoustic music, there are a plethora of digital techniques now that allow an extremely fine-grained analysis of a sound's spectral and temporal structure, and then either a direct modification of certain of these parameters or even a modeling of their behavior and a subsequent resynthesis with changes in the model parameters. This kind of sonic elaboration is a class of transformation that remains quite close to the timbre and sound source identity of the materials to be transformed. Compelling examples of this approach can be found in Trevor Wishart's *Vox 5* (1986), Jean-Claude Risset's *Sud* (1985), and Jonathan Harvey's *Ritual Melodies* (1990). In particular, in the Harvey piece models of various kinds of non-Western instruments and vocal styles are used to create interpolations between the sound universes of each, including not only the sound qualities themselves, but also the styles of playing (e.g., forms of vibrato, tremolo, and portamento).

By far the most widely used type of variation involves change in the surface level structure while some invariant is retained at a higher hierarchical level. This kind of technique is quite common in theme-and-variations forms in music of the common practice period. The notions of elaboration and reduction have been formalized to a large extent by Lerdahl and Jackendoff,[12] although their applicability to contemporary music as proposed by Lerdahl[13] has been questioned.[14] Indeed, Dibben suggests that since structurally important events that "stand for" sets of subordinated events in tonal music do not have the same kind of representation in atonal music, it may be necessary to consider more associational structures to explain formal coherence in this latter kind of music, as will be discussed at the end of this paper. Evidence for similarity among musical materials related by a common reduced structure in tonal/metric music is clear (see, e.g., Refs. 15–17) and will not be belabored here. However, notions of elaboration also exist in contemporary music, although it is not clear to what extent commonalities of mental processing exist between the different musical systems.

Some of the psychological questions that arise from this brief and extremely schematic consideration of the possibilities of musical variation may include the following: How far can one transform original musical material before it becomes completely new, and perceptually unrelated—that is, what are the bounds of perceptual similarity? What does the answer to this question tell us about the representations of musical materials and the nature of the psychologically realistic processes of transformation that can operate on them?

## SIMILARITY AND INVARIANCE

At the crux of the issue of musical variation are the notions of perceptual similarity and invariance, *if* the varied material appears to have at least some relation to the original material. What we actually mean by "similarity" is not easy to pin down operationally. Let us just assume that it is some degree of match between the properties and features of two materials. The experimental question is thus to understand what properties contribute to perceived similarity and how the degree of match between these (to-be-determined) properties determines its strength. We will speculate on the nature of various kinds of properties and levels of abstraction that might contribute to musical similarity as judged directly by listeners.

Properties that contribute to similarity may be of various levels of abstraction, moving from the most concrete (specific values of attributes of individual events in the sequence) to the most abstract (higher-level reductions of the hierarchical event structure, if such exists for the material being heard, or harmonic contours that provide a kind of skelelton for a tonal work). The values and relations thus analyzed from the musical sequence may also be explicitly tied to the temporal order of the musical sequence or experienced under transformations of temporal order or even as distributions with reference to specific order.

Concerning surface values, the surface content of a material and its transformation (e.g., specific pitches, durations, loudnesses, and timbres) may have more or fewer events in common along the various auditory dimensions. Also, the temporal order of these events may be the same or may be transformed to a greater or lesser extent. Exact repetition would, of course, have the highest similarity, and completely different values would have no similarity. However, under some situations the exact same number of events with specific attributes in completely different orders (a random rearrangement of the events, for example) may still be experienced as similar due to identical statistical distributions of material properties. As mentioned above, however, musical material can also be represented in terms of relations among attributes in a more or less precise way: exact intervals, relations within a musical system such as a scale, contours, and so on. The perception of similarity would thus be based on a match between relations abstracted from the material attributes rather than directly on the values of the attributes themselves. A melodic contour in Béla Bartók's *Music for Strings, Percussion and Celesta* (1936), for example, might undergo an expansive transformation in which all of the intervals are increased, and yet the whole stays within a certain pitch system, and the pitch contour is preserved. Since the rhythmic material is relatively similar between the versions, listeners may hear a strong link between this variation and the original.

Finally, as also mentioned previously, research[16,17] has shown that materials that are quite different in surface structure (affecting specific surface values and the relations between adjacent events) may still be perceived as similar if they share an underlying structure, such as the reduced structure proposed by Lerdahl and Jackendoff.[12] In these cases, it is necessary to postulate that the representation of the event structure of the material is organized to some extent in a hierarchical fashion and that it is some level of this hierarchy that is compared between versions. The experimental question here concerns the depth of the hierarchical representation that can be used to derive a sense of similarity between materials that differ drastically in surface structure. It would seem that it is this kind of structural abstraction that would be one of the main bases of the musical invariance underlying theme-and-variations forms.

From these considerations, we propose that there are (at least) three types of similarity that should be taken into consideration, all the while understanding that they are overlapping and non-mutually exclusive categories. The first is relatively abstract and concerns similarity of the statistical distribution of surface values and their derivatives as well as of surface relations of first and higher orders, including perhaps even transitional probabilities from one value or relation to another. This kind of similarity would seem to be involved in the perception of similar textures, where it is not the exact values that count but the probability of occurrence and of transition that define the texture. The second kind concerns figural similarity, specific patterns of attributes (perhaps associated intimately across auditory dimensions) that give perceptual landmarks within an extended material and that allow its recognition when a transformed version still contains the same identifying pattern, or something very similar to it. This notion of figural similarity is akin to the notion of imprint proposed by Deliège.[18] The third kind of similarity is structural and relies on the abstraction of structural invariants related to the event hierarchy, often defined by underlying harmonic and metric templates that are maintained constant in cases of elaborative or reductive transformations of thematic materials.

## FORM-BEARING DIMENSIONS

A framework has been developed concerning the psychological constraints on form-bearing dimensions in music.[1] Among these psychological constraints, some seem particularly cogent for the present concerns. Form-bearing dimensions are those along which invariant relations can be configured. For them to be useful in musical discourse these relations must be easily discriminable. Further, in evaluating the degree of similarity between thematic materials and their transformations, time has necessarily passed between the two, often several tens of seconds in real music. The configurations must thus be memorizable to some degree of precision either in their absolute form or in some abstracted relative form.

In real music, several dimensions may be varied for a given material. For example, transposing a melody in pitch that is played by the same instrument maintains the same rhythmic structure, and while globally the timbre may change slightly, it remains roughly constant. One psychological constraint on form-bearing dimensions that merits extensive consideration is the resistance of perceived invariance on one dimension to independent variation on other dimensions. In other words, to what extent are the dimensions processed independently or do they interact? This raises the modularity question that has been addressed in both theoretical and experimental terms.[19–21] Both the psychological and the musical issues are extremely complex here, as it becomes clear that one may have varying degrees of identity and difference on different dimensions, and the question would be to know how strong the sense of similarity was under certain conditions, or even whether similarity could be maintained. The perception of such transformations would depend on several factors. At the level of the stimulus, it would depend on the way the composer used variation across dimensions and on the types of interactions between dimensions, which may indeed by asymmetric. For example, in Olivier Messiaen's *Mode de Valeurs et d'Intensités* for piano (1950), many dimensions are used: pitch, dynamics, articulation, duration. However, there is a coupling between the values of one and those of the others. Krumhansl has shown[22] that listeners pick up on these interdimensional correlations and can judge that new material fits or does not fit with rules as developed by the composer. In integral serialism, on the other hand, the composer may have series of pitches, dynamics, articulations, and durations, each series having a different number of elements, and they thus become uncorrelated with one another, as in Pierre Boulez' *Structures* for two pianos (1952). It would be interesting to repeat Krumhansl's experiment on such material. However, she and her colleagues have shown that listeners do hear different transformations of row forms (retrograde, inversion, and retrograde inversion) as having a similarity relation to the original series in twelve-tone rows of Schoenberg.[23]

A psychological factor that needs to be considered is the resistance of perceived invariance/similarity to perturbation by intervening material. We often presume as score gazers that recognition of the return of a theme is unaffected by new intervening material. However, work in auditory short-term memory has shown that relations of attribute similarity between the to-be-remembered information and the intervening material can degrade recognition and discrimination performance. This kind of auditory memory does seem to be somewhat modular, however, in that intervening material that is different on the criterial dimension but similar on other dimensions does not necessarily degrade performance.[24–28] At a more complex level, it is unclear at present what kinds of contextual effects may exist that would render some intervening material more or less perturbing of the memory trace of the original material and thus hinder its recognition. If such were the case, then the "success" of a given theme in contributing to a sense of coherence in a musical form would certainly be compromised.

Given that materials can vary on different dimensions and that there have been strong tendencies in many cultures to focus primarily on pitch and duration, another psychological question that comes to the fore is the extent to which listeners can learn to focus on changes along a dimension not usually used in one's past experience as a structuring dimension in musical discourse, such as timbre or space, for example. That listeners do not have experience in listening for such structures would probably show up in psychological testing, but we could not conclude from such results that the use of these dimensions was impossible or doomed to failure from the outset. So the question of perceptual learning will need to be addressed within the framework of contemporary musical experience.

Finally, one of the crucial problems of the perception of musical transformations in relation to their original material concerns perceptual coherence. Variations that affect stream formation or change the availability of properties of the material in working memory may, indeed, render what is perceived unsimilar to the original material, as relations among perceptual attributes would appear to be computed on events belonging to the same auditory stream.[6,29]

## EXPERIMENTS ON MUSICAL SIMILARITY

Having drawn with broad strokes a rough and largely intuitive account of what musical similarity might be and what its role in music might be, let us examine the few studies that have specifically addressed this question, focusing in particular on some of our own recent work that takes a few small steps toward the answer. Several experiments have studied the implication of musical training, musical memory, pitch hierarchy, and musical context in the perception of tonal and nontonal transformations and their similarity ratings.[8,23,30–33] For example, similarity perception is not a symmetrical phenomenon: the similarity perceived between a melody A and a melody B does not automatically predict the similarity perceived between B and A.[30] Finally, although participants show an ability to recognize the type of transformations applied to a nontonal material presented isochronously, this ability tends to decline when transformations are multidimensional—that is, when both pitch and rhythm dimensions vary simultaneously.[8,23,31] All these facts, however, do not explore or explain the computational basis of melodic similarity perception. Is it plausible to imagine that melodic similarity perception could rely on general mental processes that are common to all listeners, independently from the melody's morphology (that is, for example, number of pitches, rhythmic figures, and metronomic speed) and the listener's musical aptitude?

In recent work, we asked the following questions: What is the nature of musical transformation space, and how is it constrained by the type of musical material and musical training? To what extent do pitch and duration components of musical materials contribute independently to global similarity judgments and discrimination focused on each? The first question was addressed in Experiment 1 and the latter in Experiment 2. The full details of the experiments are found in References 2 and 3.

## EXPERIMENT 1

Three tonal melodies were used as reference material. Following Serafine *et al.*[16] a lure melody was constructed for each one such that the surface characteristics were very similar, while the underlying reduced structure was different. For each melody, 16 transformations were composed that differed from the original in pitch content, rhythmic content, or both. The transformations could be either elaborations or reductions of the original material, reduction being taken in the sense of a simplified abstract representation of a musical structure that keeps structural information such as important notes in the tonal and metric hierarchies.[12] Further, the pitch and rhythmic transformations either respected tonal/metric syntax (reasonable variations within the scale and meter) or departed from it (unlikely chromatic and unmetrical variations). Corresponding structural transformations were applied to the lure melodies as well. The melodies were verified with respect to the desired properties by a practising composer (Joshua Fineberg) working at IRCAM.

Musician and nonmusician listeners heard one of the three reference melodies followed by a test melody. The test melodies could be the reference or its corresponding lure or any of their transformations. The listeners made a similarity rating on a scale of 1 to 9, with the high end corresponding to maximum similarity.

The experiment was designed to test several hypotheses:

(1) Musicians should be able to hear similarity to greater degrees of transformation if the transformations respect the syntactical rules.
(2) If listeners are sensitive to commonalities at certain levels of hierarchical reduction, transformations that respect the reduction should be more similar than those that violated it (including the lure transformations).
(3) Drastic transformations in both pitch and rhythm should render the variation as dissimilar to the reference as unrelated material, represented by the lure transformations.
(4) The same types of transformation should have similar effects on the three reference melodies, although local morphological differences may prevail due to figural properties of the melodies.

The data revealed no systematic differences between the two populations of listeners. This last finding is concordant with previous studies that have specifically examined effects of musical training on musically realistic tasks.[34] Transformations that respected the reduced structure were judged significantly more similar to the reference than those that did not. This effect was particularly strong for chromatic transformations and for combined chromatic and nonmetric transformations, many of which were judged as dissimilar as the lure melody and its transformations. Finally, the three reference melodies were different with regard to the perceived effects of the transformation types, demonstrating the importance of local surface features in the perception of melodic similarity and the fact that a transformational algorithm must necessarily interact complexly with the to-be-transformed material. However, rank correlations between reference melodies were very high, indicating similar qualitative effects of the set of transformations applied to all three melodies.

## EXPERIMENT 2

The answer to one question that cannot be easily gleaned from the results of the first study is the degree to which pitch and rhythmic materials interact in the perception of similarity. The second experiment was designed to address this problem more explictly. We, therefore, introduced a simple computational model based on a modular hypothesis that could represent the basis of the mental processing underlying melodic similarity perception. If the computational model turns out to be reliable for tonal melodies, can it also predict similarity perception of nontonal melodies and thus indicate general processing mechanisms independently of the musical system? One may wonder whether or not a modular approach is relevant to the study of music perception in general and of similarity perception in particular. Examples taken from vision, language, audition, and neuropsychology lead us to consider the modular hypothesis in order to clarify the specificities of similarity perception. Interestingly, aside from modular considerations in music theory,[19] the only consequential empirical research done within an explicitly modular perspective comes from the neuropsychological work of Peretz and collaborators.[20,35,36]

Based on previous findings, which argue in favor of independent pitch and duration processing structures,[33,37] the main hypothesis posited here is that a similarity judgment made between a reference melody and a transformation of that melody could be understood to a certain extent as an additive integration of two similarity coefficients calculated by the two hypothetical pitch and duration modules. While such an additive approach is certainly a simplistic way to consider the relations between the two modules, such a general principle should be applicable to any musical system without regard to the nature of the underlying musical grammar, the tonal and twelve-tone systems in our case.

To test the independence, a paradigm similar to that of Garner and Morton[38] was used in which discriminations of changes on one dimension were made in the presence or absence of changes along another dimension and vice versa by different groups of listeners. A third group was required to make both discriminations. We coupled this discrimination task with a similarity judgment task. This paradigm was used in order to investigate whether the listeners could recognize a pitch or duration pattern independently from the values taken by the other dimension. Note that since the judgment is focused on a single dimension, at times in the face of variation on the other dimension, perfect discrimination performance across all conditions indicates an ability to ignore the other dimension, implying perceptual independence.

Four reference melodies were composed: one major, one minor, and two twelve-tone melodies. In addition to the original pitch or rhythm patterns, three variations of each dimension were created. These variations were designed to be progressively less similar to the reference pattern. The pitch and rhythm patterns were crossed, giving 15 transformations of each reference. The melodies were all composed to achieve the desired properties by a practising composer (Jacopo Baboni-Schilingi) working at IRCAM. The transformations could thus differ only in pitch, only in rhythm, or in both. Listeners were divided into three groups: pitch discrimination, rhythm discrimination, or both. They first made discrimination judgments on the pitch and/or temporal dimension and then judged globally on a 10-point scale the strength of similarity for each type of transformation in relation to the reference melody.

The discrimination results globally corroborate the modularity hypothesis, although complete perceptual independence was not observed. In particular, the pitch and duration modules do not exhibit the same processing properties, as duration processing seems to be less permeable to pitch information than pitch is to duration (at least for tonal melodies). This is reflected when comparing participants' discrimination performance on the two musical dimensions for the two musical systems. For all melodies, duration configurations were better discriminated than pitch configurations; and, in particular, identical pitch patterns were judged as being progressively different when the rhythm pattern was, indeed, more and more removed from the reference, even though the pitch pattern was identical. By deduction this finding is coherent with existing empirical data in melodic similarity perception, where it has been shown that the duration dimension was the major one used by participants for establishing similarity judgments.[39] The pitch permeability to duration configuration was limited to tonal melodies, suggesting a functional interaction between the two in that musical system that is decoupled in the twelve-tone system.

Variations among listeners' similarity judgments showed that tonal and nontonal transformations were globally perceived in different ways. While the tonal transformations were hierarchically distinguished, this was not the case for the perception of the nontonal transformations, where the pitch and duration modalities were not hierarchically differentiated. Further, whereas the perception of tonal transformations appeared to be perceived in a systematic way over repetitions and for the two reference melodies, this was not the case for the nontonal transformations, where the range of variation of listeners' judgments was found to be much greater. Consequently, it seems reasonable to infer that similarity perception of tonal and nontonal melodies depended at least partly on different mental processing mechanisms.

Under the perceptual independence hypothesis, we claimed that similarity perception could be approximated by an elementary additive model. The model decomposes the listeners' similarity ratings into two numerical values, basic mean pitch and duration values for each level of transformation of that dimension without change in the other dimension. The additive combination of the mean values for pitch and duration gives the model prediction. For three of the four reference melodies (including both tonal melodies and the first nontonal melody), listeners' mean similarity judgments for the nine two-dimensional transformations were highly correlated with model predictions. So although some differences between the tonal and nontonal melodies exist, these melodies also seem to have certain commonalities with regard to processing in rhythmic and pitch similarity. There are, nonetheless, deviations of the data from model predictions for each transformation, and these will need to be examined in detail to determine their cause, particularly for the second nontonal melody. There are perhaps salient surface or structural features not taken into account by this simplistic model that are playing a role in the perceived similarity.

## CONCLUSIONS/QUESTIONS

From the data, it is clear that the space of possible variation of thematic material that still maintains a link of perceptual similarity to the original is limited. Although conditions of direct comparison are different from those in real music, where intervening material or competing material would often be present, a number of factors can be targeted provisionally for more contextually realistic work in the future. A transformation that maintains a majority of exact values or relations among values in the correct temporal order will be more strongly similar than one that changes all values and all relations. Although this has not been explicitly tested, it may be that maintaining the same distribution of values or relations could still elicit a sense of similarity if compared with material that had a dramatically different distribution of attribute values. Further relations of similarity can exist at more abstract levels of a hierarchical reduction, indicating something about the hierarchical nature of representation of the mental representation of musical material in both tonal and nontonal music. From examining the similarity judgments on elaborative and reductive transformations, it would seem that these kinds of transformations retain similarity to the extent that the transformations do not violate the "grammaticality" of the musical system within which the material has been conceived, suggesting a close link between comprehension of the musical system and the categorization processes that one may hypothesize to operate in the recognition of variations of an original theme.

Another issue of major import for the notion of musical similarity and invariance is the notion of perceptual independence between patterns on different musical dimensions. While there seems to be some degree of independence between pitch and duration structures at a global level, there are also asymmetrical interactions in which duration changes more strongly affect the perceived similarity of identical pitch patterns than is the case for the reverse. More refined experimentation will be necessary to probe the nature of the interaction between the dimensions that gives rise to such asymmetries. Also, future research should clearly study more thoroughly different musical systems, including those of other cultures, in an attempt to distinguish the more general cognitive processes from those that are specific to a particular musical grammar.

Finally, future research should examine the role that the perception of musical similarity plays in the apprehension or experience of musical form. The process of similarity perception necessarily involves comparison across time of different manifestations of musical material and thus involves memory processes. One might imagine that the recognition of similarity could thus establish some kind of associative link between different points in the piece and thus contribute in some way to a sense of formal cohesion over longer time spans. Further, one not only recognizes the link between two manifestations, but can also appreciate the difference between the two, a perception that perhaps contributes to a sense of the kind of change or the trajectory of change that has taken place. This trajectory may contribute in itself to a sense of musical development over longer time spans and also reinforce the integration of materials at different points in time into a greater whole. There are numerous theoretical and methodological obstacles to be overcome in the design of appropriate experiments to test these contributions in ongoing music listening.

## ACKNOWLEDGMENTS

This work has benefited from fruitful discussions with several composers (Jacopo Baboni-Schilingi, Benjamin Carson, Joshua Fineberg, and Roger Reynolds) and a fellow psychologist (Emmanuel Bigand), to whom we express our gratitude.

## REFERENCES

1. McAdams, S. 1989. Psychological constraints on form-bearing dimensions in music. Contemp. Music Rev. **4**(1): 181–198.
2. Matzkin, D. 1998. La perception de similarité musicale peut-elle être étudiée dans le cadre de la théorie de la modularité et de l'indépendance perceptive? Mémoire de DEA, EHESS. Paris.
3. Matzkin, D. & S. McAdams. Does similarity perception in music involve systematic mental processing regardless of the nature of the musical system? Submitted for publication.
4. Reynolds, R. 2001. Form and Method: Composing Music. Contemporary Music Studies. Harwood Academic Press. London.
5. Schoenberg, A. 1978. Theory of Harmony. (Original German publication 1911). University of California Press. Berkeley, CA.
6. McAdams, S. & A.S. Bregman. 1979. Hearing musical streams. Comput. Mus. J. **3**(4): 26–43.

7. DOWLING, W.J. 1986. Context effects on melody recognition: scale-step versus interval representations. Music Percept. **3:** 281–296.
8. DOWLING, W.J. 1972. Recognition of melodic transformations: inversion, retrograde, and retrograde inversion. Percept. Psychophys. **12:** 417–421.
9. EHRESMAN, D. & D.L. WESSEL. 1978. Perception of timbral analogies. IRCAM, Report 13. Paris.
10. MCADAMS, S. & J.C. CUNIBILE. 1992. Perception of timbral analogies. Philos. Trans. R. Soc. Lond. Ser. B. **336:** 383–389.
11. ROADS, C. 1985. Granular synthesis of sound. *In* Foundations of Computer Music. C. Roads & J. Strawn, Eds: 145–159. MIT Press. Cambridge, MA.
12. LERDAHL, F. & R. JACKENDOFF. 1983. The Generative Theory of Tonal Music. MIT Press. Cambridge, MA.
13. LERDAHL, F. 1989. Atonal prolongational structure. Contemp. Music Rev. **4:** 65–87.
14. DIBBEN, N. 1994. The cognitive reality of hierarchical structure in tonal and atonal music. Music Percep. **12:** 1–25.
15. STOFFER, T.S. 1985. Representation of phrase structure in the perception of music. Music Percept. **3:** 191–220.
16. SERAFINE, M.L., N. GLASSMAN & C. OVERBEEK. 1989. The cognitive reality of hierarchic structure in music. Music Percept. **6:** 397–430.
17. BIGAND, E. 1990. Abstraction of two forms of underlying structures in a tonal melody. Psychol. Music **18:** 45–60.
18. DELIÈGE, I. 1989. A perceptual approach to contemporary musical forms. Contemp. Music Rev. **4:** 213–230.
19. TEMPERLEY, D. 1995. Motivic perception and modularity. Music Percept. **13:** 141–170.
20. PERETZ, I. & J. MORAIS. 1989. Music and modularity. Contemp. Music Rev. **2**(1): 279–294.
21. FODOR, J. 1983. The modularity of mind. MIT Press. Cambridge, MA.
22. KRUMHANSL, C.L. 1991. Memory for musical surface. Mem. Cognit. **19:** 401–411.
23. KRUMHANSL, C.L., G.J. SANDELL & D.C. SERGEANT. 1987. The perception of tone hierarchies and mirror forms in twelve tone serial music. Music Percept. **5:** 31–78.
24. STARR, G. & M. PITT. 1997. Interference effects in short-term memory for timbre. J. Acoust. Soc. Am. **102:** 486–494.
25. SEMAL, C. & L. DEMANY. 1991. Dissociation of pitch from timbre in auditory short-term memory. J. Acoust. Soc. Am. **89:** 2404–2410.
26. SEMAL, C. & L. DEMANY. 1993. Further evidence for an autonomous processing of pitch in auditory short-term memory. J. Acoust. Soc. Am. **94:** 1315–1322.
27. DEUTSCH, D. 1970. Tones and numbers: specificity of interference in immediate memory. Science **168:** 1604–1605.
28. DEUTSCH, D. 1972. Mapping of interactions in the pitch memory store. Science **175:** 1020–1022.
29. BREGMAN, A.S. 1990. Auditory Scene Analysis: The Perceptual Organization of Sound. MIT Press. Cambridge, MA.
30. BARTLETT, J.C. & W.J. DOWLING. 1988. Scale structure and similarity of melodies. Music Percept. **5:** 286–313.
31. FRANCÈS, R. 1972. La perception de la musique. Vrin. Paris.
32. HALPERN, A.R., J.C. BARTLETT & W.J. DOWLING. 1998. Perception of mode, rhythm, and contour in unfamiliar melodies. Effect of age and experience. Music Percept. **15:** 335–355.
33. PITT, M.A. & C.B. MONAHAN. 1987. The perceived similarity of auditory polyrhythms. Percept. Psychophys. **41:** 534–546.
34. BIGAND, E. 1990. Perception et compréhension des phrases musicales. Ph.D. thesis. Université de Nanterre (Paris X). Nanterre, France.
35. PERETZ, I. 1993. Auditory agnosia: a functional analysis. *In* Thinking in Sound: The Cognitive Psychology of Human Audition. S. McAdams & E. Bigand, Eds: 199–230. Oxford University Press. Oxford.
36. PERETZ, I. & R. KOLINSKY. 1993. Boundaries of separability between melody and rhythm in music discrimination: a neuropsychological perspective. Q. J. Exp. Psychol. **46A:** 301–325.

37. PALMER, C. & C.L. KRUMHANSL. 1987. Independent temporal and pitch structures in the determination of musical phrases. J. Exp. Psychol. Hum. Percept. Perform. **13:** 116–126.
38. GARNER, W.R. & J. MORTON. 1969. Perceptual independence: definition, models, and experimental paradigms. Psychol. Bull. **72:** 233–259.
39. MONAHAN, C.B. & E.C. CARTERETTE. 1985. Pitch and duration as determinants of musical space. Music Percept. **1:** 1–32.

# Tonal Cognition

CAROL L. KRUMHANSL AND PETRI TOIVIAINEN

*Department of Psychology, Cornell University, Ithaca, New York 14853, USA*

*Department of Music, University of Jyväskylä, Jyväskylä, Finland*

**ABSTRACT: This article presents a self-organizing map (SOM) neural network model of tonality based on experimentally quantified tonal hierarchies. A toroidal representation of key distances is recovered in which keys are located near their neighbors on the circle of fifths, and both parallel and relative major/minor key pairs are proximal. The map is used to represent dynamic changes in the sense of key as cues to key become more or less clear and modulations occur. Two models, one using tone distributions and the other using tone transitions, are proposed for key-finding. The tone transition model takes both pitch and temporal distance between tones into account. Both models produce results highly comparable to those of musically trained listeners, who performed a probe tone task for ten nine-chord sequences. A distributed mapping of tonality is used to visualize activation patterns that change over time. The location and spread of this activation pattern is similar for experimental results and the key-finding model.**

**KEY WORDS: Music; Tonality; Cognition; Probe tone**

## INTRODUCTION

Tonality induction refers to the process through which the listener develops a sense of the key of a piece of music. The concept of tonality is central to Western music, but eludes definition. From the point of view of musical structure, tonality is related to a cluster of features, including musical scale (usually major or minor), chords, the conventional use of sequences of chords in cadences, and the tendencies for certain tones and chords to suggest or be "resolved" to others. From the point of view of experimental research on music cognition, tonality has implications for establishing hierarchies of tones and chords, and for inducing certain expectations in listeners about how melodic and harmonic sequences will continue. One method for studying the perception of tonality is the probe tone method, which quantifies the tonal hierarchy. When applied to unambiguous key-defining contexts, it provides a standard for determining key strengths when more ambiguous and complex musical materials are presented. In addition to experimental studies, considerable effort has been spent developing computational models. This effort has produced various symbolic and neural network models, including a number that take musical input and return a key identification, sometimes called key-finding models.

Address for correspondence: Professor Carol L. Krumhansl, Department of Psychology, Uris Hall, Cornell University, Ithaca, NY 14853. Voice: 607-255-6351; fax. 607-255-8433. clk4@cornell.edu

## PROBE TONE METHODOLOGY

An experimental method introduced to study tonality is sometimes referred to as the probe tone method.[1] It is best illustrated with a concrete example. Suppose you hear the tones of the ascending C major scale: C D E F G A B. There is a strong expectation that the next tone will be the tonic, C, first, because it is the next logical tone in the series and, second, because it is the tonic of the key. In the experiment, the incomplete scale context was followed by the tone C (the probe tone), and listeners were asked to judge how well it completed the scale on a numerical scale (1 = very bad, 7 = very good). As expected, the C received the maximal rating. Other probe tones, however, also received fairly high ratings, and they were not necessarily those that are close to the tonic C in pitch. For example, the most musically trained listeners also gave high ratings to the dominant, G, and the mediant, E. In general, the tones of the scale received higher ratings than the nonscale tones, C♯ D♯ F♯ G♯ A♯. This suggested that it is possible to get quantitative judgments of the degree to which different tones are perceived as stable, final tones in tonal contexts.

A subsequent study[2] used this method with a variety of musical contexts at the beginning of the trials. They were chosen because they are clear indicators of a key. They included the scale, the tonic triad chord, and chord cadences in both major and minor keys. These were followed by all possible probe tones in the chromatic scale, which musically trained listeners were instructed to judge in terms of how well they fit with the preceding context in a musical sense. Different major keys were used, as were different minor keys. The results for contexts of the same mode were similar when transposed to a common tonic. Also, the results were similar independent of which particular type of context was used. Consequently, the data were averaged over these factors. We call the resulting values the K-K profiles, which can be expressed as vectors. The vector for major keys is: K-K major profile = <6.35, 2.23, 3.48, 2.33, 4.38, 4.09, 2.52, 5.19, 2.39, 3.66, 2.29, 2.88>. The vector for minor keys is: K-K minor profile = <6.33, 2.68, 3.52, 5.38, 2.60, 3.53, 2.54, 4.75, 3.98, 2.69, 3.34, 3.17>.

We can generate K-K profiles for 12 major keys and 12 minor keys from these. If we adopt the convention that the first entry in the vector corresponds to the tone C, the second to C♯/D♭, the third to D, and so on, then the vector for C major is: <6.35, 2.23, 3.48, 2.33, 4.38, 4.09, 2.52, 5.19, 2.39, 3.66, 2.29, 2.88>, the vector for D♭ major is: <2.88, 6.35, 2.23, 3.48, 2.33, 4.38, 4.09, 2.52, 5.19, 2.39, 3.66, 2.29>, and so on. The vectors for the different keys result from shifting the entries the appropriate number of places to the tonic of the key.

## TRACING THE DEVELOPING AND CHANGING SENSE OF KEY

The probe tone method was then used to study how the sense of key develops and changes over time.[2] Ten nine-chord sequences were constructed, some of which contained modulations between keys. Musically trained listeners did the probe tone task after the first chord, then after the first two chords, then after the first three chords, and continued until the full sequence was heard. This meant that 12 (probe tones) × 9 (chord positions) × 10 (sequences) = 1080 judgments were made by each listener. Each of the 90 sets of probe tone ratings was compared with the ratings made for the unambiguous key-defining contexts. That is, each set of probe tone ratings was correlated with the K-K profiles for the 24 major and minor keys. For some of the sets of probe tone ratings (some probe positions in some of the chord sequences), a high correlation was found, indicating a strong sense of key. For other sets of probe tone ratings, no key was highly correlated, which was interpreted as an ambiguous or weak sense of key.

As should be obvious from the above, the probe tone task requires an intensive empirical effort to trace how the sense of key develops and changes, even for short sequences. In addition, the sequence needs to be interrupted, and the judgment is made after the sequence has been interrupted. For these reasons, the judgments may not faithfully mirror the experience of music in time. Hence, we[3] are currently testing an alternative form of the probe tone methodology. In this method, which we call the concurrent probe tone task, the probe tone is presented continuously while the music is played. The complete passage is sounded together with a probe tone. Then the passage is sounded again, this time with another probe tone. This process is continued until all probe tones have been sounded. Preliminary results suggest this methodology produces interpretable results, at least for musically trained participants. Our focus here, however, is on how the sense of key can be represented, whether the input to the representation is from a probe tone task or from a model of key-finding as described later.

## A GEOMETRIC MAP OF KEY DISTANCES FROM THE TONAL HIERARCHIES

The K-K profiles generated a highly regular and interpretable geometric representation of musical keys.[2] The basic assumption underlying this approach was that two keys are closely related to each other if they have similar tonal hierarchies. That is, keys were assumed to be closely related if tones that are stable in one key are also relatively stable in the other key. To measure the similarity of the profiles, a product-moment correlation was used. It was computed for all possible pairs of major and minor keys, giving a $24 \times 24$ matrix of correlations showing how similar the tonal hierarchy of each key was to every other key. To give some examples, C major correlated relatively strongly with A minor (0.651), with G major and F major (both 0.591), and with C minor (0.511). C minor correlated relatively strongly with E♭ major (0.651), with C major (0.511), with A♭ major (0.536), and less strongly with F minor and G minor (both 0.339).

A technique called multidimensional scaling was then used to create a geometric representation of the key similarities. The algorithm locates 24 points (corresponding to the 24 major and minor keys) in a spatial representation to best represent their similarities. It searches for an arrangement such that points that are close correspond to keys with similar K-K profiles (as measured by the correlations). In particular, nonmetric multidimensional scaling seeks a solution such that distances between points are (inversely) related by a monotonic function to the correlations. A measure (called "stress") measures the amount of deviation from the best-fitting monotonic function. The algorithm can search for a solution in any specified number of dimensions. In this case, a good fit to the data was found in four dimensions.

The four-dimensional solution located the 24 keys on the surface of a torus (generated by one circle in dimensions 1 and 2, and another circle in dimensions 3 and 4). Because of this, any key can be specified by two values: its angle on the first circle and its angle on the second circle. Thus, the result can be depicted in two dimensions as a rectangle where it is understood that the left edge is connected to the right edge, and the bottom edge is connected to the top edge. The locations of the 24 keys were interpretable in terms of music theory. There was one circle of fifths for major keys (...F♯/G♭, D♭, A♭, E♭, B♭, F, C, G, D, A, E, B, F♯/G♭...) and one circle of fifths for minor keys (...f♯, c♯, g♯, d♯/e♭, b♭, f, c, g, d, a, e, b, f♯...). These wrap diagonally around the torus such that each major key is located near both its relative minor (for example, C major and A minor) and its parallel minor (for example, C major and C minor). (See FIGS. 1 and 2 below.)

## REPRESENTING THE SENSE OF KEY ON THE TORUS

The continuous spatial medium containing the 24 major and minor keys affords representing the sense of key in a graphical form,[2] using a technique called multidimensional unfolding. It is a method that is closely related to multidimensional scaling. Multidimensional unfolding begins with a multidimensional scaling solution, in this case the torus representation of the 24 major and minor keys. This solution is considered fixed. The algorithm then finds a point in the multidimensional scaling solution to best represent the sense of key at each point in time. Let $P_1$ be the probe tone ratings after the first chord in a sequence; it is a 12-dimensional vector of ratings for each tone of the chromatic scale. This vector was correlated with each of the 24 K-K vectors, giving a 24-dimensional vector of correlations. The unfolding algorithm finds a point to best represent these correlations. Suppose $P_1$ correlates highly with the K-K profile for F major and fairly highly with the K-K profile for D minor. Then the unfolding algorithm will produce a point near these keys and far from the keys with low correlations. Then the vector of correlations was computed for $P_2$, giving a second point. This process continues until the end of the sequence.

In this manner, each of the ten nine-chord sequences[2] generated a series of nine points on the torus representation of keys. For nonmodulating sequences, the points remained in the neighborhood of the intended key. For the modulating sequences, the first points were near the initial intended key, then shifted to the region of the second intended key. Modulations to closely related keys appeared to be assimilated more rapidly than those to distantly related keys. That is, the points shifted to the region of the new key earlier in sequences containing close modulations than in sequences containing distant modulations.

## MEASUREMENT ASSUMPTIONS OF THE MULTIDIMENSIONAL SCALING AND UNFOLDING METHODS

The above methods make a number of assumptions about measurement, only some of which will be noted here. The torus representation is based on the assumption that correlations between the K-K profiles are appropriate measures of interkey distance. It further assumes that these distances can be represented in a relatively low-dimensional space (four dimensions). This latter assumption was supported by the low stress value (high goodness-of-fit value) of the multidimensional scaling solution. It was further supported by a subsidiary Fourier analysis of the K-K major and minor profiles, which found two relatively strong harmonics.[4] In fact, plotting the phases of the two Fourier components for the 24 key profiles was virtually identical to the multidimensional scaling solution. This supports the torus representation, which consists of two orthogonal circular components. Nonetheless, it would seem desirable to see whether an alternative method with completely different assumptions recovers the same toroidal representation of key distances.

The unfolding method also adopts correlation as a measure of distances from keys, this time using the ratings for each probe position in the chord sequences and the K-K vectors for the 24 major and minor keys. The unfolding technique finds the best-fitting point in the four-dimensional space containing the torus. It does not provide a way of representing cases in which no key is strongly heard, because it cannot generate points outside the space containing the torus. Thus, an important limitation of the unfolding method is that it does not provide a representation of the strength of the key or keys heard at each point in time. For this reason, we sought a method that is able to represent both the region of the key or keys that are heard, together with their strengths.

In addition, the unfolding approach assumes that spatial locations between the 24 points for major and minor keys are meaningful. An intermediate position would result from a blend of tonal hierarchies of nearby keys. However, other sets of probe tone ratings might also map to the same position. Thus, the identification of points between keys is not necessarily unique. This motivated an alternative model which explicitly specifies the meaning of positions between keys.

## THE SELF-ORGANIZING MAP (SOM) OF KEYS

The self-organizing map (SOM)[5] is an artificial neural network that simulates the formation of ordered feature maps. The SOM consists of a two-dimensional grid of units, each of which is associated with a reference vector. Through repeated exposure to a set of input vectors, the SOM settles into a configuration in which the reference vectors approximate the set of input vectors according to some similarity measure; the most commonly used similarity measures are the Euclidean distance and the direction cosine. The direction cosine between an input vector **x** and a reference vector **m** is defined by

$$\cos\theta = \frac{\sum_i x_i m_i}{\sqrt{\sum_i x_i^2}\sqrt{\sum_i m_i^2}} = \frac{\mathbf{x}\cdot\mathbf{m}}{\|\mathbf{x}\|\,\|\mathbf{m}\|}. \tag{1}$$

Another important feature of the SOM is that its configuration is organized in the sense that neighboring units have similar reference vectors. For a trained SOM, a mapping from the input space onto the two-dimensional grid of units can be defined by associating any given input vector with the unit whose reference vector is most similar to it. Because of the organization of the reference vectors, this mapping is smooth in the sense that similar vectors are mapped onto adjacent regions. Conceptually, the mapping can be thought of as a projection onto a nonlinear surface determined by the reference vectors.

We trained the SOM with the 24 K-K profiles. The SOM was specified in advance to have a toroidal configuration, that is, the left and the right edges of the map were connected to each other as were the top and the bottom edges. Euclidean distance and direction cosine, when used as similarity measures in training the SOM, yielded identical maps. (The resulting map is displayed in FIGS. 1 and 2 below.) The map shows the units with reference vectors that correspond to the K-K profiles. The SOM configuration is highly similar to the multidimensional scaling solution[2] and the Fourier-analysis-based projection[4] obtained with the same set of vectors. Unlike those maps, however, all locations in the map are explicitly associated with a reference vector so that they are uniquely identified.

## REPRESENTING THE SENSE OF KEY ON THE SOM

A distributed mapping of tonality can be defined by associating each unit with an activation value. For each unit, this activation value depends on the similarity between the input vector and the reference vector of the unit. Specifically, the units whose reference vectors are highly similar to the input vector have a high activation, and vice versa. The activation value of each unit can be calculated, for instance, using the direction cosine of Equation 1. Dynamically changing data from either probe tone experiments or key-finding models can be visualized as an activation pattern that changes over time. The location and spread of this activation pattern provides information about the perceived key and its strength. More specifically, a focused activation pattern implies a strong sense of key and vice versa.

## KEY-FINDING MODELS

A variety of key-finding algorithms have been proposed. The objective of these models is to assign a key to an input sample of music (such as the fugue subjects of J. S. Bach's *Well-Tempered Clavier*). Symbolic models, reviewed elsewhere,[4] have taken into consideration a number of factors in assigning key, including scales that contain the tones of the sample, and the presence of such cues to key as the tonic-fifth, tonic-third, and tonic–leading-tone intervals, characteristic tone sequences, and cadences. Some effort has been made to take phrasing, melodic accent, and rhythm into account. More recently proposed symbolic models[6,7] have made advances in both computational and music-analytic sophistication.

Concurrently, neural network models have provided an alternative subsymbolic approach.[8,9] In these, the input sample typically gives rise to activation levels of units associated with different keys. Thus, these models return graded measures of key strength. The problem of representing these in a way that takes into account the distances between keys, however, remains unsolved. In addition, little has been done to compare the output of key-finding algorithms to perceptual judgments, which may change over time as the cues to key become more or less clear and as the music may modulate to other keys. For the most part, the model's output has been compared with the composer's key signature.

Both these issues were addressed by a key-finding algorithm developed using the K-K profiles.[4] The input to this algorithm was a 12-dimensional vector specifying the total durations of the twelve chromatic scale tones in the musical selection to which the key was to be assigned. One application was to Bach's *C Minor Prelude* from Book II of the *Well-Tempered Clavier.* It was treated on a measure-by-measure basis. Each input vector was projected onto the toroidal map of keys using the phases of the two strong Fourier components. Two musical experts gave quantitative judgments of key strength in each measure of the piece, which were also projected onto the map. The key-finding algorithm followed the same pattern of modulations found by the experts, suggesting the approach lends itself to tracing dynamic changes in key as modulations occur. One obvious limitation of this approach is that it ignores the order of the tones in the input sample. This potentially important cue for key led to the development of the tone transition model described below.

## TONE TRANSITIONS AND KEY-FINDING

The order in which tones are played may provide additional information that is useful for key-finding. This is supported by studies on both tone transition probabilities[10,11] and perceived stability of tone pairs in tonal contexts.[4,13] In samples of compositions by Bach, Beethoven, and Webern, only a small fraction of all the possible tone transitions were actually used (the fractions were 23, 16, and 24 percent, respectively).[10] Furthermore, in a sample of 20 songs by Schubert, Mendelssohn, and Schumann, there was an asymmetry in the transition frequencies in the sense that certain tone transitions were used more often than the same tones in the reverse temporal order.[11] For instance, the transition B-C was used 93 times, whereas the transition C-B was used only 66 times. A similar asymmetry was found in studies on perceived stability of tone pairs in a tonal context.[4,13] After the presentation of a tonal context, tone pairs that ended with a tone that was high in the tonal hierarchy were given higher ratings than the reverse temporal orders. For instance, in the context of C major, the ratings for the transitions B-C and C-B were 6.42 and 3.67, respectively.

Determining tone transitions in a piece of polyphonic music is not a trivial task, especially if one aims at a representation that corresponds to perceptual reality. Even in a monophonic piece, the transitions can be ambiguous in the sense that their perceived strengths may depend on the tempo. Consider, for example, the tone sequence C4-G3-D4-G3-E4, where all the tones have equal durations. When played slowly, this sequence is heard as a succession of tones oscillating up and down in pitch. With increasing tempi, however, the subsequence C4-D4-E4 becomes increasingly prominent. This is because these tones segregate into one stream due to the temporal and pitch proximity of its members, separate from G3-G3. With polyphonic music, the ambiguity of tone transitions becomes even more obvious. Consider, for instance, the sequence consisting of a C major chord followed by a D major chord, where the tones of each chord are played simultaneously. In principle, this passage contains nine different tone transitions. Some of these transitions are, however, perceived as stronger than others. For instance, the transition G-A is, due to pitch proximity, perceived as stronger than the transition G-D.

It seems, thus, that the analysis of tone transitions in polyphonic music should take into account principles of auditory stream segregation.[14] Furthermore, it may be necessary to code the presence of transitions on a continuous instead of a discrete scale. In other words, each transition should be associated with a strength value instead of just coding whether that particular transition is present or not. Below, a dynamic system that embraces these principles is described. In regard to the evaluation of transition strength, the system bears a resemblance to a proposed model applying the concept of apparent motion to music.[15]

## TONE TRANSITION MODEL

Let the piece of music under examination be represented as a sequence of tones, where each tone is associated with pitch, onset time, and duration. The main idea of the model is the following: given any tone in the sequence, there is a transition from that tone to all the tones following that particular tone. The strength of each transition depends on three factors: pitch proximity, temporal proximity, and the duration of the tones. More specifically, a transition between two tones has the highest strength when the tones are proximal in both pitch and time and have long durations. These three factors are included in the following dynamic model.

### *Representation of Input*

The pitches of the chromatic scale are numbered consecutively. The onset times of tones having pitch $k$ are denoted by $t_{ki}^{n}$, $i = 1, \ldots, n_k$, and the offset times by, $t_{ki}^{f}$, $i = 1, \ldots, n_k$, where $n_k$ is the total number of times the $k^{\text{th}}$ pitch occurs.

### *Pitch Vector* $\mathbf{p}(t) = (p_k(t))_k$

Each component of the pitch vector has nonzero value whenever a tone with the respective pitch is sounding. It has the value of 1 at each onset at the respective pitch, decays exponentially after that, and is set to zero at the tone offset. The time evolution of $\mathbf{p}$ is governed by the equation

$$\dot{p}_k = -p_k/\tau_p + \sum_{i=1}^{n_k} \delta\left(t - t_{ki}^{n}\right) - p_k \sum_{i=1}^{n_k} \delta\left(t - t_{ki}^{f}\right), \tag{2}$$

where $\dot{p}_k$ denotes the time derivative of $p_k$ and $\delta(\cdot)$ the Dirac delta function (unit impulse function). The time constant $\tau_p$ has the value of $\tau_p = 0.5$ sec. With this value, the integral of $p_k$ saturates at about 1 sec after tone onset, thus approximating the durational accent as a function of tone duration.[16]

### *Pitch Memory Vector* $\mathbf{m}(t) = (m_k(t))_k$

The pitch memory vector provides a measure of both the perceived durational accent and the recency of notes played at each pitch. In other words, a high value of $m_k$

indicates that a tone with pitch $k$ and a long duration has been played recently. The dynamics of **m** are governed by the equation

$$\dot{\mathbf{m}} = \mathbf{p} - \mathbf{m}/\tau_m, \tag{3}$$

The time constant $\tau_m$ determines the dependence of transition strength on the temporal distance between the tones. In the simulations, the value of $\tau_m = 3$ sec has been used, corresponding to typical estimates of the length of the auditory sensory input.[17–19]

### *Transition Strength Matrix* S(*t*) = $(s_{kl}(t))_{kl}$

The transition strength matrix provides a measure of the instantaneous strengths of transitions between all pitch pairs. More specifically, a high value of $s_{kl}$ indicates that a long tone with pitch $k$ has been played recently and a tone with pitch $l$ is currently sounding. The temporal evolution of **S** is governed by the equation

$$s_{kl} = m_k \frac{1 + \operatorname{sgn}(p_l - p_k)}{2} p_l e^{-(k-l)^2/\alpha^2}, \tag{4}$$

In this equation, the nonlinear term $(1 + \operatorname{sgn}(p_l - p_k))/2$ is used for distinguishing between simultaneously and sequentially sounding pitches. This term is nonzero only when $p_l > p_k$, that is, when the most recent onset of pitch $l$ has occurred more recently than that of pitch $k$. The term $e^{-(k-l)^2/\alpha^2}$ weights the transitions according to the interval size. For the parameter $\alpha$, the value $\alpha = 6$ has been used. With this value a perfect fifth gets a weight of about 0.37 times the weight of a minor second.

### *Dynamic Tone Transition Matrix* N(*t*) = $(n_{kl}(t))_{kl}$

The dynamic tone transition matrix is obtained by temporal integration of the transition strength matrix. At a given point of time, it provides a measure of the strength and recency of each possible tone transition. The time evolution of **N** is governed by the equation

$$\dot{\mathbf{N}} = \mathbf{S} - \mathbf{N}/\tau_N, \tag{5}$$

where the time constant $\tau_N$ is equal to $\tau_m$, that is, $\tau_N = 3$ sec.

To examine the role of tone transitions in key-finding, we developed two key-finding models. Model 1 is based on pitch class distributions. Model 2 is based on tone transition distributions. Below, a brief description of the models is given.

## KEY-FINDING MODEL 1

Model 1 is based on pitch class distributions only. Like the earlier algorithm based on the K-K profiles,[4] it does not take tone transitions into account. However, it has a dynamic character in that both the pitch vector and the pitch memory vector depend on time. It uses a pitch class vector $\mathbf{p}_c(t)$, which is similar to the pitch vector $\mathbf{p}(t)$ used in the dynamic tone transition matrix, except that it ignores octave information. Consequently, the vector has 12 components that represent the pitch classes.

The pitch class memory vector $\mathbf{m}_c(t)$ is obtained by temporal integration of the pitch class vector according to the equation

$$\dot{\mathbf{m}}_c = \mathbf{p}_c - \mathbf{m}_c/\tau_d . \tag{6}$$

Again, the time constant has the value $\tau_d = 3$ seconds. To obtain estimates for the key, vector $\mathbf{m}_c(t)$ is correlated with the K-K profiles for each key. Alternatively, or in addition, the vectors $\mathbf{m}_c(t)$ can be projected onto the toroidal key representation using activation values as described earlier. Both approaches will be taken here.

## KEY-FINDING MODEL 2

Model 2 is based on tone transitions. Using the dynamic transition matrix **N**, it calculates the octave-equivalent transition matrix $\mathbf{N}' = (n'_{ij})_{ij}$ according to

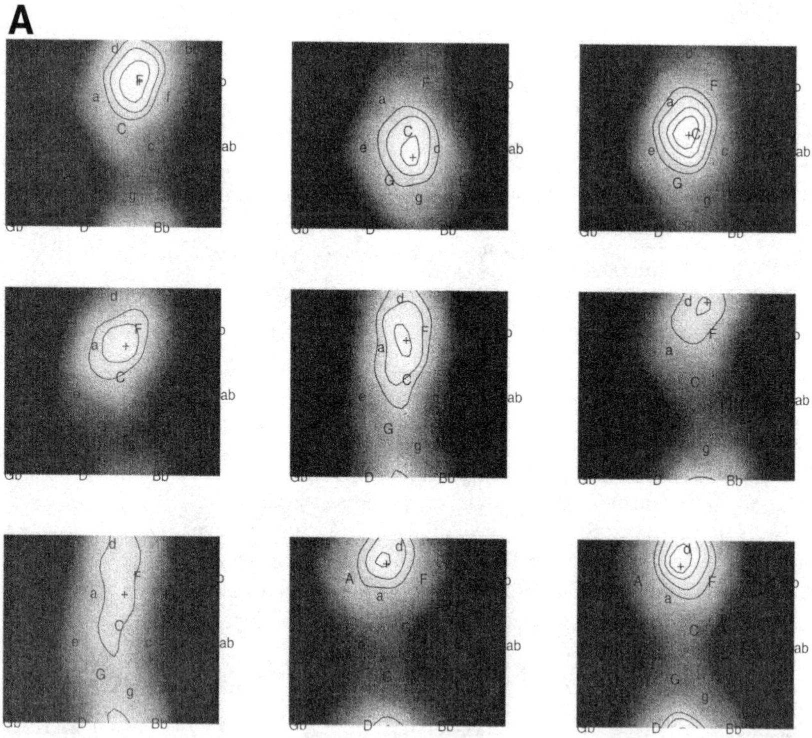

**FIGURE 1.** **(A)** Distributed mapping of tonality on the SOM for probe tone judgments for the nine-chord sequence modulating from C major to D minor. The *top row* shows the results for chords 1, 2, and 3 (*left to right*), the *second row* for 4, 5, and 6 (*left to right*), and the *bottom row* for chords 7, 8, and 9 (*left to right*).

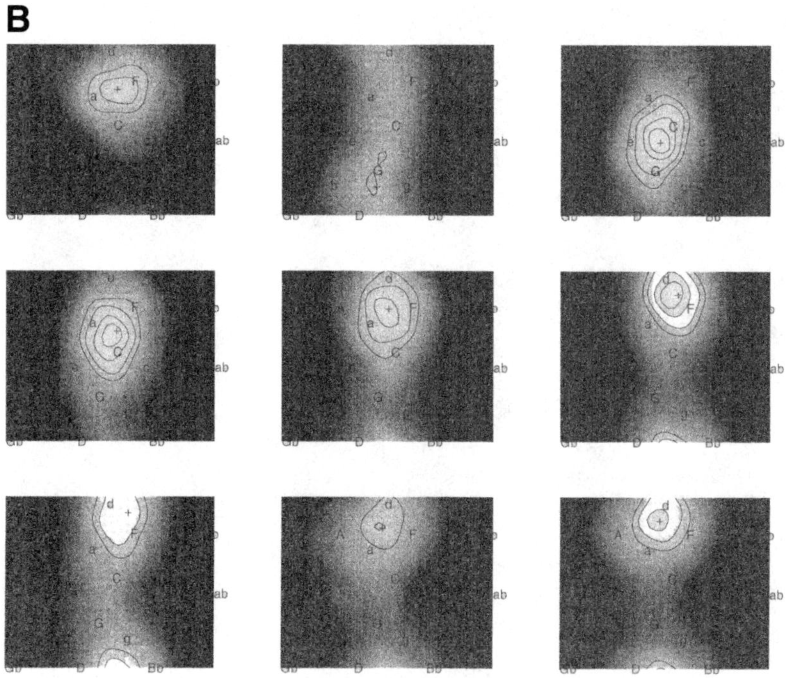

FIGURE 1. (B) Distributed mapping of model 1 for the same sequence.

$$n_{ij}'(t) = \sum_{\substack{i = p \bmod 12 \\ j = p \bmod 12}} n_{pq}(t) . \quad (7)$$

In other words, transitions whose first and second tones have identical pitch classes are considered equivalent, and their strengths are added. Consequently, the melodic direction of the tone transition is not taken into account. To obtain estimates for the key, the pitch class transition matrix is correlated with the matrices representing the perceived stability of two-tone transitions for each key.[4]

## APPLICATION OF MODELS 1 AND 2 TO THE CHORD SEQUENCE DATA

The probe tone data for the ten chord sequences described earlier[2] were compared with the results of models 1 and 2 applied to those sequences. An interchord onset time of one second was used. The perceptual judgments were correlated with the K-K profiles, and these were compared with the key correlations of the two models. Model 1 results correlated highly ($r(2158) = 0.87$, $p < 0.0001$) with the experimental data. Model 2 also correlated highly ($r(1918) = 0.86$, $p < 0.0001$). (Model 2 produces

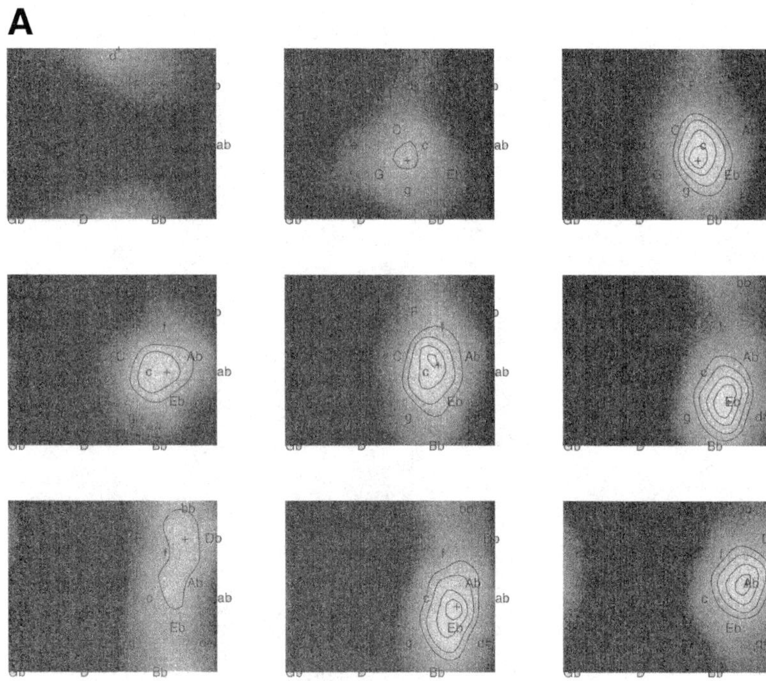

**FIGURE 2.** **(A)** Distributed mapping of tonality on the SOM for probe tone judgments for the nine-chord sequence modulating from C minor to A♭ minor.

an output only after the first two chords.) Model 2 contributed additional precision when combined with model 1, as shown by a multiple regression predicting the experimental data from both models ($R(2, 1917) = 0.89$, $p < 0.0001$, with both models contributing significantly at $p < 0.0001$; the standard coefficient was 0.50 for model 1 and 0.43 for model 2). Thus, the two models generally matched well the experimental results, and modeled slightly different aspects of the listeners' responses.

FIGURES 1 and 2 show the distributed mapping of tonality on the SOM for the listeners (FIGS. 1a and 2a) and model 1 (FIGS. 1b and 2b) for two illustrative chord sequences. (Because of issues about how best to visualize the results of model 2, we show only model 1 here.) In these representations, a single contour outlines correlations 0.70 or above, double contours outline correlations 0.80 or above, triple contours outline correlations 0.90 or above, and quadruple contours outline correlations 0.95 or above.

The sequence depicted in FIGURE 1 consists of the chords F G C F d B♭ e$^{o}$ A d, containing a relatively distant modulation from C major to D minor. After the first chord, F, appeared, the listeners and the model had a clear focus on the key of F major in which the chord is I. When the second chord, G, was sounded, listeners apparently interpreted this as a IV–V progression in C major, resulting in a focus near that

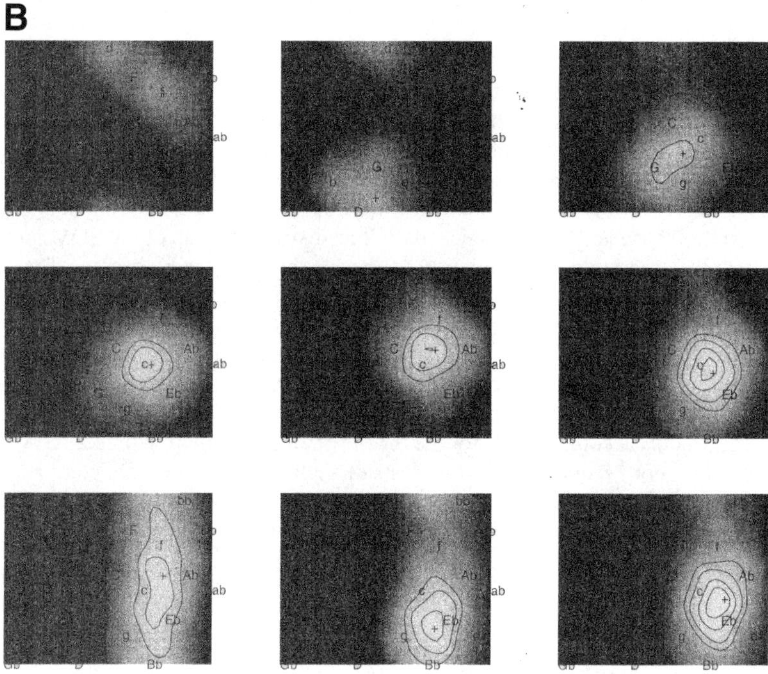

**FIGURE 2.** (**B**) Distributing mapping of model 1 for the same sequence.

key. The model, however, did not find this focus until the tonic triad, C, in the third position. For the fourth and fifth chords, F and d, the focus of both listeners and model remained in the region of C major but was shifted somewhat toward F major and D minor. The sixth chord, B♭, which is contained in D minor but not C major, shifted the focus farther toward D minor. The seventh chord, e$^{o}$, greatly weakened the sense of key for both listeners and model; a diffuse elongated region of weak activation was found. With the last two chords, A and d, the final focus on D minor (in which these are V–I) was arrived at.

The sequence depicted in FIGURE 2 consists of the chords d$^{o}$ G c A♭ F D♭ b♭ E♭ A♭, containing a modulation from C minor to A♭ major, a modulation to what is considered a relatively close key. Both listeners and model yielded an extremely weak activation pattern for the first two chords, d$^{o}$ and G, although the focus for listeners after the second chord was weakly near C minor. The third chord, c, clarified this focus near C minor for both listeners and model, where it remained for the fourth and fifth chords, A♭ and f. The sixth chord, E♭, which is the tonic triad in the relative major of C minor (E♭ major) produced a strong focus on E♭ major for listeners and a shift in that direction for the model. The seventh chord, B♭, diffused this focus for both listeners and model before the new key of A♭ major clarified with the E♭ and A♭ chords in the last two positions (which are V–I in the A♭ major key).

## CONCLUSION

Experimental studies suggest that the sense of tonality undergoes dynamic and subtle changes when a listener hears music. The sense of key develops and strengthens as certain cues appear, then may weaken or shift to a new key as subsequent events are sounded. An important step in understanding this process is a suitable means of representing such changes. Toward this end, we have developed a spatial representation based on psychological data. The distributed map of the SOM provided a visually accessible representation of these subtle dynamic changes. The key-finding models described here suggested that listeners' sense of key can be modeled quite well using tone distributions and tone transitions. Two models were formulated to incorporate various psychological phenomena, including durational accent as a function of tone duration, the duration of sensory memory, temporal-order asymmetries, and pitch streaming as a function of the distances between tones in pitch and time. The models accounted well for the results of an experiment with relatively simple and short chord sequences. Further tests of such models with more complex, extended, and musically realistic materials may point to additional factors, such as rhythm, meter, phrasing, and form, that may also influence the sense of tonality.

## REFERENCES


1. Krumhansl, C.L. & R.N. Shepard. 1979. Quantification of the hierarchy of tonal functions within a diatonic context. J. Exp. Psychol. Hum. Percept. Perform. **5:** 579–594.
2. Krumhansl, C.L. & E.J. Kessler. 1982. Tracing the dynamic changes in perceived tonal organization in a spatial representation of musical keys. Psychol. Rev. **89:** 334–368.
3. Krumhansl, C.L. & P. Toiviainen. 2000. Dynamics of tonality induction: a new method and a new model. *In* Proceedings of the Sixth International Conference on Music Perception and Cognition. C. Woods, G. Luck, R. Brochard, F. Seddon & J.A. Sloboda, Eds.: 1504–1513. Keele University. Keele, UK.
4. Krumhansl, C.L. 1990. Cognitive Foundations of Musical Pitch. Oxford University Press. New York.
5. Kohonen, T. 1997. Self-organizing Maps. Springer-Verlag. Berlin.
6. Vos, P.G. & E.W. Van Geenen. 1996. A parallel-processing key-finding model. Music Percept. **14:** 185–223.
7. Vos, P.G. & M. Leman, Eds. 2000. Tonality Induction. Special issue of Music Percept.
8. Leman, M. 1995. A model of retroactive tone-center perception. Music Percept. **12:** 439–471.
9. Bharucha, J.J. 1999. Neural nets, temporal composites, and tonality. *In* The Psychology of Music. 2nd edit. D. Deutsch, Ed.: 413–440. Academic Press. San Diego, CA.
10. Fucks, W. 1962. Mathematical analysis of the formal structure of music. I R E Trans. Information Theory **8:** 225–228.
11. Youngblood, J.E. 1958. Style as information. J. Music Theory **2:** 24–35.
12. Knopoff, L. & W. Hutchinson. 1978. An index of melodic activity. Interface **7:** 205–229.
13. Krumhansl, C.L. 1979. The psychological representation of musical pitch in a tonal context. Cognit. Psychol. **11:** 346–374.
14. Bregman, A.S. 1990. Auditory Scene Analysis. M.I.T. Press. Cambridge, MA.
15. Gjerdingen, R.O. 1994. Apparent motion in music? Music Percept. **11:** 335–370.
16. Parncutt, R. 1994. A perceptual model of pulse salience and metrical accent in musical rhythms. Music Percept. **11:** 409–464.

17. Darwin, C.J., M.T. Turvey & R.G. Crowder. 1972. An auditory analogue of the Sperling partial report procedure: evidence for brief auditory storage. Cognit. Psychol. **3:** 255–267.
18. Fraisse, P. 1982. Rhythm and tempo. *In* The Psychology of Music. 1st edit. D. Deutsch, Ed.: 149–180. Academic Press. San Diego, CA.
19. Treisman, A.M. 1964. Verbal cues, language, and meaning in selective attention. Am. J. Psychol. **77:** 206–219.

# Neurobiological Foundations for the Theory of Harmony in Western Tonal Music

MARK JUDE TRAMO,[a,b,c] PETER A. CARIANI,[a,b] BERTRAND DELGUTTE,[b,c] AND LOUIS D. BRAIDA[c]

[a]*Department of Neurology, Harvard Medical School and Massachusetts General Hospital, Boston, Massachusetts 02114-2696, USA*

[b]*Eaton-Peabody Laboratory of Auditory Physiology, The Massachusetts Eye and Ear Infirmary, Boston, Massachusetts 02114, USA*

[c]*Research Laboratory of Electronics, Massachusetts Institute of Technology, Cambridge, Massachusetts 02139-4307, USA*

**ABSTRACT: Basic principles of the theory of harmony reflect physiological and anatomical properties of the auditory nervous system and related cognitive systems. This hypothesis is motivated by observations from several different disciplines, including ethnomusicology, developmental psychology, and animal behavior. Over the past several years, we and our colleagues have been investigating the vertical dimension of harmony from the perspective of neurobiology using physiological, psychoacoustic, and neurological methods. Properties of the auditory system that govern harmony perception include (1) the capacity of peripheral auditory neurons to encode temporal regularities in acoustic fine structure and (2) the differential tuning of many neurons throughout the auditory system to a narrow range of frequencies in the audible spectrum. Biologically determined limits on these properties constrain the range of notes used in music throughout the world and the way notes are combined to form intervals and chords in popular Western music. When a harmonic interval is played, neurons throughout the auditory system that are sensitive to one or more frequencies (partials) contained in the interval respond by firing action potentials. For consonant intervals, the fine timing of auditory nerve fiber responses contains strong representations of harmonically related pitches implied by the interval (e.g., Rameau's fundamental bass) in addition to the pitches of notes actually present in the interval. Moreover, all or most of the partials can be resolved by finely tuned neurons throughout the auditory system. By contrast, dissonant intervals evoke auditory nerve fiber activity that does not contain strong representations of constituent notes or related bass notes. Furthermore, many partials are too close together to be resolved. Consequently, they interfere with one another, cause coarse fluctuations in the firing of peripheral and central auditory neurons, and give rise to perception of roughness and dissonance. The effects of auditory cortex lesions on the perception of consonance, pitch, and roughness, combined with a critical reappraisal of published psychoacoustic data on the relationship between consonance and roughness, lead us to conclude that consonance is first and foremost a function of the pitch relationships among notes. Harmony in the vertical dimension is a positive phenomenon, not just a negative phenomenon that depends on the absence of**

Address for correspondence: Dr. Mark Jude Tramo, MGH EDR-405, 55 Fruit Street, Boston, MA 02114-2696. Voice: 617-726-5409; fax: 617-726-5457.
mtramo@hms.harvard.edu

**roughness—a view currently held by many psychologists, musicologists, and physiologists.**

**KEYWORDS: Consonance; Dissonance; Harmony, musical; Intervals, musical; Perception of harmony; Psychoacoustics of harmony**

## INTRODUCTION

Why do some combinations of simultaneous tones sound more harmonious than others? Pythagoras's curiosity about the nature of harmony inspired some of the earliest experiments relating mathematics and physics to perceptual phenomena.[1] Approaching the problem from the perspective of neurobiology, we ask: Are there physiological and anatomical properties of the auditory system and related cognitive systems that determine the degree to which simultaneous notes sound harmonious?

We restrict our consideration of harmony to basic tenets articulated by Piston,[2] among others. Harmony has a vertical dimension and a horizontal dimension. The vertical dimension encompasses the relationships among simultaneous notes. By convention, *note* refers to a pitch in the musical scale, and *harmonic interval* refers to two notes sounded simultaneously (FIG. 1A–D). When a note is played on a musical instrument, its pitch corresponds to the fundamental frequency (F0) of the complex tone generated by the instrument. Some synthesizers and other types of equipment are capable of generating pure tones, in which case the pitch of the note corresponds to the frequency of the pure tone. Harmonic intervals are a type of *dyad*. Three or more notes played simultaneously make up a *chord*. Chords with three notes are called *triads*. The time window over which acoustic information is integrated in the vertical dimension spans about a hundredth of a second to a few seconds (e.g., sixteenth notes to tied whole notes at a tempo of 120 beats per min). The horizontal dimension encompasses successive tones (melodic intervals and melodic progressions) and successive harmonic intervals and chords (harmonic progressions). Certain intervals and chords are treated as consonant (e.g., fifths, major triads) and others as dissonant (e.g., minor seconds, diminished triads). Acknowledging that different psychologists have attached different perceptual attributes and meanings to the terms *consonance* and *dissonance*, we nonetheless find considerable agreement among music texts and dictionaries that consonant means harmonious, agreeable, and stable, and that dissonant means disagreeable, unpleasant, and in need of resolution.[2–5] Psychoacoustic experiments bear out semantic overlap among the terms *consonant, pleasant, beautiful,* and *euphonious*.[6,7]

These basic concepts apply to a wide range of musical styles enjoyed by people throughout much of the industrialized world: contemporary pop and theater (including rock, rhythm and blues, country, and Latin-American), European music from the Baroque, Classical, and Romantic eras (1600–1900), children's songs, and many forms of ritualistic music (e.g., church songs, processionals, anthems, and holiday music). The overlap in their harmonic structure incorporates commonalities in musical phonology and syntax.[8] In our view, the widespread popularity of Western pop taps into (1) universal competence in auditory functions needed to extract the pitch of a note and to analyze the harmonic relationships among different pitches, and (2) universal competence in cognitive functions that parse acoustic information and as-

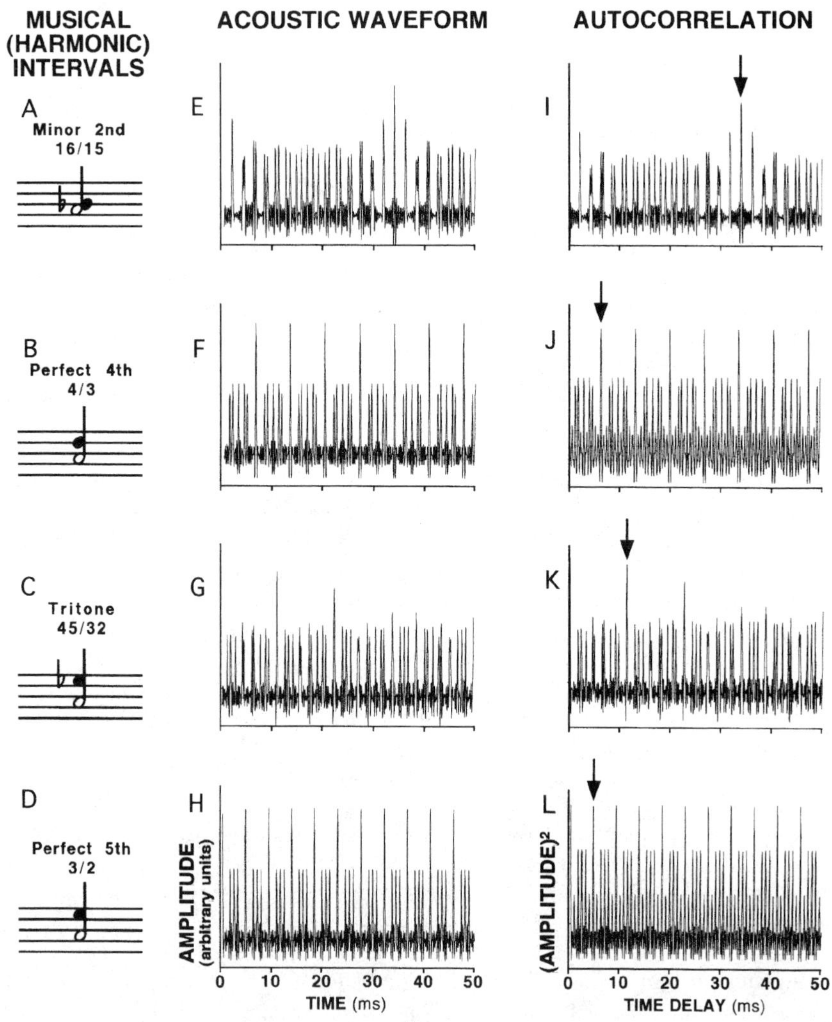


**FIGURE 1.** Acoustic representations of musical (harmonic) intervals in the time domain. *Left column* (**A–D**): Musical interval stimuli depicted in standard notation (G clef) below the name of each interval (e.g., minor second) and the F0 ratio of its notes (16/15). *Middle column* (**E-H**): Acoustic waveforms of each interval. *Right column* (**I-L**): Autocorrelations of the acoustic waveforms. *Arrows* indicate the peaks corresponding to the period of each interval's "missing" F0.

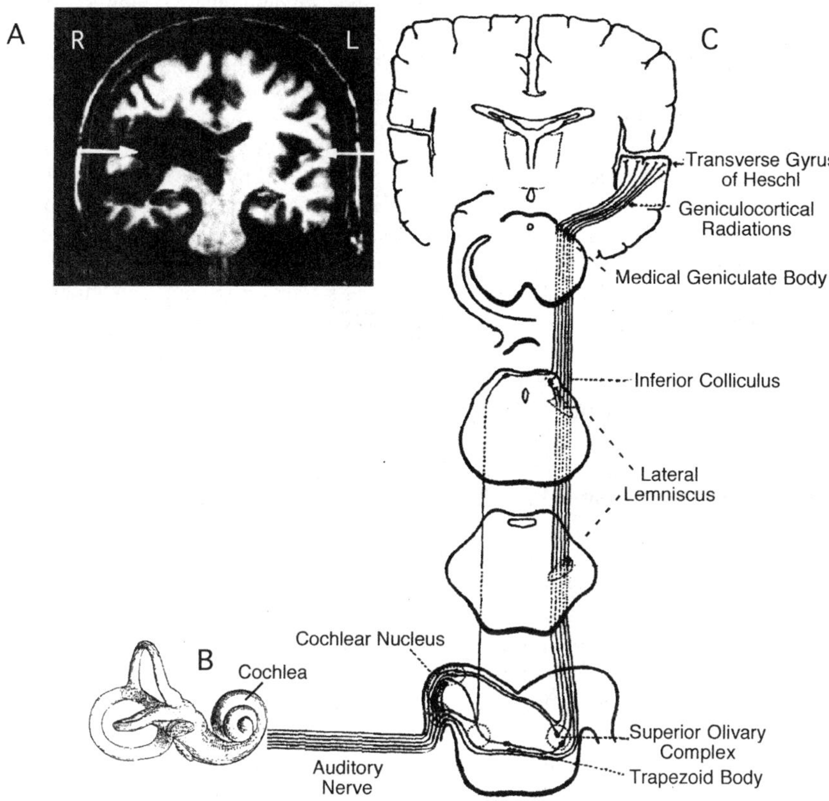


**FIGURE 2.** The auditory system. (**A**) Magnetic resonance image (coronal section) through the transverse gyri of Heschl and the superior temporal gyri of case MHS, who suffered bilateral infarction nine years earlier. *White arrows* point to the region of the transverse gyrus of Heschl in the right (*R*) and left hemispheres (*L*). The results of psychoacoustic experiments performed by MHS are shown in FIG. 7. (**B**) The inner ear and the auditory nerve, the obligatory pathway from the cochlea to the cochlear nuclei in the lower brain stem. (**C**) Schematic of the central auditory pathway showing the main relay stations and projection patterns. Adapted from Tramo *et al.*[15] (**A**), Helmholtz[16] (**B**), and Davis[17] (**C**).

sociate perceptual attributes with emotion and meaning. Experimental results suggest that similar perceptual attributes can be associated with similar emotions and social contexts across different cultures.[9,10] Moreover, listeners from different cultures often use similar cognitive schemata to structure the processing of pitch-sequences.[11,12]

Much has been written in the psychology literature about the terms *harmony*, *consonance*, and *dissonance*. At present, many psychologists and musicologists subscribe to Terhardt's[13] two-component model of *musical consonance*, which subsumes all these terms. One component is *sensory consonance*, the absence of annoying features, such as roughness, in both musical and nonmusical sounds. The other component, *harmony*, is based on music-specific principles that govern pitch relationships in melodic and harmonic progressions. Terhardt asserts that "sensory consonance...dominates the evaluation of single isolated chords...whereas harmony does not enter into the subject's response" (p. 282).[13]

With respect to terminology, we adhere in this discussion to the simple distinction between the vertical and horizontal dimensions of harmony. We restrict use of the terms *consonance* and *dissonance* to the vertical dimension and keep the term *harmony* supraordinate to them. We make no assumptions about the level of auditory processing (e.g., sensory, peripheral) where the perceptual attribute of consonance takes shape. We consider it likely that a listener's implicit (or explicit) knowledge about harmony in the horizontal dimension[14] bears on harmony perception in the vertical dimension.

In this paper, we present neurophysiological, neurological, and psychoacoustic evidence to support our contentions that (1) pitch relationships among tones in the vertical dimension influence consonance perception and (2) consonance cannot be explained solely by the absence of roughness. First, we review terminology and basic psychoacoustics pertinent to our subsequent discussion of experimental results. Second, we demonstrate that the harmonic relationships of tones in musical intervals are represented in the temporal discharge patterns of auditory nerve fibers. Third, we critically reevaluate the psychoacoustic literature concerning the consonance of isolated intervals and chords, paying particular attention to (1) the relationships among interval width, roughness detection thresholds, and consonance ratings; and (2) the predictions of roughness-based computational models about relative consonance as a function of spectral energy distribution. Finally, we discuss evidence that impairments in consonance perception following auditory cortex lesions are more likely to result from deficits in pitch perception than to deficits in roughness perception. This evidence highlights the dependence of so-called low-level perceptual processing on the integrity of the auditory cortex, the highest station in the auditory nervous system (FIG. 2).

## PSYCHOACOUSTICS AND NEUROPHYSIOLOGY OF HARMONY

For authoritative reviews of the psychoacoustics of harmony, we refer the reader to Krumhansl,[14] Parncutt,[18] and Deutsch.[19] Here, only basic concepts and terminology pertinent to our subsequent discussion of psychoacoustic and neurophysiological experiments are covered.

Let us consider a modern restatement of Pythagoras's observation: The degree to which two simultaneous notes (a harmonic interval) sound consonant is determined by the simplicity of the ratio $x$:$y$, where $x$ is the F0 associated with one tone and $y$ the F0 of the other, lower tone. In musical terms, $y$ is the root of the interval. $X$ and $y$ can take on any value between about 25 Hz and about 5 kHz. This upper limit coincides with (1) the upper F0 of notes on a piccolo (~4500 Hz); (2) the upper F0 for which octave similarity can be reliably judged;[20,21] and (3) the upper F0 of strong phase locking by auditory nerve fibers—that is, the highest frequency at which neurons can fire in time with amplitude fluctuations in the acoustic waveform.[22,23] This convergence of facts from music, psychoacoustics, and physiology suggests that limitations in the phase-locking capacity of neurons in the auditory periphery constrain the range of note F0s that are used in music and the way they are combined in the vertical dimension of harmony. Other authors have discussed the relationships among the temporal discharge patterns of auditory nerve fibers, fundamental pitch perception, octave equivalence, and the consonance of intervals formed by simple integer ratios.[24–29]

By convention, notes in popular Western music are tuned to the scale of equal temperament, which chunks the F0 continuum into octaves (i.e., doublings of F0) and each octave into twelve discrete, equal, logarithmic steps. Each step within the octave is called a semitone, and the F0s of adjacent semitones differ by a factor of $2^{1/12}$, or about 6 percent. The chromatic scale is made up of all twelve tones in the octave, whereas the major and minor (diatonic) scales are made up of partially overlapping sets of seven tones in the octave. Harmonic intervals and melodic intervals are named according to the scale relationship of the upper note to the lower note. Thus when the fifth note in the major or minor scale sits atop the first note on the scale, the interval is called a *fifth* (FIG. 1D). The note named $A_4$ is assigned an F0 value of 440 Hz. The letter name of each note corresponds to one of the twelve notes in the octave; the number of each note indicates the octave the note is in, with increments at each occurrence of C along the scale (e.g., ...$A_4$—$A\sharp_4$—$B_4$—$C_5$...). The frequency range over which pure tones are audible to humans extends from approximately $E_0$ (20.6 Hz) to $E_{10}$ (21.1 kHz). For the purposes of this discussion, we will set *y*, the F0 of the root of a harmonic interval, equal to 440 Hz ($A_4$).

In music theory, the interval formed by notes that are an octave apart (e.g., $A_4$ and $A_5$) is the most consonant interval, followed by the fifth ($A_4$ and $E_5$), fourth ($A_4$ and $D_5$), major third ($A_4$ and $C\sharp_5$), and minor third ($A_4$ and $C_5$). In the scale of just intonation, these intervals correspond to $x{:}y$ ratios of 2:1, 3:2, 4:3, 5:4, and 6:5, respectively, consistent with Pythagoras's claim that the simplicity of the integer ratio correlates with perceived consonance. Combinations of some other notes on the scale between $A_4$ and $A_5$ have more complex $x{:}y$ ratios and sound dissonant. For example, the minor second and the tritone (also known as the augmented fourth, which, in equal-tempered tuning, is equivalent to the diminished fifth) have $x{:}y$ ratios of 16:15 and 45:32 (or approximately 7:5), respectively. The dependence of consonance on the simplicity of F0 ratios tolerates small deviations from perfect integer relationships. For example, because the scale of equal temperament is based on equal logarithmic steps within each octave, a major third in this scale has a ratio of 5.04:4, not 5:4. This deviation amounts to 0.8 percent. Even highly practiced listeners participating in psychoacoustic experiments under ideal listening conditions cannot reliably detect a mistuned lower harmonic embedded within a harmonic complex tone if the deviation is less than 0.9% (harmonics 1–12 at 60 dB SPL [sound pressure level] and isophase, $F0 = 200$ Hz, duration $\leq 410$ ms).[30] Moreover, conservatory students who excel at interval identification cannot reliably judge whether a mistuned major third composed of two harmonic complex tones has been stretched or compressed away from a perfect 5:4 ratio if the deviation is less than 1.2% (each tone with harmonics 1–20 in isophase, first harmonic at 80 dB SPL, higher harmonics at a 6 dB decrease per octave, F0 between 260 Hz and 525 Hz, and duration = 1000 ms).[31]

All experimental studies that have used stimuli consisting of single, isolated, harmonic intervals formed by two complex tones (as would be the case if the intervals were sung or played on guitar or piano) show that listeners consistently perceive the fifth and fourth as more consonant than the minor second and tritone.[6,32–35] This convergence of results across study populations from different countries (USA, Germany, Japan), generations (1909–1969), and musical backgrounds, combined with results obtained in infants from the USA[36] and European starlings,[37] motivates the hypothesis that common, basic auditory mechanisms underlie perceptual categorization of harmonic intervals as consonant or dissonant. However, there is disagreement about the nature of the underlying neural mechanisms, and few physiological experiments have systematically analyzed responses to harmonic intervals at any level of the auditory nervous system.[25,27,38–41] Still, a large body of data is available on the responses of neurons to other types of complex tones in the auditory nerve,[42] cochlear nucleus,[43] inferior colliculus,[44,45] medial geniculate nucleus,[46] and auditory cortex[47–49] (FIG. 2; only a few of the many available papers are cited here; for review see Ehret and Romand[50]).

## NEURAL CODING OF PITCH RELATIONSHIPS AS A PHYSIOLOGICAL BASIS OF HARMONY

We synthesized simultaneous complex tones forming four musical intervals: the minor second (F0 ratio = 16/15), perfect fourth (4:3), tritone (45:32), and perfect fifth (3:2, FIG. 1). Each of the two complex tones in the interval contained the first six harmonics with equal amplitude (60 dB SPL re: 20 μPa) and equal phase (cosine, FIG. 3). Each interval had a duration of 200 ms (a bit shorter than an eighth note at a tempo of 120 beats per minute), including 5-ms rise and fall times. These stimuli are acoustically similar to the inputs into the computational models used by Plomp and Levelt,[51] Kameoka and Kuriyagawa,[35] and Hutchinson and Knopoff[52] to predict the consonance of complex-tone intervals on the basis of psychoacoustic data on pure-tone intervals.

FIGURE 1 illustrates two time-domain representations of our stimuli: the acoustic waveform, which plots sound pressure amplitude as a function of time (FIG. 1E–H); and the autocorrelation of the waveform (FIG. 1I–L). In the acoustic waveform of the most consonant interval, the perfect fifth, we see a regular pattern of major and minor peaks (FIG. 1H). The pattern with one major peak and three minor peaks repeats every 4.55 ms (1/x = F0 = 220 Hz). This periodicity corresponds to the missing F0 of a harmonic series containing energy at the second harmonic (440 Hz, $A_4$) and third harmonic (660 Hz, $E_5$), the F0s of the notes actually present in the stimulus. Rameau's concept of the "basse fondametale" (fundamental bass) in his *Treatise on Harmony*[53] is related to the missing F0 of a harmonic interval.

Autocorrelation functions provide another representation of temporal regularities and irregularities embedded in acoustic waveforms (FIG. 1I–L). Autocorrelation functions are computed by multiplying the waveform with a delayed copy of itself and integrating over time. A large value at a given delay indicates the presence of a dominant periodicity in the waveform whose period equals the delay. Like pitch percepts, but unlike acoustic waveforms, autocorrelation functions are stable despite changes in the relative phases of frequency components. In the autocorrelation functions plotted in FIGURE 1I–L, the periodicity at the upper limit of the $x$ axis (50 ms) corresponds to 20 Hz, the lowest audible frequency.

In the autocorrelation function of the perfect fifth (FIG. 1L), the first major peak again corresponds to the missing F0, $A_3$ (220 Hz). The second major peak occurs at

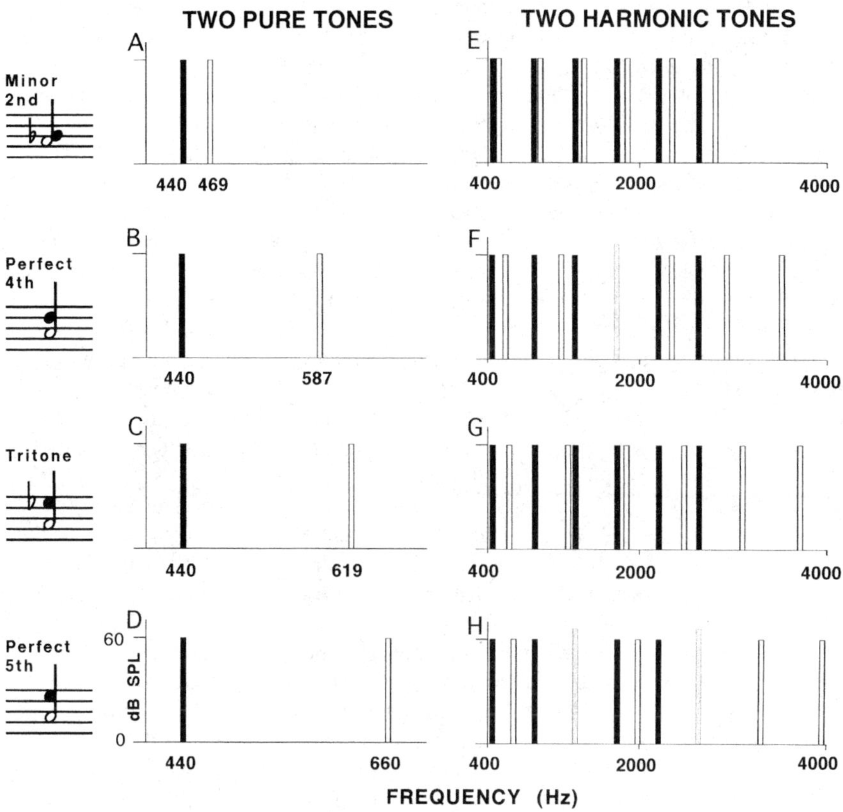


**FIGURE 3.** Line amplitude spectra of the four musical intervals shown in FIG. 1. *Black* indicates the root and its harmonics, *white* the interval and its harmonics, and *gray* the frequencies at which harmonics of the root and interval overlap. (**A–D**) A minor second, perfect fourth, tritone, and perfect fifth composed of two pure tones. (**E–H**) The same intervals composed of two harmonic complex tones containing the first six harmonics at equal amplitude. These are the spectral domain representations corresponding to the time domain representations in FIG. 1E–L.

9.09 ms, which corresponds to $A_2$ (110 Hz), the bass note an octave below. In fact, all the major peaks up to 50 ms correspond to the fundamental bass and its subharmonics (undertones) at $A_2$, $D_2$, $A_1$, and on down to $A_0$, the lowest note on the piano (F0 = 27.5 Hz).

In between the major peaks is a set of three, evenly spaced, minor peaks. The first of these minor peaks occurs at 1.51 ms, which corresponds to $E_5$ (660 Hz), the upper note of the interval. The second minor peak occurs at 2.27 ms, which corresponds to $A_4$ (440 Hz), the root of the interval. The third minor peak occurs at 3.03 ms, which corresponds to $E_4$ (330 Hz), the octave below the fifth and the fifth above the fundamental bass at $A_3$.

Temporal regularities are also seen in the waveform and autocorrelation of the perfect fourth, the other consonant interval in our stimulus set (FIG. 1F and J). Here the major peaks are at 6.82 ms ($D_3$, F0 = 146.7 Hz) and 13.6 ms ($D_2$, F0 = 73.3 Hz). Thus, in addition to a representation of the fourth, there is a representation of its inversion as a fifth with the implied root at $D_3$ and the fundamental bass at $D_2$. The autocorrelation function of the fourth is a bit more complicated than that of the fifth, as there are two more peaks between each pair of major peaks. In the first set of minor peaks, the following notes are represented: $A_4$ (the root), $D_4$ (the interval), $A_3$ (the octave below the root), and $G_3$ (the fifth below $D_4$, and the fourth of an interval rooted at $D_3$). Thus, we find representations of notes that function as fourths and fifths in the major and (all) minor scales of A and D.

In summary, the temporal fine structure of the perfect fifth and fourth contains representations of the two notes constituting the interval, plus harmonically related bass notes that are implied by the interval. In music, these bass notes support the deep structure of harmony. Parncutt[18] demonstrated experimentally that listeners associate major triads with pitches that are harmonically related to note F0s, including the fundamental bass, plus the pitches of note F0s actually in the stimulus. These pitches cannot be accounted for simply on the basis of combination tones (for review, see Wightman and Green[54]). Houtsma and Goldstein[55] showed that musicians can use missing F0 pitches to identify melodic intervals (major and minor seconds and thirds), even when two upper harmonics are presented separately (dichotically) to each ear.

For the dissonant intervals in our stimulus set, the minor second and tritone, we find no such temporal regularity in the acoustic waveform and autocorrelation function. For the minor second (FIG. 1I), the largest peak in the autocorrelation function occurs at 34.1 ms, which corresponds to a frequency of 29.3 Hz, and it decays rapidly into the background. This periodicity lies outside the range associated with strong periodicity pitch percepts (for review, see Moore[56]). In addition, we find multiple peaks between zero and the maximum peak. The largest of these is the first peak at 2.20 ms, which corresponds to mean of the note F0s, 455 Hz. This pitch does not correspond to any of the notes in any scale that has $A_4$ in it. In short, there is no strong representation of any pitch below the note pitches in the interval, and the dominant pitch is off the scale. Both of these factors contribute to the dissonance of the minor second.

Likewise, the autocorrelation function of the tritone (FIG. 1K) does not show a simple, regular pattern of peaks. The largest peak occurs at 11.4 ms, which corresponds to F0 = 88 Hz, and it, too, decays into the background. This periodicity corresponds to a near coincidence between the fifth subharmonic of the root (440 Hz divided by 5) and the seventh subharmonic of the tritone (618.7 Hz divided by 7). It also lies close to $F_2$, which is related to the fundamental bass of an F dominant-seventh chord in its first inversion. Thus the autocorrelation function of the tritone implies a chord that is, in music theory, less consonant than the chords implied by the autocorrelation functions of the fifth and fourth.

To summarize, for the consonant intervals (the fifth and fourth), the pattern of major and minor peaks in the autocorrelation is perfectly periodic, with a period related to the fundamental bass. This pattern is obtained because these stimuli have a unique, clearly defined fundamental period. By contrast, for dissonant intervals (the minor second and tritone), no true periodicity is seen in the autocorrelation function.

While some peaks occasionally stand out at specific delays, indicating a pseudoperiod, either there are no consistent peaks at multiples of this pseudoperiod, or the amplitudes of these peaks decay rapidly with increasing multiples of the pseudoperiod.

These observations suggest that the consonance of harmonic intervals reflects regularities in their temporal fine structure in the range of tenths to tens of milliseconds. Do neurons in the auditory system represent this information using a time code? Galileo,[57] who wrote about consonance while he was under house arrest for his work on the solar system, may have been the first to postulate that temporal coding in the auditory periphery was the physiological basis for consonance:

> Agreeable consonances are pairs of tones which strike the ear with a certain regularity; this regularity consists in the fact that the pulses delivered by the two tones, in the same interval of time, shall be commensurable in number, so as not to keep the ear drum in perpetual torment, bending in two different directions in order to yield to the ever-discordant impulses....The unpleasant sensation produced by [dissonances] arises, I think, from the discordant vibrations of two different tones which strike the ear out of time. Especially harsh is the dissonance between notes whose frequencies are incommensurable;...this yields a dissonance similar to the augmented fourth or diminished fifth [*tritono o semidiapente*]." (Galileo,[57] 1638, pp. 103–104.)

To investigate the neural coding of consonance in the auditory periphery, we analyzed the responses of over 100 cat auditory nerve fibers to the minor second, perfect fourth, tritone, and perfect fifth. Auditory nerve fibers are the central axons of spiral ganglion cells that synapse on cochlear nucleus neurons in the brain stem (FIG. 2; for review, see Pickles[58]). In humans, each auditory nerve contains about 30,000 auditory nerve fibers. Spiral ganglion cells also have peripheral axons that synapse on sensory receptors in the cochlea—the inner hair cells that ride atop the basilar membrane. Virtually all information about sound is transmitted from the ear to the brain via trains of action potentials fired by auditory nerve fibers.

When a minor second or some other interval is sounded, an auditory nerve fiber will increase the number of action potentials it fires only if it is sensitive to the frequencies present in the interval (FIG. 4A). The time between consecutive action potentials in the train is called an *interspike interval* (ISI), and a plot of the number of times each ISI occurs in the spike train is called an *ISI histogram* (FIG. 4B–E). We measured all the ISIs between all possible pairs of spikes (FIG. 4A, $ISI_1$, $ISI_2$...$ISI_N$) with a precision of approximately one microsecond. The corresponding plot is called an *all-order ISI histogram*, which is equivalent to the autocorrelation of the spike train. The spike train of each fiber in the auditory nerve can be analyzed in this way, and the resultant ISI histograms can be combined to show the ISI distribution in the entire population of auditory nerve fibers. Single-unit physiology experiments[59–61] and computational models[62] have shown that the first among the major peaks in the all-order ISI histogram computed from the entire auditory nerve fiber ensemble (the population ISI distribution) matches the fundamental period of complex tones and thus their periodicity pitch. This is essentially the time-domain equivalent of Terhardt's[63] spectrally based subharmonic sieve for virtual pitch extraction.

FIGURE 4B–E illustrates the population ISI distributions embedded in the spike trains fired by over 50 auditory nerve fibers in response to the minor second, fourth, tritone, and fifth. In the response to the fifth (FIG. 3E), we see major peaks corresponding to the fundamental bass ($A_3$, 4.55 ms) and its subharmonics, just as we did in the acoustic waveform (FIG. 1H) and the autocorrelation of the waveform (FIG.

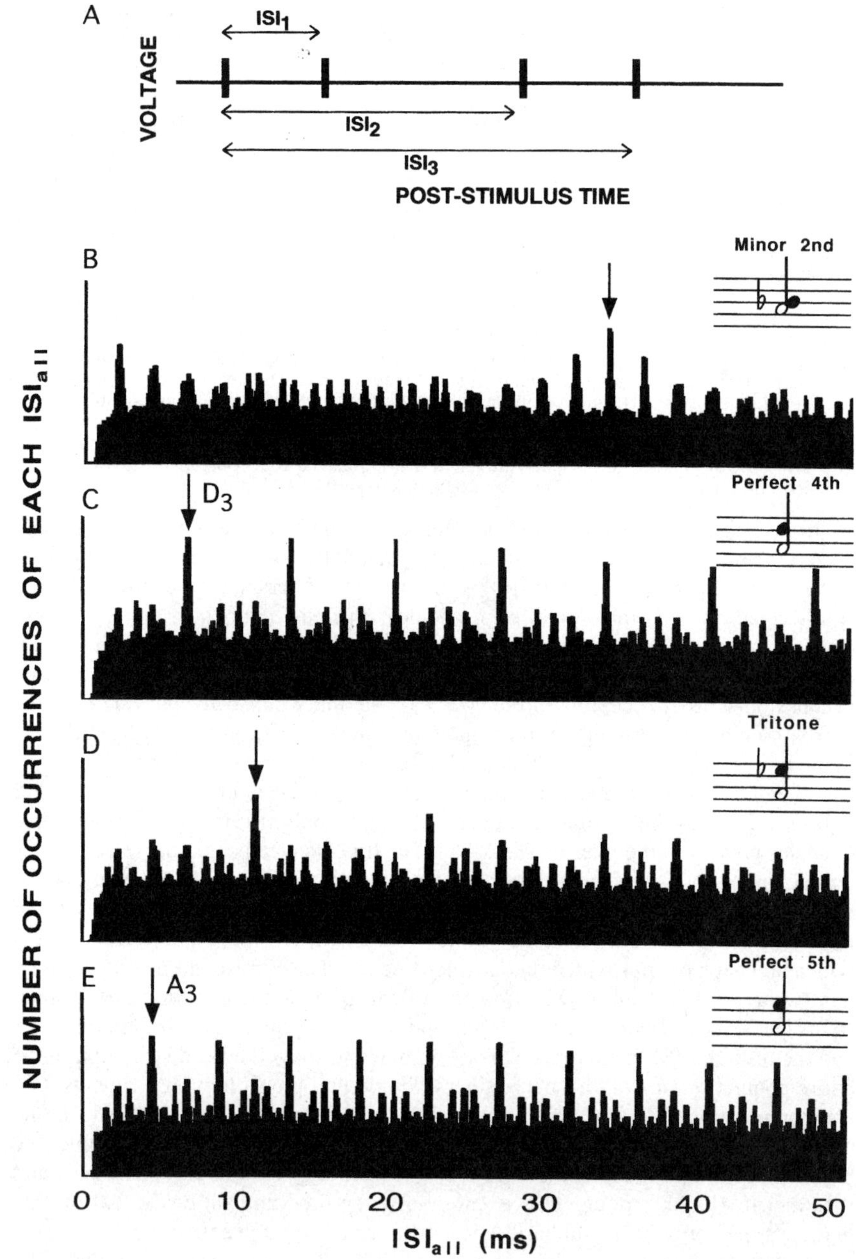


**FIGURE 4.** Interspike interval (ISI) distributions embedded in the responses of axons throughout the auditory nerve during stimulation with four musical intervals. (**A**) Schematic of a train of action potentials (or "spikes," *vertical lines*) fired by an auditory nerve fiber when a musical interval is played. *Double arrows* demarcate some of the ISIs between pairs

1L). Indeed, for all four intervals, the autocorrelation histograms of neural responses (FIG. 4B–E) mirror the fine structure of acoustic information in the time domain (FIG. 1).

Thus the peaks in the population ISI distribution evoked by consonant intervals reflect the pitches of each note, the fundamental bass, and other harmonically related pitches in the bass register. By contrast, the dissonant intervals (the minor second and tritone) are associated with population ISI distributions that are irregular. These contain little or no representation of pitches corresponding to notes in the interval, the fundamental bass, and related bass notes.

To obtain a physiological measure of the strength of the fundamental pitch of each interval relative to other pitches, we measured the number of intervals under the peak in the all-order ISI distribution corresponding to the missing F0 (arrows in FIG. 3B–E; bin width = 300 μs). We then divided that value by the value of $y$ in each $x$ bin from $x = 0$ ms – 50 milliseconds. We found a high correlation ($r = 0.96$) between our physiological measure of fundamental pitch strength and previous psychoacoustic measures of the "clearness" of musical intervals composed of two complex tones.[35]

In summary, the all-order ISI distribution embedded in auditory nerve fiber firing patterns contains representations of the pitch relationships among note F0s that influence the perception of musical intervals as consonant or dissonant. The neural coding mechanisms that provide representations of these pitch relationships form part of the neurobiological foundation for the theory of harmony in its vertical dimension.

## NEURAL CODING OF ROUGHNESS AS THE PHYSIOLOGICAL BASIS FOR HARMONY PERCEPTION

Whereas pitch-based accounts treat consonance as a positive perceptual phenomenon associated with the presence of highly structured temporal information, roughness-based accounts treat consonance as a negative phenomenon associated with the absence of annoying perceptual attributes. Terhardt's notion that the consonance of isolated intervals and chords depends on the absence of roughness[13] echoes one of the main points in Helmholtz's[16,64] monumental work, *On the Sensations of Tone as a Physiological Basis for the Theory of Music*:

> As long as several simple tones of a sufficiently different pitch enter the ear together, the sensation due to each remains undisturbed in the ear, probably because entirely different bundles of [auditory] nerve fibers are affected. But tones of the same, or of nearly the same pitch, which therefore affect the same nerve fibers, do not produce a sensation which is the sum of the two they would have separately excited, but new and peculiar phenomena arise which we term interference...and beats[16] (p. 160)....Rapidly beating tones are jarring and rough...the sensible impression is also unpleasant[16]

---

of spikes in the spike train. $ISI_1$ refers to the first-order ISI, $ISI_2$ and $ISI_3$ to higher-order ISIs. All possible ISIs are in "all-order" ISI histograms. (**B–E**) All-order ISI histograms showing ISIs embedded in the responses of 50 auditory nerve fibers to musical intervals composed of two complex tones (FIG. 1E–H and FIG. 3E–H). *Arrows* mark peaks in the ISI pattern that correspond to the missing F0 of the interval. If the peak corresponds to a note on the scale, the name of the note is given.

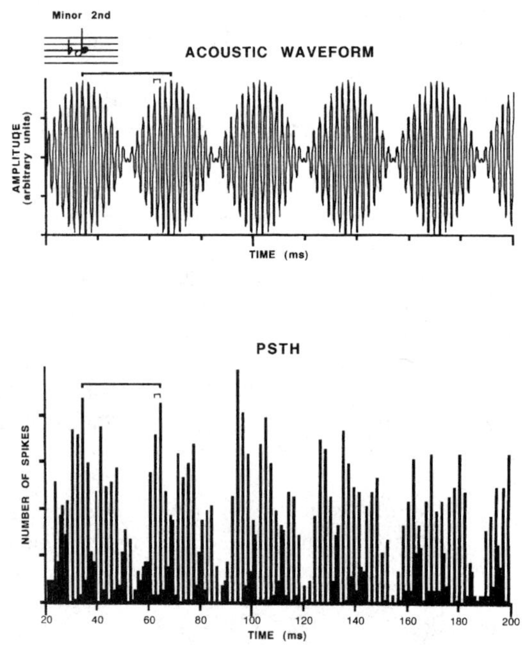


**FIGURE 5.** (*Top*) Acoustic waveform of a minor second composed of two pure tones with the root at $A_4$. *Thick bars* show the period of envelope fluctuations that render the minor second rough ($P = 1/\Delta F = 34.1$ ms). *Thin bars* show the period of fluctuations under the envelope that corresponds to the mean frequency of the tones and the pitch of the interval ($P = 2.20$ ms). (*Bottom*) Poststimulus time histogram (PSTH) showing the number of spikes fired by a single auditory nerve fiber during the steady state portion of its response to the minor second. Note that the global and local fluctuations in firing rate mirror those seen in the acoustic waveform of the minor second. This fiber was sensitive to frequencies at both the root and the interval at 60 dB SPL. Bin width = 1 ms. Number of stimulus repetitions = 100.

> (p. 168). Consonance is a continuous, dissonance an intermittent tone sensation. The nature of dissonance is simply based on very fast beats. These are rough and annoying to the auditory nerve.[64] (*Helmholtz* 1863,[64] 1885.[16])

Because frequency selectivity throughout the auditory nervous system is finite, simultaneous pure tones that are separated by small frequency differences (ΔFs), such as a minor second (FIG. 3A), cannot be separated or "filtered out" from one another. Consequently, their waveforms are effectively summed, and the pitch of the tone combination matches their mean frequency.[65] The envelope of the summed waveform contains periodic amplitude fluctuations whose frequency equals ΔF (FIG. 5, *top*). If these envelope fluctuations fall in the range of 20–200 Hz (the precise values depend on the frequencies of the two tones, FIG. 6A), interruptions in continuous tone sensation are perceptible. These interruptions make the tone combination sound "rough," analogous to the interruptions one feels on the fingertips when touching coarse sandpaper. At smaller frequency differences, and thus slower amplitude modulations, one perceives a single, continuous tone that is slowly fluctuating in loudness, or "beating." Auditory nerve fibers[38,39] (FIG. 5, *botttom*), inferior colliculus neurons,[40,41] and

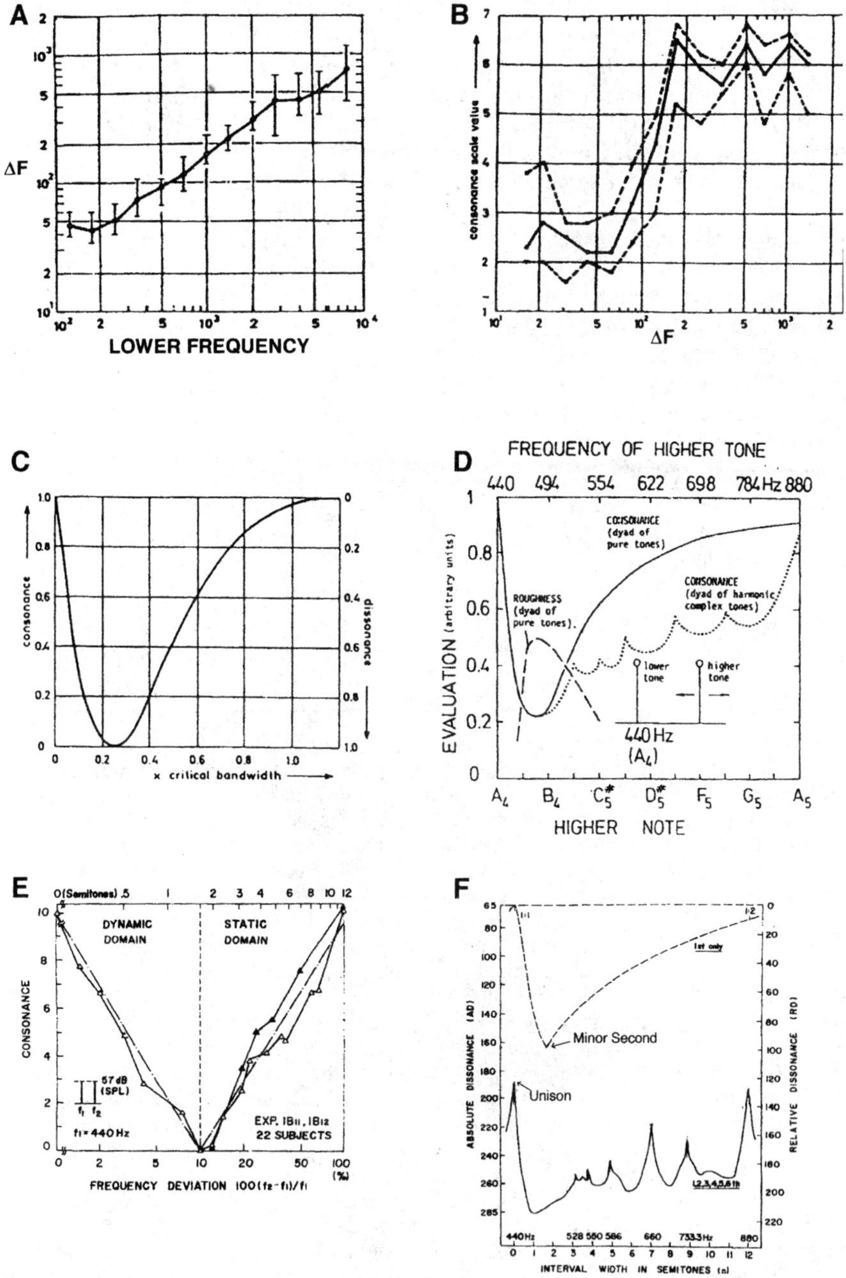


**FIGURE 6.** *See following page for caption.*

populations of primary auditory cortex neurons[48] can fire in synchrony with amplitude fluctuations in the ΔF range associated with perception of roughness and beats.

The concept of *critical bandwidth* refers to the limits of ΔF over which frequency selectivity operates in the auditory system. Critical bandwidth has been estimated psychoacoustically in several ways that have yielded somewhat different results depending on the method (for reviews, see Greenwood[66] and Moore[56]). One estimate of critical bandwidth is based on the ΔF above which roughness disappears.

When musical intervals are composed of two complex tones (FIG. 3E–H), the partials may interfere with one another and produce amplitude fluctuations at the corresponding ΔF. There is more interference between adjacent partials in the minor second and tritone (FIG. 3E and G) than in the fourth and fifth (FIG. 4F and H). Several computational models of consonance[16,35,51,52] assume that (1) the roughness generated by all the partials in the interval are added together (presumably by a central processor in the auditory brain stem or cortex), and (2) this total roughness determines the degree to which the interval is perceived as consonant.

FIGURE 6A shows Plomp and Steeneken's data on the relationships among the ΔF between two pure tones, the frequency of the lower tone (or root), and just-noticeable roughness (line).[67] The data at $x = 500$ Hz would thus apply to the case of a harmonic interval composed of two pure tones with the root at $B_4$ ($x = 494$ Hz). The fifth of $B_4$, $F\sharp_5$, has a frequency of 741 Hz, so the ΔF between the root and the fifth is 247 Hz. According to FIGURE 6A, at $x = 500$ Hz, roughness disappears for ΔF values above 90 Hz–125 Hz. Therefore, the fifth should not be associated with roughness. Experimental studies agree that an isolated fifth composed of pure tones in this frequency register sounds "consonant"[7,51] or "pleasant."[6,33] The pure-tone fifth is thus

**FIGURE 6.** (**A**) Just-noticeable roughness (*line*) as a function of the frequency difference between two pure tones (ΔF) and the lower frequency of the tones. Tones were presented monaurally at 60 phons. The *bars* show the interquartile range of 20 subjects (musical background not given). (Adapted from Plomp and Steeneken,[67] FIG. 1, p. 883.) (**B**) Plomp and Levelt's data on consonance ratings as a function of ΔF for simultaneous pure tones with a mean frequency of 1000 Hz. Tones were presented in free field at 65 dB SPL. The *solid line* shows the mean consonance ratings of 10 subjects (musical background not given). *Dashed lines* show the interquartile range. (Adapted from Plomp and Levelt,[51] FIG. 6, p. 554.) (**C**) Plomp and Levelt's idealized plot of the relationship between consonance and critical bandwidth. The *y* axis is in units of consonance (*left*) and dissonance (*right*). (From Plomp and Levelt[51] [FIG. 10, p. 556]. Reproduced by permission.) (**D**) Terhardt's idealized plot showing the relationship of consonance and roughness to interval width (root at $A_4$, 440 Hz). The *solid line* shows consonance vs. ΔF when the interval is composed of two pure tones. The *dotted line* shows consonance vs. ΔF0 when the interval is composed of two complex tones, each containing several lower harmonics. The *dashed line* shows the roughness of two pure tones as a function of ΔF. (From Terhardt[13] [FIG. 1, p. 281]. Reproduced by permission.) (**E**) Kameoka and Kuriyagawa's data on consonance as a function of ΔF for simultaneous pure tones with the root at $A_4$ and intensities of 57 dB SPL. The *dashed line* is the mean performance of 22 audio engineers who performed the task twice (*solid lines* with *black* and *white triangles*). (From Kameoka and Kuriyagawa[72] [FIG. 1, p. 1452]. Reproduced by permission.) (**F**) Kameoka and Kuriyzgawa's plot of dissonance as a function of interval width. The *dashed line* at the *top* shows the function when the interval is composed of two pure tones. The *solid line* at the *bottom* gives the calculated dissonance of an interval composed of two complex tones, each containing the first six harmonics at 57 dB SPL. (Adapted from Kameoka and Kuriyagawa,[35] FIG. 8, p. 1465.)

associated with the absence of roughness and with strong pitches associated with temporal regularities in its acoustic waveform and autocorrelation (similar to those described in the preceding section for a fifth composed of two harmonic complex tones; FIG. 1H and L). The same set of observations applies to a perfect fourth composed of two pure tones.

Now consider the case of a minor second composed of two pure tones with the root at $B_4$ (494 Hz). Here, $\Delta F$ (between $B_4$ and $C_5$) is 33 Hz. This falls well within the range of noticeable roughness (FIG. 6A). In fact, it lies near the $\Delta F$ associated with maximal roughness[67–69] (not shown). Experimental studies agree that a pure-tone minor second (and other tone combinations close to it) sounds "dissonant"[7,51] or "unpleasant."[6,33]

The case of a tritone composed of two pure tones provides an interesting test of the roughness hypothesis. The $\Delta F$ between a tritone at $F_5$ and a root at $B_4$ is 201 Hz, well above the $\Delta F$ for just-noticeable roughness. Thus the fourth, tritone, and fifth are all above the roughness range. Does that mean they all have the same consonance?

FIGURE 6B shows Plomp and Levelt's data on consonance ratings as a function of $\Delta F$ for two pure tones whose mean frequency is 1000 Hz.[51] Because the *y* axis is an ordinal scale, not an interval or ratio scale, it is inappropriate to assume that equal distances reflect equal differences in consonance. It follows that $4 > y > 4$ is not to be taken as the categorical boundary for dissonance and consonance, respectively. In addition, because Plomp and Levelt intentionally avoided using standard intervals like the fourth and tritone (they were concerned that interval recognition would influence consonance ratings), it is difficult to estimate where on the curve these intervals would fall. These caveats aside, it is clear that all pure-tone combinations with $\Delta F$s above approximately 150 Hz are consonant. This would apply to the fourth, tritone, and fifth with their roots in the vicinity of $A_5$. Tone combinations with $\Delta F0 =$ 20 to 80 Hz are dissonant; this would apply to a minor second with the root near $A_5$. Superficially, it would appear that we have a convergence between the disappearance of roughness at $\Delta F0 = 150$–$250$ Hz (FIG. 6A, interquartile range for a lower frequency of 1 kHz) and a steep increase in consonance ratings at $\Delta F0 > 80$ Hz (at and above a minor third, FIG. 6B). However, the $\Delta F$ associated with the highest mean consonance rating ($\Delta F \sim 180$ Hz) is within the range of noticeable roughness for many of Plomp and Steeneken's subjects.[67]

Beyond about $\Delta F = 180$ Hz, mean consonance ratings vary by only one rating point or less, but they are not perfectly flat (FIG. 6B). One can discern alternating peaks and valleys out to about $\Delta F = 1200$ Hz. We estimated where the minor second, perfect fourth, tritone, and perfect fifth might fall on the interpolated lines drawn by Plomp and Levelt[51] (FIG. 6B), and we estimated the frequency ratios and intervals that correspond to the peaks and valleys beyond $\Delta F = 1000$ Hz. The first peak is near 6:5, which would correspond to a minor third. The second peak falls close to 5:3, which would correspond to a major sixth (or inverted minor third). The ratio 5:3 could also be thought of as the fifth and third harmonics of a harmonic series corresponding to the third and first notes of a major triad in its second inversion. The third peak is at or close to a ratio of 3:1, which corresponds to the interval of a twelfth, that is, a root and a fifth in the octave above. High consonance ratings for the twelfth are also found in Plomp and Levelt's data for two pure tones with a mean frequency of 500 Hz[51] (not shown). The first valley between the first two peaks is near the tritone, and the second valley appears to be a mistuned octave, with a frequency ratio near 2.03:1.

FIGURE 6C shows Plomp and Levelt's idealized plot of the relationship between consonance and critical bandwidth[51] (the latter is defined here by loudness summation[70]). Note that the curve reaches an asymptote near the end of the *x* axis, at about one critical bandwidth. Thus a critical band account of consonance as the absence of roughness *cannot* apply to pure-tone intervals that are wider than a minor third or so.

Yet Terhardt's[13] idealized plot (FIG. 6D) of the relationships among roughness, consonance, and pure-tone ΔF shows a monotonic increase in perceived consonance all the way out to the octave, well beyond the ΔF at which (1) roughness disappears (FIG. 6A and D), (2) consonance ratings plateau (FIG. 6B), and (3) loudness summation and masking effects are observed (FIG. 6C; for review see Yost[71]). The representation of the psychoacoustic literature summarized in FIGURE 6D appears to draw upon Kameoka and Kuriyagawa's data showing increases in consonance well beyond the ΔF associated with disappearance of roughness[72] (FIG. 6E). Although it is generally accepted that Kameoka and Kuriyagawa's work supports the idea that consonance is a function of roughness and critical bandwidth, comparisons of FIGURE 6A, C, and E reveal that their data actually argue against it, at least for musical interval widths greater than the ΔF for just-noticeable roughness.

The disagreement may arise from two sources. First, Kameoka and Kuriyagawa's Japanese audio engineers were instructed to judge tones for *sunda* (which they translate as "clearness" in English) and *nigotta* ("turbidity").[72] Consequently, these listeners may have been rating different perceptual attributes than Plomp and Levelt's Dutch subjects, who were instructed to judge how "consonant" [or *mooi* ("beautiful") or *welluidend* ("euphonious")] the intervals sounded.[51] Second, Kameoka and Kuriyagawa used an incomplete paired comparison paradigm—incomplete because only three or four adjacent intervals were paired for comparisons of relative consonance,[72] a much more restricted ΔF range than the one Plomp and colleagues used in their one-interval consonance rating paradigm.[7,51] When Kameoka and Kuriyagawa tried the method of magnitude estimation, presumably using all possible pairings, the task turned out to be "rather difficult" for "naive subjects"[72] (p. 1453), and they dropped it in favor of incomplete pairings. Comparing only adjacent intervals may have biased subjects to focus on differences they would not have otherwise attended to if all intervals had been paired with one another. The pattern of results suggests that pitch height, rather than absence of roughness, influenced consonance judgments beyond a minor third or so. Since the authors used these data to calculate the consonance of musical intervals formed by two complex tones (FIG. 6F), they may have confounded roughness and pitch height in their predictions.

We reviewed previous studies that used isolated minor seconds, fourths, tritones, and fifths composed of two pure (or nearly pure) tones as experimental stimuli. Kaestner,[6] who used a *Stimmgabelklangen* to generate tones that were "poor in overtones," found that subjects judged the fourth to be slightly more "pleasant" than the tritone. Malmberg,[32] who used tuning forks, found a more marked preference for the fourth over the tritone for judgments of "blending," "purity," and "smoothness." Pratt,[73] who used a Stern variator that may have produced weak overtones, found that the fourth was judged to be more "pleasant," "smoother," and more "unitary" than the tritone. Brues,[74] using a Stern variator that produced weak energy at the first overtone, found the fourth, tritone, and fifth were similar with respect to "unitariness." Guthrie and Morrill[75] used a Stern variator that produced "very faint traces of the third partial" and reported that the fourth was judged to be more "pleasant" than the tritone and of equal "consonance." Guernsey,[33] who used tuning forks, reported that nonmusicians, amateur musicians, and professional musicians found the fourth more "pleasant" and "smooth" than the tritone. Schellenberg and Trehub's recent experiments with nine-month-old babies are also relevant here.[76] When the upper pure tone of a repeating harmonic interval was flattened by one-fourth of a semitone, infants could detect the change if the interval was a fourth but not if it was a tritone. Their findings indicate that fourths provided a more stable background against which changes in tuning could be detected. All in all, these results indicate that the fourth, even when it is composed of two pure tones, is often perceived as more consonant than the tritone.

Another challenge for roughness-based accounts of consonance arises when we compare the consonance of pure-tone intervals and the consonance of complex-tone intervals. In FIGURE 6D, Terhardt[13] plots consonance ratings for pure-tone and complex-tone intervals on the same scale. In fact, Kameoka and Kuriyagawa's psychoacoustic data and calculations put them on different scales[35] (FIG. 6F). Likewise, Plomp and Levelt use a dissonance scale from one to zero for their pure-tone data (FIG. 6C) and a dissonance scale from six to zero for their complex-tone calculations (not shown).[51] Kameoka and Kuriyagawa's calculations predict that a minor second composed of two pure tones is more consonant than the unison of two complex tones with the first six harmonics at isoamplitude[35] (FIG. 6F). Intuitively, this notion is untenable; however, direct comparisons between pure-tone intervals and complex-tone intervals have not been reported in the literature. We synthesized a pure-tone minor second and the unison of two complex tones (with the acoustic parameters specified by Kameoka and Kuriyagawa[35]) and asked several of our students to judge which of these two stimuli sounded more "consonant." These and Huron's (personal communication) informal observations raise the possibility that a combination of pitch height effects and loudness (shrillness), rather than or in addition to roughness, accounts for Kameoka and Kuriyagaw's predictions. For example, in the case of unison at $A_4$ (FIG. 6F), spectral energy extends all the way up to 2200 Hz (fifth harmonic) and 2640 Hz (sixth harmonic), so there are high-frequency components that are greater in sensation level than the note F0s. At the same time, $\Delta F$ (440 Hz) is higher than the highest $\Delta F$ associated with just-noticeable roughness when the root is at 2000 Hz (FIG. 6A, interquartile range for $\Delta F \sim 250$–400 Hz[67]).

In summary, the neural coding mechanisms that provide representations of roughness form part of the neurobiological foundation for the theory of harmony in its vertical dimension. However, our reappraisal of the psychoacoustic literature leads us to conclude that the dependence of consonance on the absence of roughness is overstated. We believe pitch relationships, as well as roughness, influence the perception of intervals and chords as consonant or dissonant in the vertical dimension.

## EFFECTS OF AUDITORY CORTEX LESIONS

Another approach to assessing the relative contributions of pitch and roughness to consonance perception might be to determine whether impairments in consonance perception caused by brain lesions are associated with deficits in one, the other, or both.

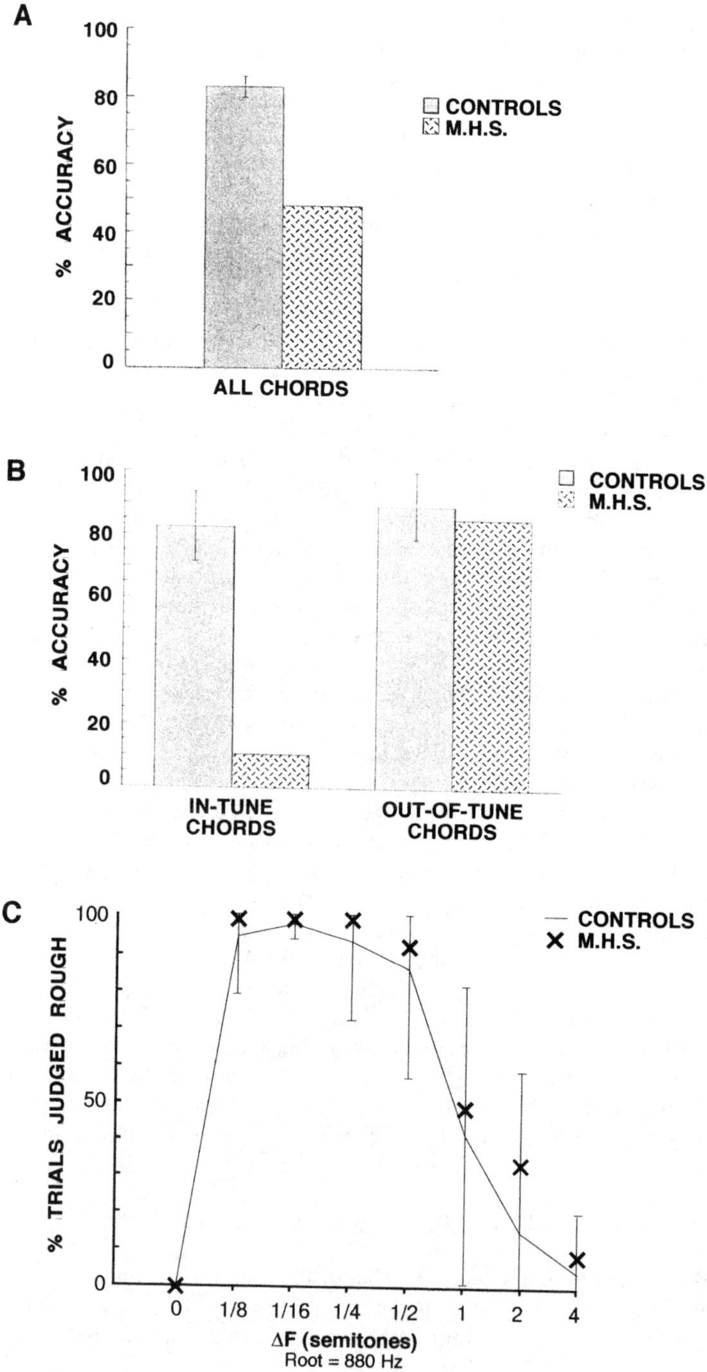


**FIGURE 7.** *See following page for caption.*

Consonance perception has been reported to be severely impaired following bilateral lesions of the auditory cortex.[15] In an experiment employing a one-interval, two-alternative, forced-choice paradigm, two types of stimuli were presented: a major triad, and a triad whose fifth was flattened by a fraction of a semitone. In each trial of the experiment, a young stroke patient, MHS, was asked if a single, isolated chord sounded "in tune" or "out of tune." His response accuracy was 56%, better than chance ($p < 0.05$), but more than two standard deviations below the mean of thirteen normal controls (FIG. 7A). Magnetic resonance imaging revealed that MHS's infarcts involved the primary auditory cortex in both hemispheres, all or almost all of the auditory association cortex in the right hemisphere, and about 20% of the posterior auditory association cortex in the left hemisphere (FIG. 2A). His pure-tone audiograms were within normal limits. Speech perception was impaired.

We subsequently compared MHS's performance on in-tune trials versus out-of-tune trials. If roughness perception were impaired, then MHS might make more errors in in-tune trials than out-of-tune trials. If frequency selectivity were coarsened and pitch perception impaired, then MHS might make more errors on in-tune trials. Consistent with the latter possibility, we found a marked response bias for out-of-tune judgments (FIG. 7B). Two possible interpretations follow: (1) MHS was having difficulty extracting the pitches of chord frequency components and analyzing their harmonic relationships; and/or (2) he heard more roughness in the chords than normals because his effective critical bandwidths were wider.

To assess whether MHS was having difficulty with frequency discrimination, we examined his performance on the Pitch Discrimination subtest of the Seashore Measures of Musical Talents.[77,78] This test uses the method of constant stimuli to measure one's ability to judge whether the second of two pure tones is higher or lower in pitch than the first tone. The $\Delta$F between the tones gets smaller over successive blocks of trials. The tones were centered at 500 Hz, 600 ms in duration, and 600 ms apart, with an intensity of 35–40 dB above sensation level. Overall, MHS scored in the 15th percentile. His error pattern was again revealing. He performed poorly in the last third of the test, where the $\Delta$Fs between the tones were smallest. We subsequently measured pure-tone frequency difference thresholds for pitch discrimination using an adaptive procedure and a two-interval, two-alternative, forced-choice paradigm.[79] Whereas normal controls and patient controls had frequency difference thresholds corresponding to Weber fractions ($\Delta$F/mean frequency) of around 1%, MHS's Weber fractions were over 10 percent. In short, his ability to judge the direction of a pitch change was markedly impaired. A similar deficit has since been reported in patients with surgical lesions of the right primary auditory cortex and right

---

**FIGURE 7.** Perception of consonance (**A** and **B**) and roughness (**C**) in a patient, MHS, with bilateral lesions of auditory cortex (FIG. 2A). (**A**) Data from Tramo *et al.*[15] showing response accuracy on a task that required MHS to determine whether a major triad was "in tune" or "out of tune." (**B**) MHS's error rate as a function of stimulus condition. (**C**) Percentage of trials in which two simultaneous pure tones were judged to be fluctuating and rough (as opposed to steady and smooth) as a function of the $\Delta$F between the tones. The mean (*line*) and standard deviation (*bars*) show the combined performance of 10 normal subjects and 8 neurological patients who did not have bilateral lesions of auditory cortex. Three of the patients had chronic unilateral lesions of superior temporal cortex (left or right). and one had a lesion of the right inferior colliculus.

anterior auditory association cortex.[80] We also found that perception of the missing F0 of harmonic complex tones was impaired,[81] again consistent with observations in most right temporal lobectomy patients with partial or complete excisions involving the primary auditory cortex.[82] Taken together, these findings are consistent with the hypothesis that MHS's impaired consonance perception was related to deficits in pitch processing.

Was this increase in ΔF thresholds for pure-tone pitch discrimination associated with higher ΔF values for disappearance of roughness? We assessed MHS's ability to judge two simultaneous pure tones as "steady and smooth" versus "fluctuating and rough" using a one-interval, two-alternative, forced-choice paradigm and the method of constant stimuli. The lower tone was fixed at either 220 Hz ($A_3$) or 880 Hz ($A_5$), and the upper tone was above the root by a variable number of semitones: 0, 1/16, 1/8, 1/4, 1/2, 1 (a minor second), 2 (a major second), or 4 (a major third). FIGURE 7C shows the results when the root was at 880 Hz. When the tones were between 1/16 to 1/2 semitone apart, MHS, like controls, judged the combination as rough on >80% of trials. When the tones were zero, two, or four semitones apart, MHS and controls judged the combination to be rough on less than 20% of trials. At one and two semitones apart, MHS's performance fell near the mean of controls, but there is too much variability in the normal data to meaningfully assess MHS's performance. Still, these observations mitigate the possibility that consonance perception was impaired because he heard more roughness in chords than normals.

In summary, MHS's bias to hear major triads as mistuned appears to be associated with impairments in pitch perception but not roughness perception. Consistent with our physiological data and review of the psychoacoustic literature, this pattern of lesion effects indicates that pitch relationships influence harmony perception in the vertical dimension.

## CONCLUSIONS

Basic physiological and anatomical properties of auditory and cognitive systems determine why some combinations of simultaneous tones sound more harmonious than others. Distinctive acoustic features of consonant and dissonant intervals are translated into distinctive patterns of neural activity. A faithful representation of temporal regularities in the acoustic structure of consonant intervals exists in the population interspike interval (ISI) distribution of auditory nerve fibers. The most common ISIs in the distribution correspond not only to the pitches of note F0s actually present in the consonant intervals, but also to the pitches of harmonically related notes in the bass register, such as the fundamental bass. By contrast, for dissonant intervals, the most common ISIs in the distribution do not correspond to one of the note F0s, nor do they correspond to harmonically related notes. The relative strength of the missing F0 in the population ISI distribution predicts the relative consonance of the minor second, perfect fourth, tritone, and fifth. Limits on the temporal precision and frequency selectivity of neurons throughout the auditory system constrain the range of note F0s we can hear as strong pitches and how they are combined into intervals and chords. Implicit knowledge about the hierarchical relationships of pitches in a given tonal system is likely to exert cognitive influences on the degree to which intervals and chords sound consonant or dissonant, even when they are heard in isolation.

Representations of roughness exist in temporal patterns of neural activity at several levels of the auditory system. For the minor second, fourth, tritone, and fifth, the amount of 20- to 200-Hz temporal fluctuations in the firing patterns of auditory nerve fibers inversely correlates with perceived consonance. These representations of roughness are multiplexed with pitch representations in the spike trains of auditory nerve fibers. These two neural time codes operate over different time regimes. The fine timing of action potential firing with precision in the submillisecond range carries information about fundamental pitch. Periodic fluctuations in discharge rate with precision in the range of milliseconds to tens of milliseconds carry information about roughness.

Bilateral lesions of primary auditory cortex and auditory association cortex can lead to severe impairments in consonance perception, with a bias to judge well-tuned chords as out of tune. In our patient, MHS, impaired consonance perception was associated with severely impaired pitch perception, but roughness perception appeared to be normal or near normal.

We interpret our findings and the results of previous psychoacoustic experiments as evidence in favor of the hypothesis that harmony in the vertical dimension, like harmony in the horizontal dimension, is principally a function of the pitch relationships among tones, with roughness playing a secondary role. In light of these observations, and in view of the likelihood that cognitive representations of pitch hierarchies influence harmony perception in the vertical dimension, we urge that the terms *sensory consonance* and *sensory dissonance* be reconsidered.

## ACKNOWLEDGMENTS

This work was supported by NIH DC03382 (M.J. Tramo), DC03054 (P.A. Cariani), DC02258 (B. Delgutte), and DC00117 (L.D. Braida); and the McDonnell-Pew Program in Cognitive Neuroscience (M.J. Tramo). We gratefully acknowledge the instructive comments of David Cohen and Kay Shelemay of the Harvard Music Department; the technical support of Leslie Liberman of EPL, and David Lum, Andrew Grant, and Danielle Lemay of RLE; the graphics and manuscript assistance of Adda Kridler, Janelle Mallett, and Monique James; and, especially, our collaborators in the experimental studies cited here: Jamshed Bharucha, Martin McKinney, Michael Gazzaniga, Frank Musiek, Robert Zatorre, Carla Kovacs, Gaurav Shah, Eiling Yee. Special thanks go to Mark Smith and his family.

## REFERENCES

1. COHEN, M.R. & I.E. DRABKIN. 1948. A Source Book in Greek Science. McGraw-Hill Book Company, Inc. New York.
2. PISTON, W. 1941/1987. Harmony, 5th edit. Revised and Expanded by Mark DeVoto. W. W. Norton & Company, Inc. New York.
3. APEL, WILLI, Ed. 1972. Harvard Dictionary of Music, 2d edit. Belknap Press of Harvard University Press. Cambridge, MA.
4. RANDEL, D.M. 1986. The New Harvard Dictionary of Music, 3d edit. Belknap Press of Harvard University Press. Cambridge, MA.

5. TYRRELL, J. & S. SADIE, Eds. 2001. The New Grove Dictionary of Music and Musicians, 2d edit. Grove's Dictionaries, Inc. New York.
6. KAESTNER, G. 1909. Untersuchungen uber den Gefuhlseindruck unanalysieter Zweiklange. Psychol. Studien. **4:** 473–504.
7. VAN DE GEER, J.P., W.J.M. LEVELT & R. PLOMP. 1962. The connotation of musical consonance. Acta Psychologia **20:** 308–319.
8. BERNSTEIN, L. 1976. The Unanswered Question. Harvard University Press. Cambridge, MA.
9. GUNDLACH, R.H. 1935. Factors determining the characterization of musical phrases. Am. J. Psychol. **47:** 624–643.
10. BALKWILL, L. & W.F. THOMPSON. 1999. A cross-cultural investigation of the perception of emotion in music: psycho physical and cultural cues. Music Percept. **17:** 43–64.
11. CASTELLANO, M.A, J.J. BHARUCHA & C.L. KRUMHANSL. 1984. Tonal hierarchies in the music of north India. J. Exp. Psychol. Gen. **113:** 394–412.
12. KESSLER, E.J., C. HANSEN & R.N. SHEPARD. 1984. Tonal schemata in the perception of music in Bali and in the West. Music Percept. **2:** 131–165.
13. TERHARDT, E. 1984. The concept of musical consonance: a link between music and psychoacoustics. Music Percept. **1:** 276–295.
14. KRUMHANSL, C.L. 1990. Cognitive Foundations of Musical Pitch. Oxford University Press. New York.
15. TRAMO, M.J., J.J. BHARUCHA & F.E. MUSIEK. 1990. Music perception and cognition following bilateral lesions of auditory cortex. J. Cognit. Neurosci. **2:** 195–212.
16. HELMHOLTZ, H. 1877/1954. On the Sensations of Tone as a Physiological Basis for the Theory of Music. Translated by A.J. Ellis, 1885. Dover Publications, Inc. NewYork.
17. STEVENS, S.S. & H. DAVIS. 1938/1983. Hearing: Its Psychology and Physiology. American Institute of Physics, Inc. New York.
18. PARNCUTT, R. 1989. Harmony: A Psychoacoustical Approach. Springer-Verlag. Berlin.
19. DEUTSCH, D., Ed. 1999. The Psychology of Music, 2nd edit. Academic Press. San Diego, CA.
20. DEMANY, L. & C. SEMAL. 1990. Harmonic and melodic octave templates. J. Acoust. Soc. Am. **88:** 2126–2135.
21. DEMANY, L., C. SEMAL & R.P. CARLYON. 1991. On the perceptual limits of octave harmony and their origin. J. Acoust. Soc. Am. **90:** 3019–3027.
22. KIANG, N.Y.-S., T. WATANABE, E.C. THOMAS & L.F. CLARK. 1965. Discharge Patterns of Single Fibers in the Cat's Auditory Nerve. M.I.T. Press. Cambridge, MA.
23. ROSE, J.E., J.F. BRUGGE, D.J. ANDERSON & J.E. HIND. 1967. Phase-locked response to low frequency tones in single auditory nerve fibers of the squirrel monkey. J. Neurophysiol. **30:** 769–793.
24. BOOMSLITTER, P. & W. CREEL. 1961. The long pattern hypothesis in harmony and hearing. J. Music Theory **5:** 2–31.
25. BRUGGE J.F., D.J. ANDERSON, J.E. HIND & J.E. ROSE. 1969. Time structure of discharges in single auditory nerve fibers of the squirrel monkey in response to complex periodic sounds. J. Neurophysiol. **32:** 386–401.
26. MOORE, B.C.J. 1980. Neural interspike intervals and pitch. Audiology **19:** 363–365.
27. ROSE, J.E. 1980. Neural correlates of some psychoacoustic experiences. *In* Neural Mechanisms in Behavior. D. McFadden, Ed.: 1–37. Springer-Verlag. New York.
28. OHGUSHI, K. 1983. The origin of tonality and a possible explanation of the octave enlargement phenomenon. J. Acoust. Soc. Am. **73:** 1695–1697.
29. PATTERSON, R.D. 1986. Spiral detection of periodicity and the spiral form of musical scales. Psychol. Music **14:** 44–61.
30. MOORE, B.C.J., R.W. PETERS & B.R. GLASBERG. 1985. Thresholds for the detection of inharmonicity in complex tones. J. Acoust. Soc. Am. **77:** 1861–1867.
31. VOS, J. 1982. The perception of pure and mistuned musical fifths and major thirds: thresholds for discrimination, beats, and identification. Percept. & Psychophys. **32:** 297–313.
32. MALMBERG, C.F. 1918. The perception of consonance and dissonance. Psychol. Monogr. **25:** 93–133.
33. GUERNSEY, M. 1928. The role of consonance and dissonance in music. Am. J. Psychol. **40:** 173–204.

34. BUTLER, J.W. & P.G. DASTON. 1968. Musical consonance as musical preference: a cross-cultural study. J. Gen. Psychol. **79:** 129–142.
35. KAMEOKA, A. & M. KURIYAGAWA. 1969. Consonance theory part II: consonance of complex tones and its calculation method. J. Acoust. Soc. Am. **45:** 1460–1469.
36. ZENTNER, M.R. & J. KAGAN. 1998. Infants' perception of consonance and dissonance in music. Infant Behav. & Dev. **21:** 483–492.
37. HULSE, S., D.J. BERNARD & R.F. BRAATEN. 1995. Auditory discrimination of chord-based spectral structures by European starlings (*Sturnus vulgaris*). J. Exp. Psychol. Gen. **124:** 409–423.
38. TRAMO, M.J., P. CARIANI & B. DELGUTTE. 1992. Representation of tonal consonance and dissonance in the temporal firing patterns of auditory nerve fibers. Soc. Neurosci. Abstr. **18:** 382.
39. TRAMO, M.J., M.C. MCKINNEY, P.A. CARIANI & B. DELGUTTE. 2000. Physiology of tonal consonance and dissonance. Assoc. Res. Otolaryngol. Abstr. **23:** 275–276.
40. MCKINNEY, M.F., M.J. TRAMO & B. DELGUTTE. 2001. Neural correlates of musical dissonance in the inferior colliculus. *In* Physiological and Psychophysical Bases of Auditory Function. A.J.M. Houtsma *et al.*, Eds.: 71–77. Shaker Publishing. Maastricht, the Netherlands.
41. MCKINNEY, M.F., M.J. TRAMO & B. DELGUTTE. 2001. Neural correlates of the dissonance of musical intervals in the inferior colliculus. Assoc. Res. Otolaryngol. Abstr. **24:** 54–55.
42. JAVEL, E. 1980. Coding of AM tones in the chinchilla auditory nerve: implications for the pitch of complex tones. J. Acoust. Soc. Am. **68:** 133–146.
43. RHODE, W.S. 1995. Interspike intervals as a correlate of periodicity pitch in cat cochlear nucleus. J. Acoust. Soc. Am. **95:** 2414–2429.
44. SCHREINER, C.E. & G. LANGNER. 1988. Coding of temporal patterns in the central auditory nervous system. *In* Auditory Function Neurobiological Bases of Hearing. G.M. Edelman *et al.*, Eds: 337–361. John Wiley & Sons. New York.
45. KRISHNA, B.S. & M.N. SEMPLE. 2000. Auditory temporal processing: responses to sinusoidally amplitude-modulated tones in the inferior colliculus. J. Neurophysiol. **84:** 255–273.
46. PREUSS, A. & P. MULLER-PREUSS. 1990. Processing of amplitude modulated sounds in the medial geniculate body of squirrel monkeys. Exp. Brain Res. **79:** 207–211.
47. BIESER, A. & P. MULLER-PREUSS. 1996. Auditory responsive cortex in the squirrel monkey: neural responses to amplitude-modulated sounds. Exp. Brain Res. **108:** 273–284.
48. FISHMAN, Y.I., D.H. RESER, J.C. AREZZO & M. STEINSCHNEIDER. 2000. Complex tone processing in primary auditory cortex of the awake monkey. I. Neural ensemble correlates of roughness. J. Acoust. Soc. Am. **108:** 235–246.
49. FISHMAN, Y.I., D.H. RESER, J.C. AREZZO & M. STEINSCHNEIDER. 2000. Complex tone processing in primary auditory cortex of the awake monkey. II. Pitch versus critical band representation. J. Acoust. Soc. Am. **108:** 247–262.
50. EHRET, G. & R. ROMAND, Eds. 1997. The Central Auditory System. Oxford University Press. New York.
51. PLOMP, R. & W.J.M. LEVELT. 1965. Tonal consonance and critical bandwidth. J. Acoust. Soc. Am. **38:** 548–560.
52. HUTCHINSON, W. & L. KNOPOFF. 1978. The acoustic component of Western consonance. Interface **7:** 1–29.
53. RAMEAU, J.-P. 1722/1971. Treatise on Harmony. Translated by P. Gossett. Dover Publications, Inc. New York.
54. WIGHTMAN, F.L. & D.M. GREEN. 1974. The perception of pitch. Am. Sci. **62:** 208–215.
55. HOUTSMA, A.J.M. & J.L. GOLDSTEIN. 1971. The central origin of the pitch of complex tones: evidence from musical interval recognition. J. Acoust. Soc. Am. **51:** 520–529.
56. MOORE, B.C.J. 1997. An Introduction to the Psychology of Hearing. Academic Press. San Diego, CA.
57. GALILEO GALILEI. 1638/1954. Dialogues Concerning Two New Sciences. Translated by H. Crew & A. de Salvio, 1914. Dover Publications. New York.
58. PICKLES, J.O. 1988. An Introduction to the Physiology of Hearing. Academic Press, Harcourt Brace Jovanovich, Publishers. London.

59. CARIANI, P.A. & B. DELGUTTE. 1996. Neural correlates of the pitch of complex tones. I. Pitch and pitch salience. J. Neurophysiol. **76:** 1698–1716.
60. CARIANI, P.A. & B. DELGUTTE. 1996. Neural correlates of the pitch of complex tones. II. Pitch shift, pitch ambiguity, phase-invariance, pitch circularity, and the dominance region for pitch. J. Neurophysiol. **76:** 1717–1734.
61. CARIANI, P. 1999. Temporal coding of periodicity pitch in the auditory system: an overview. Neural Plast. **6:** 147–172.
62. MEDDIS, R. & M.J. HEWITT. 1991. Virtual pitch and phase sensitivity of a computer model of the auditory periphery. II. Phase sensitivity. J. Acoust. Soc. Am. **89:** 2883–2894.
63. TERHARDT, E. 1974. Pitch, consonance, and harmony. J. Acoust. Soc. Am. **55:** 1061–1069.
64. HELMHOLTZ, H.L.F. 1863/1913. Die Lehre von der Tonempfindungen als physiologische Grundlage fur die Theorie der Musik, 6th edit. F. Vieweg. Brauschweig. Quote translated by E. Terhardt,[13] pp. 283–284.
65. DAI, H. 1993. On the pitch of two-tone complexes. J. Acoust. Soc. Am. **94:** 730–734.
66. GREENWOOD, D.D. 1991. Critical bandwidth and consonance in relation to cochlear frequency-position coordinates. Hear. Res. **54:** 164–208.
67. PLOMP, R. & H.J.M. STEENEKEN. 1968. Interference between two simple tones. J. Acoust. Soc. Am. **43:** 883–884.
68. MAYER, A.M. 1874. Researches in acoustics. Am. J. Sci. Arts, 3rd Ser. **8:** 241–255.
69. CROSS, C.R. & H.M. GOODWIN. 1891. Some considerations regarding Helmholtz's theory of consonance. Am. J. Sci. **58:** 1–12.
70. ZWICKER, E., G. FLOTTORP & S.S. STEVENS. 1957. Critical band width in loudness summation. J. Acoust. Soc. Am. **29:** 548–557.
71. YOST, W.A. 2000. Fundamentals of Hearing. Academic Press. San Diego, CA.
72. KAMEOKA, A. & M. KURIYAGAWA. 1969. Consonance theory part I: consonance of dyads. J. Acoust. Soc. Am. **45:** 1451–1459.
73. PRATT, C.C. 1921. Some qualitative aspects of bitonal complexes. Am. J. Psychol. **32:** 490–515.
74. BRUES, A.M. 1927. The fusion of non-musical intervals. Am. J. Psychol. **38:** 624–638.
75. GUTHRIE, E.R. & H. MORRELL. 1928. The fusion of non-musical intervals. Am. J. Psychol. **40:** 624–625.
76. SCHELLENBERG, E.G. & S.E. TREHUB. 1996. Natural musical intervals: evidence from infant listeners. Psychol. Sci. **7:** 272–277.
77. TRAMO, M.J. 1990. Impaired perception of relative pure tone pitch following bilateral lesions of auditory cortex in man. Soc. Neurosci. Abstr. **16:** 580.
78. SEASHORE, C.E., D. LEWIS & J.G. SAETVEIT. 1960. Seashore Measures of Musical Talents. The Psychological Corporation. New York.
79. TRAMO, M.J., A. GRANT & L.D. BRAIDA. 1994. Psychophysical measurements of frequency difference limens for relative pitch discrimination reveal a deficit following bilateral lesions of auditory cortex. Soc. Neurosci. Abstr. **20:** 325.
80. JOHNSRUDE, I.S., V.B. PENHUNE & R.J. ZATORRE. 2000. Functional specificity in the right human auditory cortex for perceiving pitch direction. Brain **123:** 155–163.
81. BHARUCHA, J.J., M.J. TRAMO & R.J. ZATORRE. 1993. Abstraction of the missing fundamental following bilateral lesions of auditory cortex. Soc. Neurosci. Abstr. **19:** 1687.
82. ZATORRE, R.J. 1988. Pitch perception of complex tones and human temporal-lobe function. J. Acoust. Soc. Am. **84:** 566–572.

# Intracerebral Evoked Potentials in Pitch Perception Reveal a Functional Asymmetry of the Human Auditory Cortex

CATHERINE LIÉGEOIS-CHAUVEL, KIMBERLY GIRAUD, JEAN-MICHEL BADIER, PATRICK MARQUIS, AND PATRICK CHAUVEL

*INSERM EPI 9926, Laboratoire de Neurophysiologie et Neuropsychologie, Faculté de médecine, 13005 Marseille, France*

**ABSTRACT: One acoustic feature that plays an important role in pitch perception is frequency. Studies on the processing of frequency in the human and animal brain have shown that the auditory cortex is tonotopically organized: low frequencies are represented laterally whereas high frequencies are represented medially. To date, the study of the functional organization of the human auditory cortex in the processing of frequency has been limited to the use of either scalp-recorded auditory evoked potentials (AEPs), which have relatively poor spatial resolving power, or functional imagery techniques, which have poor temporal resolving power. The present study uses intracerebrally recorded AEPs to explore this topic in the primary and secondary auditory cortices of both hemispheres of the human brain. Recordings were carried out in 45 adult patients with drug-resistant partial seizures. In the right hemisphere, clear spectrally organized tonotopic maps were observed with distinct separations between different frequency-processing regions. AEPs for high frequencies were recorded medially, whereas AEPs for low frequencies were recorded laterally. In the left hemisphere, however, this tonotopic organization was less evident, with different regions involved in the processing of a range of frequencies. The hemisphere-related difference in the processing of tonal frequency is discussed in relation to pitch perception.**

**KEYWORDS: Pitch perception; Auditory cortex; Neural processing**

The ability to process pitch is of major importance to music and language perception. Although the acoustic cues necessary to the perception of pitch are mainly of a temporal nature, human subjects describe pitch as having a spatial dimension (i.e., musical scale). Traditionally, the literature on the processing of pitch in the human brain has been dominated by two main theories: one that postulates an exclusively tonotopic coding and another that is in favor of an exclusively "temporal" coding.[1–3] Recently, however, Langner[4,5] has suggested that temporal information, coded at the subcortical level as spikes synchronized to the periodicity of the acoustic signal, is

Address for correspondence: Catherine Liégeois-Chauvel, INSERM EPI 9926, Laboratoire de Neurophysiologie et Neuropsychologie, Faculté de médecine, 27 Bd. Jean Moulin, 13005 Marseille, France. Voice: +33 4 91 32 44 63; fax: +33 4 91 78 99 14.

catherine.liegeois@medecine.univ-mrs.fr

subsequently converted to spatial information coded in the form of tonotopic maps in the auditory cortex. This hypothesis is consistent with the finding that the temporal resolution of the auditory nerve[6] and inferior colliculus[7,8] may be higher than at the cortical level, where the temporal resolution of auditory information is possible only at frequencies below 100 Hz for the majority of neurons.[9,10] At the cortical level, pitch appears to be coded according to a frequency spectrum (i.e., tonotopically).

Experimental studies using single-unit recording techniques indicate that for most auditory neurons, a "best frequency" (or BF) can be determined at which a low-intensity auditory stimulus evokes the greatest electrophysiological response in a given region. The auditory cortex is said to be tonotopically organized because an ordered change in BFs occurs as one moves along the cortex. The auditory cortex is organized according to a frequency gradient, with low frequencies represented laterally and high frequencies represented medially.[11–18]

The tonotopic organization of the auditory cortex is best understood in nonhuman animals. The cytoarchitecture of the auditory cortex is not always uniform,[14] however, rendering the mapping of different tonotopic regions difficult. Merzenich and Brugge[12] have found that the auditory cortex in the superior temporal plane of the macaque can be subdivided into at least six distinct fields on the basis of cytoarchitectonic characteristics and BFs. Likewise, Reale and Imig[16] observed a tonotopic organization in six different auditory areas of the cat, and a similar tonotopic organization has been described in the albino rat.[18]

In humans, the use of noninvasive techniques such as magnetoencephalography (MEG) and positron emission tomography (PET) has shed some light on the functional organization of the auditory cortex with respect to frequency. In their PET study, Lauter *et al.*[19] observed that changes in regional cerebral blood flow induced by a high-frequency tone (4 kHz) were located deeper and more posteriorly in the temporal lobe than those induced by a low-frequency tone (500 Hz). In a similar vein, MEG studies on the tonotopic organization of N1 generators have shown that their estimated depth can vary as a function of frequency: higher frequencies involve more medial regions of the Heschl's gyrus (HGs), whereas lower frequencies involve more lateral areas.[20–24]

In a previous study,[25] using depth electrodes implanted in the HG of human subjects with medically intractable epilepsy, we identified four separable components of middle-latency auditory evoked potentials (MLAEPs), which peaked at approximately 30, 50, 60, and 80 ms. The sources of these components were definable, although some overlap was observed. In a recent MEG study, Pantev *et al.*[23] explored the sources of different middle-latency components as well as their distribution. In all subjects and for all the frequencies studied, only two components were consistently observed: the 30-ms "Pam" component and the N1m. Also, a tonotopic organization was observed in the primary auditory cortex that mirrored that observed in the secondary cortex. The localization of Pam's source moved more superficially with higher frequencies, while the localization of the N1m's source moved progressively deeper with higher frequencies. The authors, however, did not consistently observe the 50-ms component, which is thought to be generated in the lateral part of the primary auditory cortex in close proximity to the neighboring secondary areas.

The present paper explores the tonotopic organization of the human auditory cortex using intracerebrally recorded evoked potentials studied as a function of anatomical recording site. The sensitivity of a neuronal population to a given frequency was determined from fluctuations in AEP amplitude between different recording sites in the primary auditory cortex and surrounding secondary areas like the planum temporale. Indeed, previous data suggest that the auditory cortex is tonotopically organized for frequency and that even several different tonotopic maps may be described for different cortical regions. But is this tonotopic organization the same in both left and right cerebral hemispheres? A substantial body of data suggests that the perception and processing of pitch is specific to a single hemisphere,[26] and hemisphere-related differences in the functional organization of the auditory areas underlying such processing are not inconceivable. The main objective of this study was to investigate the tonotopic organization of the human auditory cortex in both cerebral hemispheres.

## PATIENTS AND METHODS

Intracerebral AEPs were recorded from 45 adult patients (aged 20 to 56 years) with medically intractable partial epileptic seizures of right and left temporal lobe origin (in 31 and 14 patients, respectively). These patients were undergoing stereoelectroencephalographic (SEEG) evaluation for the surgical relief of their seizures (for a description of this procedure, see Refs. 27 and 28). Recordings were carried out in either the left or the right HG and right planum temporale, during which time subjects laid comfortably on a bed and remained vigilant and cooperative. All patients had previously given informed consent to participate in the study.

The anatomical localization of each implanted electrode was based on a stereotaxic procedure described elsewhere.[29–31] The three-dimensional coordinates of each electrode lead were defined and measured with respect to the temporal branch of the middle cerebral artery and allowed us to distinguish the primary from the secondary auditory areas in the different patients.[25,32]

The visualization of electrode tracks using stereotaxic MRI represents a means of confirming the anatomical localization of implanted sites and provides additional information on the anatomy of the HG in each subject (e.g., the presence of one or two gyri[33]). Such information can often be helpful in the interpretation of AEP data. MRI sections were cut along three spatial planes using bicommissural coordinates. FIGURE 1 shows the position of one electrode implanted orthogonally in the lateral part of the left HG.

### *Auditory Stimulation*

Auditory stimuli were tone bursts (with a 0.3-ms rise and decay time and a total duration of 30 ms) of different frequencies (ranging from 250 Hz to 4 kHz). These stimuli were generated using an SM 700 NICOLET auditory stimulator and presented at a rate of 0.7/s via TDH 39 earphones to the ear contralateral to the implanted site. Stimuli were presented in random order at an intensity of 70 dB SL (sensation level) for each subject.

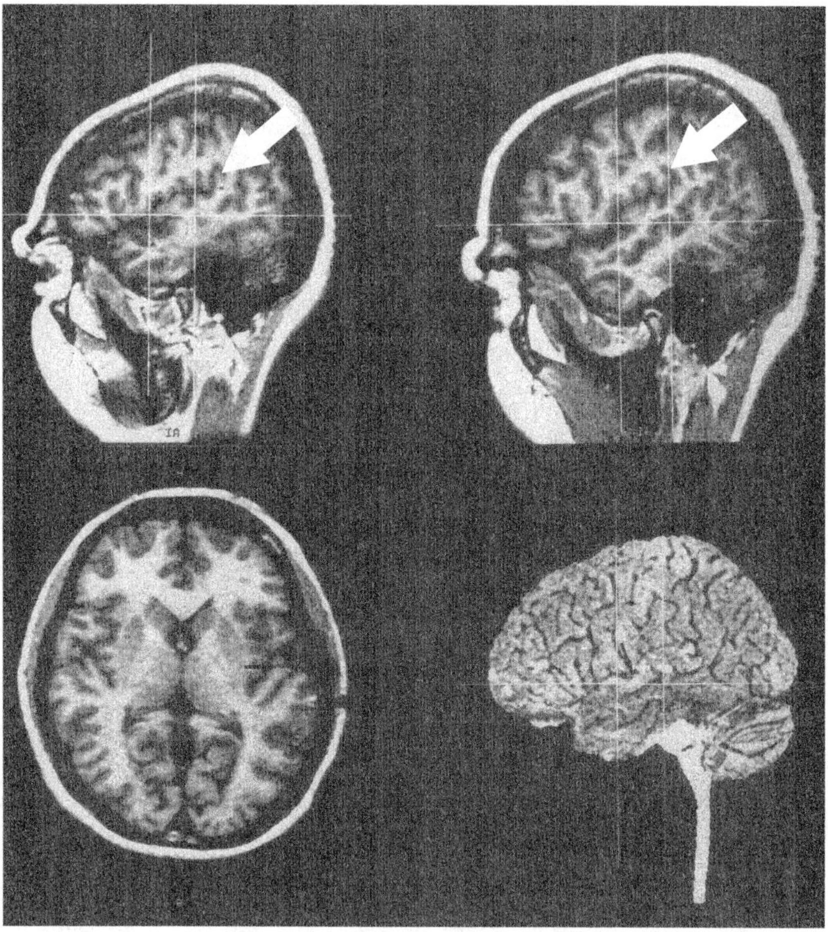

**FIGURE 1.** *Top*: Sagittal MRI sections show the trajectory of the Sylvian fissure in the left (**A**) and right (**B**) temporal lobes. Note here the distinct shape of the end of the lateral sulcus (Sylvian fissure) on both sides (*white arrows*) and the longer length of the left planum temporale in this right-handed patient. The *three axes* of the bicommissural reference system (AC-PC, VAC, VPC) have been drawn. *Left bottom*: The verticofrontal section shows the left Heschl's gyrus, and the *black arrow* indicates the trace of the electrode recorded from the auditory cortex. *Right bottom*: On the surface view, the *arrow* shows the impact point of the same electrode.

### *Depth AEP Recording and Averaging*

Intracerebral AEPs were recorded from all leads of an electrode (using an extracephalic reference looped around the neck) and were simultaneously amplified and bandpass filtered 1 Hz to 1.5 kHz. Recording epochs lasted 350 ms with a prestimulus time of 35 ms, 10% of the epoch. Two trials of 100 samples were administered to each subject. Variations (i.e., polarity inversions and/or changes in amplitude in

AEPs) recorded from different leads were studied for each patient. Tonotopic maps were based on a comparison of the results obtained from patients with different anatomical recording sites.

The acquisition system was a Nicolet Pathfinder II with 16 acquisition channels and an 8-bit–resolution digital converter. Negative potentials were displayed as up and positive as down. Recordings of brainstem evoked potentials carried out before SEEG recordings confirmed that all the patients studied had normal afferent pathway conduction.

### *Temporospectral Mapping*

Data were stored on the hard drive of the computer and transferred via an asynchronous line to an IBM PC/AT. Off-line processing consisted of the construction of temporospectral maps, using a technique adapted from Rémond,[34] to depth recordings, particularly appropriate to the depth electrode geometry.[35,36]

## RESULTS

Primary and secondary areas were identified on the basis of component latency (for a description, see Ref. 25). The selectivity of an auditory area's response (and hence its BF was determined from frequency-related fluctuations in a component's amplitude and/or polarity inversions.

### *Recordings from the Primary Auditory Cortex*

#### *Right Hemisphere*

FIGURE 2 shows AEPs recorded from 10 contiguous leads of an electrode implanted in the right auditory cortex of one patient. The first four leads recorded activity from the primary auditory cortex; the other six recorded activity from secondary areas. Tone bursts at different frequencies were presented to the patient's left ear. A triphasic N30/P50/N80 complex was recorded from the first four leads for which a polarity inversion was observed between leads 4 and 5, as well as a shift in latency between leads 6 and 10. AEPs recorded from lead 4 were highly sensitive to tonal frequency. For tones higher than 1 kHz, the amplitude of the 50- and 80-ms components progressively decreased until an isopotential response (note the polarity inversion between leads 3 and 5) was observed for tones at 4 kHz, suggesting that the cortical region at lead 4 has a high sensitivity for frequencies in the 4 kHz range. AEPs did not differ in terms of tonal frequency on any of the other leads.

It should be noted, of course, that the amplitude changes observed were never an all-or-none phenomenon, but were observed as increasing or decreasing from lead to lead in progressive steps. Also, unlike the 50- and 80-ms components, the amplitude of the 30-ms component did not appear to be frequency dependent.

FIGURE 3 is a temporospectral representation of AEPs recorded for different tonal frequencies from a second patient. In this figure, the amplitudes of the same three components (30, 50, and 80 ms) as for the first patient are shown for the three most responsive leads (2, 3, and 6 for different tonal frequencies). In each case, frequency is represented along the x-axis and latency along the y-axis. On lead 2, amplitude

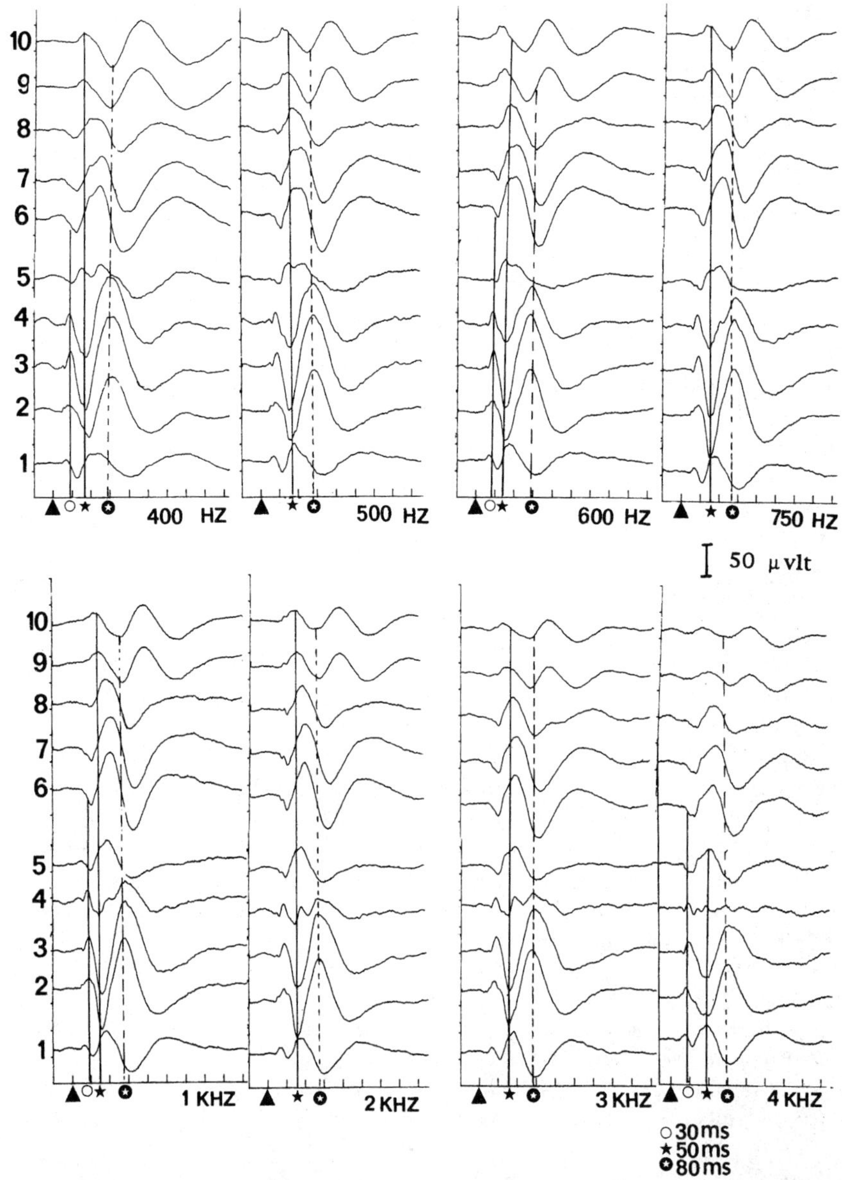


**FIGURE 2.** Recordings from the medial part of the primary auditory cortex in right HG according to different tonal frequencies. Each response corresponds to a monopolar recording of each lead of the electrode that is referenced to an extracephalic site. The leads are numbered from 1 to 10, from the medial to lateral part of the auditory cortex. Note the amplitude variations of the 50-ms component (indicated by the *star*) according to frequency (see comments in the text). The *arrow* indicates the onset of the stimulus. Recording time ends at 350 ms.

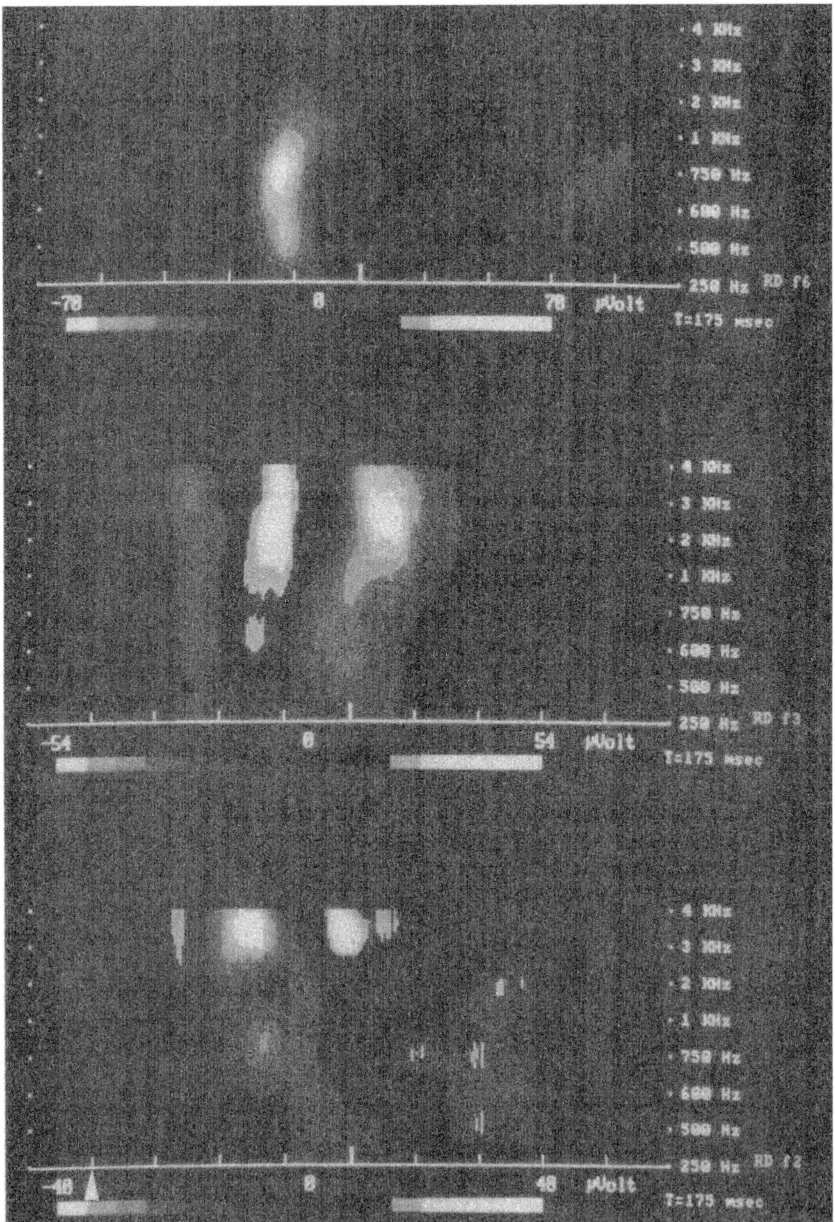


**FIGURE 3. See Plate 1 in color section.** Temporospectral maps of amplitude variations of potentials recorded from the right primary auditory cortex (leads 2, 3, and 6). Time, related in abscissae, is reduced to 175 ms, full scale. On the ordinate axis, each *yellow dot* represents the frequency. Amplitude is represented by color. A cubic spline interpolation is used in order to obtain a continuous potential line between the different frequencies. Each spatial /reconstruction gives a continuous map both in time and in frequency. *Dark colors* are values close to the neutral potential, and *bright colors* are used for the activity—*blue* represents negativity and *red* represents positivity. The *white arrow* indicates the beginning of the sound. (For comments see text.)

was maximal for tone bursts with the highest frequencies (3–4 kHz), although frequency-dependent fluctuations in amplitude were greater for the N50 and P80 components than for the N30. This was observed to an even greater extent on lead 3, where the amplitude of the N30 did not vary significantly according to tonal frequency, in spite of the highly significant fluctuations in P50 and N780 amplitude. On lead 3, the P50 and N80 were less sharply tuned and their amplitude was maximal at frequencies of 2–3 kHz. Lead 6 recorded neuronal activity in a region between the lateral part of the primary cortex and the medial part of the secondary cortex and yielded responses characterized, this time, by an almost complete absence of the 30-ms component. The N50 was the only component for which amplitude varied with tonal frequency. Its amplitude was maximal at 750 Hz.

*Left Hemisphere*

The left side of FIGURE 4 shows AEPs recorded from the left primary auditory cortex (leads 4 and 5) for different tonal frequencies. As was the case in the right primary cortex, a triphasic N30/P45/N100 complex was observed. P45 and N100 amplitude recorded from the left hemisphere, however, did not vary as a function of frequency, as shown on the right in the same figure. For these components, amplitude was maximal at a range of frequencies between 600 Hz and 2 kHz.

### *Recordings from the Secondary Auditory Cortex in the Left and Right Hemispheres*

The secondary auditory areas studied were confined to an area anterior and lateral to the primary cortex, as previously identified in physiological[25,32] and morphological studies.[37,38] In this secondary area, frequency-dependent fluctuations in amplitude were observed for the 80-ms component. FIGURE 5 shows the localization of electrodes implanted in the right (left side of figure) and left (right side of figure) secondary auditory areas in 11 patients (6 right; 5 left). The sagittal views (top) show the localization of each electrode, while the coronal views (below sagittal views) show the localization of those electrode leads yielding the greatest frequency-dependent fluctuations in amplitude. In the right hemisphere, a mediolateral as well as an anteroposterior distribution of BFs was observed for the 80-ms component. Neuronal populations sensitive to high frequencies (graphs A and E) 80-ms poststimulus onset were localized in the posterolateral part of right HG, while those sensitive to low frequencies (graph B) were localized in the anterolateral part of right HG. For the 80-ms component, high frequencies were also represented ventrally in an area adjacent to the lateral part of the primary cortex in the middle of right HG, overlapping the area in which low frequencies were represented for the 50-ms component.

The profile of amplitude curves for the right hemisphere was noticeably different from that for the left hemisphere. In the dorsolateral part of the left HG, the generators of the 80-ms component were most responsive over a range of high frequencies (1 kHz to 4 kHz; graph B) or to two isolated sets of frequencies (500 Hz and 1–2 kHz; graph C); in the anterolateral part of left HG, this component was sensitive either to a range of low-to-medium frequencies (500 Hz to 1 kHz) or to two isolated sets of frequencies (400 Hz and 1 kHz; graph D).

FIGURE 6 summarizes the main results. In the right primary cortex, the 30-ms component was not clearly tonotopic. A first tonotopic map is observed for the 50-

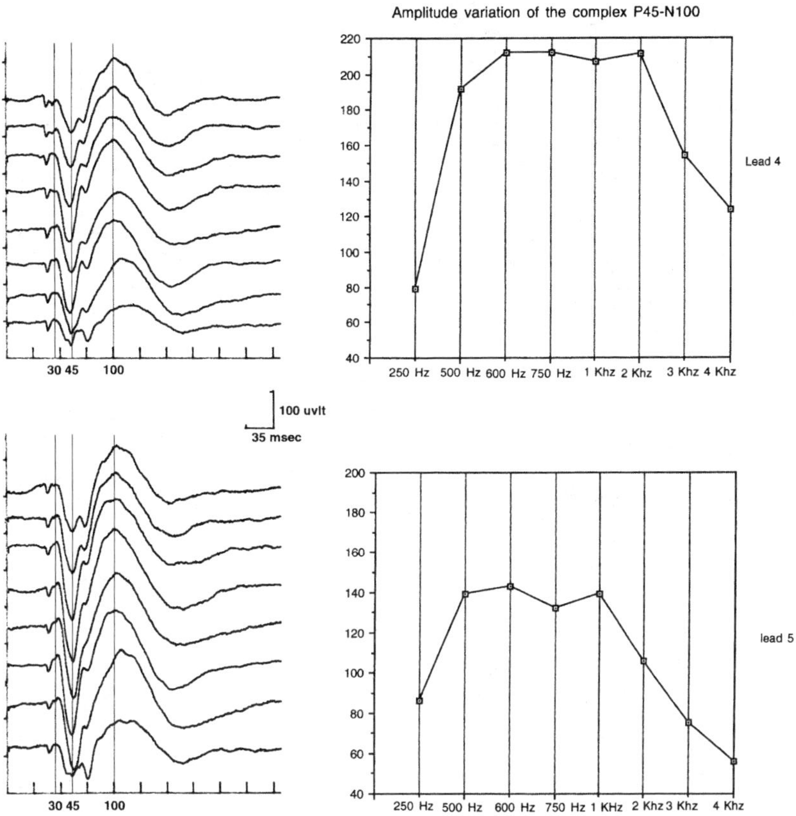


**FIGURE 4.** *Left*: Evoked responses recorded from the medial part of the primary auditory cortex (leads 4 and 5) in the left HG. *Right*: Graphs representing amplitude variations for the P45/N100 complex according to different tonal frequencies (ordinate axis: absolute amplitude expressed in μV; abscissa axis: tonal frequency). The frequencies vary from 250 Hz (*first line*) to 4 kHz (*last line*) (see comments in the text).

ms component. A strict organization displaying high frequencies medially and low frequencies laterally is observed. A second map is observed for the 80- or 100-ms components generated in the secondary cortex. Neurons responsive to low frequencies were lateral to those responding to higher frequencies, but they were less finely tuned (broken line). Another map was observed for the 100-ms component generated in the planum temporale. High frequencies seemed to be more represented. On the left hemisphere, in the primary as well in the secondary auditory cortex, the neurons were most sensitive to a range of frequencies (low or high). Nevertheless, the same tonotopic organization of the frequencies seems to be respected—high frequencies medially and low frequencies laterally represented.

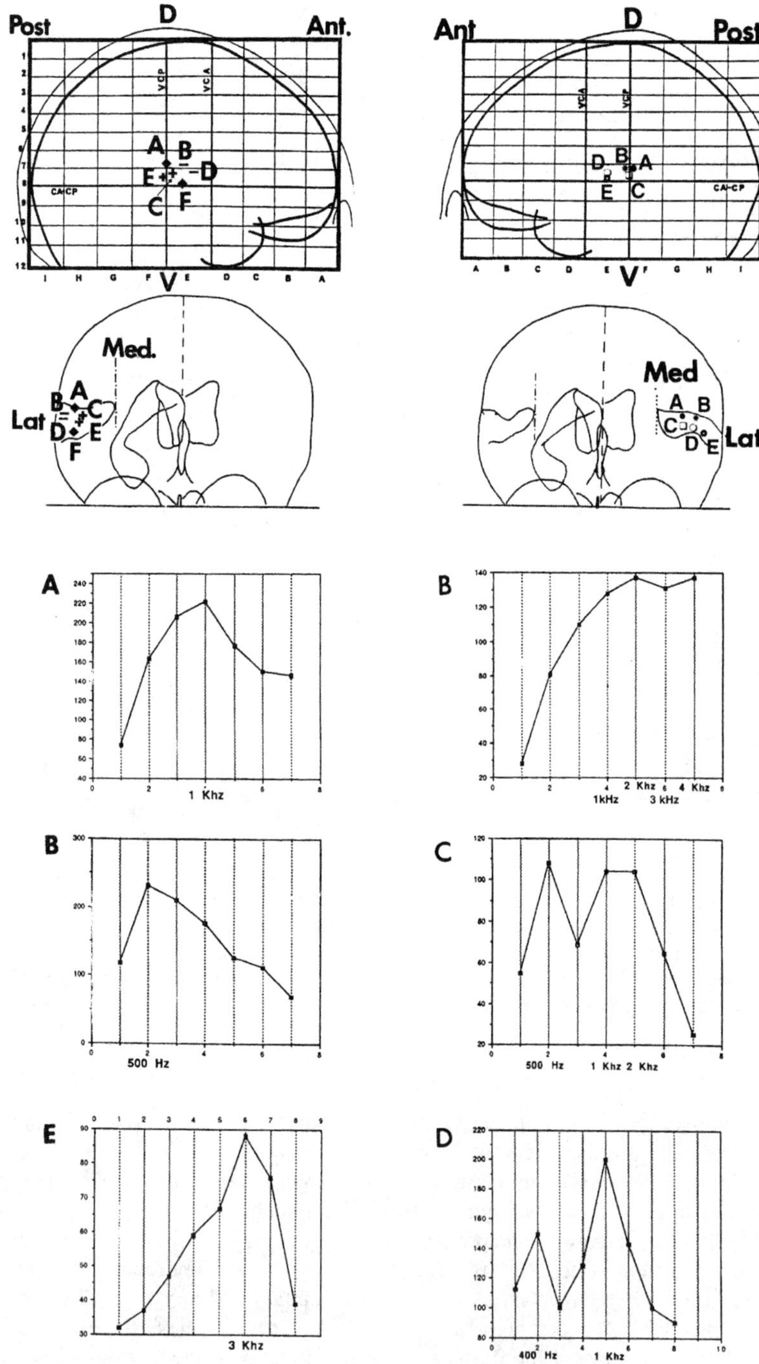


**FIGURE 5.** *See following page for caption.*

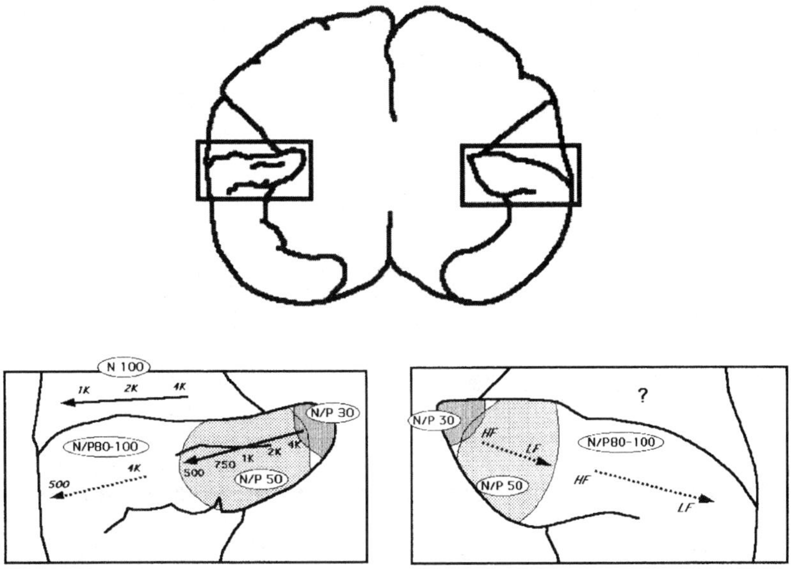


**FIGURE 6.** Schematic representation of the tonotopic maps recorded from the right and left auditory areas (see comments in the text).

## DISCUSSION

The present study suggests that the auditory cortex is composed of frequency-dependent tonotopic maps and that these maps are more complex than previously reported in surface EP studies.[20,21,23,39] Although the previously reported mediolateral tonotopic organization was also observed in this study, it was not robust for all AEP components recorded and was hemisphere specific.

In the primary auditory cortex, no frequency-dependent changes in amplitude or latency were observed for the earliest components (N13/P17) in either hemisphere. For the N30, only slight variations in amplitude were seen in response to tonal frequency, and most often its amplitude covaried with that of the 50-ms component. The amplitude of the 50-ms component was maximal for high frequencies (4 kHz) in medial regions of the primary cortex and for low frequencies (400–500 Hz) in lateral regions. Both mediolateral and anteroposterior organizations observed were

---

**FIGURE 5.** Mapping of amplitude variations for the 80-ms component (ordinate axis: absolute amplitude expressed in μV; abscissa axis: frequency) recorded from different loci of the lateral part of the right and left HG referenced to Talairach's proportional standard grid on the lateral view. On the frontal view, the depth of the most responsive lead of each patient is reported in the Heschl's gyrus. The frequencies vary from 250 Hz (*first line*) to 4 kHz (*last line*). The *peak of the curve* corresponds to the CF (see comments in the text).

marked by a progressive decrease in BFs toward the boundaries of the primary cortex.

The present data suggest the existence of an orderly, tonotopic representation of frequencies in the human primary auditory cortex, similar to that described in nonhuman animals.[13,40] In humans, BFs vary from 4 kHz medially to 400 Hz laterally for the 50-ms component. This range roughly corresponds to the frequency range useful for speech (this latter range actually begins near 250 Hz—a frequency at which a clear evoked response was difficult to obtain in this study). Studies exploring the auditory cortex of nonhuman primates have shed a great deal of light on the functional organization of the human auditory cortex, because its organization in primates appears to be similar to that in humans. In primates, BFs are represented in an orderly pattern in the primary auditory cortex and range from below 100 Hz to about 32 kHz, nearly the full frequency hearing range for primates. The lowest BFs are represented rostrally and laterally, whereas the highest BFs are found caudally and medially. Such an organization is thought to cover almost the entire auditory cortex, with all frequencies (except those at the extreme ends of the frequency hearing range) being represented.[2,16,41,42] The BFs observed in the present study were highly stable from patient to patient, in spite of the intersubject variability in localization and orientation of the primary auditory cortex.

In a recent study, Pantev *et al.*[23] observed a tonotopic organization of the human auditory cortex for the Pam component (corresponding to the 30-ms intracerebral AEP component); unlike the results of the present study, however, higher frequencies were represented more laterally than were lower frequencies. This discrepancy may be explained by the fact that, in the present study, only slight frequency-dependent fluctuations in amplitude were observed for the 30-ms component, compared to those observed for the P/N 50. Furthermore, the P1m (corresponding to the P/N 50 in this study), generated in the primary auditory cortex, was not systematically discernible in Pantev *et al.*'s study and has been shown in several MEG[44–46] and EEG studies to have highly variable scalp topography. With intracerebrally recorded AEPs, on the other hand, the P/N 50 is robust and can reach amplitudes of up to 150 μV.[25] It is also probable that two different overlapping tonotopic maps exist, each corresponding to either the 30- or the 50-ms component, and that different investigation techniques have differential capacities to explore these maps.

Frequency-dependent fluctuations in amplitude were also observed for the 80-ms component, which is generated in secondary auditory areas anterior and lateral to the primary cortex.[25] In the right hemisphere, a mediolateral as well as an anteroposterior tonotopic organization was seen, with high frequencies represented posteromedially and low frequencies represented anterolaterally. High frequencies were also represented ventrally in an area adjacent to the lateral part of the primary cortex, which was also the area in which low frequencies were represented for the 50-ms component. While the distribution of BFs was roughly similar for the 80-ms component in both hemispheres, this distribution was not as well defined in the left hemisphere, where neuronal populations were often found to respond to more than one frequency or over a broad range of frequencies. Neither Merzenich *et al.*[13] nor Reale and Imig,[16] however, found a clear tonotopic organization of the secondary cortex (AII) in their studies with nonhuman animals: several groups of neurons had multiple BFs or responded over a broad range of frequencies. Evoked responses from this area had longer latencies and were less well tuned than those from the primary cortex.[16,47] In their study, Merzenich *et al.*[14] reported that neurons in these secondary areas often responded over a broad range of frequencies and that the assignment of BFs was difficult, with two maxima often being observed. It is possible that a second harmonic distortion produced this "double range" because the pairs of maxima in this study were often harmonically related.[48]

With respect to area 22, in the present study no tonotopic organization was observed in either hemisphere. This finding is consistent with data from animal studies using single-unit recordings[49] showing that neurons in nonprimary auditory areas in the macaque are not selectively responsive to pure tones, but respond only to frequency-centered sound bursts containing many frequencies.

One of the most significant findings in this study is the difference in the response selectivity of neurons in the right and left cerebral hemispheres. Neurons in the right auditory cortex were more sharply tuned to frequency than neurons in the homologous region of the left hemisphere. In a previous study, we showed that left auditory neurons were specifically sensitive to the temporal features of auditory information.[50]

As discussed earlier in this paper, the perceptual identity of a sound depends on both a sound's temporal structure as well as its spectral content. For example, periodic modulations of a sound's amplitude evoke a pitch percept. Taken together, the results from the numerous studies on the functional organization of the human auditory cortex suggest a different functional hemispheric asymmetry in the processing of spectral and temporal information. Frequency (i.e., spectral information) may be processed in HG of the right hemisphere, while temporal information may be processed in the left hemisphere. Not all data, however, is consistent with this idea. Using MEG, Langner *et al.*[51] found that periodicity pitch and frequency are arranged orthogonally in the human auditory cortex, with no differences observed according to hemisphere.

The asymmetry observed in the tonotopic organization of the right and left auditory cortices could underlie hemispheric specialization and could account for a fair amount of behavioral data obtained in studies on human auditory function. Robin *et al.*[52] found that left temporoparietal lesions impair the perception of temporal but not spectral auditory information; lesions in homologous regions of the right hemisphere were found to have the opposite effect. The lesions referred to in their study affected only the associative auditory areas, but such lesions could also disconnect the primary cortex from the rest of the brain.[53,54] Taken together with these findings, the present study's results suggest that the frequency analysis of sounds corresponds to the neural sharpening for neurons tuned to frequency. This mechanism could be conferred on the right auditory cortex, which plays an important role in the processing of pitch.[26,55] Auditory neurons in the left hemisphere, on the other hand, have been shown to be more selective for temporal pattern[50] than for spectral pattern. Such data are consistent with the results from studies with language-disordered subjects who have left-hemispheric lesions, suggesting that many of these patients are also impaired in auditory temporal perception.[56]

In conclusion, these results suggest that the processing of spectral information is carried out primarily in the right hemisphere, and the processing of the temporal aspects of auditory stimuli in the left auditory cortex.[50] Future research in this direction could not only provide insight into the cortical processes underlying speech and music perception, but may also help us better understand the psychoacoustic sequelae of cortical lesions.

## ACKNOWLEDGMENTS

The authors are grateful to Drs. B. Devaux (Ste. Anne Hospital, Paris), A. Musolino (Clinica Neurochirurgica, Messina, Italy), and J.M. Scarabin (Pontchaillou Hospital, Rennes) for their collaboration.

## REFERENCES

1. WHITFIELD, I.C. 1970. Central nervous processing in relation to spatio-temporal discrimination of auditory patterns. *In* Frequency Analysis and Periodicity Detection in Hearing. R. Plomp & Smoorenburg, Eds.: 136–152. Sijthoff. Leiden.
2. SHAMMA, S.A. 1985. Speech processing in the auditory system II: lateral inhibition and the processing of speech evoked activity in the auditory nerve. J. Acoust. Soc. Am. **78:** 1622–1632.
3. SHAMMA, S.A. 1986. Encoding the acoustic spectrum in the spatio-temporal responses of the auditory nerve. *In* Auditory Frequency Selectivity. B.C.J. Moore & R.D. Patterson, Eds.: 289–296. Plenum. New York.
4. LANGNER, G. 1997. Neural processing and representation of periodicity pitch. Acta Otolaryngol. (Stockh.) **532:** 68–76.
5. LANGNER, G. 1998. Neuronal periodicity coding and pitch effects. *In* Central Auditory Processing and Neural Modeling. P.W. Poon & J.F. Brugge, Eds.: 31–41. Plenum. New York.
6. PALMER, A.R. 1982. Encoding of rapid amplitude fluctuations by cochlear nerve fibers in the guinea pig. Acta Otolaryngol. (Stockh.) **236:** 197–202.
7. SCHREINER, C.E. & G. LANGNER. 1984. Representation of periodicity information in the inferior colliculus of the cat. Abstr. Soc. Neurosci. **10:** 395.
8. LANGNER, G. & C.E. SCHREINER. 1988. Periodicity coding in the inferior colliculus in the cat. I: neuronal mechanisms. J. Neurophysiol. **60:** 1799–1822.
9. GOLDSTEIN, M.H., F. DE RIBAUPIERRE & G.H. YENI-KOMOSHIAN. 1971. Cortical coding of periodicity pitch. *In* Physiology of the Auditory System. M.B. Sach, Ed.: 299–306. National Education Consultants. Baltimore, MD.
10. RIBAUPIERRE, F. DE, M.H. GOLDSTEIN & G.H. YENI-KOMOSHIAN. 1972. Cortical coding of repetitive acoustic pulses. Brain Res. **48:** 205–225.
11. GOLDSTEIN, M.H., M. ABELES, M.R. DALY & J. MCINTOSH. 1970. Functional architecture in cat primary auditory cortex: tonotopic organization. J. Neurophysiol. **33:** 188–197.
12. MERZENICH, M.M. & J.F. BRUGGE. 1973. Representation of the cochlear partition on the superior temporal plane of the macaque monkey. Brain Res. **50:** 275–296.
13. MERZENICH, M.M., P.L. KNIGHT & G.L. ROTH. 1975. Representation of cochlea within primary auditory cortex in the cat. J. Neurophysiol. **38:** 231–249.
14. MERZENICH, M.M., J.H. KAAS & G.L. ROTH. 1976. Auditory cortex in the grey squirrel: tonotopic organization and architectonic fields. J. Comp. Neurol.**166:** 387–402.
15. PHILLIPS, D.P. & D.R.F. IRVIN. 1981. Responses of single neurons in physiologically defined primary auditory cortex (A1) of the cat: frequency tuning and responses to intensity. J. Neurophysiol. **45:** 16–34.
16. REALE, R.A. & T.J. IMIG. 1980. Tonotopic organization in auditory cortex of the cat. J. Comp. Neurol. **291:** 192–265.
17. PHILLIPS, D.P. & S.S. ORMAN. 1984. Responses of single neurons in posterior field of cat auditory cortex to tonal stimulation. J. Neurophysiol. **51:** 147–163.

18. SALLY, S.L. & J.K. KELLY. 1988. Organization of auditory cortex in the albino rat: sound frequency. J. Neurophysiol. **59:** 1627–1638.
19. LAUTER, J.L., P. HERSCOVITCH, *et al.* 1975. Tonotopic organization in human auditory cortex revealed by positron emission tomography. Hear. Res. **20:** 199–205.
20. ROMANI, G.L., S.J. WILLIAMSON & L. KAUFMAN. 1982. Tonotopic organization of the human auditory cortex. Science **216:** 1339–1340.
21. PANTEV, C., M. HOKE, K. LEHNERTZ, *et al.* 1988. Tonotopic organization of the human auditory cortex revealed by transient auditory evoked magnetic fields. EEG Clin. Neurophysiol. **69:** 160–170.
22. PANTEV, C., M. HOKE, K. LEHNERTZ, *et al.* 1990. Identification of sources of brain neuronal activity with high spatiotemporal resolution through combination of neuromagnetic source localization (NMSL) and magnetic resonance imaging (MRI). EEG Clin. Neurophysiol. **75:** 173–184.
23. PANTEV, C., O. BERTRAND, C. EULITZ, *et al.* 1995. Specific tonotopic organizations of different areas of the human auditory cortex revealed by simultaneous magnetic and electric recordings. EEG Clin. Neurophysiol. **94:** 26–41.
24. CANSINO, S., S.J. WILLIAMSON & D. KARRON. 1994. Tonotopic organization of human auditory association cortex. Brain Res. **663:** 38–50.
25. LIEGEOIS-CHAUVEL, C., A. MUSOLINO, J.M. BADIER, P. MARQUIS & P. CHAUVEL. 1994. Evoked potentials recorded from the auditory cortex in human: evaluation and topography of the middle latency components. EEG Clin. Neurophysiol. **92:** 204–214.
26. ZATORRE, R.J., A.C. EVANS, *et al.* 1992. Lateralization of phonetic and pitch discrimination in speech processing. Science **256:** 846–849.
27. BANCAUD, J., J. TALAIRACH, *et al.* 1965. La stéréoélectroencephalographie dans l'epilepsie: informations neurophysiopathologiques apportées par l'investigation fonctionnelle stéréotaxique. 1965. Masson. Paris.
28. BANCAUD, J., F. BRUNET-BOURGIN, P. CHAUVEL, *et al.* 1994. Anatomical origin of *déjà vu* and vivid "memories" in human temporal lobe epilepsy. Brain **117:** 71–90.
29. SZIKLA, G., G. BOUVIER, T. HORI & V. PETROV. 1977. Angiography of the human brain cortex. *In* Atlas of Vascular Patterns and Stereotactic Cortical Localization. Springer Verlag. Berlin.
30. MUSOLINO, A., J. TALAIRACH, P. TOURNOUX & O. MISSIR. 1988. Comparative study between stereotactic angiography and magnetic resonance imaging (MRI) data on spatial organization of sulci and convolutions in man. Boll. Lega Italiana contro Epilessia **62/63:** 51–56.
31. MUSOLINO, A., P. TOURNOUX, O. MISSIR & J. TALAIRACH. 1990. Methodology of *in vivo* anatomical study and stereo-electroencephalographic exploration in brain surgery for epilepsy. J. Neuroradiol. **17:** 67–102.
32. LIEGEOIS-CHAUVEL, C., A. MUSOLINO & P. CHAUVEL. 1991. Localization of primary auditory area in man. Brain **114:** 139–153.
33. STEINMETZ, H., J. RADEMACHER, *et al.* 1989. Cerebral asymmetry: Mr planimetry of the human planum temporale. J. Comput. Assist.Tomogr. **13:** 996–1005.
34. REMOND, A. 1961 Integrated and topological analysis of the EEG Clin. Neurophysiol. (Suppl.) **20:** 64–67
35. CHAUVEL, P., P. BUSER, J.M. BADIER, *et al.* 1987. La "zone épileptogène" chez l'homme: représentation des évènements intercritiques par cartes spatio-temporelles. Rev. Neurol. (Paris) **143**(5): 443–450.
36. BADIER, J.M. 1991. Etude de la localisation des sources cérébrales d'activité paroxystique par cartographie. Thèse de Doctorat. Université de technologie de Compiègne. Bertrand *et al.*
37. BRAAK, H. 1978. Architectonics of the human telencephalic cortex. Springer Verlag. Berlin.
38. GALABURDA, A.M. & F. SANIDES. 1980. Cytoarchitectonic organization of the human auditory cortex. J. Comp. Neurol. **190:** 597–610.
39. BERTRAND, O., F. PERRIN & J. PERNIER. 1971. Evidence for a tonotopic organization of the auditory cortex observed with auditory evoked potentials. Acta Otolaryngol. (Stockh.) (Suppl.) **491:** 116–123.

40. SCHREINER, C.E. & J.R. MENDELSON. 1990. Functional topography of cat primary auditory cortex: distribution of integrated excitation. J. Neurophysiol. **64:** 1442–1459.
41. IMIG, T.J., M.A. RUGGERO, L.M. KITZES, *et al.* 1977. Organization of auditory cortex in the owl monkey. J. Comp. Neurol. **171:** 111–128.
42. AITKIN, L.M., M.M. MERZENICH, D.R.F. IRVINE, *et al.* 1986. Frequency representation in auditory cortex of the common marmoset. J. Comp. Neurol. **252:** 175–185.
43. IMIG, T.J. & R.A. REALE. 1980. Patterns of cortico-cortical connections related to tonotopic maps in cat auditory cortex. J. Comp. Neurol. **192:** 293–332.
44. LUETHKE, L.E., L.A. KRUBITZER & J.H. KAAS. 1988. Cortical connections of electrophysiologically and architectonically defined subdivisions of auditory cortex in squirrels. J. Comp. Neurol. **268:** 181–203.
45. REITE, M., P. TEALE, J. ZIMMERMAN, *et al.* 1988 Source location of a 50 Msec latency auditory evoked field component. EEG Clin. Neurophysiol. **70:** 490–498.
46. MAKELA, J.P., M. HAMALAINEN, R. HARI & L. MCEVOY. 1994. Whole-head mapping of middle-latency auditory evoked magnetic field. EEG Clin. Neurophysiol. **92:** 414–422.
47. SCHREINER, C.E. & M.S. CYNADER. 1984. Basic functional organization of second auditory cortical field (AII) of the cat. J. Neurophysiol. **51:** 1284–1305.
48. GOLDSTEIN, M.H., J.L. HALL & B.O. BUTTERFIELD. 1968. Single unit activity in the primary auditory cortex of unanesthetized cats. J. Acoust. Soc. Am. **43:** 444–455.
49. RAUSCHECKER, J.P., B.TIAN, *et al.* 1997. Serial and parallel processing in rhesus monkey auditory cortex. Science **268:** 111–114.
50. LIEGEOIS-CHAUVEL, C., J. DE GRAAF, V. LAGUITTON, *et al.* 1999. Specialization of left auditory cortex for speech perception in man depends on temporal coding. Cereb. Cortex **9:** 484–496.
51. LANGNER, G., M. SAMS, *et al.* 1997. Frequency and periodicity are represented in orthogonal maps in the human auditory cortex evidence from magnetoencephalography. J. Comp. Physiol. **181 (6):** 665–676.
52. ROBIN, D.A., D. TRANEL & H. DAMASIO. 1990. Auditory perception of temporal and spectral events in patients with focal left and right cerebral lesions. Brain Lang. **39:** 539–555.
53. PANDYA, D.N. & H.G.J.M. KUYPERS. 1969. Cortico-cortical connections in the rhesus monkey. Brain Res. **13:** 13–36.
54. PANDYA, D.N. & B. SELTZER. 1982. Association areas of the cerebral cortex. Trends Neurosci. **5:** 386–390.
55. ZATORRE, R.J. 1988. Pitch perception of complex tones and human temporal-lobe function. J. Acoust. Soc. Am. **84:** 566–572.
56. TALLAL, P. & P. NEWCOMBE. 1978. Impairment of auditory perception and language comprehension in dysphasia. Brain Lang. **5:** 137–152.

# The Neural Processing of Complex Sounds

TIMOTHY D. GRIFFITHS

*Departments of Neurology and Physiological Sciences, Newcastle University Medical School, Newcastle-upon-Tyne NE2 4HH, United Kingdom*

**ABSTRACT: This paper considers the temporal processing of complex sounds relevant to musical analysis. Functional imaging studies, using positron emission tomography (PET), functional magnetic resonance imaging (fMRI), and magnetoencephalography (MEG), and the psychophysical assessment of patients with lesions allow two different approaches to this. Functional imaging allows the determination of structures *normally involved* in temporal analysis, while patient studies allow inference about the *necessary* structures for temporal analysis. Both approaches suggest a hierarchal organization in the brain corresponding to the processing of music. The features of individual notes are analyzed in the pathway up to and including the auditory cortices, while higher-order patterns formed by those features are analyzed by distributed networks in the temporal lobe and frontal lobes distinct from the auditory cortices.**

**KEYWORDS: Neural processing; Complex sounds; Musical analysis; Functional imaging**

## INTRODUCTION

This paper might have been called "The Neural Processing of Complex Sound *Features*." *All* sound is processed by mechanisms for the analysis of simple acoustic features (intensity, frequency, onset), complex acoustic features (such as patterns of these simple features as a function of time), and semantic features (learned association of sound patterns and meanings). Here, I concentrate on the analysis of complex features likely to be important in musical analysis, below the level of semantic and affective processing. Such analysis is represented in the middle box in FIGURE 1.

For the analysis of music, analysis of temporal features is likely to be particularly important, although I would certainly not dismiss the role of spectral or spatial analysis. Temporal analysis can be considered at different levels that I will call fine temporal structure and higher-order temporal structure. Such a categorical distinction works well for music as a sound with a highly segmented structure. Fine temporal structure corresponds to temporal structure at the level of milliseconds or tens of milliseconds; this is a "window" for the processing of the temporal regularity within individual notes that is relevant to complex pitch perception (see Refs. 1 and 2 for discussion of mechanisms of pitch perception). Higher-order temporal structure cor-

Address for correspondence: Timothy D. Griffiths, Departments of Neurology and Physiological Sciences, Newcastle University Medical School, Framlington Place, Newcastle-upon-Tyne NE2 4HH, UK.Voice: +44 (191) 222-6953; fax: +44 (191) 222-6706.

t.d.griffiths@ncl.ac.uk (e-mail preferred for all correspondence.)

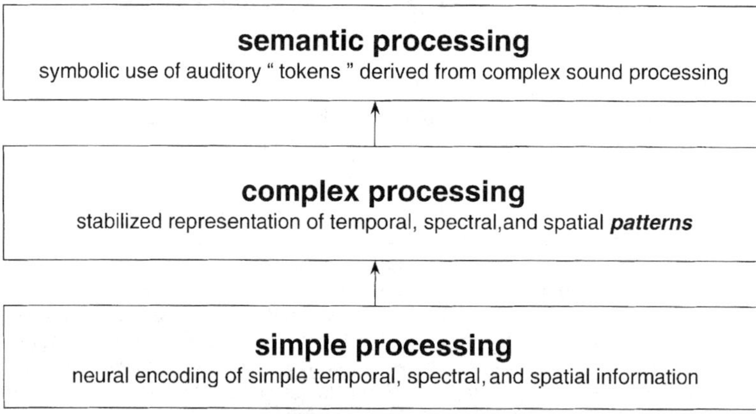


**FIGURE 1.** Modular representation of complex sound perception relevant to music.

responds to temporal structure at the level of patterns of the pitch, onset time, or duration of notes. This corresponds to a "window" at the level of hundreds of milliseconds or seconds. Higher-order structure might itself be subdivided into "local patterns" (for example, single intervals) and "global patterns" (for example, contour) as used by Peretz and others.[3,4] For present purposes, higher-order structure simply means any structure formed by patterns of the features in individual notes. This categorical distinction of temporal structure is convenient when considering processing relevant to music. This is not the only way of considering the temporal processing of sound. In particular, I would point out approaches based on the modulation transfer function. Here, responses to modulation of the sound amplitude or frequency are considered as *continuous* functions of modulation rate. This approach has been used in human psychophysical work,[5–8] evoked-potential work,[9,10] and functional imaging work.[11,12]

In the human auditory system much work on mechanisms of temporal processing, particularly in the case of music, has focused on the auditory cortex and cortical networks involving the auditory cortices. However, the ascending auditory pathway (FIG. 2) affords an extensive mechanism for the processing of complex signals before the cortex is reached. Subcortical mechanisms for the processing of temporally complex sound in the form of continuous modulation have been investigated in animals[13–15] and in man.[11] Little work has been carried out on the processing of segmented sound closer to music in the ascending pathway, especially in man. Because of the extensive cross-connections in the auditory pathway, bilateral lesions of the ascending pathway are needed to produce temporal deficits. Such lesions are rarely compatible with life, but do occur, for example, in multiple sclerosis.[16] Imaging of human auditory brain stem processing is technically limited in functional imaging techniques, including positron emission tomography (PET), functional magnetic resonance imaging (fMRI) and magnetoencephalography (MEG). I point this out primarily to declare an open mind about the existence of ascending processing systems for sound sequences such as music. The concentration in this paper on cortical

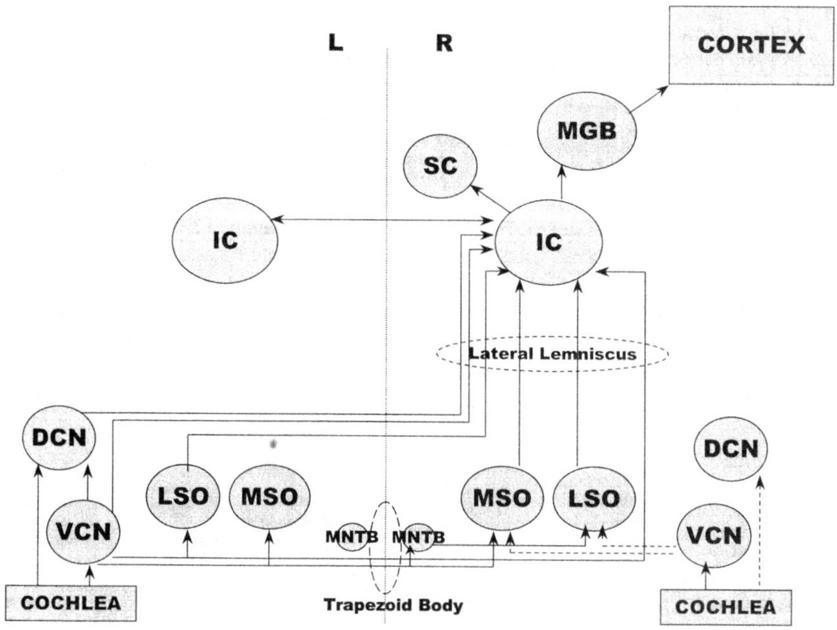


**FIGURE 2.** Simplified representation of ascending auditory pathway to the auditory cortex. This is based on animal and human studies (e.g., Refs. 18 and 34) and demonstrates extensive processing of auditory information before the cortex is reached. For clarity, only the convergence of the input from both ears at the right-sided brain stem auditory nucleii is shown. A dorsal pathway, via the dorsal cochlear nucleus (DCN) and inferior colliculus (IC) is shown, which is involved in animals in the analysis of spectrally complex waveforms. The diagram also shows a ventral pathway, via the ventral cochlear nucleus (VCN), trapezoid body and medial superior olive (MSO), thought to be important in animals for the analysis of spatial information. LSO = lateral superior olive, MNTB = medial nucleus of the trapezoid body, MGB = medial geniculate body, SC = superior colliculus.

mechanisms simply reflects a lack of data, rather than any a priori limitation on such processing in the ascending pathway. Another area of future interest is cortical "molding" of auditory processing in the ascending pathway that has been demonstrated in other species such as the bat.[17] The human auditory system[18] also contains a rich descending system that could mold behaviorally significant sounds such as music.

This paper considers temporal processing in humans relevant to musical processing. I will consider functional imaging (using PET, fMRI, and MEG) and the psychophysical assessment of patients with lesions. Functional imaging allows the determination of structures *normally involved* in temporal analysis, and patient studies allow inference about the *necessary* structures for temporal analysis. For discussion of the extension of this approach above the level of acoustic pattern analysis, I refer the reader to the paper by Zatorre in this volume.

## LESION STUDIES

### *Fine Temporal Structure*

There have been a number of studies suggesting that the detection of sounds of different frequencies (as measured in the routine pure tone audiogram) depends on an intact ascending auditory pathway to at least one of the auditory cortices. The rare condition of cortical deafness occurs following lesions to both superior temporal lobes (see Ref. 19 for review). The perception of pitch, as opposed to the detection of a pure tone at a given frequency, has been investigated using complex stimuli where the pitch is not immediately determined by the spectral characteristics of the sound. Zatorre[20] used a stimulus where the pitch is determined by the "missing fundamental" of the spectral representation. The stimulus in this experiment, considered in the frequency domain, is a harmonic series where the lower harmonics including the fundamental stimulus are removed. The pitch of this stimulus is determined by the missing fundamental, against the idea that for a pitch to be heard there needs to be a representation of that pitch in the stimulus spectrum. The determination of pitch by any frequency domain model requires a form of template fitting;[21] the spectrum still contains harmonics that are *spaced* by an amount equal to the fundamental frequency. Alternatively, the pitch of this stimulus might be explained by temporal domain models; the waveform of this stimulus has a periodicity equal to the reciprocal of the fundamental, both before and after removal of the lower harmonics. Using the missing fundamental stimulus Zatorre investigated the perception of this complex pitch in subjects with well-characterized lesions affecting the superior temporal lobes. Subjects with right lesions involving the primary auditory cortex had deficits in perception of this pitch, but not subjects with right superior temporal lobe lesions sparing the primary auditory cortex, or subjects with lesions of the left primary auditory cortex. This suggests a particular involvement of the right primary auditory cortex in complex pitch perception.

Patterson, Yost, and others have developed stimuli containing regular intervals where the pitch properties are most parsimoniously explained by the fine temporal rather than the spectral characteristics of the sound.[22,23] Of course, both characteristics are interchangeable at the stimulus level, but these stimuli are manipulated in such a way that spectral template models are difficult to apply because of the filtering properties of the auditory system. Iterated rippled noise (IRN) is one such type of regular interval sound produced by taking bandpass noise and adding it to itself in a delay-and-add cycle. With increasing delay-and-add cycles (or gain in each cycle) the stimulus becomes more regular, as shown by the increasing first peak in the autocorrelation spectrum. In parallel with this temporal change, a pitch emerges from the noise with a signal-to-noise ratio that parallels the degree of temporal regularity of the auditory signal. In contrast, the spectrum of the auditory signal retains its bandpass properties. It is therefore easier to infer in the case of this signal (than for the missing fundamental) that changes in perceived properties are due to the fine temporal structure rather than the spectral structure in the auditory signal (see Ref. 2 for more detailed discussion). Using this stimulus I investigated one of Peretz's patients, I.R., with apperceptive amusia following bilateral temporal lobe lesions[24] (see the paper by Peretz, this volume, for further discussion of amusia). This patient has normal appreciation of pitch but is unable to perceive musical sequences. She has an intact pathway up to and including the right primary auditory cortex, but effective "isolation" of that cortex from other areas of cortex that are lesioned. Testing this patient with IRN showed that she had a normal threshold for the detection of this stimulus presented to both ears. Testing each ear separately with contralateral masking noise showed that her normal performance was entirely accounted for by the performance at the left ear, the predominant input to the intact right auditory cortex.

Taken together these two studies are consistent with the pathway up to and including the primary auditory cortex being a necessary and sufficient substrate for the perception of complex pitch. I have discussed these studies under the heading fine temporal structure; both studies can be interpreted in terms of a need for primary cortex (particularly on the right) to perceive fine temporal structure, but the missing fundamental work is equally consistent with a spectral template model. These data would be consistent with activity of the human primary cortex being a neural correlate of the conscious perception of complex pitch.[25] Such processing is important for the perception of the pitch of the individual notes produced by most musical instruments.

### *Higher-Order Temporal Structure*

Lesion studies can also allow inference about the detection of higher-order temporal patterns of segmented sounds, including music. A number of studies have looked at very simple tests of segmented sound processing in patients with auditory agnosias including amusia (see Ref. 19 for review). The tests used include gap detection and click-fusion threshold. Many studies have shown deficits, and been interpreted in terms of a temporal processing deficit, but conventional gap detection paradigms are confounded by the presence of spectral artifact in the stimulus (this is overcome in more refined paradigms such as those developed by Phillips[26]). Many of the patients in these studies had large bilateral lesions involving the superior temporal cortices, consistent with a role for the secondary and association areas of auditory cortex in this region being needed for the perception of this type of temporal pattern.

A deficiency in the evaluation of patients with auditory agnosia is the lack of a systematic battery of tests of sound sequence processing above the level of gap detection but below the level of actual musical sequences. Such a battery would have the potential to demonstrate "generic" sound sequence mechanisms that might be relevant to a number of different sound processing disorders without higher-level confounds. A crude test of sequence discrimination in a patient, H.V., with amusia[27] suggested a loss of the perception of *rapid* but not slow sequences. This paralleled the phenomenology in a patient who had lost the ability to recognize and enjoy music with a high tempo. That patient had a unilateral lesion affecting the posterior superior temporal lobe and inferior parietal lobe, and the deficit in his case was less profound than in other cases reported following bilateral lesions.

Auditory agnosia can also be studied by a parallel approach based on the modulation transfer function as discussed earlier. In terms of the temporal "window," the level of higher-order structure for patterns of notes (100s of milliseconds or seconds) parallels the level of low modulation rates (of tens of Hz or less). The patient with mild amusia discussed above (H.V.[27]) had a mild deficit in the detection of sinusoidal frequency modulation (FM) and amplitude modulation (AM) at a rate of 2 Hz.

The dissociation between this and his profound deficit in the discrimination of rapid sequences of segmented sound was striking, and cautions against any universal approach in these patients based on modulation characterization. In contrast, the patient I.R., with a more marked amusia due to bilateral lesions, had a striking deficit in the detection of both AM and FM, particularly at low modulation rates (her threshold for 2-Hz sinusoidal AM was 11 times normal). Apart from demonstrating auditory deficits linked to the phenomenology, her case allows the inference that a single intact pathway to the right auditory cortex is not a sufficient substrate for the normal perception of modulation. The presence of normal processing of IRN by I.R. provides an important control to allow the exclusion of other possible causes of the deficit. The IRN task demanded similar attention and working memory for simple acoustic features using a similar two-alternative forced-choice paradigm. The presence of such dissociated deficits is critical to these studies,[28] and studies showing a general deficit in performance in a range of tasks do not allow such interpretation in terms of distinct perceptual processes.

In terms of the psychophysical examination of higher-order temporal processing in subjects with lesions, these studies suggest that the deficits are produced by temporal lobe lesions that involve superior temporal lobe areas beyond the primary auditory cortex. Additionally, subject I.R. has a large right frontal lesion implicating this area in the perception of sound sequences and low-rate modulation. More studies of such patients using detailed modern structural imaging and systematic psychoacoustic testing will characterize the necessary structures for higher-order temporal processing. Such studies will never be supplanted by functional imaging, from which *sufficient* processing mechanisms cannot be immediately inferred.

## FUNCTIONAL IMAGING

### *Fine Temporal Structure*

The missing fundamental stimulus has been used in MEG experiments[29,30] where a mapping of perceived pitch has been demonstrated. For reasons discussed above, this map cannot be interpreted unambiguously in terms of the temporal structure of the stimulus, as template models could also explain the perceived pitch. But the mapping of perceived complex pitch to auditory cortex, derived by whatever means, is convergent with the lesion data discussed above. In the study of Langner, distinct representations of the frequency and temporal domains were suggested, with different orientations in the superior temporal planum. This would allow extensive processing in auditory cortex of spectrotemporal properties such as pitch as well as spectral (timbre) and temporal (attack) instrument characteristics. These are all features of individual notes.

IRN has also been used in functional imaging experiments, with the aim of identifying brain areas where the brain activity varies as a function of the fine temporal regularity of the stimulus.[2] In an initial study with PET in collaboration with Pattterson, such a relationship was demonstrated in both auditory cortices. On the right, a region of activation was demonstrated in the medial part of Heschl's gyrus, containing the primary auditory cortex. On the left, the activation was situated more laterally and may correspond to secondary auditory cortex; these asymmetries are of interest in view of the lesion data suggesting a particular involvement of the right primary auditory cortex in the processing of temporal regularity. The activity shown in the PET study was argued to correspond to neural activity occurring after temporal regularity is converted into a more stable code in the ascending pathway, a process called temporal integration. The study is critically dependent on this assumption, but not on the exact mechanism by which such temporal regularity is stabilized, the simplest possibility being based on autocorrelation as mentioned above. Models for this temporal integration mechanism[31] allow it to occur at an early stage in the auditory pathway, perhaps as early as the cochlear nucleus, and ongoing work is being carried out with fMRI to further delimit it to brain stem structures.

The work using IRN therefore also supports the existence of a representation of temporal regularity in the auditory cortex, with closer control of the auditory spectrum than in the case of the missing fundamental.

### *Higher-Order Temporal Structure*

In the IRN experiment,[2] higher-order structure was also investigated. The individual notes with pitches determined by temporal regularity were presented as two pitch patterns: as a regular pitch "staircase" and as novel diatonic melodies. We hypothesized that areas involved in the analysis of sound sequences would show greater activation as a function of the pitch strength of the individual notes for the melody condition than for the staircase condition. Sound sequence analysis represents an emergent property of the detection of the pitch of individual sounds, and areas involved in sequence analysis will show a more marked dependence on pitch strength for more complex pitch sequences like the melody condition. This was demonstrated in four areas, distinct from the primary auditory cortex, in both anterior temporal lobes near the poles and in the posterior superior temporal lobes. These areas are shown in FIGURE 3. A scheme is hypothesized, consistent with the data, where the temporal properties of the individual notes are extracted in the region of the primary auditory cortex before higher-order temporal properties of the stimulus sequence are analyzed in the distinct network shown in FIGURE 3.

A subsequent study used PET to demonstrate areas involved in the processing of simple sequences that are not musical.[32] Here, simple six-element atonal sequences were presented to nonmusical subjects in order to demonstrate areas involved in the analysis of both pitch sequences and sequences of notes of different duration. Pairs of sequences were presented to subjects where either the pitch sequence, the duration sequence, neither, or both were varied between the two. During any scan, subjects were required to detect either changes in pitch sequence or duration sequence. The results showed, again, a network of activation distinct from the primary auditory cortex (in addition to activation in the primary auditory cortex in the principal comparison between the sound conditions and rest). Bilateral activation was demonstrated in the posterior superior temporal planum in the region of the planum temporale, and in both frontal opercula, with greater spatial extent of the clusters of activity demonstrated on the right. Highly significant activation was also demonstrated in both lobes of the cerebellum. The activation during the pitch sequence and duration sequence tasks was strikingly similar, and direct comparison of the two conditions showed no significant differences in activation. This would be consistent with a common mechanism for the two tasks at some level, consistent with some work in the musical domain.

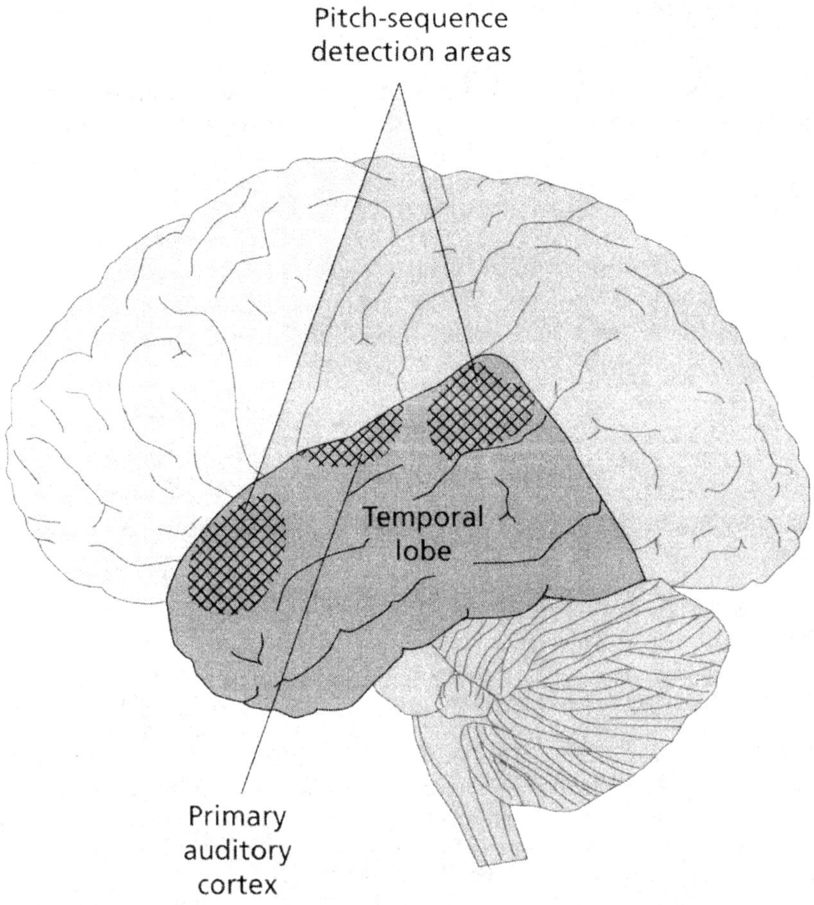


**FIGURE 3.** Hypothesized basis for hierarchal processing of temporal structure in the temporal cortices suggested by the experiment using iterated rippled noise. The temporal structure within individual notes is processed in the pathway up to and including the primary auditory cortex, while pitch sequences are processed in the distinct areas shown. (Artist: ©1998, Bob Crimi. Reprinted with permission.)

These studies using "submusical" stimuli, therefore, suggest the existence of networks for processing of higher-order structure in both temporal lobes that are distinct from the primary auditory cortices. The planum temporale is consistently activated in a number of studies, with less consistency in activation anterior to the primary auditory cortex. Activation anterior to the auditory cortex occurs in the anterior temporal lobe in the IRN study and frontal opercula in a number of other studies. The patterns of activation shown are similar to those demonstrated by studies using actual musical stimuli[33] (see also the paper by Zatorre, this volume), but show less striking lateralization. This suggests the possibility that the imposition of higher levels of cognitive processing in more musical studies contributes to the lateralization during musical processing.

## CONCLUSION

Both studies of humans with brain lesions and functional imaging provide convergent evidence for the existence of a neural substrate for the processing of sound sequences that is hierarchal in organization. The pathway up to and including the primary auditory cortices may provide a sufficient mechanism for the processing of spectrotemporal features of individual sounds. Subsequent processing of higher-order temporal patterns formed by those individual sounds depends on processing in bilateral temperofrontal networks. These networks provide a first stage in the neural processing of music. Outstanding issues relate to the relationship between the elements of these networks and the perceptual attributes of the stimulus, the type of subject, and the type of task, to name but three.

## REFERENCES

1. Hartmann, W.M. 1997. Signals, Sound, and Sensation. American Institute of Physics Press. New York.
2. Griffiths, T.D., C. Buechel, R.S.J. Frackowiak, *et al.* 1998. Analysis of temporal structure in sound by the human brain. Nat. Neurosci. **1:** 421–427.
3. Peretz, I. 1990. Processing of local and global musical information by unilateral brain-damaged patients. Brain **113:** 1185–1205.
4. Liegeois-Chauvel, C., I. Peretz, M. Babai, *et al.* 1998. Contribution of different cortical areas in the temporal lobes to music processing. Brain **121:** 1853–1867.
5. Zwicker, E. 1952. Die Grenzen der Horbarkeit der Ampiltudenmodulation und der Frequenzmodulation eines Tones. Acustica Akust. Beih. **3:** 125–133.
6. Kay, R. 1982. Hearing of modulation in sounds. Physiol. Rev. **62:** 894–975.
7. Moore, B.C.J. & A. Sek. 1995. Effects of carrier frequency, modulation rate, and modulation waveform on the detection of modulation and the discrimination of modulation type (amplitude modulation versus frequency modulation). J. Acoust. Soc. Am. **97:** 2468–2478.
8. Wojczak, M. & N.F. Viemeister. 1999. Adaptation produced by amplitude modulation. Abstr. 59. Association for Research in Otolaryngology. St. Petersburg, FL.
9. Rees, A., G.G.R. Green & R.H. Kay. 1986. Steady-state evoked responses to sinusoidally amplitude-modulated sounds recorded in man. Hear. Res. **23:** 123–133.
10. Stefanatos, G.A., G.G.R. Green & G.G. Ratcliff. 1989. Neurophysiological evidence of auditory channel anomalies in developmental dysphasia. Arch. Neurol. **46:** 871–875.
11. Harms, M.P., J.R. Melcher & R. Weisskoff. 1998. Time courses of fMRI signals in the inferior colliculus, medial geniculate body, and auditory cortex show different dependencies on noise burst rate [abstract]. Neuroimage **7:** S365.
12. Giraud, A.L., C. Lorenzi, J. Wable, *et al.* 1999. Temporal envelope representation in the human auditory cortex [abstract]. Neuroimage **9:** S787.
13. Moller, A.R. 1974. Response of units in the cochlear nucleus to sinusoidally amplitude modulated tones. Exp. Neurol. **45:** 104–117.
14. Rees, A. & A.R. Moller. 1983. Responses of neurons in the inferior colliculus of the rat to AM and FM tones. Hear. Res. **10:** 301–330.
15. Langner, G. & C.E. Schreiner. 1988. Periodicity coding in the inferior colliculus of the cat. I. Neuronal mechanisms. J. Neurophysiol. **60:** 1799–1822.
16. Quine, D.B., D. Regan, K.I. Beverly, *et al.* 1984. Patients with multiple sclerosis experience hearing loss for shifts of tone frequency. Arch. Neurol. **41:** 506–507.

17. YAN, W. & N. SUGA. 1998. Corticofugal modulation of the midbrain frequency map in the bat auditory system. Nat. Neurosci. **1:** 54–85.
18. MOORE, J.K. 1994. The human brainstem auditory pathway. *In* Neurotology. R.K. Jackier & D.E. Brackmann, Eds.: 1–17. Mosby. St. Louis, MO.
19. GRIFFITHS, T.D., A. REES & G.G.R. GREEN. 1999. Disorders of human complex sound processing. Neurocase **5:** 365–378.
20. ZATORRE, R. 1988. Pitch perception of complex tones and human cerebral lobe function. J. Acoust. Soc. Am. **84:** 566–572.
21. COHEN, M.A., S. GROSSBERG & L. WYSE. 1995. A spectral network model of pitch perception. J. Acoust. Soc. Am. **98:** 862–879.
22. PATTERSON, R.D., S. HANDEL, W.A. YOST, *et al.* 1996. The relative strength of the tone and the noise components in iterated rippled noise. J. Acoust. Soc. Am. **100:** 3286–3294.
23. YOST, W.A., R. PATTERSON & S. SHEFT. 1996. A time domain description for the pitch strength of iterated rippled noise. J. Acoust. Soc. Am. **99:** 1066–1078.
24. GRIFFITHS, T.D., V. PENHUNE, I. PERETZ, *et al.* 2000. Frontal processing and auditory perception. NeuroReport **11:** 919–922.
25. FRITH, C., R. PERRY & E. LUMER. 1999. The neural correlates of conscious experience: an experimental framework. Trends Cognit. Sci. **3:** 105–114.
26. PHILLIPS, D.P., T.L. TAYLOR, S.E. HALL, *et al.* 1997. Detection of silent intervals between noises activating different perceptual channels: some properties of "central" auditory gap detection. J. Acoust. Soc. Am. **101:** 3694–3705.
27. GRIFFITHS, T.D., A. REES, C. WITTON, *et al.* 1997. Spatial and temporal auditory processing deficits following right hemisphere infarction: a psychophysical study. Brain **120:** 785–794.
28. ELLIS, A. & A.W. YOUNG. 1998. Human Cognitive Neuropsychology. Lawrence Erlbaum Associates. Hove, Sussex, UK.
29. PANTEV, C., M. HOKE, B. LUTKENHONER, *et al.* 1989. Tonotopic organisation of the auditory cortex: pitch versus frequency representation. Science **242:** 486–488.
30. LANGNER, G., M. SAMS, P. HEIL, *et al.* 1997. Frequency and periodicity are represented in orthogonal maps in the human auditory cortex: evidence from magnetoencephalography. J. Comp. Physiol. **181:** 665–676.
31. PATTERSON, R.D., M.H. ALLERHAND & C. GIGUERRE. 1995. Time-domain modeling of peripheral auditory processing: a modular architecture and a software platform. J. Acoust. Soc. Am. **98:** 1890–1894.
32. GRIFFITHS, T.D., I. JOHNSRUDE, J.L. DEAN, *et al.* 1999. A common neural substrate for the analysis of pitch and duration pattern in segmented sound? Neuroreport **18:** 3825–3830.
33. ZATORRE, R.J., A.C. EVANS & E. MEYER. 1994. Neural mechanisms underlying melodic perception and memory for pitch. J. Neurosci. **14:** 1908–1919.
34. IRVINE, D.R.F. 1986. The Auditory Brainstem. A Review of the Structure and Function of the Auditory Brainstem Processing Mechanisms. Springer-Verlag. Berlin.

# Music and the Neurologist

## A Historical Perspective

JOHN C. M. BRUST

*Department of Neurology, Harlem Hospital Center and Columbia University College of Physicians and Surgeons, New York, New York 10037, USA*

**ABSTRACT: Neurological disorders affecting musical function can produce either positive or negative symptoms. Positive phenomena include musicogenic epilepsy (seizures triggered by music), musical partial seizures (hallucinated music as the expression of the seizure), musical release hallucinations (nonepileptic musical hallucinations, usually associated with impaired hearing), and synesthesia (hallucinated colors triggered by musical tones). Negative phenomena comprise the amusias, which can be receptive, expressive, or both, and can selectively involve particular components of musical processing, including pitch, interval, contour, rhythm, meter, timbre, and emotional response. Amusia is often accompanied by aphasia, but each can occur in the absence of the other. Neurological disorders provide evidence that musical processing is multimodal and widely distributed in both cerebral hemispheres.**

**KEYWORDS: Musicogenic seizures; Musical partial seizures; Release hallucinations; Amusia; Aphasia**

## INTRODUCTION

In nearly all right-handed people the left cerebral hemisphere is "dominant" for propositional or analytic processing (exemplified by language), whereas the right hemisphere is more responsible for appositional or gestalt processing (exemplified by spatial manipulation and aspects of emotional expression or response). How the brain processes music has intrigued neurologists and psychologists for over a century. A straightforward answer would not be expected, for the components of music—for example, pitch, timbre, duration, loudness, and rhythm—are likely processed through separate circuits. Music can be both linear (e.g., melody) and nonlinear (e.g., chords), and different kinds of music make different intellectual and emotional demands on different kinds of listeners or performers.

The biological survival value of music remains an enigma. Darwin believed that musical "calling systems" evolved into speech,[1] whereas Herbert Spencer believed that music evolved as a stylized form of speech.[2] Steven Pinker notes, "Of [all] mental faculties...music...shows the clearest signs of not being [adaptive],"[3] yet in all

Address for correspondence: John C.M. Brust, M.D., Department of Neurology, Columbia University College of Physicians and Surgeons, Harlem Hospital Center, New York, NY 10037. Voice: 212-939-4244; fax: 212-939-4245.
jcb2@columbia.edu

cultures most people enjoy music. On the other hand, it takes severe mental retardation to prevent language acquisition, whereas musical aptitude ranges from tone deafness to Mozartian genius independently of other measures of intelligence.[4] Estimates of absolute pitch in the general population range from 1 in 1500 to 1 in 10,000.[5]

Clues as to how the brain processes music come from clinical case reports of patients with deranged musical function, from psychological testing of either normal or brain-damaged subjects, and, more recently, from studies using functional imaging (positron emission tomography [PET] and functional magnetic resonance imaging [fMRI]). Neurological symptoms can be positive or negative. Positive phenomena include musicogenic epilepsy, musical partial seizures, musical release hallucinations, and synesthesia. Negative phenomena include the amusias.

## POSITIVE PHENOMENA

Musicogenic seizures are triggered by hearing music. Attacks typically begin with subjective distress, agitation, tachycardia, and tachypnea and may or may not generalize to grand mal. Victims are often musically talented but do not have to be. The stimulus can be quite specific—for example, classical versus popular music, or a particular piece. Reported triggers include the second movement of Beethoven's Symphony No. 5 and a sequence of church bells with a frequency range of less than one octave.[6,7] In one case seizures occurred only if the patient actively listened to the music. In another, seizures could be precipitated by background music of which the patient was barely aware. Seizures have been triggered by the voices of particular singers.[8] A two-year-old boy's seizures were precipitated by his own singing.[9] In another case seizures were triggered by playing "Now Thank We All Our God" on the organ but not by hearing it, singing it, silently reading it, or silently playing the notes on the keyboard.[10] Lesions causing musicogenic epilepsy are usually in the temporal lobe, right more often than left.[11]

Musical partial seizures consist of hallucinated music, which can be of any type but tends to be stereotypic for a particular epileptic subject. Such patients are less likely than those with musicogenic epilepsy to be musically talented. In a reported series of 666 temporal lobe epileptics, 16% had auditory hallucinations, the great majority of which were unformed (e.g., banging, ticking).[12] Formed hallucinations could be verbal, musical, or both. In a more detailed description of hallucinatory seizures precipitated by cortical electrical stimulation, subjective phenomena included a male chorus and orchestra singing "White Christmas," a radio jingle, organ music, and an orchestra playing in a music hall, which was simultaneously visually hallucinated.[13] Seizures most predictably followed stimulation of the superior temporal gyrus, right more often than left. Interestingly, spoken words also followed either left or right hemispheric stimulation, and there was no topographical separation between areas precipitating voices and those precipitating music.

Visual or auditory hallucinations associated with impaired vision or deafness ("Charles Bonnet syndrome," release hallucinations) can consist of music; they most often affect elderly people with presbycusis. Neither dementia nor psychiatric disease need be present, and the hallucinations are often pleasurable rather than frightening or threatening. Some subjects sing along with the music, and some report that they can switch to another tune by consciously thinking about it. In some cases the pleasurable aspects of the music deteriorate over time, becoming distorted or too loud, or degenerating into tinnitus.[14,15] The pathophysiological basis of such hallucinations is uncertain; plausibly, reduced sensory input results in disinhibition of perception-bearing circuits, and perceptual traces are "released."

Synesthesia is an involuntary perception produced by stimulation of another sense. Thus, sounds produce the perception of colors. Geometric patters can occur, but more elaborate formed hallucinations do not. In some cases particular pitches produce the same color; for example, C produced red, D green, and E blue in a boy with poor vision and musical talent (but not absolute pitch).[16] In addition to deafferentation, synesthesia is associated with hallucinogenic drug use (e.g., LSD) and, in one case, with a brain stem neoplasm. Many cases are idiopathic and noted in early childhood; some are familial. Women and left-handers are overrepresented, as are people who have difficulties with spatial relations.[17] Alexander Scriabin and Nicholai Rimsky-Korsakoff are believed to have had colored-hearing synesthesia. The disorder, if it can be called that, is an example of unusually strong cross-modal association, but the physiological basis is otherwise unclear.

## NEGATIVE PHENOMENA: AMUSIA AND APHASIA

Amusia is an acquired impairment of musical processing. As with aphasia—a comparable impairment of language—the problem can be expressive, receptive, or both, and the disorder cannot be explained by damage either to the articulatory apparatus or to primary receptor mechanisms. In other words, tongue paralysis is not expressive amusia, and deafness is not receptive amusia. Amusia and aphasia can occur together or each can occur in the absence of the other.

Since the nineteenth century, clinicians investigating the amusias have tended to adopt either a localizationist/anatomical or a holistic/psychological approach. Among the localizers, Henschen classified amusia as either motor or sensory and then identified "centers" in the left cerebral hemisphere: singing resided in the pars triangularis of the third frontal convolution, musical reception in the temporal pole, note reading near the angular gyrus, and instrument playing at the foot of the second frontal convolution (with a separate center for violinists).[18,19] Similarly, Kleist described motor and sensory amusia, each affecting either tone or melody. His anatomical localization included separate centers for singing and for whistling.[20]

Opposed to such rigid diagramming was Feuchtwanger, who considered music too complex to localize or even to restrict to one cerebral hemisphere.[21] Ustvedt noted the heterogeneity of amusic patients and the lack, in most reports, of premorbid baseline information or of standardized tests.[22] A review of such reports reveals that heterogeneity is indeed the rule and that the presence or absence of aphasia does not predict the type of amusia.

The first report of aphasia and relatively preserved musical ability was in 1745; the patient, who had right hemiparesis and speech limited to the word "yes," could sing hymns if someone else sang along.[23] Similar cases include two children with nonfluent aphasia, one of whom could sing with words and the other without words;[24] a right-handed man with largely expressive aphasia following left hemispherectomy who could sing familiar songs with some errors in articulation but with little hesitation;[25] and a woman with transcortical aphasia (her verbal expression was limited to a few recurrent phrases, such as "Hi, Daddy," but she had compulsively preserved speech repetition with echolalia) who could sing both old and newly taught songs.[26] Of 24 patients with Broca's aphasia and severely restricted speech output, 21 could sing (6 excellently), some sang only without words, some sang with paraphasic errors, and some needed help to get started.[27] A musically naïve patient with transcortical aphasia had hyperprosodic speech, whistled much of the time (including responses to questions), and listened to music much more often than before his illness.[28]

A professional pianist with Wernicke's aphasia, including alexia and agraphia for language, continued to play publicly; he could play unfamiliar melodies sung to him; reproduce heard chords; recognize dynamics, instruments, and intentional errors; and read a score nearly correctly.[29] A music professor with aphasia that included impaired speech comprehension and severe agraphia continued to write music.[30] The composer Shebalin developed Wernicke's aphasia with alexia and agraphia yet continued composing pieces "which other musicians considered to be up to standard," including songs, quartets, choruses, and a symphony.[31] A conductor with global aphasia and ideomotor apraxia continued to conduct his orchestra.[32] A blind organist and composer who had written and read music in Braille developed Wernicke's aphasia and could no longer read or write language in Braille; yet he continued to read and write music in Braille.[33]

Cases of amusia plus aphasia are similarly diverse. A patient with anomia heard music as dissonant and voices, including his own, as too high pitched; he could neither sing nor reproduce rhythmic patterns.[34] Following surgery for a right frontal glioma, an accordionist had left hemiparesis and mild aphasia. Thereafter he sang "wrong" and could no longer play his instrument (although he held it properly); he could correctly sing notes sung to him, but hearing them played on his accordion, he sang them a fourth too high or too low, and he could not imitate rhythm unless it was bound to a melody.[35] A violinist with absolute pitch had a stroke with right hemiparesis and Wernicke's aphasia, after which he had difficulty recognizing intervals, pitches, and rhythmic patterns; imitating heard rhythm; singing a melody previously unknown to him; and writing music spontaneously or to dictation.[36]

A man with probable Broca's aphasia was unable to identify popular songs or to sing melodies, but he could sing single tones and sequences without rhythm; rhythm recognition and production were completely lost whether presented visually, aurally, or tactily.[37] A musician with primary progressive aphasia, unable any longer to read music, could improvise but did so repetitively and without completion; he could not write or copy music or reproduce rhythm, but he could reproduce fragments of melodies with a sense of meter.[38] A woman with computerized tomographic evidence of right perisylvian atrophy developed aprosodic speech and could not recognize the prosody (e.g., sad, glad, mad) of others; she could no longer sing even single notes or identify melodies, although she could tell which of two notes was higher, and she could read and write music.[39]

Maurice Ravel had progressive aphasia with alexia, agraphia, and ideomotor apraxia; "musical thinking" was said to be intact—he recognized melodies and errors of rhythm and pitch and continued to listen to music critically—but he could no longer write or dictate music and thus could not compose the pieces he heard "in his head."[40,41] Following excision of a meningioma beneath the left temporal lobe, a young voice student had transcortical aphasia, alexia, and agraphia; she also had musical alexia and agraphia, yet she could still identify pitch, interval, melody, harmony, timbre, rhythm, and meter, and in a singing recital that included works in five languages, there were no apparent musical errors (although there may have been occasional paraphasic errors in the foreign languages).[42]

Cases of amusia without language disturbance are also heterogeneous; not surprisingly, in right-handers the lesions most often involve the right cerebral hemisphere. Reported symptoms include difficulty recognizing sounds as musical,[43] loss of a sense of rhythm,[44] hearing musical sounds "out of tune,"[45] and hearing both voices and music as monotonal.[46] Following a right temporoparietal infarct a man could no longer play the organ, yet other musical capabilities such as singing and imitating rhythm were intact ("expressive instrumental amusia").[47] Following a right temporal lobe infarct a man lost appreciation of timbre, including nonmusical sounds and people's voices; music sounded "low and dull" with a kind of echo overlapping the sounds, and although he could recognize a guitar, a trumpet, or a drum, he could not recognize a violin, a piano, or an organ.[48]

A composer with right frontoparietotemporal infarction lost his emotional response to music and could no longer compose meaningfully; he could still compose serial music, "...but I have never been awed by the musical content of this style of composition."[49] A man with a right temporoparietal vascular malformation had "difficulty understanding the nuances of words and the inflections of sentences" and in understanding music; the relationship between a soloist and an accompanist became blurred, and there was a loss of aesthetic pleasure, yet there was no auditory agnosia, and he correctly recognized pitch, melody, timbre, harmony, and rhythm.[50]

A piano teacher with bilateral posterior cerebral atrophy of uncertain cause developed object-use apraxia and could no longer write or play the piano with two hands; her playing and singing were melodically disorganized and discontinuous, with poor rhythm and meter, and she read music note for note, with loss of "global processing of pitch sequences."[38] Following surgical clipping of bilateral middle cerebral artery aneurysms, a woman could no longer recognize or hum well-known tunes, and although she could tell if piano tones were the same or different, she could not tell if melodies were the same or different, whether melodic contour was preserved or violated, or whether a melody ended on the tonic or a nontonic; rhythm recognition and production were intact, however.[51]

## STUDIES OF BRAIN-DAMAGED POPULATIONS

Studies involving groups of patients with either left or right cerebral hemispheric disruption have also been inconsistent. Subjects receiving right carotid artery amobarbital injection could not sing properly, with pitch more distorted than rhythm, but they recognized their errors and identified songs sung to them.[52] Patients who had undergone either left or right temporal lobectomy were studied using the Seashore Test of Musical Abilities, which measures discrimination of pitch, loudness, rhythm, duration, timbre, and tonal memory; right but not left lobectomy resulted in impaired recognition of timbre and tonal memory and a lesser degree of impairment involving pitch and rhythm.[53] A different testing instrument, the Musical Aptitude Profile, which involves tonal imagery (melody and harmony), rhythm (tempo and meter), and "musical sensitivity" (phrasing, balance, and style) was used to study subjects with no musical training who had undergone either left or right temporal lobectomy; right-sided but not left-sided surgery resulted in impairment involving rhythm and sensitivity, not tonal imagery.[54]

Patients with either left or right temporal lobectomy were presented with spoken words, tunes without words, and words and tunes combined as an unfamiliar song. Left lobectomy resulted in impaired recognition of words alone, and right lobectomy resulted in impaired recognition of tunes alone, but either left or right lobectomy resulted in impaired recognition of melody combined with words.[55]

Several studies have compared local and global aspects of musical information processing in brain-damaged subjects. (For melody, pitch intervals are local, whereas melodic contour is global; for temporal patterns, rhythm is local, whereas meter is global.) In one study of patients with either right or left cerebral stroke, right hemispheric damage produced difficulty recognizing changes in either interval or contour, whereas left hemispheric damage produced difficulty only with intervals; either right or left hemispheric damage produced difficulty with rhythm but not with meter.[56] By contrast, although a similar study of patients following right or left cerebral corticectomy found comparable abnormalities involving interval and contour, either right or left hemispheric damage produced difficulty with meter, not rhythm.[57] In another study of patients with cerebral strokes, of eight patients with right-sided damage, five had impaired musical processing; of 12 patients with left-sided damage, nine had impaired musical processing. Right hemispheric damage caused abnormal processing of both contour and interval and of both meter and rhythm, whereas left hemispheric damage caused much more variable abnormalities: contour and interval could be affected alone or in combination, and meter and rhythm could be affected alone or in combination.[58]

Normal subjects recognize pitch even when the fundamental is missing if overtones are retained. This ability is lost following right temporal lobe damage that includes Heschl's gyrus.[59] Another study assessed the ability to recognize timbre derived either from spectral envelopes or from onset-offset times, and it was found that right temporal lobectomy produced greater impairment in either ability than did left temporal lobectomy.[60]

Following right but not left temporal lobectomy, patients had difficulty reproducing auditory but not visual rhythm; lesions producing such impairment encroached on the primary auditory cortex.[61]

In normal subjects the harmonic context established by a prime chord influences judgments of target chord intonation; in patients who had undergone corpus callostomy, this normal interaction was preserved in the right cerebral hemisphere but not the left.[62]

## STUDIES OF NORMAL SUBJECTS

Paralleling clinical case reports and studies of brain-damaged subjects have been studies of musical processing in normal people. Space does not allow a detailed review of such investigations, the findings of which have often been as inconsistent as those of clinical reports. Early studies used the technique of dichotic listening, in which the two ears simultaneously receive different auditory information. In one such investigation musically naïve subjects appeared to have left ear advantage—that is, right cerebral hemispheric dominance—listening to melodies, whereas musically sophisticated subjects had right ear advantage; the findings were considered to reflect right hemispheric gestalt and emotional processing on the part of naïve listeners, in contrast to left hemispheric detailed analysis on the part of the musicians.[63] Other investigators, however, found no differences between musically sophisticated and musically naïve subjects,[64,65] and in some reports differences were found that seemed to depend on inherent musical aptitude,[66] age,[67] sex,[68] or right-ear advantage for higher pitch.[69,70] In one study the right hemisphere was used more for melody recognition when the tones had a complex timbre with rich overtones, again perhaps reflecting a right hemispheric advantage in dealing with simultaneous or gestalt, rather than linear or analytic, information.[71] It should be noted, however, that many investigators consider dichotic listening a highly unreliable technique for defining hemispheric functional "dominance."[72]

In normal subjects, functional imaging—positron emission tomography (PET) and functional magnetic resonance imaging (fMRI)—allows different components of musical processing to be assessed in real time. Such studies demonstrated that pitch recognition selectively activates that right prefrontal cortex, whereas active retention of pitch requires interaction of right frontal and right temporal cortices.[73,74] In another study, musical sight-reading, unlike reading words, activated the left occipitoparietal cortex, perhaps reflecting the dependence of musical reading on spatial location, in contrast to the dependence of word reading on feature detection.[75] In another study subjects listened to the same piece but were instructed to attend to different aspects of the music—for example, melody recognition, pitch, rhythm, or how many instruments were playing; even though the auditory information did not change with repeated trials, brain activation patterns were sharply altered by such changes in selective attention.[76]

As with brain-damaged patients, studies of normal subjects involve particular aspects of musical experience—auditory (pitch, harmony, timbre, interval, contour), visual (notational score, which includes real words; notes designated by letter but represented spatially, with duration indicated by appearance; purely musical symbols, such as rests, fermata, clef, crescendo, staccato, and repeat; and display of both simultaneous and sequential sounds), motor (performance, dance); memory; and emotion. Musical experience, however, is more than the sum of these parts. The multimodal components of musical processing cannot be sharply localized to one part of the brain or even to one cerebral hemisphere. The psychological whole that emerges from their interaction is even more widely distributed.

## REFERENCES

1. Darwin, C. 1871. The Descent of Man, and Selection in Relation to Sex. Appleton. New York.
2. Spencer, H. 1857/1951. The origin and function of music. *In* Literary Style and Music. Philosophical Library. New York.
3. Pinker, S. 1997. How the Mind Works. W.W. Norton & Co. New York.

4. GARDNER, H. 1983. Musical Intelligence. *In* Frames of Mind. The Theory of Multiple Intelligences: 99–127. Basic Books, Inc. New York.
5. PROFITA, J. & T.G. BIDDER. 1988. Perfect pitch. Am. J. Med. Genet. **29:** 763–771.
6. CRITCHLEY, M. 1937. Musicogenic epilepsy. Brain **60:** 13–27.
7. POSKANZER, C., E. BRAWN & H. MILLER. 1962. Musicogenic epilepsy caused by only a discrete frequency band of church bells. Brain **85:** 77–92.
8. BRIEN, S.E. & T.J. MURRAY. 1984. Musicogenic epilepsy. Can. Med. Assoc. J. **131:** 1255–1258.
9. HERSKOWITZ, J., N.P. ROSMAN & N. GESCHWIND. 1984. Seizures induced by singing and recitation. A unique form of reflex epilepsy in childhood. Arch. Neurol. **41:** 1102–1103.
10. WIESER, H.G., H. HUNGERBÜHLER, A.M. SIEGEL & A. BUCK. 1997. Musicogenic epilepsy: review of the literature and case report with ictal single photon emission computed tomography. Epilepsia **38:** 200–207.
11. SUTHERLING, W.W., L.M. HERSHMAN, J.Q. MILLER, *et al.* 1980. Seizures induced by playing music. Neurology **30:** 1001–1004.
12. CURRIE, S., K.W.G. HEATHFIELD, R.A. HENSON & D.F. SCOTT. 1971. Clinical course and prognosis of temporal lobe epilepsy. A survey of 666 patients. Brain **94:** 173–190.
13. PENFIELD, W. & P. PEROT. 1963. The brain's record of auditory and visual experience. Brain **86:** 595–696.
14. ROSS, E.D., P.B. JOSSMAN, B. BELL, *et al.* 1975. Musical hallucinations in deafness. J. Am. Med. Assoc. **231:** 620–621.
15. HAMMEKE, T.A., M.P. MCQUILLEN & B.A. COHEN. 1983. Musical hallucinations associated with acquired deafness. J. Neurol. Neurosurg. Psychiatry **46:** 570–572.
16. RIZZO, M. & P.J. ESLINGER. 1989. Colored hearing synesthesia: an investigation of neural factors. Neurology **39:** 781–784.
17. CYTOWIC, R.E. 1989. Synesthesia and mapping of subjective sensory dimensions. Neurology **39:** 849–850.
18. HENSCHEN, S.E. 1925. Clinical and anatomical contributions on brain pathology. Arch. Neurol. Psychiatry **13:** 226–249.
19. Henschen, S.E. 1926. On the function of the right hemisphere of the brain in relation to the left in speech, music, and calculation. Brain **49:** 110–123.
20. KLEIST, K. 1928. Gehirnpathologie und localisatorische Ergebnisse über Horstorungen, Gerauschtaubheiten und Amusien. Monatsschr. Psychiatr. Neurol. **68:** 853–860.
21. FEUCHTWANGER, E. 1930. Amusie. Studien zur pathologischen Psychologie der akustischen Wahrnehmung und Vorstellung undihrer Strukturgebiete, besonders in Musik und Sprache. Julius Springer. Berlin.
22. USTVEDT, H.J. 1937. Ueber die unersuchung der musikalischen funktionen bei patienten mit gehirnleiden, besonders bei patienten mit aphasie. Acta Med. Scand. (Suppl.) **86:** 1–737.
23. DALIN, O. 1745. Berättelse om en dumbe, som kan siumga. Kung Sven Vetensk Acad Handlinger **6:** 114–115. Cited in Benton, A.K. & R.J. Joynt. 1960. Early descriptions of aphasia. Arch. Neurol. **3:** 205–222.
24. JACKSON, J.H. 1871. Singing by speechless (aphasic) children. Lancet **2:** 430–431.
25. SMITH, A. 1966. Speech and other functions after left (dominant) hemispherectomy. J. Neurol. Neurosurg. Psychiatry **29:** 467–471.
26. GESCHWIND, N., F.A. QUADFASEL & J.M. SEGARRA. 1968. Isolation of the speech area. Neuropsychologia **6:** 327–340.
27. YAMADORI, A., Y. OSUMI, S. MASUHARA & M. OKUBO. 1977. Preservation of singing in Broca's aphasia. J. Neurol. Neurosurg. Psychiatry **40:** 221–224.
28. JACOME, D.E. 1984. Aphasia with elation, hypermusia, musicophilia, and compulsive whistling. J. Neurol. Neurosurg. Psychiatry **47:** 308–310.
29. ASSAL, G. 1973. Aphasie de Wernicke sans amusie chez un pianiste. Rev. Neurol. **129:** 251–255.
30. ASSAL, G. & J. BUFFET. 1983. Agraphie et conservation de l'écriture musicale chez un professeur de piano bilingue. Rev. Neurol. **139:** 569–574.

31. LURIA, A.R., L.S. TSVETKOVA & D.S. FUTER. 1965. Aphasia in a composer. J. Neurol. Sci. **2:** 288–292.
32. BASSO, A. & E. CAPITANI. 1985. Spared musical abilities in a conductor with global aphasia and ideomotor apraxia. J. Neurol. Neurosurg. Psychiatry **48:** 407–412.
33. SIGNORET, J.L., P. VAN EECKHOUT, M. PONCET, *et al.* 1987. Aphasie sans amusie chez un organiste aveugle. Rev. Neurol. **143:** 172–181.
34. PÖTZL, O. & H. UIBERALL. 1937. Zur Patholgie der Amusie. Wien Klin. Wochenschr. **50:** 770–775.
35. BOTEZ, M. & N. WERTHEIM. 1959. Expressive aphasia and amusia following right frontal lesion in a right-handed man. Brain **82:** 186–202.
36. WERTHEIM, N. & M. BOTEZ. 1961. Receptive amusia: a clinical analysis. Brain **84:** 19–30.
37. MAVLOV, L. 1980. Amusia due to rhythm agnosia in a musician with left hemispheric damage: a non-auditory supramodal defect. Cortex **16:** 331–338.
38. POLK, M. & A. KERTESZ. 1993. Music and language in degenerative disease of the brain. Brain Cogn. **22:** 98–117.
39. CONFAVREUX, C., B. CROISILE, P. GARASSUS, *et al.* 1992. Progressive amusia and aprosody. Arch. Neurol. **49:** 971–976.
40. ALAJOUANINE, T. 1948. Aphasia and artistic realization. Brain **71:** 229–241.
41. HENSON, R.A. 1988. Maurice Ravel's illness: a tragedy of lost creativity. Br. Med. J. **296:** 1585–1588.
42. BRUST, J.C.M. 1980. Music and language. Musical alexia and agraphia. Brain **103:** 367–392.
43. PITTRICH, H. 1956. Sensorische Amusie mit Paramusie nach rechtsseitger Temporalverletzung. Cited by Spreen, O., A.L. Benton & R.W. Fincham. 1965. Auditory agnosia without aphasia. Arch. Neurol. **13:** 84–92.
44. PÖTZL, O. 1939. Zur Pathologie der Amusie. Z. Neurol. Psychiatr. **165:** 187–195.
45. JELLINECK, A. 1956. Amusia. On the phenomenology and investigation of central disorders of the musical functions. Folia Phoniatr. **8:** 124–149.
46. SIDTIS, J.J. & B.T. VOLPE. 1988. Selective loss of complex-pitch or speech discrimination after unilateral lesion. Brain Lang. **34:** 235–245.
47. MCFARLAND, H.R. & D. FORTIN. 1982. Amusia due to right temporoparietal infarct. Arch. Neurol. **39:** 725–727.
48. MAZZUCCHI, A., C. MARCHINI, R. BUDAI, *et al.* 1982. A case of receptive amusia with prominent timbre perception defect. J. Neurol. Neurosurg. Psychiatry **45:** 644–647.
49. JUDD, T.L., A. ARSLANIAN, L. DAVIDSON, *et al.* 1979. A right hemisphere stroke in a composer. Presented at International Neuropsychological Society. February 2. New York.
50. MAZZONI, M., P. MORETTI, L. PARDOSSI, *et al.* 1993. A case of musical imperception. J. Neurol. Neurosurg. Psychiatry **56:** 322.
51. PERETZ, I. & R. KOLINSKY. 1993. Boundaries of separability between melody and rhythm in music discrimination: a neuropsychological perspective. Q. J. Exp. Psychol. **46A:** 301–325.
52. GORDON, H.W. & J.E. BOGEN. 1970. Hemispheric lateralization of singing after intracarotid sodium amylobarbitone. J. Neurol. Neurosurg. Psychiatry **37:** 737–738.
53. MILNER, B. 1962. Laterality effects in audition. *In* Interhemispheric Relations and Cerebral Dominance. V.B. Mountcastle, Ed.: 177–195. John Hopkins University Press. Baltimore.
54. KESTER, D.B., A.J. SAYKIN, M.R. SPERLING, *et al.* 1991. Acute effect of anterior temporal lobectomy on musical processing. Neuropsychologia **29:** 703–708.
55. SAMSON, S. & R.J. ZATORRE. 1991. Recognition memory for text and melody of songs after unilateral temporal lobe lesions: evidence for dual encoding. Q. J. Exp. Psychol: Learn Mem. Cogn. **17:** 793–804.
56. PERETZ, I. 1990. Processing of local and global musical information by unilateral brain-damaged patients. Brain **113:** 1185–1205.
57. LIÉGOIS-CHAUVEL, C., I. PERETZ, M. BABAI, *et al.* 1998. Contribution of different cortical areas in the temporal lobes to music processing. Brain **121:** 1853–1867.

58. SCHUPPERT, M., T.F. MÜNTE, B.M. WIERINGA, *et al.* 2000. Receptive amusia: evidence for cross-hemispheric neural networks underlying music processing strategies. Brain **123:** 546–559.
59. ZATORRE, R.J. 1988. Pitch perception of complex tones and human temporal lobe function. J. Acoust. Soc. Am. **84:** 566–572.
60. SAMSON, S. & R.J. ZATORRE. 1994. Contribution of the right temporal lobe to musical timbre discrimination. Neuropsychologia **32:** 231–240.
61. PENHUNE, V.B., R.J. ZATORRE & W.H. FEINDEL. 1999. The role of auditory cortex in retention of rhythmic patterns as studied in patients with temporal lobe removals including Heschl's gyrus. Neuropsychologia **37:** 315–331.
62. TRAMO, M.J. & J.J. BHARUCHA. 1991. Musical priming by the right hemisphere post-callosectomy. Neuropsychologia **29:** 313–325.
63. BEVER, T.G. & R.J. CHIARELLO. 1974. Cerebral dominance in musicians and non-musicians. Science **185:** 537–539.
64. PERETZ, I. & J. MORAIS. 1980. Modes of processing melodies and ear asymmetry in non-musicians. Neuropsychologia **18:** 477–489.
65. ZATORRE, R.J. 1979. Recognition of dichotic melodies by musicians and non-musicians. Neuropsychologia **17:** 607–617.
66. GAEDE, S.E., O.A. PARSONS & J.H. BERTERA. 1978. Hemispheric differences in music perception: aptitude vs. experience. Neuropsychologia **16:** 369–373.
67. WAGNER, M.T. & R. HANNON. 1981. Hemispheric asymmetries in faculty and student musicians and non-musicians during melody recognition tasks. Brain Lang. **13:** 379–388.
68. SHANNON, B. 1984. Asymmetries in musical aesthetic judgements. Cortex **20:** 567–573.
69. DEUTSCH, D. 1974. An auditory illusion. Nature **251:** 307–309.
70. DEUTSCH, D. 1992. Paradoxes of musical pitch. Sci. Am. **267:** 88–95.
71. SIDTIS, J.J. 1980. On the nature of cortical function underlying right hemisphere auditory perception. Neuropsychologia **18:** 321–330.
72. SERGENT, J. 1993. Mapping the musician brain. Hum. Brain Map **1:** 20–38.
73. ZATORRE, R.J., A.C. EVANS, E. MEYER, *et al.* 1992. Lateralization of phonetic and speech discrimination in speech processing. Science **256:** 846–849.
74. ZATORRE, R.J., A.C. EVANS & E. MEYER. 1994. Neural mechanisms underlying melodic perception and memory for pitch. J. Neurosci. **14:** 1908–1919.
75. SERGENT, J., E. ZUCK, S. TERRIAH, *et al.* 1992. Distributed neural network underlying musical sight-reading and keyboard performance. Science **257:** 106–109.
76. PLATEL, H., C. PRICE, J.-C. BARON, *et al.* 1997. The structural components of music perception. A functional anatomic study. Brain **120:** 229–243.

# Brain Specialization for Music

## New Evidence from Congenital Amusia

ISABELLE PERETZ

*Department of Psychology, University of Montreal, Montreal, Quebec, H3C 3J7, Canada*

**ABSTRACT: Brain specialization for music refers to the possibility that the human brain is equipped with neural networks that are dedicated to the processing of music. Finding support for the existence of such music-specific networks suggests that music may have biological roots. Conversely, the discovery that music may have systematic associations with other cognitive domains or variable brain organization across individuals supports the view that music is a cultural artifact. Currently, the evidence favors the biological perspective. There are numerous behavioral indications that music-specific networks are isolable in the brain. These neuropsychological observations are briefly reviewed here with special emphasis on a new condition, that of congenital amusia (also commonly referred to as tone deafness).**

**KEYWORDS: Music agnosia; Congenital amusia; Music-specific neural networks; Tone-deafness**

The notion that music might have biological foundations has only recently gained legitimacy. Over the past 30 years, music has mostly been studied as a cultural product. Musicologists have been analyzing each musical system in the context of its specific culture. Neuroscientists and psychologists have considered music as a convenient window to the general functioning of the human brain. However, neuropsychological observations have consistently and recurrently suggested that music might well be distinct from other cognitive functions, in being subserved by specialized neural networks. As such, music might be viewed as pertaining more to biology than to culture. The goal of the present paper is to review the neuropsychological evidence that supports the biological perspective.

## SPECIALIZED NEURAL NETWORKS FOR MUSIC PROCESSING

If music is biologically determined, then music should enjoy functional and neuroanatomical specialization. That is, music should be supported by neural networks that are dedicated to its processing and that are unresponsive or inadequate for deal-

Address for correspondence: Isabelle Peretz, Department of Psychology, University of Montreal, C.P. 6128, succ. Centre-ville, Montreal, Quebec, H3C 3J7, Canada. Voice: 514-343-5840; fax: 514-343-5787.

isabelle.peretz@umontreal.ca

ing with nonmusical input. Presently, support for the existence of such specialized neural networks is compelling. Most evidence in this regard derives from the functional examination of individuals whose brain conditions are disturbed in highly selective aspects. The brain anomaly may either disturb or spare musical abilities exclusively. These disorders are classified as three types of neurological conditions: acquired disorders, congenital disorders, and brain stimulation.

### *Acquired Disorders*

Acquired disorders refer to sequelae of a brain accident. Such injuries may be diverse but most frequently result from a cerebrovascular accident. In such cases, disorders are expected to reveal brain organization due to the fact that the disorders are constrained by the organization of the undamaged system, which was fully stabilized prior to the accident. Thus, studying brain-damaged individuals with the objective of uncovering the functional organization of the normal brain is an endeavor that is akin to the process of reverse engineering.

This research strategy is facilitated by the recurrent observation that brain damage does not affect cognition in its entirety, but rather, in particular aspects. Cognitive disorders can be highly selective. The selectivity of the disorder can take spectacular forms such as in brain-damaged composers who may lose their language and yet remain able to maintain their musical activities at their prior professional level.

The most famous case is probably that of Shebalin, the Russian composer, who, following successive vascular accidents occurring in his left hemisphere, suffered from severe disturbances of his language abilities. He remained aphasic for the rest of his life; he could neither understand nor speak intelligibly. Nevertheless, he continued to compose, notably completing his Fifth Symphony, which Shostakovitch considered to be one of his most brilliant and innovative works.[1] Therefore, Shebalin displayed severe language deficits yet retained his musical skills to a remarkable degree.

This dissociation between language and music cannot simply be explained by the fact that these professional musicians were "abnormally musical" from the start. Indeed, the reverse dissociation can also be observed, even in ordinary listeners. That is, persons devoid of any special talent, linguistic or musical, can experience spectacular losses of musical abilities, like losing the ability to recognize one's national anthem, without accompanying language difficulties. I was fortunate to be able to study in detail three such cases.[2,3]

One of them, I.R., was a restaurant manager who suffered bilateral brain damage as a result of successive surgeries in both sides of the brain for the clipping of ruptured aneurysms at the age of 28. Ten years postonset, I.R. still suffers from severe and irreversible deficits in musical perception and memory as a consequence to her brain condition.[3] Prior to her brain surgeries, music had great value to her. She was raised in a musically inclined family, her only brother being a professional musician. Fortunately, I.R. did not lose her language skills. She understands speech perfectly and remains verbally fluent, being able to express herself quite effectively (see her poem recently published in French[3]).

Such musical disorders—called amusia in neurological terms—are not rare and are dissociable from disorders of language—that is, aphasia, as reviewed by Brust (this volume). This recurrent finding suggests that most processing components that underlie language and music are not shared and are neuroanatomically separate.

**TABLE 1. Case reports of selective impairment and of sparing in the recognition of music**

| | Domain | | | | |
|---|---|---|---|---|---|
| Reports | Music | Speech (not prosody) | Environmental sounds | Voices | Lesions |
| Peretz *et al.*[2], C.N., and | –[a] | + | + | + | Bilateral temporal lobes |
| G.L. | – | +[b] | + | + | Bilateral temporal lobes |
| Peretz *et al.*[3], I.R. | – | + | + | + | Bilateral temporal lobes and right frontal lobe |
| Griffith *et al.*[6] | – | + | + | | Posterior right temporal lobe |
| Laignel-Lavastine *et al.* [10] | + | – | – | | Right temporal lobe |
| Godefroy *et al.*[9,b] | + | – | – | | Right posterior hemisphere |

[a] +, normal recognition; –, impaired recognition.
[b] After or during recovery.

Although dissociation between music and language has been reported in various spheres of activities, the two domains are rarely compared in analogous contexts. Hence, significant association between language and music might have escaped scrutiny. Thus, it is important to examine music and language in comparable tasks. One such simple task that the auditory system performs constantly and without effort, for both music and language, is sound pattern recognition. Yet, it has been shown that this task is not performed by a single shared auditory device, but by several, each being specialized for its particular domain.[4,5]

For instance, auditory perception of music appears to recruit mechanisms that are not implicated in speech recognition or in environmental sound recognition. Indeed, there are brain-damaged persons whose unique symptom is the loss of the ability to recognize and memorize music. The patients retain the ability to recognize and understand speech as well as to identify common environmental sounds normally.[2,3,6,7] The deficit can be remarkably selective. For example, C.N. was unable to recognize hummed melodies coming from familiar songs. Yet, she could perfectly recognize the lyrics accompanying the melodies that she failed to recognize.[7] Moreover, C.N. was able to recognize the voice of speakers[2] and the intonation of speech.[8] The existence of such a specific problem with music alongside normal functioning of other kinds of auditory abilities, including speech comprehension, suggests damage to processing components that are not only specific to the musical domain but that are also essential to the normal process of music recognition. We refer to this condition, which is a particular form of amusia, as *music agnosia.*[7]

The reverse condition that corresponds to a selective sparing of music recognition has been reported (see TABLE 1) but not yet well documented. Nevertheless, in the few available cases,[9,10] the lesion spared music processing relative to both speech comprehension and environmental sound recognition, which were both severely disturbed. Such cases suggest isolated sparing of music recognition abilities, hence complementing the music-specific deficits described above. Thus, the current evidence, summarized in TABLE 1, is indicative of the presence of specialized brain circuits for music recognition. These circuits are damaged in the cases of music agnosia and spared in the few cases of verbal agnosia and associated agnosia for environmental sounds.

### *Congenital Disorders*

Further neuropsychological evidence that is indicative of brain specialization for music comes from congenital disorders. These disorders refer to unexpected failures or preservation in a particular cognitive domain in comparison to the general level of intellectual and socioemotional functioning. These deficiencies are termed congenital since their presence can be detected very early in development. One such well-known condition corresponds to the "music-savant syndrome" that can often be observed in autistic individuals. The etiology of autism is not yet known; however, its incidence is relatively high, with 1 to 2 cases out of every 1000 births (about the same rate as Down syndrome). Autism is currently viewed as deriving from some brain anomaly because of its frequent association with another brain defect (e.g., epilepsy), its genetic transmission, and its atypical cerebral functioning as measured by brain imaging studies.[11] More interestingly from our perspective, 1 to 10% of autistic individuals might be considered musicians.[12] In effect, autistic subjects are generally more apt in the area of music than in other domains, such as language.[13] Several even become *musical savants*, a term that refers to the observation of high achievements in musical competence in individuals who are otherwise socially and mentally handicapped.

A well-known case (described in greater detail by L. Miller[12]) is that of "Blind Tom." Blind Tom was a young blind slave who gave piano concerts at the White House and all around the world. Although his language repertoire consisted of less than 100 words, his musical repertoire contained more than 5000 musical pieces. This "music savant" was sold along with his mother to Colonel Bethune in 1850 during a slavery sale in Georgia. Until age 5, he did not say a word and manifested no other sign of intelligence than his remarkable interest for the musical performance given by the colonel's daughters. At age 4, he was playing Mozart sonatas, which he had heard. At age 6, he was able to improvise; and at age 7, he gave his first recital. In 1862, despite the fact that he did not know how to read music, he was able to play without error 14 pages of an original composition that he had heard just once. Blind Tom gave recitals until age 53 when, following the colonel's death, he had to end his career.

The mirror image of this condition corresponds to individuals who are totally musically inept, despite normal exposure to music, normal intelligence, and social adaptation. Such individuals are sometimes called *tone-deaf* (see Grant-Allen[14] for the first report).

Che Guevara was known to be "tone-deaf."[15] He was well aware of his handicap, as the following anecdote illustrates. At a party, by prior arrangement, Alberto, his best friend, was required to give a poke to Che every time a tango was played. At some point during the party, the orchestra played an agitated Brazilian shora that had been Alberto's favorite. Alberto wished to share his enthusiasm with Che, but Che, with his eyes on a woman across the room, believed Alberto's nudge to be a tango signal and took to the floor, dancing a slow and passionate tango with everyone else jiggling to the shora. Realizing something was wrong, Che Guevara came over to ask Alberto for advice, who was too convulsed with laughter to be able to explain.

These individuals have a rare condition called *congenital amusia*. The term *amusia* is preferable to *tone-deafness* to reflect the likelihood that there might be as many forms of musical disorders that characterize congenital musical deficits as there are multiple forms of acquired musical disorders (see Brust, this volume, for an illustration of the heterogeneity of acquired amusias).

Unlike other developmental disorders, such as dysphasia and dyslexia, congenital amusia has not received much scientific attention. One obvious reason for this neglect is that attention is directed primarily at learning disabilities that affect language because of their wide educational implications. Moreover, music educators are reluctant to suspect the presence of congenital disorders because such a diagnosis may mean discontinuation of musical studies.[16] The other reasons are less pragmatic. As mentioned at the outset, many scientists conceive music as the product of a general-purpose brain organization. In that context, amusia is not a developmental disorder that is expected to occur in isolation. Rather, amusia is expected to result from intellectual and/or socioemotional dysfunctioning. However, as the study of music savants suggests, music proficiency does not seem to depend on the normal development of the cognitive and affective system. Musical proficiency can be achieved while sociocognitive functioning is globally deficient.

Therefore, in an effort to learn more about congenital amusia, we actively advertised and searched for three years for people who were tone-deaf. The early discovery of a textbook case (called Monica, reported in Peretz *et al.*[34] and represented in FIG. 1) who closely matched the two case descriptions available in the literature (i.e., Grant-Allen's case[14] and Geshwind's case[17]) has contributed greatly to the advancement of the study. Since then, the presence of amusia has been confirmed in 16 adults (Ayotte *et al.*, in preparation). All are self-declared tone deaf. Yet, self-report of music deficiency requires verification, as this is a frequent complaint in ordinary listeners. Therefore, we selected only subjects who exhibited clear-cut performance deficits in laboratory tests. Furthermore, in order to exclude extraneous causalities, we carefully selected participants who had no psychiatric or neurological history and who possessed a solid level of education. To ensure adequate stimulation and to exclude lack of interest or motivation, only volunteers who had experienced unsuccessful attempts at learning music during childhood were considered.

For all selected persons with congenital amusia, results from a musical screening battery of tests originally designed for brain-damaged patients gave behavioral evidence for the disorder. This battery has proved to be very effective in identifying adult nonmusicians with a deficit in music perception and memory.[18–20] By experience, we know that when a score obtained in one of the six tests falls below the lowest performance obtained by neurologically intact subjects who have reported no particular musical disorder, the outlier score is indicative of a genuine deficit. In the present study, we adopted a more conservative criterion and decided to consider participants as amusic if their performance fell below three standard deviations from the mean of a group of 60 unselected nonmusicians, on at least two tests. Their scores are represented in FIGURE 1 with two tests that allow comparisons of results on the melodic and temporal dimensions because the tests are very close in structure. One test serves to evaluate the use of melodic contour and the other to assess the use of rhythmic contour in the discrimination of two successive short musical sequences as "same or different." Each test comprises 30 trials made of two successive synthesized melodies to be compared. The comparison melody is repeated in half the trials. When different, the comparison melody contains one altered note that modifies the

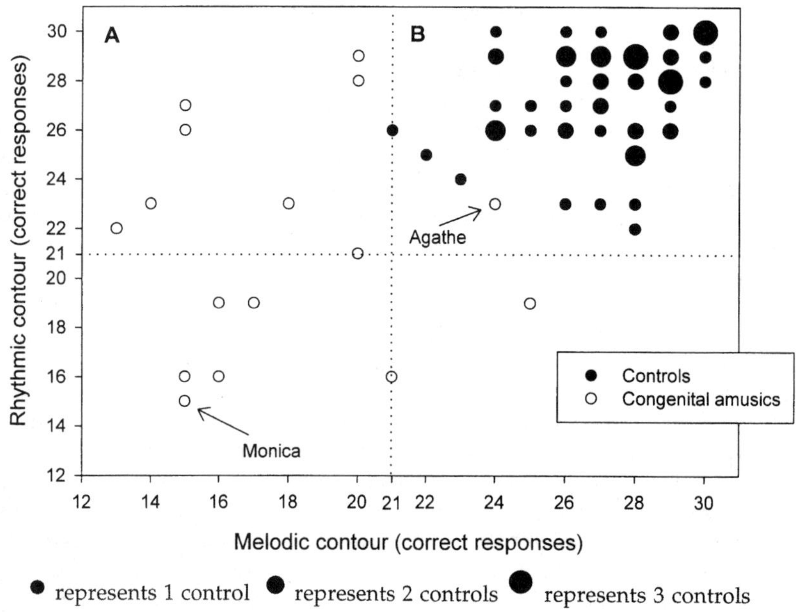


**FIGURE 1.** Number of correct responses obtained by 16 congenital amusics relative to the normal distribution (constituted by 60 nonmusicians, age range: 14–74 yr; education range: 7–20 yr) in two screening tests of our musical battery[17–19] assessing melodic contour discrimination (abscissa) and rhythmic contour discrimination (ordinate axis). The *dashed line* represents 3 standard deviations below the mean of controls.

pitch directions of the surrounding intervals in the melodic contour condition. In the rhythmic contour test, the duration of two successive tones was inverted so as to create a different rhythmic grouping in the comparison sequence. The original set of melodies served to construct both tests. The melodies obey the rules of Western music and were composed specifically for the battery.

Examination of the data in FIGURE 1 is informative in several aspects. First, most controls are confined to the right top quarter (B) and show very little overlap with amusics' performance, with one exception. Agathe was unimpaired on these two tests but was nonetheless classified as amusic because her performance was below the cut-off scores in two other tests that are not illustrated here. Her condition points to the possible heterogeneity of the congenital amusia syndrome. Another interesting trend is reflected by the subgroup of amusics located in the left top quarter (A). These amusics perform close to chance on the melodic dimension, whereas they perform normally on the rhythmic dimension. Such dissociation, between impaired melody and spared rhythm, is relatively frequent after brain damage.[20,21] Congenital amusia seems to fractionate along similar lines. Note, however, that the large majority of amusics (88%) have a deficit on the melodic dimension. This deficit seems to lie on a continuum and may result from a basic auditory problem related to pitch discrimination (Hyde, Peretz, and Ayotte *et al.*, in preparation). Further psychoacoustical work is currently under way with the objective to circumscribe the pitch deficit experienced by these unmusical individuals.

It is theoretically appealing to consider congenital amusia as arising from a basic pitch perception defect. Such an account would square rather nicely with the idea that some disorders in specific cognitive domains can be explained in terms of low-level auditory mechanisms. Currently a highly popular view of language-specific disorders involves a basic deficiency in fine temporal resolution.[22] Ascribing some form of congenital amusia to a faulty analysis of pitch variations constitutes an interesting complementary disorder. The pitch-based account is also reasonable because fine-grained discrimination of pitch is probably required by most musical systems, but not by speech intonation patterns. Intonation uses rather large variations in pitch to convey meaningful information. For example, the question, *Elle est devant la gare?* has a final rising pitch that is about half an octave (5 to 6 semitones) higher than the final pitch in the assertion *Elle est devant la gare*. By contrast, melodies mostly use small pitch intervals (on the order of one-twelfth or one-sixth of an octave).[23] Therefore, a degraded pitch perception system may well compromise music perception but leave speech intonation unaffected. The available data are consistent with this prediction, since our group of amusics does not exhibit difficulties discriminating questions from assertions (Ayotte *et al.*, in preparation).

Whatever the exact origin of congenital amusia, it is important to emphasize that, indeed, the syndrome appears to be music specific. First, all subjects had reached a high level of education. Thus, if they were having general auditory learning deficiencies, they should have had problems learning speech, for example, which, in turn, should have interfered with normal education, but it did not. To demonstrate this in a more rigorous fashion, we used another set of tests devised for music agnosic patients.[7] These tests are standard memory recognition tasks that differ only by the domain from which the test items are taken. In the music memory test, subjects are presented with 20 tunes (taken from familiar songs) to memorize. The melodies are then represented among 20 unstudied (but equally familiar) melodies that are randomly mixed. The subject is requested to indicate which melodies were heard in the study phase. For comparison purposes, in the lyrics and the environmental sound tests, subjects are given similar opportunities to learn and recognize 20 spoken lyrics (taken from the same familiar songs) and 20 environmental sounds (e.g., a barking dog), respectively. The three tests are performed in different sessions. As can be seen in FIGURE 2, amusics perform at chance level in the music test, whereas they score fairly high and within normal limits for nonmusical material, be they song lyrics or environmental sounds. The results clearly show the music specificity of the learning disability.

In conclusion, congenital amusia seems to be a true disorder. Affected individuals appear to be born without the essential wiring elements for developing a normally functioning system for music, while achieving a high degree of proficiency in their professional and social lives. This impairment seems to result, at least in some cases, from a more elemental problem that makes the individual unable to hear the pitch-relevant variations in music. Although it seems unlikely that a single defective mechanism, such as a pitch defect, is responsible for all processing components underlying musical functions, it may be that a single essential defective component brings the development of the musical system to a halt even if it needs many properly functioning components to work normally.

The existence of congenital amusia coupled with the music-savant syndrome strongly suggests the presence of early pressures for the development of normal neu-

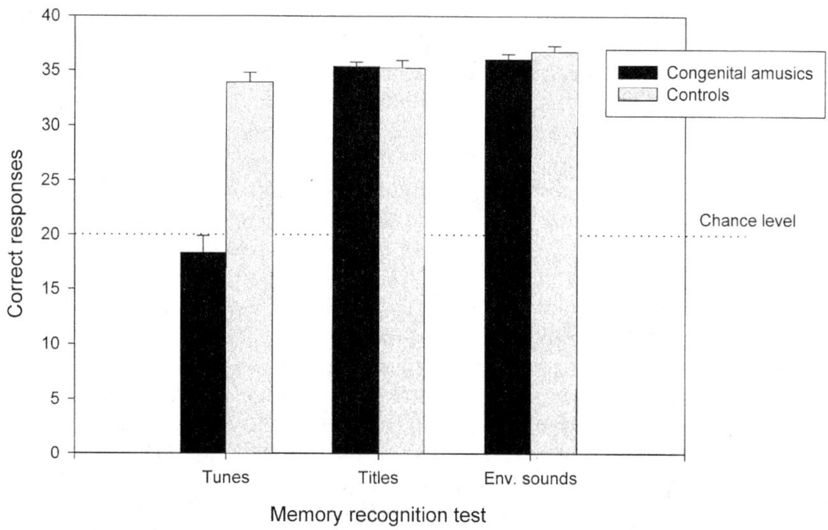


**FIGURE 2.** Number of correct responses obtained by 12 amusics and 20 controls in the memory recognition of studied melodies, lyrics, and environmental sounds. *Error bars* represent standard error of the mean.

ral networks that are dedicated to music. These predispositions would not find a suitable neural space in the amusic's brain, whereas they would find adequate wiring in the autistic child's brain. Identification of these neural bases as well as of their associated heritability should provide crucial elements in the debate surrounding the existence of biological foundations for musicality.

### *Brain Stimulation*

Investigations with epileptic patients provides the third source of evidence that speaks for the existence of neural networks that are dedicated to music. In a few individuals, music will be the exclusive trigger of the pathological firing of neurons conductive to the epileptic crisis. This form of epilepsy is called *musicogenic epilepsy*[24] (see also Brust, this volume) and suggests that the epileptogenic tissue lies in a neural region that is tied to music processing.

The British neurologist Macdonald Critchley[25] describes one regular visit of an epileptic patient to report progress and to collect medication. The patient explained to him that hearing music provoked her attacks exclusively. The patient specified that music of the popular type had no effect, only classical music did, while confessing she had no particular preference for classical music. She was subsequently admitted to the hospital and, despite her discomfort, was presented with various kinds of music. The only record of classical music possessed by the experimenters, which contained "The Walse of the Flowers" by Tchaikovsky, immediately provoked a seizure with generalized convulsive movements, frothing at the mouth, and cyanosis.[25]

During musicogenic epileptic seizures, abnormalities in electrical activity (recorded from the scalp) are generally observed at the temporal lobes, with a slight bias toward the right one (see Wieser *et al.*[24] for a recent review). Thus, some processing component that is exclusively related to music must be located in those regions.

Musicogenic epilepsy is in some sense a natural accident that can be simulated by direct application of electrical stimulation of the brain. *In vivo* electrical stimulation of particular areas of the auditory associative cortex of awake patients may produce highly vivid musical hallucinations.[26] These provoked hallucinations suggest that the stimulation has been applied to circuits that contain memories of musical experiences.

> Penfield and Perot report stimulating a particular region of the first right temporal circumvolution. The patient then said: "I hear music." The experimenters then repeated the stimulation, without telling the patient, who immediately said, "I hear the music again. It is like the radio." When asked what tune she was hearing, she said she did not know but it was familiar. The stimulation was again repeated and the patient shouted: "I hear it." The electrode was kept in place and the patient was asked to describe what she heard. The patient hummed the tune quite distinctly. The song came out so clearly that one of the nurses recognizes "Rolling along together." The patient agreed that this sounded like the words in the song (Penfield and Perot,[26] case 5, p. 620).

The left and, slightly more often, right temporal regions are prone to evoke such musical experiences. The fact that musical memories can be exclusively elicited in individuals with no musical training underscores once again brain specialization for music.

In summary, the patient-based approach converges on one precise point: neuronal networks that are situated in or close to the superior temporal gyrus participate in music perception and memory in a decisive and exclusive manner. Presently, there is little evidence for music specificity coming from the study of normal brains (but see Halpern, this volume, for recent data).

## LOCALIZATION OF THE MUSIC-SPECIFIC NETWORKS

Localization of the music-specific networks is essential when biological determinism is at issue. Biological functions are expected to be associated with a fixed neural organization across individuals. However, finding brain specialization for music is necessary but not sufficient. Showing brain specialization means that music possesses a neurofunctional substrate by itself and hence is not a parasite or a by-product of a more important brain function. Yet, brain specialization does not entail prewiring.

Brain specialization for music may result from recruitment of any free neural space in the infant's brain. Music would simply modify that space and adjust it to its processing needs. This opportunistic scenario of brain organization for music may respond to cultural pressure and not biological forces. If this were true, then a highly variable localization and distribution of the musical networks should be observed across individuals. Depending on the moment, quality, and quantity of exposure, various brain spaces might be mobilized. Thus, if music is a "squatter" in the brain, localization should vary capriciously across members of the same culture.

By contrast, a prewired organization is expected to exhibit consistency in localization. For example, the primary auditory areas (the Heschl's gyri) are systematically buried in the sylvian fissure; this holds for all humans. If development of musical networks is under the guidance of early predispositions, then these networks are expected to have a *fixed* arrangement. That is, brain implementation of music networks should be similar in the vast majority of humans, nonmusicians and musicians alike. Moreover, this organization is not expected to vary as a function of the musical culture considered. Musical functions are expected to be similarly implemented in the brain of an isolated Pacific Islander, a Chinese opera singer, and a Western fan of rap music. This prediction is quite strong and perfectly suited to the exploitation of the new brain imagery techniques.

Although clear and straightforward, the demonstration of a similar brain organization for music in all humans remains elusive. Localization of the brain substrates underlying music has been an enduring problem for more than a century. In my view, the only consensus that has been reached today about the cerebral organization underlying music concerns pitch contour processing. The vast majority of studies point to the superior temporal gyrus and frontal regions on the right side of the brain as the responsible areas for processing pitch contour information.[27] However, it remains to be determined if this mechanism is music specific, since the intonation patterns of speech seem to recruit similarly located, if not identical, brain circuitries.[8,28]

Similarly, assuming comparable brain organization in musicians and nonmusicians may sound controversial, since there is strong evidence that musical training has a sizable effect on cortical morphology (see Schlaug, this volume) and activity (see both Pantev and Pascual-Leone, this volume). It remains to be determined to what extent these changes due to musical training are merely quantitative or rather represent qualitative modifications. In other words, it will be important to qualify the effect of training on brain organization. Presently, music training does not seem to distribute musical modules differently in the neural space of the expert compared to the ordinary listener, as the parasitic scenario would predict. Yet, the relevant evidence is still scarce.

Clearly, what is needed at the present stage is a grid that allows specification of the processing mechanisms that are essential for music appreciation, an ability shared by all humans. Once these essential ingredients have been identified, their respective localization may be tracked down in the brains of musicians and nonmusicians of different musical cultures. The research agenda involved is dense and will only be briefly sketched in the next section.

## WHAT IS THE CONTENT OF THE MUSIC-SPECIFIC NEURAL NETWORKS?

The music-specific neural networks should correspond to a common core of musical abilities that are acquired by all individuals of the same culture and that form the essence of musical competence acquired by members of all cultures. This common core can probably be reduced to a few essential processing components that represent the germ of brain specialization for music. In this perspective, there is no need for all musical abilities to have initial specialization. Brain specialization for a few mechanisms that are essential to the normal development of musical skills should suffice.

I am proposing that the two anchors of brain specialization for music are the encoding of pitch along musical scales and the ascribing of a regular beat to incoming events. The notion that a special device exists for tonal encoding of pitch has been developed in previous papers[29] and will thus not be elaborated further. Similarly, the notion that regularity might be fundamental to music appreciation is slowly emerging[30] (e.g., Drake, this volume), although its specificity to music is rarely addressed.

Universality of musical scales and of pulse regularity is another issue that has received very little attention from ethnomusicologists. As mentioned earlier, ethnomusicologists cautiously avoid generalization across musical cultures. Yet, the plausibility of considering pitch scales and regularity as universals of music has increased in recent years.[31]

By contrast, developmental psychologists have made significant progress in identifying plausible musical universals. Studies with infants have largely confirmed the presence of precocious sensitivity to musical scales and to temporal synchronicity in auditory processing. For example, six- to nine-month-old infants process consonant intervals better than dissonant intervals[32] and exhibit learning preferences for musical scales.[33] In most musical cultures, musical scales make use of unequal pitch steps. Infants already show a sensitivity bias toward musical scales, since they have been shown to be better at detecting a small pitch change in an unequal-step scale than in an equal-step scale (see Trehub, this volume). On the time dimension, infants prefer music that is subject to an isochronous temporal pulse. For instance, like adults, four-month-old infants are biased toward perceiving regularity; they exhibit sensitivity to slight disruptions of temporal isochrony[30] (Drake, this volume). All of these aspects of auditory pattern processing suggest the presence of innate learning preferences.

In conclusion, appreciation of music fits well with the product of a specialized cortical arrangement that is present and functional early in human development and that is shared by the vast majority of ordinary listeners. Hence, music does not seem to be a mere game for the mind, for the neurons, or for the senses. Music seems to serve needs that are so important to humans that their brain has dedicated some neural space to its processing. It remains to demonstrate that these music-specific networks are fulfilling needs that are not optional but that have adaptive value.

## ACKNOWLEDGMENTS

Parts of the work summarized in this article were done in collaboration with Julie Ayotte and Krista Hyde. I thank Simone Dalla Bella for editing the figures and making insightful comments on the paper and Krista Hyde for language editing. The paper is based on studies supported by grants from the Medical Research Council of Canada.

## REFERENCES

1. LURIA, A., L. TSVETKOVA & J. FUTER. 1965. Aphasia in a composer. J. Neurol. Sci. **2:** 288–292.
2. PERETZ, I., R. KOLINSKY, M. TRAMO, *et al.* 1994. Functional dissociations following bilateral lesions of auditory cortex. Brain **117:** 1283–1301.

3. PERETZ, I., S. BELLEVILLE & S. FONTAINE. 1997. Dissociations entre musique et langage après atteinte cérébrale: un nouveau cas d'amusie sans aphasie. Rev. Can. Psychol. Exp. **51:** 354–367.
4. PERETZ, I. 1993. Auditory agnosia: a functional analysis. *In* Thinking in Sound. The Cognitive Psychology of Human Audition. S. McAdams & E. Bigand, Eds: 199–230. Oxford University Press. New York.
5. POLSTER, M. & S. ROSE. 1998. Disorders of auditory processing: evidence for modularity in audition. Cortex **34:** 47–65.
6. GRIFFITHS, T., A. REES, C. WITTON, *et al.* 1997. Spatial and temporal auditory processing deficits following right hemisphere infarction: a psychophysical study. Brain **120:** 785–794.
7. PERETZ, I. 1996. Can we lose memories for music? The case of music agnosia in a non-musician. J. Cognit. Neurosci. **8:** 481–496.
8. PATEL, A.D., I. PERETZ, M. TRAMO & R. LABRECQUE. 1998 Processing prosodic and musical patterns: a neuropsychological investigation. Brain Lang. **61:** 123–144.
9. GODEFROY, O., D. LEYS, A. FURBY, *et al.* 1995. Psychoacoustical deficits related to bilateral subcortical hemorrhages: a case with apperceptive auditory agnosia. Cortex **31:** 149–159.
10. LAIGNEL-LAVASTINE, M. & T. ALAJOUANINE. 1921. Un cas d'agnosie auditive. Rev. Neurol. **37:** 194–198.
11. BAILEY, A., W. PHILLIPS & M. RUTTER. 1996. Autism: towards an integration of clinical, genetic, neuropsychological, and neurobiological perspectives. J. Child Psychol. Psychiatry **37:** 89–126.
12. MILLER, L. 1989. Musical savants. Exceptional skill in the mentally retarded. Erlbaum. Hillsdale, NJ.
13. HEATON, P., B. HERMELIN & L. PRING. 1998. Autism and pitch processing: a precursor for savant musical ability? Music Percept. **15:** 291–305.
14. GRANT-ALLEN. 1878. Note-deafness. Mind **10:** 157–167.
15. TAIBO, P.I., II. 1996. Ernesto Guevara, también conocido como el Che. Planeta. Buenos Aires.
16. KAZEZ, D. 1985. The myth of tone deafness. Music Educators J. **71:** 46–47.
17. GESCHWIND, N. 1984. The brain of a learning-disabled individual. Ann. Dyslexia **34:** 319–327.
18. AYOTTE, J., I. PERETZ, I. ROUSSEAU, *et al.* 2000. Patterns of music agnosia associated with middle cerebral artery infarcts. Brain **123:** 1926–1938.
19. LIÉGEOIS-CHAUVEL, C., I. PERETZ, M. BABAÏ, *et al.* 1998. Contribution of different cortical areas in the temporal lobes to music processing. Brain **121:** 1853–1867.
20. PERETZ, I. 1990. Processing of local and global musical information by unilateral brain-damaged patients. Brain **11:** 1185–1205.
21. PERETZ, I. & R. KOLINSKY. 1993. Boundaries of separability between melody and rhythm in music discrimination: a neuropsychological perspective. Q. J. Exp. Psychol. **46:** 301–325.
22. WRIGHT, B., L. LOMBARDINO, W. KING, *et al.* 1997. Deficits in auditory temporal and spectral resolution in language-impaired children. Nature **387:** 176–177.
23. VOS, P. & J. TROOST. 1989. Ascending and descending melodic intervals: statistical findings and their perceptual relevance. Music Percept. **6:** 383–396.
24. WIESER, H.G., H. HUNGERBÜHLER, A. SIEGEL & A. BUCK. 1997. Musicogenic epilepsy: review of the literature and case report with ictal single photon emission computed tomography. Epilepsia **38:** 200–207.
25. CRITCHLEY, M. 1977. Musicogenic epilepsy. *In* Music and the Brain. M. Critchley & M. Henson, Eds: 344–353. W. Heinemann Medical Books. London.
26. PENFIELD, W. & P. PEROT. 1963. The brain's record of auditory and visual experience. Brain **86:** 595–696.
27. PERETZ, I. 2001. Music perception and recognition. *In* The Handbook of Cognitive Neuropsychology. B. Rapp, Ed.: 521–540. Psychology Press. Hove, UK.
28. ZATORRE, R.J., A.C. EVANS, E. MEYER & A. GJEDDE. 1992. Lateralization of phonetic and pitch processing in speech perception. Science **256:** 846–849.
29. PERETZ, I. & J. MORAIS. 1989. Music and modularity. Contemp. Music Rev. **4:** 277–291.

30. DRAKE, C. 1998. Psychological processes involved in the temporal organization of complex auditory sequences: universal and acquired processes. Music Percept. **16:** 11–26.
31. AROM, S. 2000. Prolegomena to a biomusicology. *In* The Origins of Music. N. Wallin, B. Merker & S. Brown, Eds.: 27–29. MIT Press. Cambridge.
32. SCHELLENBERG, E.G. & S. TREHUB. 1996. Natural musical intervals: evidence from infants' listeners. Psychol. Sci. **7:** 272–277.
33. TREHUB, S., G. SCHELLENBERG & S. KAMENETSKY. 1999. Infants' and adults' perception of scale structure. J. Exp. Psychol. Hum. Percept. Perform. **25:** 965–975.
34. PERETZ, I., J. AYOTTE, J. MEHLER, *et al.* Congenital amusia: a disorder of fine-grained pitch discrimination. Submitted for publication.

# Cerebral Substrates for Musical Temporal Processes

SÉVERINE SAMSON,[a,b] NATHALIE EHRLÉ,[a,b] AND MICHEL BAULAC[b]

[a]*Université de Lille 3, URECA, UFR de Psychologie, 59653 Villeneuve d'Ascq cedex, France*

[b]*Epilepsy Unit, Salpêtrière Hospital and Laboratoire de Neurosciences Cognitives et Imagerie (CNRS UPR 640), Paris, France*

**ABSTRACT: Music as well as language consists of a succession of auditory events in time, which require elaborate temporal processing. Although several lines of evidence suggest that the left dominant hemisphere is predominantly involved in the processing of rapid temporal changes of speech, very little is known about the cerebral substrates underlying such auditory temporal processes in music. To investigate this issue, we examined epileptic patients with either left (LTL) or right (RTL) temporal lobe lesions as well as normal control subjects (NC) in two different tasks involving the processing of time-related (temporal) information. By manipulating the interonset interval (IOI) in a psychophysical task, as well as in a task of detection of rhythmic changes in real tunes, we studied the processing of temporal microvariations in music. The first task assessed anisochrony (or irregularity) discrimination of sequential information according to different presentation rates (between 80 and 1000 ms IOI). For all subjects, an effect of tempo was obtained; thresholds were lower for the 80 ms IOI than for longer IOIs. Furthermore, there was a specific impairment of rapid anisochronous discrimination (80 ms IOI) for LTL patients as compared to RTL and NC subjects, but no deficit was observed for longer IOIs. These findings suggest the specialization of left temporal lobe structures in processing rapid sequential auditory information. The second task involved the detection of IOI increments in familiar monodic tunes. Performance was measured for two increments (easy vs. difficult to detect according to cognitive expectation) to assess the effect of cognitive expectation using a forced-choice paradigm (changed vs. unchanged melody). The results showed that LTL patients but not RTL were impaired as compared to NC subjects in the increment detection. However, all groups showed differences between the two levels of difficulty, suggesting that top-down processing remains functional. These findings suggest that left temporal lobe structures are predominantly involved in perceiving time-related perturbations in familiar tunes as well as in isochronous sequences, extending to the musical domain findings previously reported in speech.**

**KEYWORDS: Brain; Music; Musical temporal processes**

Address for correspondence: Séverine Samson, Université de Lille 3, URECA, UFR de Psychologie, BP 149, 59653 Villeneuve d'Ascq cedex, France. Voice: +33 3 2041 6443; fax: +33 3 2041 6324.

samson@univ-lille3.fr

## INTRODUCTION

Music as well as language consists of a succession of auditory events in time, which require elaborate temporal processing. In the present study, we will consider the temporal coding of musical information that plays an important role in performing and perceiving rhythm. This time-related processing concerns a wide range of time frames that may implicate different forms of information processing. In the neuropsychological domain, simple auditory sequences and musical patterns have been used to investigate the cerebral structures underlying musical temporal processes. Studies reported in the literature will be reviewed first. Then, experimental findings that we obtained in two different studies will be reported to clarify the role of the left as opposed to the right temporal lobe structures in processing subtle temporal variations within a range of ten to hundreds of milliseconds.

### *Temporal Processing in Simple Auditory Sequences*

Several studies have explored time-related processing by using simple auditory sequences. In this domain, several paradigms classically used in experimental psychology have been adapted to neuropsychology. In a seminal paper, Efron[1] investigated the ability of brain-damaged patients to judge the temporal order of two tones of different frequencies separated by a silent interval. The results showed that aphasic patients with left-hemisphere lesions required longer intervals (between 140 and 400 ms) to discriminate temporal order than nonaphasic patients with right-hemisphere lesion or normal subjects (75-ms interval). It was therefore suggested that the left-hemisphere structures generally thought to be involved in language processing may also contribute to the temporal analysis of fast auditory sequential information. Subsequently, Tallal and Newcombe[2] provided convergent evidence by demonstrating that selective damage to the left but not to the right hemisphere disrupted the ability to process two tones separated by a short interval (300 ms or less between the tones). Importantly, neither left- nor right-hemisphere–damaged subjects were affected when longer intervals were used. This finding has been reproduced with similar paradigms and generalized to other tasks involving gap detection or perception of simultaneity and succession.[3–7] One case report of an amusic patient (H.V.) seems to contradict previous results, since a deficit in temporal order processing of rapid patterns was observed in the presence of a right-hemisphere lesion.[8] However, the authors noted that the cortical lesion of this patient was associated with an underlying bilateral white matter lesion, which may have caused bilateral cortical deafferentation. Such a cerebral dysfunction, which presumably involved the left hemisphere, might have been responsible for the disruption of rapid temporal processing, thus explaining the apparent contradiction between the results.

However, inconsistent results have been reported in studies investigating the perception of duration. Indeed, results of different studies have shown that impairments in this perceptual ability have been observed in patients with unilateral lesions implicating the right or the left hemisphere[9] in a case of auditory agnosia following a bilateral cerebral dysfunction,[10] as well as in patients with right temporal lobe lesions.[11] These findings suggest that the perception of continuous signals, as opposed to discrete events, depends on different processing involving a distinct neural substrate.

Except for the perception of duration, the results previously reviewed suggest that left-hemisphere structures are predominantly involved in the processing of rapid sequential information. Recent electrophysiological results indicate that this function can be linked to auditory cortices of the left temporal lobe. By recording intracerebral evoked potentials to syllables in the right and the left human auditory cortices, Liégeois-Chauvel and her collaborators[12] demonstrated a specialization of the left auditory cortex for speech perception that depends on rapid temporal coding (within a few tens of milliseconds). If the left temporal cortex is dominant for fine-grained time-related processing of language, it seems plausible to hypothesize that it would also be important for such temporal processing in music.

### *Temporal Processing in Musical Sequences*

Parallel to the previously reported studies using simple auditory sequences, several studies investigated temporal processes in musical sequences. It has been suggested that temporal processing, of which musical rhythm would be an example, is best performed by the functions of the dominant left hemisphere. However, the evidence reported in the literature provides little support for this hypothesis. The few studies investigating the perception of dichotically presented stimuli in normal listeners usually report a right-ear advantage, which is supposed to reflect left-hemisphere predominance for the perception of temporally complex nonspeech stimuli,[13–16] but this perceptual asymmetry has not always been obtained.[17] Similarly, studies carried out in brain-damaged subjects have not systematically documented a deficit in rhythmic tasks in the presence of a left-hemisphere lesion. Although evidence supporting left-hemisphere involvement in rhythm has been reported in a few studies carried out in unilateral brain-damaged subjects,[18,19] other studies have demonstrated the contribution of right-hemisphere structures and/or a bilateral cerebral involvement in rhythm discrimination.[20–25] Two studies have even failed to report deficits in rhythm discrimination after unilateral temporal lobe resection.[11,26] However, the use of musical stimuli whose familiarity and rhythmical complexity are extremely variable make the comparison between studies difficult. In light of these seemingly contradictory results obtained in lesion studies, it seems impossible to conclude that left-hemisphere structures are predominantly involved in musical rhythm.

One functional neuroimaging study investigated the temporal processing of musical patterns,[27] but interpretation of the results in this study is made difficult by methodological factors. In normal subjects, an increase of cerebral blood flow was obtained by positron emission tomography in left inferior Broca's area and in left insula when rhythmic and pitch judgment conditions were compared. Although this finding was interpreted by the authors as evidence supporting left-hemisphere superiority for temporal processing, it seems impossible to attribute these foci of activation to a purely temporal processing. Indeed, many other nontemporal aspects related to the task demands and to the stimuli that differ between the two compared conditions might have contributed to these metabolic changes.

In cognitive psychology, it has been proposed that the perception of rhythmic grouping, consisting of a sequential organization of relative durations of tones and silent intervals, should be differentiated from the perception of meter, referring to the underlying perceived beat marking off equal duration units.[28–31] Neuropsychological studies have demonstrated a dissociation between these two components involved in subjective organization of temporal patterns,[23,26,32] but this dissociation has not been systematically observed.[20] Moreover, no indication in favor of a lateralized deficit in metrical processing has been reported.

On the basis of this review, we can assume that the left dominant hemisphere and more specifically the temporal lobe plays an important role in the processing of simple auditory sequences requiring rapid timing variations ranging from tens to hundred of milliseconds. However, this left hemisphere superiority has not been consistently demonstrated in studies using musical sequences. Although precise timing information is not usually reported in such studies, the description of the stimuli indicates that the manipulated temporal information refers to slower changes resulting, for example, from the permutation of two notes. As previously emphasized, left-hemisphere damage affected the processing of fast but not slow temporal information, therefore explaining why left cerebral lesions do not systematically interfere with time-related judgment involved in musical sequences (which usually concerned longer temporal information greater than 200 ms[33]).

To try to clarify the role of the left temporal lobe in fast temporal coding in music, we examined the consequences of unilateral temporal lobe dysfunction in processing rapid sequential information. In this paper, we present evidence supporting the hypothesis that left temporal lobe structures are predominantly involved in auditory temporal processes within the tens to hundreds of milliseconds time frame. For this purpose, interonset interval (IOI) was manipulated in a psychophysical task of anisochronous discrimination by using simple auditory sequences, as well as in a detection task involving rhythmic changes in real tunes.

## PERCEPTION OF INTERONSET INCREMENT IN SIMPLE AUDITORY SEQUENCES: ANISOCHRONOUS DISCRIMINATION

As frequently suggested, an important component of the rhythmic subjective structure is its underlying isochrony (i.e., a regular beat), which corresponds to musical meter. In its simplest form, the meter can be expressed as a series of regular sounds. In the present study, we designed an experiment based on anisochronous perception, which refers to the ability to differentiate a regular from an irregular sequence of sounds. The goal of this experiment was to test the effect of tempo on anisochronous (or irregularity) discrimination in patients with unilateral temporal lobe dysfunction, using an adaptive procedure. The rate of presentation or tempo (defined by the IOIs separating the sounds in the sequence) was systematically manipulated to compare anisochronous discrimination using different tempi.

The anisochronous paradigm is very adequate for looking at temporal processing. As indicated by the results of experiments carried out in normal subjects,[34] anisochronous discrimination remains unaffected by duration or number of tones, but it is influenced by tempo or presentation rate. Using sequences presented at various tempi, ranging from 80 ms (fast) to 1000 ms (slow) IOI, allowed us to compare the perception of fast to slow sequential information without changing the task demands. Results obtained in normal listeners also showed that the discrimination thresholds for the anisochronous sequences were proportional to the size of the IOI for tempi between 300 and 1400 ms, whereas sensitivity deteriorated for IOIs of 80 ms as compared to slower tempi. This finding corroborates the assumption of differ-

**FIGURE 1.** Example of the two sequences presented at one trial in the anisochronous discrimination task.

ent processes for temporal stimuli less than or equal to 200–300 ms compared to longer ones.[35–37]

Furthermore, the paradigm that we designed uses an adaptive procedure that offers the opportunity to determine precise and fine individual thresholds that represent sensitive measures of perceptual abilities. In particular, the use of such a method in neuropsychology improves the chance of documenting subtle perceptual difficulties in brain-injured patients that could not be detected by classical paradigms involving fixed levels of stimulation common to all subjects.

Two successive sequences of five tones with the same frequency, intensity, duration, and attack and decay time were presented on each trial (one regular and one irregular). The subject had to decide if the irregular sequence was in the first or second position (see FIG. 1). Five conditions with an IOI of 80, 300, 500, 800, or 1000 ms were prepared in order to test the effect of tempo on anisochronous discrimination. The anisochrony was introduced by delaying either the second or the fourth sound of the isochronous sequence. To obtain reliable measures of anisochronous discrimination, a psychophysical procedure developed by Levitt[38] was used to determine the minimum temporal shift necessary for each individual to discriminate the regular from the irregular sequence. The size of the first shift corresponded to 10% of the base interval (i.e., 50 ms for an IOI of 500 ms). The threshold of the shift detection was determined by an adaptive procedure in which the subject's response on one trial determined the size of the shift on the next trial. The perceptual threshold was then expressed as a percentage of the base IOI to allow comparison with other differential thresholds reported in the literature. Further details about the methods are provided in another paper.[39]

Eighteen patients with medically intractable epilepsy, candidates for surgical treatment at La Salpêtrière Hospital (Paris), as well as a group of normal control subjects ($n = 6$), participated in this study. They all presented medial temporal lobe epilepsy associated with lateralized hippocampal sclerosis as identified by magnetic resonance imaging (MRI). None of them suffered from language disturbances, and language function was lateralized in the left hemisphere for all subjects. These patients were divided into two groups: those with right hippocampal atrophy ($n = 8$) and those with left hippocampal atrophy ($n = 10$).

According to the literature, we hypothesized that the thresholds of patients with left-temporal lobe lesions would be significantly higher than the thresholds of patients with right-temporal lobe lesions or normal control subjects for tempi with IOIs at or below 300 ms. By contrast, no specific deficit was predicted for slower tempi with IOIs greater than 300 milliseconds. Finally, we predicted that for all subjects the

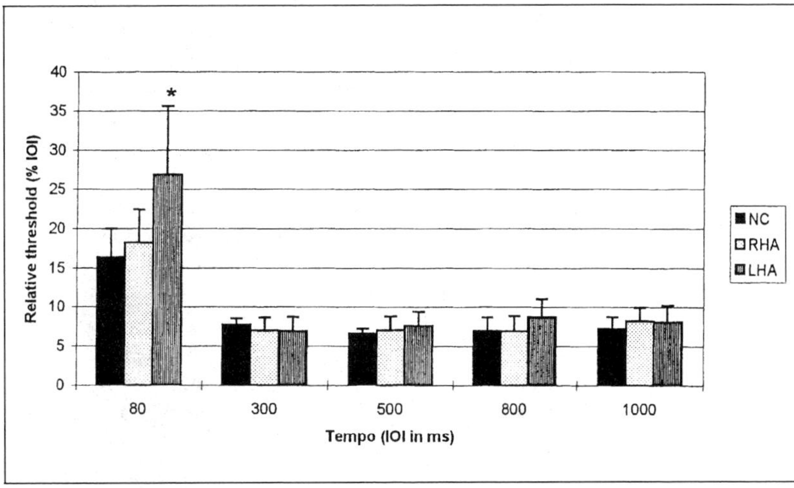


**FIGURE 2.** Relative thresholds of the three groups of subjects (LHA, left hippocampal atrophy; RHA, right hippocampal atrophy; NC, normal controls) for interonset intervals (IOI) of 80, 300, 500, 800, and 1000 ms. The *bars* represent the standard error of the mean.

thresholds obtained for the fastest tempo (80 ms) would be higher than those obtained for slower tempi (≥ 300 ms), as indicated by previous results in normal listeners.

In keeping with our predictions, the results displayed in FIGURE 2 clearly show that anisochronous discrimination thresholds for the rapid tempo (80 ms IOI) were significantly higher for the patients with left temporal lobe dysfunction (mean threshold: 27.5%) than for the patients with right temporal lobe dysfunction (17.7%) and for the normal control subjects (16.4%). However, there were no differences between the groups for the slower tempi (from 300 ms to 1000 ms IOI). This finding is compatible with results of aphasic patients with left hemisphere lesion[1,2,4] and provides strong evidence suggesting that fast auditory sequential information processing depends specifically on the integrity of left temporal lobe structures.

In addition, an effect of tempo was systematically obtained for all the subjects. The thresholds for very rapid sequences (80 ms IOI = 21.43%) were always higher than the thresholds obtained with slower tempi, ranging from 300-ms to 1000-ms IOI (7.6% on average), for which the different thresholds remained proportional to IOI. This finding confirmed results of a previous study carried out in healthy subjects[34] and is in agreement with the assumption that brief auditory information is processed differently from longer information.[35–37]

In a previous paper,[39] we proposed that the decreased sensitivity observed for the fastest tempo could be related to an auditory memory saturation. According to echoic memory studies[40] (see Ref. 41 for a review), acoustic properties of sounds necessitate a 200- to 300-ms delay following the presentation of a tone to be adequately retained. This time delay was obviously shorter for the fastest tempo (80 ms IOI). Indeed, the interval between the two sequences to be discriminated was twice the base interval, in order to maintain isochrony across the trial. Thus, the intersequence interval for the fastest tempo (80 ms IOI) was 160 ms, whereas a minimum of 600 ms separated the sequences for the other tempi. Therefore, a saturation of echoic memory could have been responsible for the observed difference between the thresholds of the faster and slower tempi, because the intersequence interval would be too short to allow consolidation of the first sequence. If such an explanation is validated in the future, the deficit documented in patients with left temporal lobe dysfunction in fast sequential auditory perception could be interpreted as an echoic memory dysfunction. Although an eventual relationship between a temporal processing dysfunction and an auditory memory deficit in dysphasic children has already been suggested by Tallal and collaborators,[42] the possible link between an echoic memory dysfunction and an anisochronous discrimination deficit of fast auditory information in patients with left temporal lobe lesion remains to be clarified.

It is also possible that the decreased temporal sensitivity for the rapid tempo is related to a difficulty in determining if the irregular sequence was the first or the second one, suggesting a deficit in temporal order processing. Although this judgment was not affected when slower tempi were presented, it becomes more difficult to make such a judgment in rapid sequences (subjects sometimes having trouble separating the sequences per se). This additional constraint might have penalized patients with left temporal lobe dysfunction more strongly than the other subjects.

In previous studies, rapid temporal processing deficits were usually reported in language-impaired subjects (see Ref. 43 for a review). Results of the present investigation extend this finding to patients presenting a limited left temporal lobe lesion without associated massive language disorders, indicating that time-related disturbances are not necessarily associated with language disturbances. However, the present results differ from the data obtained in language-impaired subjects, since, in this latter case, subjects presented a deficit with a slower tempo (300 ms IOI), while the subjects with left temporal lobe dysfunction did not present any difficulties in this condition. It seems that the severity of the sequential auditory deficit observed in patients with or without verbal deficit varies. Although these divergent results could be explained by methodological differences between the studies, it is also possible that the severity of the temporal deficits could be correlated to the severity of verbal deficits. The patients tested in the present study might be situated at one end of the continuum, showing limited temporal processing deficits and relatively preserved verbal abilities, while dysphasic or dyslexic subjects could be located at the other end, with more severe temporal deficits and more severe language deficits. Conversely, it is also possible that rapid sequential temporal processing can be disturbed without affecting language functions, which would suggest that they depend on distinct neural substrates. Future research will be necessary to determine whether the same underlying mechanism is responsible for the deficit observed in epileptic and dysphasic patients.

## PERCEPTION OF INTERONSET INCREMENT IN MUSICAL SEQUENCES: TEMPORAL MICROSTRUCTURE OF FAMILIAR TUNES

The goal of this study was to extend to the musical domain the role of the left temporal lobe in processing small timing variations. As emphasized in the introduction, it seems that the perception of small timing differences within a musical sequence has not been investigated in neuropsychology. To explore this issue, we designed a task involving the detection of IOI increments in familiar monodic tunes, improving therefore the ecological validity of the task.

Several descriptive studies showed that performers modify considerably the temporal structure as noted in the score when interpreting a piece of music (for review, see Ref. 44), a piece never being played exactly as it is written in the score. However, music played in this way seems perfect to the listeners, suggesting that we are expecting these temporal changes. Such temporal microvariations occur at specific locations and are supposed to reflect the subjective temporal structure of music. It has been demonstrated that, consequently, a listener is expecting these musical microvariations in perceiving musical excerpts.[45] In an experimental study using classical pieces of music (microstructural expectations), Repp showed that increment detection accuracy was correlated with the temporal microstructure profiles of expert performances, suggesting that temporal increments of an interval that is usually lengthened are more difficult to detect than temporal increments of intervals for which no microvariations were introduced. In the cognitive literature, two hypotheses that are not mutually exclusive have been proposed to explain variations in the accuracy detection. A top-down hypothesis suggests that listeners' expectations reflect expressive performance of temporal microvariations,[45] whereas a bottom-up hypothesis indicates that some expectations may be due to psychoacoustical characteristics of the stimulus.[44]

Based on these observations, we designed an experiment involving the detection of IOI increments introduced in musical sequences to test perception of these temporal microvariations in patients with unilateral temporal lobe lesions. The musical excerpts used were very familiar in order to generate high expectations and to prevent bias due to each listeners' musical background.

In the experimental task, half of the trials consisted of the presentation of a score version (played exactly as it is written in the score) of a familiar tune, whereas the other half consisted of a modified version of the tune in which one IOI was increased by 25%, producing an increment of 10 to 140 ms (mean = 85 ms; standard deviation = 62 ms) depending on the size (or duration) of the modified interval within the tune. The subject's task was to decide whether the presented trial corresponded to the mechanical or to the modified version of the excerpt using a two-alternative forced-choice paradigm. Two types of IOIs were introduced. One type corresponded to expected increments, and the other one corresponded to unexpected increments. The level of expectation was determined by results of previous studies detailed elsewhere.[46] Basically, the recording of pianists' performances were compared to a score (or computerized) version of each selected tune to allow the analysis of temporal microstructure. The results of this experiment allowed the identification of one interval that was systematically lengthened in pianists' recordings as compared to the metrical version, and another interval that was systematically preserved. IOI increments were thus introduced at these specific locations, and it was hypothesized that temporal increments located at "lengthened intervals" would be more difficult to detect than temporal increments located at "preserved intervals" since listeners would be expecting temporal increments in the first but not in the second condition. The ability to distinguish these two types of interonset increments was subsequently tested in a perceptual task, confirming therefore the relevance of this temporal manipulation in normal nonmusician listeners. Based on these results, we designed an

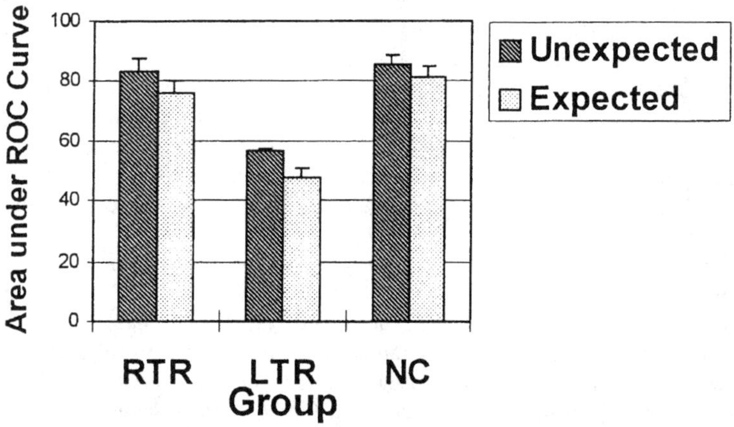


**FIGURE 3.** Detection accuracy derived from the area under the receiver operating curve (ROC) of the three groups of subjects (LTR, left temporal resection; RTR, right temporal resection; NC, normal controls) for expected and unexpected interonset increments. The *bars* represent the standard error of the mean.

experimental task to evaluate the ability to detect subtle temporal changes by taking into account implicit knowledge of the rhythmical structure underlying musical listening. By manipulating temporal increments and cognitive expectations, especially with regard to the temporal (or rhythmic) dimension, it was possible to assess bottom-up as well as top-down processing involved in musical rhythmic perception.

Twenty-two patients who had undergone a right ($n = 11$) or a left ($n = 11$) temporal lobe resection for the relief of medically intractable epilepsy, as well as 14 normal control subjects, were tested in this experiment. None of the patients presented language disorders or suffered from extra temporal lesions. Language was lateralized on the left side in all subjects. As opposed to the previous study, the patients were tested postoperatively. The temporal resection includes the medial temporal lobe structures (hippocampal and surrounding cortex), and in some patients the excision involves the temporal pole as well.

Considering the predominant contribution of left temporal lobe structures in processing rapid sequential information, we expected a deficit in detecting microtemporal variations in familiar tunes in patients with left temporal lobe lesion. In this study, the task of detection of temporal increments required no tonal judgment, for which the right temporal lobe is usually needed.[47–50] We therefore postulated adequate performance for patients with right temporal lobe lesion. Moreover, an effect of level of expectation of IOI was predicted for normal subjects and for patients with right temporal lobe lesion. If this effect reflects a perceptually driven processing, it should disappear in patients with left temporal lobe lesion, otherwise it might also be present in this patient group.

The results of this study are displayed in FIGURE 3, which represents detection accuracy of the three groups of subjects for expected and unexpected interonset increments. An area under the receiver operating curve (ROC) derived from the number of hits and false alarms was computed for each subject. As predicted, a significant deficit in detecting interonset increments was obtained in patients with left temporal lobe lesion as compared to subjects with right temporal lobe lesion and normal subjects, these two latter groups not being different. This deficit characterizing patients with left temporal lobe lesion cannot be attributed to difficulty in identifying the tunes since all the subjects were able to adequately recognize the melodies. The results of this study suggest that the ability to detect an interonset increment introduced in very well-known tunes depends on the integrity of left temporal lobe structures, extending to the musical domain the previously reported role of the left temporal lobe in rapid sequential processing.

Moreover, the results showed that it is more difficult to detect expected than unexpected temporal increments for all the subjects, indicating, therefore, that top-down processing remains functional for the three groups of subjects. Moreover, it means that expectations of nonmusician listeners were identical to those of pianists, extending to the simple monodic familiar tunes findings reported with polyphonic classical pieces.[45] Although patients with left temporal lobe lesion presented a clear deficit in perceiving temporal microvariations, they seemed to be able to generate temporal expectations to the same extent as healthy subjects.

To summarize, the results of this study illustrate a dissociation between a selective deficit in fine-grained temporal sensitivity, whereas temporal expectations seem to be relatively spared in the presence of left temporal lobe lesions, suggesting that top-down (or descending) processing may function independently from bottom-up (or ascending) processing. In addition, these data indicate that top-down processing involved in the temporal perception of musical information can still be preserved despite disrupted bottom-up processing.

## CONCLUSION

The experiments reported in this paper were designed to test the hypothesized role of left temporal lobe structures to musical temporal processing. We focused our interest on the perception of brief sequential information ranging from tens to hundreds of milliseconds by comparing results obtained with very simple auditory sequences and real musical excerpts. In keeping with our predictions, the data reported in this paper suggest that left temporal lobe structures are predominantly involved in perceiving IOI increments in familiar tunes as well as in isochronous sequences, extending to the musical domain findings previously reported in speech.[12,43]

The patients included in both studies had lesions located within the medial temporal lobe structures and more specifically in the hippocampus. It remains difficult to know if the deficit could be attributed to the hippocampal lesion or to a more global dysfunction of the temporal neocortical structures. Evidence from a functional cerebral imaging study (positron emission tomography with 18 fluorodeoxyglucose, FDG-PET) carried out in a similar population of epileptic subjects indicates that the temporal lobe dysfunction associated with unilateral hippocampal sclerosis is not restricted to the hippocampus but extends to include the anterior part of the temporal neocortex corresponding to the temporal pole.[51] Although the structural lesion is apparently limited to the hippocampal formation, it does not mean that the connections between the medial temporal lobe structures and the neocortex or the surrounding and external temporal cortex per se remain functional. The present data does not allow us to differentiate the role of the hippocampus from the contribution of the adjacent cortical areas in auditory sequential information processing. However, it seems relevant to suggest that abnormal metabolism reported in these surrounding cortical structures reflects a temporal lobe dysfunction or an interruption of the hippocampal-neocortical loop that may be responsible for the deficits obtained in our patients with left temporal lobe dysfunction.

In the present studies, the time-related deficit following left temporal lobe lesion was reproduced in two different tasks involving purely temporal stimuli as well as multidimensional musical sequences in the context of an adaptive or a fixed level paradigm, respectively. Despite numerous methodological differences, the results of these two studies provide strong support for the selective sensitivity of left temporal lobe structures in processing rapid time-related events. Indeed, the two methodologies require the detection of subtle temporal variations that were made possible by the use of regular sequences or very familiar tunes producing strong temporal expectations. It is therefore tempting to consider that a single mechanism, depending on the processing of very brief sequential events, is responsible for impairments obtained in both tasks.

However, it is too early to assume that such impairments can be explained by a unique deficit. We already suggested that deficit in anisochronous discrimination of rapid sequences can also result from the saturation of sensory memory or from an inability to judge the temporal order. Such a hypothesis should be tested in the future to elucidate the nature of the deficit-characterizing subjects with left temporal lobe dysfunction.

## ACKNOWLEDGMENTS

We are also indebted to Xavier Gangand for programming the experiments. This research was supported by a grant from Contrat d'objectif Région Nord–Pas de Calais: Motricité et Cognition (France), and by GIS Dysfonctionnement de la Cognition.

## REFERENCES

1. EFRON, R. 1963. Temporal perception, aphasia and déjà vu. Brain **86:** 403–424.
2. TALLAL, P. & F. NEWCOMBE. 1978. Impairment of auditory perception and language comprehension in dysphasia. Brain Lang. **5:** 13–24.
3. LACKNER, J.R. & H.L. TEUBER. 1973. Alterations in auditory fusion thresholds after cerebral injury in man. Neuropsychologia **11:** 409–415.
4. MILLS, L. & G.B. ROLLMAN. 1980. Hemispheric asymmetry for auditory perception of temporal order. Neuropsychologia **18:** 41–47.
5. ROBIN, D.A., D. TRANEL & H. DAMASIO. 1990. Auditory perception of temporal and spectral events in patients with focal left and right cerebral lesions. Brain Lang. **39:** 539–555.
6. SHERWIN, L. & R. EFRON. 1980. Temporal ordering deficits following anterior temporal lobectomy. Brain Lang. **11:** 195–203.
7. SWISHER, L. & I.J. HIRSH. 1972. Brain damage and the ordering of two temporally successive stimuli. Neuropsychologia **10:** 137–152.
8. GRIFFITHS, T.D. *et al.* 1997. Spatial and temporal auditory processing deficits following right hemisphere infarction. Brain **120:** 785–794.

9. MUSIEK, F.E., J.A. BARAN & M.L. PINHEIRO. 1990. Duration pattern recognition in normal subjects and patients with cerebral and cochlear lesions. Audiology **29:** 304–313.
10. ALBERT, M.L. *et al.* 1972. A case study of auditory agnosia: linguistic and non-linguistic processing. Cortex **8:** 427–443.
11. MILNER, B. 1962. Laterality effects in audition. *In* Interhemispheric Relations and Cerebral Dominance. V. Mountcastle, Ed.: 173–201. John Hopkins University Press. Baltimore, MD.
12. LIÉGEOIS-CHAUVEL, C. *et al.* 1999. Specialization of left auditory cortex for speech perception in man depends on temporal coding. Cereb. Cortex **9:** 484–496.
13. GORDON, H. 1978. Left-hemisphere dominance for rhythmic elements in dichotically-presented melodies. Cortex **14:** 58–70.
14. HALPERIN, Y., I. NACHSHON & A. CARMON. 1973. Shift of ear superiority in dichotic listening to temporally patterned nonverbal stimuli. J. Acoust. Soc. Am. **53:** 46–50.
15. PAPÇUN, G. *et al.* 1974. Is the left hemisphere specialized for speech, language and/or something else? J. Acoust. Soc. Am. **55:** 319–327.
16. ROBINSON, G.M. & D.J. SOLOMON. 1974. Rhythm is processed by the speech hemisphere. J. Exp. Psychol. **102:** 508–511.
17. PERETZ, I. & J. MORAIS. 1980. Modes of processing melodies and ear asymmetry in non-musicians. Neuropsychologia **18:** 447–489.
18. MAVLOV, L. 1980. Amusia due to rhythm agnosia in a musician with left-hemisphere damage: a non-auditory supramodal defect. Cortex **16:** 331–338.
19. PRIOR, D.A., G. KINSELLA & J. GIESE. 1990. Assessment of musical processing in brain-damaged patients: implications for laterality of music. J. Clin. Exp. Neuropsychol. **12:** 301–312.
20. KESTER, D. *et al.* 1991. Acute effect of anterior temporal lobectomy on musical processing. Neuropsychologia **29:** 703–708.
21. MICHEL, F., F. PERONNET & B. SCHOTT. 1980. A case of cortical deafness: clinical and electrophysiological data. Brain Lang. **10:** 367–377.
22. PENHUNE, V.B, R.J. ZATORRE & W.H. FEINDEL. 1999. The role of auditory cortex in the retention of rhythmic patterns as studied in patients with temporal lobe removals including Heschl's gyrus. Neuropsychologia **37:** 315–331.
23. PERETZ, I. 1990. Processing of local and global musical information by unilateral brain-damaged patients. Brain **113:** 1185–1205.
24. REITAN, R.M. & D. WOLFSON. 1989. The Seashore rhythm test and brain functions. Clin. Neuropsychol. **3:** 70–78.
25. SHAPIRO, B.E., M. GROSSMAN & H. GARDNER. 1981. Selective musical processing deficits in brain damaged populations. Neuropsychologia **19:** 161–169.
26. LIÉGEOIS-CHAUVEL, C., *et al.* 1998. Contribution of different cortical areas in the temporal lobes to music processing. Brain **121:** 1853–1867.
27. PLATEL, H. *et al.* 1997. The structural components of music perception. A functional anatomical study. Brain **120:** 229–243.
28. DOWLING, W.J. & D. HARWOOD. 1986. Music cognition. Academic Press. New York.
29. POVEL, D.J. 1981. Internal representation of simple temporal patterns. J. Exp. Psychol.: Hum. Percept. Perform. **7:** 3–18.
30. POVEL, D.J. & P. ESSENS. 1985. Perception of temporal patterns. Music Percept. **2:** 411–440.
31. REPP, B.H. 1995. Detectability of duration and intensity increments in melody tones: partial connection between music perception and performance. Percept. Psychophys. **57:** 1217–1232.
32. POLK, M. & A. KERTESZ. 1993. Music and language in degenerative disease of the brain. Brain Cogn. **22:** 98–117.
33. FRAISSE, P. 1982. Rhythm and tempo. *In* The Psychology of Music. D. Deutsch, Ed.: 149–178. Academic Press. New York.
34. EHRLÉ, N. & S. SAMSON. 1996. Détection de l'irrégularité temporelle: influence de la durée et du tempo. *In* Sciences Cognitives, Individus et Société. Actes du 6ème Colloque de l'Association pour la Recherche Cognitive. F. Anceaux & J.M. Coquery, Eds.: 209–214. Villeneuve d'Ascq.

35. MICHON, J.A. 1964. Studies on subjective duration: I. Differential sensitivity in the perception of repeated temporal intervals. Acta Psychol. **22:** 441–450.
36. NAKAJIMA, Y., G. TEN HOOPEN & R. VAN DER WILK. 1991. A new illusion of time perception. Music Percept. **8:** 431–448.
37. TEN HOOPEN, G. *et al.* 1995. Auditory isochrony: time shrinking and temporal patterns. Perception **24:** 577–593.
38. LEVITT, H. 1971. Transformed up-down methods in psychoacoustics. J. Acoust. Soc. Am. **49:** 467–477.
39. EHRLÉ, N., S. SAMSON & M. BAULAC. 2001. Processing of rapid auditory information in epileptic patients with left temporal lobe damage without dysphasia. Neuropsychologia. **39:** 525–531.
40. MASSARO, D.W. 1972. Preperceptual images, processing time and perceptual units in auditory perception. Psychol. Rev. **79:** 124–145.
41. COWAN, N. 1984. On short and long auditory stores. Psychol. Bull. **96:** 341–370.
42. TALLAL, P., R.L. SAINBURG & T. JERNIGAN. 1991. The neuropathology of developmental dysphasia: behavioral, morphological, and physiological evidence for a pervasive temporal processing disorder. Read. Writ. Interdiscip. J. **3:** 363–377.
43. TALLAL, P., S. MILLER & R.H. FITCH. 1993. Neurobiological basis of speech: a case for the preeminence of temporal processing. Ann. N.Y. Acad. Sci. **682:** 27–47.
44. PENEL, A. & C. DRAKE. 1998. Sources of timing variations in music performance: a psychological segmentation model. Psychol. Res. **61:** 12–32.
45. REPP, B.H. 1992. Probing the cognitive representation of musical time: structural constraints on the perception of timing perturbations. Cognition **44:** 241–281.
46. EHRLÉ, N. 1998. Traitement temporel de l'information auditive et lobe temporal. Ph.D. thesis. University of Reims, France.
47. SAMSON, S. & R.J. ZATORRE. 1988. Melodic and harmonic discrimination following unilateral cerebral excision. Brain Cogn. **7:** 348–360.
48. JOHNSRUDE, I., B.B. PENHUNE & R.J. ZATORRE. 2000. Functional specificity in the right human auditory cortex for perceiving pitch direction. Brain **123:** 155–163.
49. ZATORRE, R.J. 1988. Pitch perception of complex tones and human temporal-lobe function. J. Acoust. Soc. Am. **84:** 566–572.
50. ZATORRE, R.J. & S. SAMSON. 1991. Role of the right temporal neocortex in retention of pitch in auditory short-term memory. Brain **114:** 2403–2417.
51. SEMAH, F. *et al.* 1995. Is interictal temporal hypometabolism related to mesial temporal sclerosis? A positron emission tomography/magnetic resonance imaging confrontation. Epilepsia **36:** 447–456.

# Cerebral Substrates of Musical Imagery

ANDREA R. HALPERN

*Psychology Department, Bucknell University, Lewisburg, Pennsylvania 17837, USA*

**ABSTRACT: Musical imagery refers to the experience of "replaying" music by imagining it inside the head. Whereas visual imagery has been extensively studied, few people have investigated imagery in the auditory domain. This article reviews a program of research that has tried to characterize auditory imagery for music using both behavioral and cognitive neuroscientific tools. I begin by describing some of my behavioral studies of the mental analogues of musical tempo, pitch, and temporal extent. I then describe four studies using three techniques that examine the correspondence of brain involvement in actually perceiving vs. imagining familiar music. These involve one lesion study with epilepsy surgery patients, two positron emission tomography (PET) studies, and one study using transcranial magnetic stimulation (TMS). The studies converge on the importance of the right temporal neocortex and other right-hemisphere structures in the processing of both perceived and imagined nonverbal music. Perceiving and imagining songs that have words also involve structures in the left hemisphere. The supplementary motor area (SMA) is activated during musical imagery; it may mediate rehearsal that involves motor programs, such as imagined humming. Future studies are suggested that would involve imagery of sounds that cannot be produced by the vocal tract to clarify the role of the SMA in auditory imagery.**

**KEYWORDS: Musical imagery; Behavioral studies; Cognitive neuroscience**

## INTRODUCTION

Many people experience the sounds of music in two distinct but related manners. Listening to live or recorded music is, of course, the way we commonly think of enjoying music. However, many people also report that they can reexperience music by imagining it in their heads. This can be pleasurable or not, depending on the circumstances, but in either case, the experience appears to be a vivid one, even among people untrained in music. In fact, I am often asked how to "stop" a tune from obsessively intruding into everyday thoughts. Highly trained musicians report that they can use auditory imagery to help them in their everyday tasks, such as "hearing" music as they read musical notation.

In recent years, I have studied the characteristics of this auditory imagery experience, initially from a behavioral perspective and more recently from a cognitive neuroscientific perspective. I have been most interested in the auditory imagery experiences of untrained or moderately trained musicians, although many interesting questions derive from considering experts as well. My paradigms and experimental

Address for correspondence: Andrea R. Halpern, Psychology Department, Bucknell University, Lewisburg, PA 17837. Voice: 570-577-1295; fax: 570-577-7007.
ahalpern@bucknell.edu

logic have been derived partly from the more extensive literature in visual imagery; other approaches have been created anew as the need arose. In studying any mental imagery, the challenge is to externalize what is essentially an internal experience to examine what it means to have, in the case of musical imagery, a "tune inside the head."

## BEHAVIORAL APPROACH

In the 1980s, I carried out an extensive series of experiments with young adult college students to explore how auditory imagery of music may be characterized. I explored the mental representation of *tempo*[1,2] and *pitch*[3] in familiar songs. To study representation of tempo, I asked participants to set their preferred tempo for a familiar tune being played on a computer. The program allowed continuous adjustments of tempo until the tune sounded right to the listeners. I then asked them to imagine the same tunes and set a metronome to the imagined tempo. Tempo settings were highly correlated in the perception and imagination tasks. I also showed that people can manipulate the tempo of an imagined song, within limits, to make it "sound" very fast or very slow. Finally, I showed that musicians showed impressive consistency when asked to tap out the tempo of an imagined song on one day and again two to five days later (average standard deviation = 1.4 metronome settings over four attempts), although nonmusicians were somewhat less consistent.

In the domain of pitch, I asked people to imagine familiar songs and hum the starting note corresponding to their auditory image of the song. In a second task, I asked people to think of the starting note of a familiar song and select that note from a piano keyboard (all the tunes were popular folk or children's songs, unlikely to have been heard with any particular starting note in the past). Once again, consistency of pitch production or selection was impressive for musically unselected participants, even over a delay of several days (average standard deviation = 1.25 semitones for production, 2 semitones for selection over four attempts). Finally, I asked people to rate how similar a played note was to the opening note of a tune they had imagined and produced a few days earlier. The person's own preferred note was rated quite highly, as were starting notes a major third lower, minor third higher, and a perfect fifth higher than the preferred notes (these are all musically coherent intervals). However, subjects rated notes only one semitone higher or lower than their preferred note as dissimilar to their imagined pitch, showing a fairly acute sense of pitch representation.

These behavioral tasks seem to show a veridical representation of characteristics particular to music. However, I also explored the *extension in time* that is characteristic of almost all auditory stimuli and thus ought to be captured in auditory imagery. I explored this temporal aspect in the subsequent cognitive neuroscientific studies of auditory imagery for music.

To begin, I modified a paradigm introduced in visual imagery by Kosslyn, Ball, and Reiser.[4] In that study, they asked people to learn a map of an imaginary island. Pairs of features on the island were presented, and subjects had to "mentally scan" between them. Latencies to do so were highly correlated with actual distance between the features on the map, suggesting that the mental representation was preserving an analogue of space.

To extend this paradigm to musical imagery,[5] I selected a number of familiar songs where unique lyrics fell on specified beats of the tune. For instance, the first line of the American national anthem, "The Star Spangled Banner," is, "Oh, say can you see by the dawn's early light?" In the most relevant study in that series, I presented the title of the song, followed by one lyric from the first line of the song, followed by the second lyric. The task was mental pitch comparison: Was lyric 2 higher or lower in pitch than lyric 1? No singing or humming was allowed. The lyrics were either close together ("Oh" and "can") or far apart in the actual tune ("Oh" and "dawn's"). I found that this was a difficult task for my musically unselected subjects; nevertheless, reaction times increased nearly linearly as the separation in beats between the two lyrics increased. Subjects reported using auditory imagery to accomplish the task, even though they were not instructed to do so. The consistency of reaction time data with this report strongly suggested to me that this task had captured the extension in time of auditory imagery for music.

## COGNITIVE NEUROSCIENTIFIC APPROACH

The behavioral studies cited above have in common the logic that if responding to an imagined stimulus resembles responding to a perceived stimulus, we may conclude that imagery is a particularly vivid and veridical form of mental representation. However, this comparison between imagery and perception may be strengthened by examining the similarities in neural underpinnings of the two processes. This approach is complementary to the behavioral approach in that at least to some extent behavioral responses might be influenced by external influences such as demand characteristics or experimenter expectancies.[6] However, it is unlikely that people can influence their own brain structures or activities. We may thus look to similarities in the brain loci involved in auditory imagery and perception to gain a better perspective on the processing of similarities and differences in the two types of tasks. An argument for this approach is well articulated by Farah[7] for the visual domain.

The strongest hypothesis is of course that the brain areas would be identical in auditory imagery and perception. This may serve as a guiding null hypothesis, but we would not in reality expect this amount of overlap; people other than those hallucinating can tell the difference between imagining and hearing a song. However, the extent of overlap may tell us how similarly the brain processes hearing and imagining hearing. Brain areas uniquely active in imaging tasks can by extension inform us as to the additional or alternative processing demands imposed in imagery by having to, in effect, create as well as perceive the stimulus.

As noted above, the approach has been reasonably successful in the visual domain. For instance, Kosslyn *et al.*[8] used positron emision tomography (PET) to measure brain activity during parallel perceptual and imagined visual tasks. They found quite a few areas activated in common, even to the extent that varying the size of the presented object and varying the size of an imagined object activated similar brain areas in similar ways. Few people have looked at auditory imagery using parallel perceptual and imagery tasks. However, the neural structures responsible for some aspects of musical perception are well defined. My partner for most of the studies described in the following sections has been Robert Zatorre. When we began our first study, he and Samson, among others, had already established that a number of musical tasks are impaired after lesions in the right temporal neocortex.[9–12] Thus our initial hypothesis was that the same region would be involved in musical imagery tasks that resembled musical perceptual tasks.

What follows are brief descriptions of a series of studies using three different cognitive neuroscience techniques to investigate the cerebral substrates of musical imagery. The first is a lesion study, which can give information about the necessary involvement of some brain areas in an activity. The next two studies use PET paradigms to investigate brain areas that are active in auditory imagery for verbal and nonverbal songs, respectively. Finally, I describe a study using transcranial magnetic stimulation (TMS) that returns to the logic of the lesion studies, in that a brain area in normal people is disrupted for a brief time and the ensuing decrements in performance are measured.

## LESION STUDY

My first study with Zatorre[13] examined the effect of right temporal lobectomy on performance of a mental pitch comparison task that I described earlier.[5] Participants were patients having undergone surgical excision of the anterior portion of the right or left temporal lobe (excluding the primary auditory cortex) for relief of intractable epilepsy. We tested the patients either two weeks after surgery (approximately two-thirds of the patients), or at follow-up medical appointments a year or more after surgery. A control group consisted of age- and education-matched neurologically normal individuals. Preoperative testing insured that the patients had typical language representation. All participants were familiar with the songs we used, and they all passed a brief test of pitch discrimination ability.

The imagery task was essentially the one described earlier. Participants saw a title of a song, for instance, "Jingle Bells," followed by the first line of the song, with two words in capital letters, such as "Dashing through the SNOW, in a one-horse open SLEIGH." They decided if SLEIGH was higher or lower in pitch than SNOW, and pressed a button to answer. The parallel perception task, which was always presented first, was the same except that the song, sung with lyrics, was actually presented to participants from a digitized sound file while they made their judgment. Accuracy and reaction times were recorded. The reaction time pattern replicated the pattern I had shown earlier[5] of increasing latency with increasing distance in beats between the lyrics, although accuracy turned out to be the measure of interest here.

Accuracy results are shown in FIGURE 1. It is clear that the imagery task was more difficult than the perception task, as expected. It is also clear that the right temporal lobectomy group was impaired relative to controls on both the imagery and perception tasks, to an equal extent, whereas the left temporal lobectomy group was impaired on neither.

This pattern was consistent with our hypothesis that the right temporal lobe is an important mediator of musical imagery, as it has been shown to be for musical perception. We considered which particular aspect of the task was most likely to have been subserved by the right temporal lobe. All participants passed a simple tone discrimination task, so a deficit in pitch comparison itself was unlikely to be the cause of the group's impairment. Furthermore, this group was impaired in the perception task, when the tune did not need to be generated from memory, making memory re-

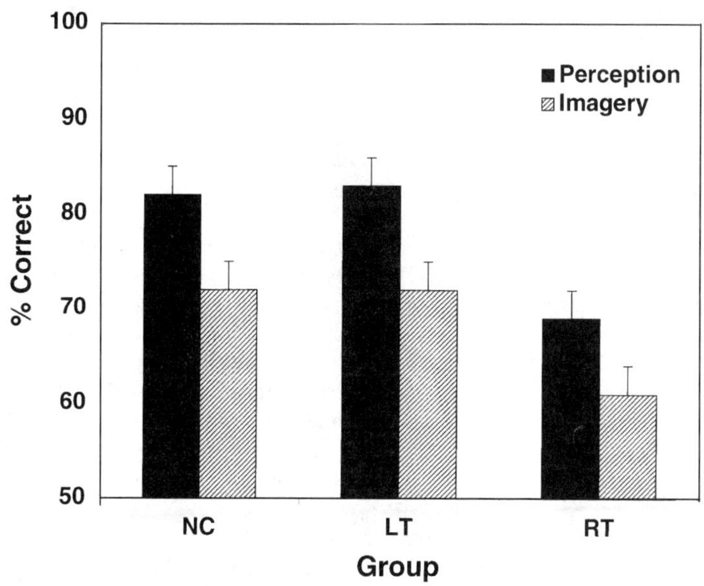


**FIGURE 1.** Mean percent correct, plus standard error, for normal control subjects (NC), patients with left temporal lobe excision (LT), and patients with right temporal lobe excision (RT) in perception and imagery tasks in the lesion study.

trieval unlikely to be the right temporal lobe's crucial contribution. Thus we concluded that the right temporal deficit is likely to be primarily due to a difficulty in remembering the (imagined) pitches to be compared.

## PET STUDIES

Lesion studies allow one to draw conclusions about the areas *crucial* to perform certain tasks. They do not, however, allow any conclusions about what brain areas are *active* in the tasks, whether or not they are crucial. Lesion studies also are confined to the particular excisions made available by accidents of nature or deliberate surgery. If we want to study the somewhat broader question of what areas throughout the brain are active during imagery and perception tasks, then functional brain imaging technology is a useful addition to our investigations. Another advantage of brain imaging techniques is that neurologically normal participants can be studied, in contrast to the neurologically abnormal people studied, by definition, in lesion work. In our next study, Zatorre and I[14] used positron emission tomography (PET) to observe the changes in cerebral blood flow (CBF) as participants performed musical imagery tasks similar to the ones already described.

In this approach, tasks are set up in a series so that more complex tasks include elements of the simpler tasks presented to subjects. Subtracting the activation observed in the simpler task from that observed in the more complex task allows one to isolate activation unique to the critical components of the more complex task. For instance, in almost any cognitive task, the brain activation due to the simple registration of auditory or visual stimuli, and the motor activation involved in pressing buttons, are not of major interest. Therefore, control tasks are often presented that involve simply seeing or hearing a stimulus and pressing a button in response. Subtracting activation due to these lower-level components gives a better indication of the mental work involved in the tasks of more interest.

In our study, we presented the imagery and perception tasks that we used previously,[14] except we presented single words instead of complete first lines from the songs. A third task was meant to control for such simpler tasks as visual perception and button pressing. This visual baseline presented pairs of words used in the imagery and perception task, but each pair member came from a different song. For each pair presented on the screen, participants had to judge which one was longer in length and press a button for their choice. Because the pairs were scrambled, we assumed listeners would not be reminded of songs; thus auditory imagery should not be activated. But over the course of the baseline condition, all words used in the main task would be seen. All songs were very familiar to the subjects, who were 12 healthy right-handed young adults. Most had some musical training but none were serious musicians.

All participants underwent a magnetic resonance imaging (MRI) scan in order to allow us to later localize CBF activity to the appropriate anatomy for each person. For the test session, the three tasks were explained and there was a short practice session. After being placed in the scanner, the subjects received an injection of radioactively labeled water ($H_2{}^{15}O$) required to index the CBF activity, and performed the visual baseline task. This was then repeated for the perception task and then the imagery task, in that order.

As expected, people were more accurate on the perception task than the imagery task, and once again reaction times in both imagery and perception increased as a function of distance between the lyrics in the real tunes. The analysis then proceeded by subtracting the activation shown in the baseline task from the perception task and also the baseline task from the imagery task. A graphic representation of the results is shown in panel I of FIGURE 2. Here we see in the upper part of panel I that the primary auditory area (located in superior temporal gyrus, or STG) is quite active when listening to sounds, as one would expect. More interesting is the fact that several areas of the STG adjacent to the primary auditory cortex (secondary auditory cortex) are also active when people were just imagining the sounds (lower part of panel I). Panel II shows several other areas of correspondence between the imagery and perception tasks: several areas in the frontal lobe were active in both tasks, as was one area in the parietal lobe. Panel III shows activity in the supplementary motor area (SMA, involved in motor planning) in both tasks, stronger in imagery than perception. When we subtracted the activity in the perception task from that in the imagery task, only four brain areas were unique to imagery, two of which are pictured in panel IV. These two areas, the thalamus and inferior frontopolar areas, are known to be involved in memory functions, and may be associated with the extensive memory demands attached to the imagery task.

The results of this study supported several of our earlier ideas but also raised new questions. Consistent with our lesion study,[13] several areas of the temporal lobe classified as secondary auditory cortex were activated when people were carrying out

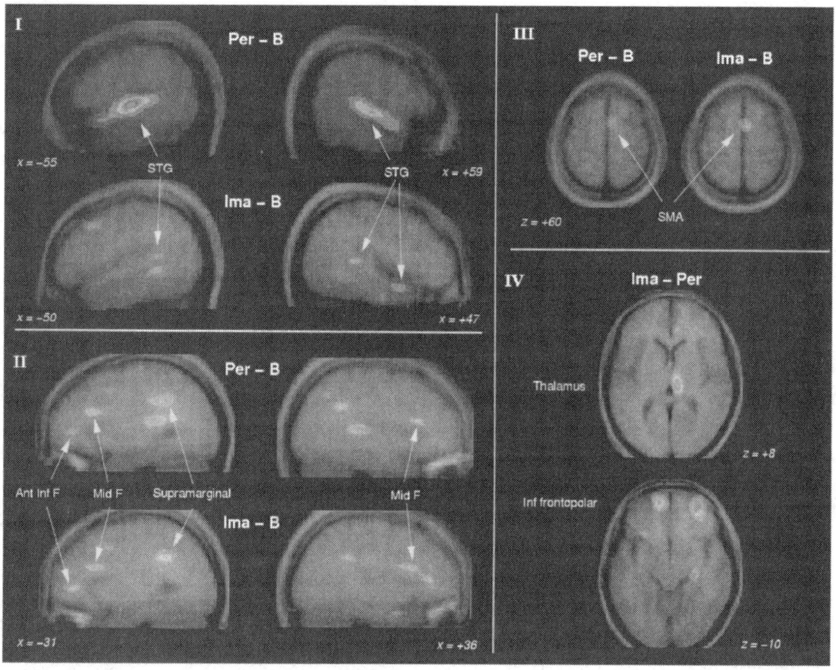


**FIGURE 2. See Plate 2 in color section.** Brain areas showing activation in the perception minus baseline, imagery minus baseline, and imagery minus perception conditions of the verbal tune PET study.

mental pitch comparisons, in the absence of any overt auditory stimulation. Thus we concluded that the right STG is both *active* during such tasks (shown by the PET study), as well as *necessary* for carrying out the task successfully (shown by the lesion study). We were also not surprised by the areas of activation in the frontal lobes. Activation in frontal areas similar to the ones seen here was found by Zatorre *et al.*[15] in a PET study in which participants had to judge relative pitches of notes in novel (perceived) tunes, and in a study where the pitches of spoken syllables were compared.[16]

Two results were more unexpected. First, contrary to the literature cited earlier that consistently showed right-sided asymmetries in brain activation during music processing,[15] and deleterious consequences of right-sided lesions for music processing,[9–13] here almost all the activated areas were bilateral, except for right-sided activation in the thalamus. A second unexpected finding was the strong activation of the SMA during imagery even though subjects were not actually producing any motor movements. Both these findings might be related to the fact that the stimulus songs all had words, which participants had to access during the tasks. The left frontal regions have been implicated in various overt verbal tasks,[17,18] so that the bilateral activation seen here might be reflecting the dual nature of the stimulus to be processed (words plus music). Activation of the SMA has been observed when subjects are asked to generate internal speech[19] and to overtly vocalize music.[20] Thus the SMA activation observed here in our imagery task may reflect a subvocal rehearsal process of either words or music to support performance on an otherwise difficult task.

In light of these issues, our next PET study[21] used only nonverbal tunes. We were interested to see whether removing words from our task would lead to more right-sided activation than we saw earlier, consistent with various findings in the music perception literature. We also wanted to see if SMA would be active even when potential rehearsal devices would not involve words. Finally, we wanted to try a different type of imagery task to see if our results would generalize over paradigms.

To this end, we developed a stimulus pool of tunes that were familiar but did not have lyrics. These included movie and television themes, classical excerpts, and miscellaneous tunes such as the Westminster (Big Ben) chimes. In our main task (cue/imagery) we played the first few notes of a theme as a cue, and asked participants to imagine the theme to the end of its first phrase (this task and the materials had been presented to people in advance of scanning to familiarize them with task parameters). To have a behavioral index of auditory imagery, the played excerpts differed in length. If subjects were carrying out instructions as we intended, latency to press the button should increase from our shortest (2.2 s on average) to longest excerpts (6.2 s on average).

We also had some control tasks. For these, we took the first few notes of each real tune, and scrambled the note order so that the cue did not elicit a memory of any real tune. In the control task for simple listening and button pressing we presented these "fake" cues and simply had people press a button after each one (control). A second control task involved imagery but no retrieval from long-term memory. In this task, we presented the fake cue and asked for people to simply reimagine it immediately after presentation (control/imagery). The subtraction of interest for current purposes is the cue/imagery minus the control task.

As previously, PET scanning was undertaken in conjunction with an MRI to provide anatomical localization of CBF activation for each person. Eight healthy, right-handed young adults participated, who had from 3 to 16 years of musical training. The conditions were presented in order of control, cue/imagery, and control/imagery.

As predicted, the average time to press a button indicating imaging of the tune was complete in the cue/imagery condition varied proportionally to the length of the tune. Thus we are confident that subjects were following our imagery instructions. The results of the cue/imagery minus control subtraction are shown in FIGURE 3. As in our previous study,[14] we found activation in the secondary auditory cortex (marked STG), although this time the activation was in the right but not left temporal lobe. We also found activation in several regions of the frontal lobe (inf F), most of which were more prominent on the right than the left side. Finally we once again found strong activation in the SMA.

We thus confirmed several findings from our study with verbal tunes:[14] areas normally concerned with processing of auditory information are recruited even when the auditory information is internally generated. This occurred even with a different behavioral task and different songs than had been used earlier. We also confirmed the activity of the SMA in our task. The fact that this area was active even though no verbal rehearsal could logically have been taking place suggests that the SMA is involved with some kind of subvocal humming strategy during the imagery generation.

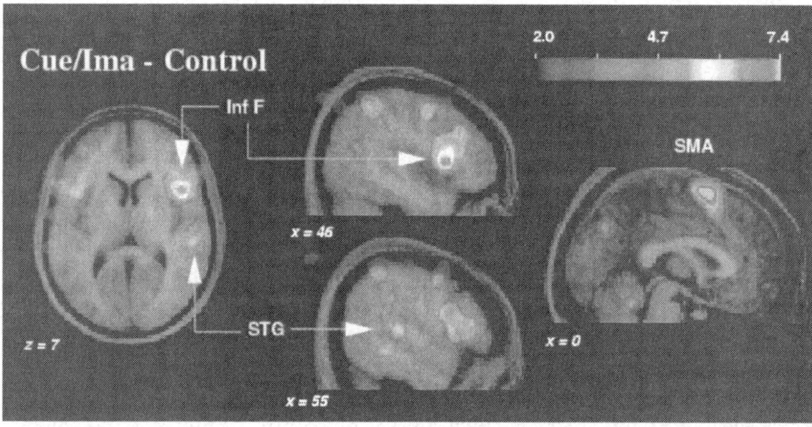


**FIGURE 3. See Plate 3 in color section.** Brain areas showing activation in the cue/imagery minus control subtraction in the nonverbal tune PET study.

The frontal areas activated here are also approximately the same as some of the frontal areas seen in our previous task. Because the current task did not require extensive involvement of working memory (mental pitch comparison was not required, merely internal scanning of the tune), it is likely that the areas in common in our two studies reflect retrieval from musical semantic memory rather than working memory.

The major new contribution here was that once words were removed from the stimulus and the task, we found prominent right-sided asymmetry in the areas active in the imagery task. The temporal lobe activation is consistent with our lesion study[13] that showed similar decrements from right-sided temporal lobectomy in both imagery and perception tasks. Thus we conclude that this area is both active in and necessary to the support of musical imagery tasks. The second right-sided asymmetry involved a region in the frontal lobes. This area, on the left, has been implicated in retrieval from verbal semantic memory,[22] suggesting that retrieval from semantic memory may be lateralized depending on the type of material. Another right-sided asymmetry was found in activation of the thalamus, a subcortical area involved in memory among other functions. In our earlier PET study,[14] that activation was above our statistical threshold; here it was just below the threshold and thus was not pictured with our main results. However, we can take this as at least suggestive evidence that a circuit involving temporal, frontal, and subcortical areas on the right are important in imagined music processing.

## TRANSCRANIAL MAGNETIC STIMULATION

The final study in the series I am reviewing is currently unpublished, and was conducted in partnership with Alvaro Pascual-Leone, Fumiko Maeda, and Gottfried Schlaug. Transcranial magnetic stimulation (TMS) is a noninvasive method of affecting underlying brain tissue by application of a magnetic pulse to the outside of the skull. The pulse is discharged via a coil that can be placed on the skull overlying the brain area of interest, using external anatomical landmarks, or in our case an MRI of each subject, to locate the areas of interest. A high-frequency series of pulses (10–20 Hz) has been shown to excite the underlying brain tissue, and low-frequency (1-Hz) sequences have been shown to inhibit the underlying tissue.[23] The inhibition is temporary, and the technique is safe for most people, although some discomfort may be experienced by the tapping sound the coil makes. Precautions are taken to exclude people with a history of seizure, as well as people who would be excluded from any study involving MRI, such as anyone with metal implants in the skull. The excitatory aspect of TMS is of interest to clinicians and researchers in the treatment of mental illness such as depression.[24] The inhibitory function of low-frequency TMS is of interest to researchers in that a temporary "lesion" can be created in otherwise normal people, and subsequent effects on behavior can be studied.

TMS has already been used to investigate visual imagery. Kosslyn, Pascual-Leone, and colleagues[25] showed that an application of 10 min of 1-Hz TMS to the visual cortex impaired later performance of both a visual perception and equivalent visual imagery task. They took this as evidence, in accord with lesion logic, that the primary visual cortex is necessary for performance of the visual imagery task, strengthening the argument that visual perception and imagery are mediated by common structures.

In our TMS study, we selected three brain areas that had been shown to be active during auditory imagery in my previous work, plus one control area. We also wanted to look at auditory imagery for both verbal and nonverbal tunes, as some different brain areas were shown to be active for these two types of tasks. To this end, we needed an auditory imagery task that would be suitable for both types of tunes. We modified a task used by Smith,[26] in which participants are given the title of a familiar tune and then asked about a pitch relationship between two notes indexed by ordinal position. In our case, we asked participants to judge whether the second note of a given tune was higher or lower than its first note. Verbal and nonverbal tunes could both be tested, and we intermixed the two types in the test.

The general outline of a session was to administer the auditory imagery task as a pretest. We then determined each person's motor threshold, or the lowest amount of energy that, when applied to the skull overlying the hand area of the motor cortex, just made the person's finger twitch. Each person received TMS at 90% of that value. Administration of 10 min of 1-Hz TMS to a designated brain area ensued, followed immediately by another auditory imagery task as a posttest (using different tunes for pretest and posttest). When more than one brain area was investigated in a single session, 30 min of rest intervened between testing different brain areas, to allow the effects of the TMS to dissipate entirely.

Participants were nine right-handed adults with average age of 34, all of whom had some musical background (four were active musicians). Each person underwent an MRI scan before the TMS session. This allowed us to locate, using anatomical measurements, the areas of interest to us. The control area we selected was primary visual cortex (coil placed at the back of the skull), which had not been implicated in auditory imagery tasks before. The three experimental areas were the SMA (top of the skull a little more than midway towards the back), and the left and the right auditory cortex (above and slightly behind the top of the ear in most people). Because

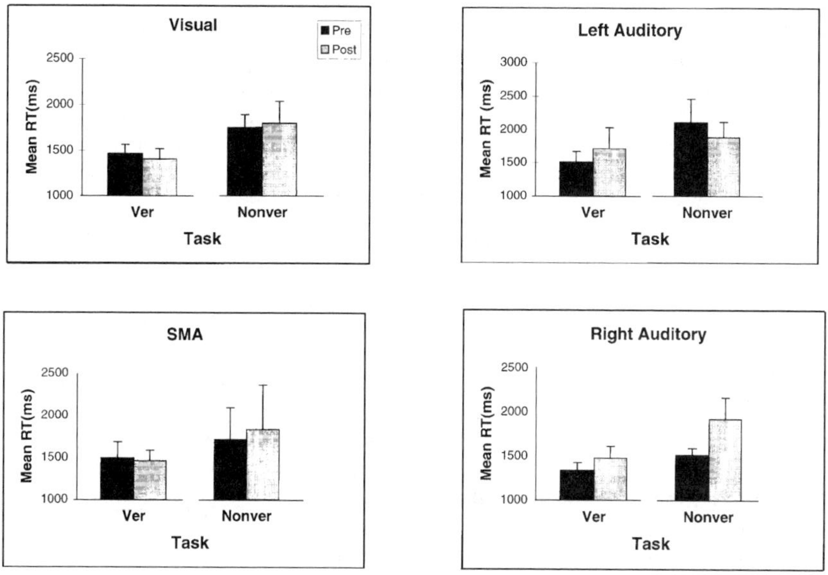


**FIGURE 4.** Mean reaction time before and after TMS for the verbal (Ver) and nonverbal (Nonver) imagery tasks, after application to the visual cortex, supplementary motor area (SMA), left auditory cortex, and right auditory cortex.

of the anatomy of the auditory cortex, we could not localize our coil placement exactly on secondary auditory cortex. We located the coil instead as near as possible on the primary auditory cortex, with the assumption that the TMS activation would also likely affect the secondary cortex that is immediately adjacent to that area. Our prediction was that after 10 min of low-frequency stimulation, performance overall (in accuracy and/or time) would decrease for the experimental but not for the control areas. We made a further prediction that both left and right auditory cortex stimulation would impair processing of verbal tunes, but that the right auditory cortex stimulation would only impair nonverbal tunes.

Brain areas were tested in different orders for each person. However, as early results seemed to indicate that stimulation to the SMA was showing no effect, three volunteers did not receive SMA stimulation, in an effort to reduce subject time and discomfort. Thus all results except those for SMA were based on nine subjects; those for SMA were based on six subjects.

Accuracy was quite high in the task, so results concentrated on reaction times. FIGURE 4 shows the results for each brain area for the verbal and the nonverbal tasks, both before and after TMS stimulation. The nonverbal tunes are a little less familiar to people, and thus they were responded to on average more slowly than the verbal tunes. Inspection of the four panels shows that in accord with the prediction, the TMS made no difference when applied to the visual cortex. But, surprisingly, TMS made no significant difference when applied to SMA and left auditory cortex. The lower right panel does show a predicted effect: TMS applied to right auditory cortex slowed down responses to the imagery task, but only for the nonverbal tunes.

Stimulation to the two auditory areas were the most comparable in terms of subjective experience: the coil clicks were rather loud in the ear, and some jaw muscles were activated by the stimulation. Thus we compared those two areas statistically in an analysis of variance, using the three factors of tune (verbal, nonverbal), area (left, right auditory) and test (pre-, post-); all factors were within subjects. The main effect of tune was significant, $F(1,8) = 6.24$, $p < 0.05$, but the crucial three-way interaction of tune, area, and test just failed to reach significance, $F(1, 8) = 3.26$, $p = 0.10$). Because of the strong prediction about these areas, we made a specific comparison of the pre- and posttests after right auditory stimulation. The comparison was not significant for verbal tunes, but was for nonverbal tunes, Newman-Keul's $t = 4.173$, $p < 0.05$.

As far as I know, this study was the first use of TMS to investigate auditory areas. As mentioned above, the coil had to be localized over primary auditory cortex, which in my previous studies was not activated in auditory imagery tasks. Thus we were relying on the somewhat weaker collateral activation to the secondary auditory cortex to show inhibition in the areas of more interest to us. Therefore, although the results were not as statistically strong as we would like, the increase in latency to nonverbal tunes after nonverbal tunes only is consistent with the pattern of data shown in our PET study with nonverbal tunes.[21] The isolation of this effect to nonverbal tunes is particularly notable, because the verbal tunes were tested in the same session with the same subjects, and did not elicit the inhibition.

The lack of effect after SMA stimulation, after strong activation of this area shown in two PET studies, is more puzzling. Null effects are always hard to interpret, but two possibilities suggest themselves. One is methodological: the neural fibers subserving SMA bend as they move ventrally from the skull surface, and it may be that our particular configuration of coil type and its placement simply did not activate the area sufficiently. On the other hand, it is possible that although SMA provides useful rehearsal mechanisms during auditory imagery tasks, it may not be an area that *must* be involved in such tasks. The task used here required retrieval of the tune once a title was given, but only the first two notes had to be retrieved. Perhaps SMA is more useful when auditory imagery tasks that extend more in time are required, such as "scanning" between two notes several beats apart[14] or "playing" several seconds of a familiar tune inside one's head.[21] Pilot versions of the TMS task, in fact, asked subjects to compare the first and third notes, which might involve SMA to a greater extent; this could be tried in the future (although accuracy rates would be lower, making interpretation of reaction times more problematic).

## CONCLUSIONS AND DIRECTIONS

The studies just reviewed seem to converge on the general idea that parts of the cortex specialized for processing actual sound are also recruited to process imagined sound. Furthermore, the particular structures processing imagined music bear some similarity to those processing heard music. Specifically, the right temporal lobe has been shown to be active in, and necessary for, adequate performance on both perceived and imagined pitch comparison tasks derived from previously known music. This structure may be especially important for processing music that does not have words, when the left temporal structures cannot be recruited for verbal processing.

The role of the SMA remains unclear, in several respects. Because of our failure to disrupt imagery processing with TMS inhibition, we are not sure if a fully functioning SMA is necessary for good performance on musical imagery tasks. The strong activation of SMA shown by PET in both verbal and nonverbal auditory imagery tasks suggests that "internal humming" may at least be helpful to maintain an auditory image over time.

This latter point can be subjected to test by devising auditory imagery tasks in which internal humming would not be helpful or even possible to support the task. One domain to investigate would be imagery for environmental sounds. Intons-Peterson[27] has shown that people can generate images of everything from volcanoes erupting to wind chimes tinkling, and make mental comparisons on subjective loudness of these sounds. As people cannot possibly produce most of these sounds, it would be useful to find a task that shows activation in secondary auditory areas (which we can, by now, use as a tentative correlate of the subjective state of auditory imagery) and see if SMA is also activated.

An alternative domain to explore might be that of timbre. Crowder[28] demonstrated in several behavioral paradigms that people asked to imagine timbres of different instruments can do so. He gave people a sine wave of a particular pitch and asked them to imagine it in an instrument timbre. A second tone was then presented in an actual timbre at the same or a different pitch and subjects had to say if the pitch was the same or different as the first note. People were slower to confirm that two pitches were the same if the imagined and perceived timbres did not match. As people cannot produce the sounds of guitars and clarinets, SMA support should not be necessary if it is providing a motor rehearsal program. Timbre imagery tasks would also have the advantage of providing us a lateralization hypothesis, as Samson and Zatorre[11] showed the importance of the right temporal lobe in timbre tasks. Thus brain imaging studies using either PET or functional magnetic resonance imaging in these domains should help us clarify the role of the SMA in auditory imagery tasks.

## ACKNOWLEDGMENTS

I thank the National Science Foundation and the McDonnell-Pew Program in Cognitive Neuroscience for support of this research.

## REFERENCES

1. HALPERN, A.R. 1988. Perceived and imagined tempos of familiar songs. Music Percept. **6:** 193–202.
2. HALPERN, A.R. 1992. Musical aspects of auditory imagery. *In* Auditory Imagery. D. Reisberg, Ed.: 1–27. Lawrence Erlbaum. Hillsdale, NJ.
3. HALPERN, A.R. 1989. Memory for the absolute pitch of familiar songs. Mem. Cognit. **17:** 572–581.
4. KOSSLYN, S.M., T.M. BALL & B.J. REISER. 1978. Visual images preserve metric spatial information: evidence from studies of image scanning. J. Exp. Psychol. Hum. Percept. Perform. **4:** 47–60.

5. HALPERN, A.R. 1988. Mental scanning in auditory imagery for tunes. J. Exp. Psychol. Learn. Mem. Cognit. **14:** 434–443.
6. INTONS-PETERSON, M.J. 1983. Imagery paradigms: how vulnerable are they to experimenters' expectations? J. Exp. Psychol. Learn. Mem. Cognit. **10:** 699–715.
7. FARAH, M.J. 1988. Is visual imagery really visual? Overlooked evidence from neuropsychology. Psychol. Rev. **95:** 307–317.
8. KOSSLYN, S.M., N.M. ALPERT, W.L. THOMPSON, *et al.* 1993. Visual mental imagery activates topographically organized visual cortex: PET investigations. J. Cognit. Neurosci. **5:** 263–287.
9. ZATORRE, R.J. 1985. Discrimination and recognition of tonal melodies after unilateral cerebral excisions. Neuropsychologia **23:** 31–41.
10. ZATORRE, R.J. 1988. Pitch perception of complex tones and human temporal-lobe function. J. Acoust. Soc. Am. **84:** 566–572.
11. SAMSON, S. & R.J. ZATORRE. 1994. Contribution of the right temporal lobe to musical timbre discrimination. Neuropsychologia **32:** 231–240.
12. ZATORRE, R.J. & S. SAMSON. 1991. Role of the right temporal neocortex in retention of pitch in auditory short-term memory. Brain **114:** 2403–2417.
13. ZATORRE, R.J. & A.R. HALPERN. 1993. Effect of unilateral temporal-lobe excision on perception and imagery of songs. Neuropsychologia **31:** 221–232.
14. ZATORRE, R.J., A.R. HALPERN, D.W. PERRY, *et al.* 1996. Hearing in the mind's ear: a PET investigation of musical imagery and perception. J. Cognit. Neurosci. **8:** 29–46.
15. ZATORRE, R.J., A.C. EVANS & E. MEYER. 1994. Neural mechanisms underlying melodic perception and memory for pitch. J. Neurosci. **14:** 1908–1919.
16. ZATORRE, R.J., A.C. EVANS, E. MEYER & A. GJEDDE. 1992. Lateralization of phonetic and pitch processing in speech perception. Science **256:** 846–849.
17. SMITH, E.E., J. JONIDES & R.A. KOEPPE. 1996. Dissociating verbal and spatial working memory using PET. Cereb. Cortex **6:** 11–20.
18. GABRIELI, J.D.E., J.E. DESMOND, J.B. DEMB, *et al.* 1996. Functional magnetic resonance imaging of semantic memory processes in the frontal lobes. Psychol. Sci. **7:** 278–283.
19. PAULESU, E., C.D. FRITH & R.S.J. FRACKOWIAK. 1993. The neural correlates of the verbal component of short-term memory. Nature **362:** 342–345.
20. PERRY, D., M. PETRIDES, B. ALIVASATOS, *et al.* 1993. Functional activation of human frontal cortex during tonal working memory tasks. Soc. Neurosci. Abst. **19:** 843.
21. HALPERN, A.R. & R.J. ZATORRE. 1999. When that tune runs through your head: a PET investigation of auditory imagery for familiar melodies. Cereb. Cortex **9:** 697–704.
22. NYBERG, L., R. CABEZA & E. TULVING. 1996. PET studies of encoding and retrieval. Psychol. Bull. Rev. **3:** 135–148.
23. WALSH, V. & M. RUSHWORTH. 1999. A primer of magnetic stimulation as a tool for neuropsychology. Neuropsychologia **37:** 125–135.
24. PRIDMORE, S. & R. BELMAKER. 1999. Transcranial magnetic stimulation in the treatment of psychiatric disorders. Psychiatry Clin. Neurosci. **53:** 541–548.
25. KOSSLYN, S.M., A. PASCUAL-LEONE, O. FELICIAN, *et al.* 1999. The role of area 17 in visual imagery: convergent evidence from PET and rTMS. Science **284:** 167–169.
26. SMITH, J.D., M. WILSON & D. REISBERG. 1995. The role of subvocalization in auditory imagery. Neuropsychologia **33:** 1433–1454.
27. INTONS-PETERSON, M.J. 1980. The role of loudness in auditory imagery. Mem. Cognit. **8:** 385–393.
28. CROWDER, R.G. 1989. Imagery for musical timbre. J. Exp. Psychol. Hum. Percept. Perform. **15:** 472–478.

# Neural Specializations for Tonal Processing

ROBERT J. ZATORRE

*Montreal Neurological Institute, McGill University, Montreal, Quebec H3A 2B4, Canada*

**Abstract: The processing of pitch, a central aspect of music perception, is neurally dissociable from other perceptual functions. Studies using behavioral-lesion techniques as well as brain imaging methods demonstrate that tonal processing recruits mechanisms in areas of the right auditory cortex. Specifically, the right primary auditory area appears to be crucial for fine-grained representation of pitch information. Processing of pitch patterns, such as occurs in melodies, requires higher-order cortical areas, and interactions with the frontal cortex. The latter are likely related to tonal working memory functions that are necessary for the on-line maintenance and encoding of tonal patterns. One hypothesis that may explain why right-hemisphere auditory cortices seem to be so important to tonal processing is that left auditory regions are better suited for rapidly changing broad-band stimuli, such as speech, whereas the right auditory cortex may be specialized for slower narrow-band stimuli, such as tonal patterns. Evidence favoring this hypothesis was obtained in a functional imaging study in which spectral and temporal parameters were varied independently. The hypothesis also receives support from structural studies of the auditory cortex, which indicate that spectral and temporal processing may depend on interhemispheric differences in grey/white matter distribution and other anatomical features.**

**Keywords: Hemispheric functional specialization; Pitch processing; Pitch patterns**

How does the nervous system analyze and make sense of information coming from the environment? This central question in contemporary neuroscience has been addressed from many perspectives in the last two decades. In the auditory domain, a great deal of effort has gone into understanding the way in which the human brain processes speech sounds. Given the importance of speech and spoken language to human communication and behavior, this is not surprising. It is interesting, however, to consider other aspects of auditory information processing that may help to round out the picture. Increasingly, music is being recognized as an important component of human activity that may help us to gain insight into the functional organization of the human brain. It may be argued that the perception, encoding, and reproduction of musical sounds requires neural mechanisms that are at least as complex as those for speech; indeed, it seems likely that speech and music must engage the most cognitively demanding aspects of auditory processing. It may further be argued that

Address for correspondence: Robert J. Zatorre, Ph.D., Montreal Neurological Institute, 3801 University St., Montreal, Quebec, Canada H3A 2B4. Voice: 514-398-8903; fax: 514-398-1338. md37@musica.mcgill.ca these two classes of stimuli are uniquely human, making it particularly relevant to study them in the context of understanding those aspects of brain organization that are most relevant to human cognition and behavior. Finally, an overarching theme of the papers presented in this volume is that music can and should be viewed as a biological phenomenon, not merely a cultural artifact; from this perspective, therefore, it is not only meaningful but even imperative to seek out the neural processes that may allow human beings to engage in the diverse array of behaviors we term music.

All of these considerations set the stage for the studies from my laboratory and from others that will be described in this paper. The recent development of brain imaging techniques, coupled with more traditional behavioral approaches, has allowed for rapid developments in the field. A great deal of closely related information on music and brain function is presented in many of the other chapters in this volume, particularly in those by Besson, Griffiths, Halpern, Pantev, Parsons, Peretz, Samson, Schlaug, and Tervaniemi. The studies to be discussed in the present chapter focus specifically on aspects of tonal processing and their neural substrates. The underlying theme of the research approach is that by studying these processes we will achieve a more complete understanding not only of the auditory nervous system, but of the human brain in general. The idea that neural systems may display functional specializations is generally well accepted in cognitive neuroscience. The work described in this paper therefore also has a goal of identifying these specializations in order to determine the extent to which musical processes depend on dedicated neural circuitry or may form part of shared neural architecture. In this context, another important feature of brain organization, hemispheric functional specialization, is also crucial, and will be important in understanding the pattern that emerges.

In order to make sense of a complex phenomenon such as music, it is necessary, at least as a starting point, to focus on a specific aspect; in the present set of studies, I have chosen to explore the processing of pitch, which is to say pitch information as used for musically relevant processes. This choice is motivated by the fact that pitch appears to be a central aspect of all music. Although other aspects of music, such as temporal organization for rhythm, for example, may play an equally important role, it is difficult to conceive of a musical system of any type that does not involve the patterning of pitches. Moreover, pitch processing is amenable to study not only because the physical parameters of sounds associated with pitch are relatively well understood and easily manipulated, but also because it affords us the opportunity to analyze different levels of processing. Thus, in what follows I will consider first of all studies examining relatively low-level aspects of pitch processing, such as pitch discrimination; second I will discuss higher-order aspects, including pitch patterns. Finally, I will present some anatomical findings that may be relevant to understanding the nature of the functional specializations that the other studies have uncovered.

## BASIC ASPECTS OF PITCH PROCESSING

To begin, let us consider a relatively simple aspect of a musically relevant function: the ability to distinguish sounds on the basis of their pitch. The neural basis for the encoding of pitch has been studied for some time. Indeed, much neurophysiological work both in the periphery and in the central nervous system has examined the processing of frequency information (see Tramo, this volume). One of the salient features of the auditory nervous system, in fact, is that a tonotopic organization exists from the earliest level of the periphery, at the basilar membrane, to many fields within the auditory cortex. This topographic organization along the frequency axis points to the importance of pitch information in auditory processing generally. However, does the cortex carry out some essential function in the representation of pitch information? Several classic studies in experimental animals with bilateral destruction of auditory cortical areas have suggested that simple pitch discrimination remains unaffected, indicating that earlier levels of processing, in the midbrain or thalamus, for example, may be sufficient.[1–3] However, the nature of the pitch processing task appears to be critical in answering this question.

In a recent study from my lab, Johnsrude *et al.*[4] explored the specificity of pitch discrimination using a behavioral-lesion technique. We tested patients who had undergone surgical excision in the auditory cortex in the right or left temporal lobe; the patients' lesions were classified according to magnetic resonance imaging (MRI) as extending or not into Heschl's gyrus (HG), which contains the primary auditory cortex (see Penhune *et al.*[5] for description of a lesion quantification procedure, and FIGURES 1 and 2). A simple psychophysical "staircase" discrimination task was used, with pairs of pure tones to determine the threshold or minimal frequency separation needed to achieve a certain level of performance. Two separate tasks were administered. In the first, the two tones were either identical or different in pitch, and the subject answered accordingly. In the second task, the two tones always differed, and the subject's task was to indicate if the second tone was higher or lower than the first. The tones themselves were identical across the two tasks. The idea here is that whereas the simple pitch discrimination task merely requires that a difference be noted, the second, pitch-direction task requires that the tones be discriminated and ordered along some underlying pitch scale.

The findings (FIG. 3) indicated several interesting dissociations. First, the patients with lesions within the left temporal lobe were quite unimpaired as compared to control participants on either task, regardless of the extent of excision with HG. Second, the patients with damage to the right temporal lobe but excluding HG also had normal thresholds. The group of subjects with right HG damage, however, showed a fourfold increase in their pitch thresholds, but only for the pitch-direction task; they were normal on the simple pitch discrimination task. The findings are thus quite specific in indicating that only a certain restricted lesion in the right temporal lobe has an effect, and then only on a particular task. Thus, one may conclude that the right primary auditory area plays a special role not simply in discriminating one pitch from another, but in some aspect of organizing the sounds according to their pitch. This aspect of pitch organization, it is argued, is essential for any musical use of pitch information. Interestingly, all but one of the patients in question were able to perform the task itself; it was only the threshold that was abnormally high. This aspect of the result suggests that a much coarser level of pitch organization still exists in these patients, which may be mediated by left auditory cortex, a point to which we shall return below.

These results are consistent with those of a much earlier study[6] in which patients with right temporal lobe excisions with encroachment onto Heschl's gyri had signif-

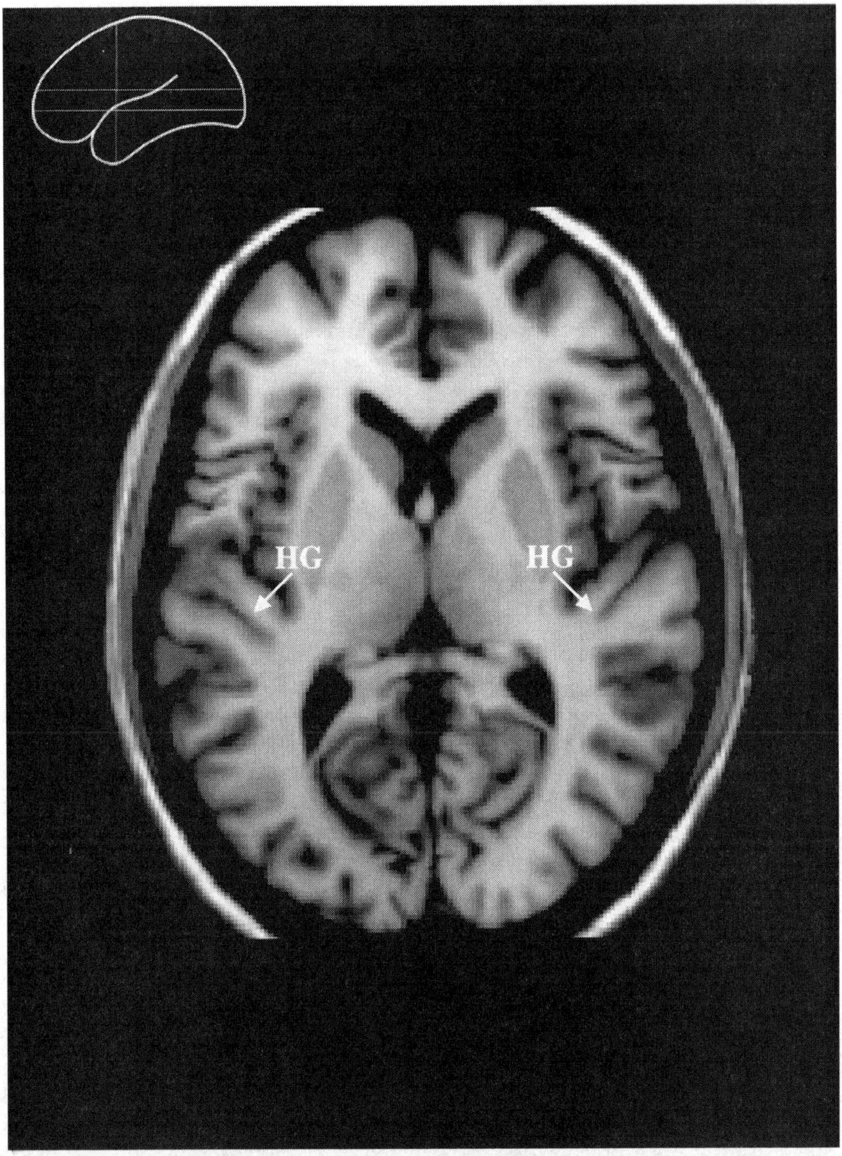


**FIGURE 1.** Magnetic resonance image of a normal human brain showing a horizontal section through the region of Heschl's gyrus, marked HG. The orientation and level of the horizontal section is indicated by the *dashed line* in the *inset*.

icant difficulties with a missing-fundamental pitch discrimination task but not with the identical task when the fundamental frequency was present. Patients with more anterior right temporal lobe damage, or those with similar damage on the left were unimpaired. Both studies thus point to a specificity for basic aspects of pitch pro-

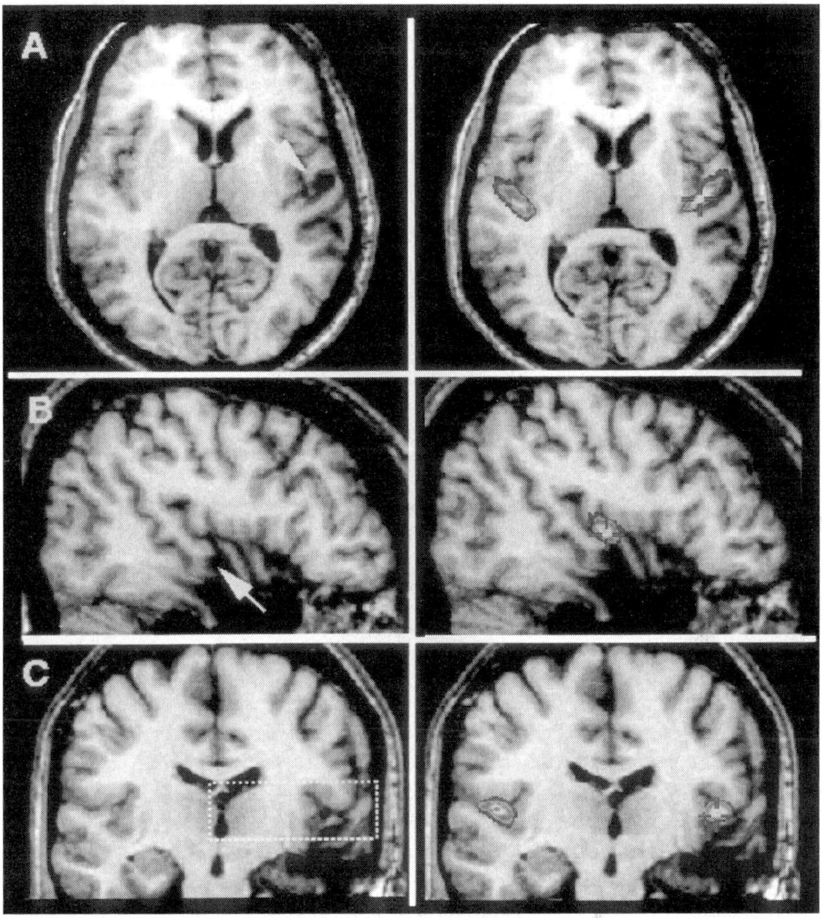


**FIGURE 2. See Plate 4A–C in color section.** MRI scan of a patient with a surgical excision in the right HG; illustrated are planes of section oriented horizontally (**A**), sagittally (**B**), and coronally (**C**). The left side of panels **A**, **B**, and **C** show the patient's scan alone, with an *arrow* indicating the region of excision/undercutting. The images on the *right side* of panels **A**, **B**, and **C** show the patient's scan coregistered with an anatomical probabilistic map of HG derived from normal individuals (Penhune *et al.*[44]). The *crosshairs* indicate the same position in the standardized space as the *arrow*. Note the correspondence between the position of HG as determined from the map and the patient's partially excised HG region. The *yellow box* in panel **C** indicates the region of the removal pictured in close-up in panel **D**, which illustrates the transition from intact, to undercut, to fully excised tissue (coronal sections taken at 3-mm intervals, posterior to anterior). *Arrows* again correspond to the *crosshairs* in the other panels and indicate the location of the HG region.

cessing, indicating a special role for areas in or adjacent to the right primary auditory cortex in these functions. Several other studies have added weight to this idea, although the degree of anatomical specificity has not always been possible; nonetheless, many authors have observed that the right temporal cortex plays a particularly important role for processing of pitch and of spectral pattern information.[7–11]

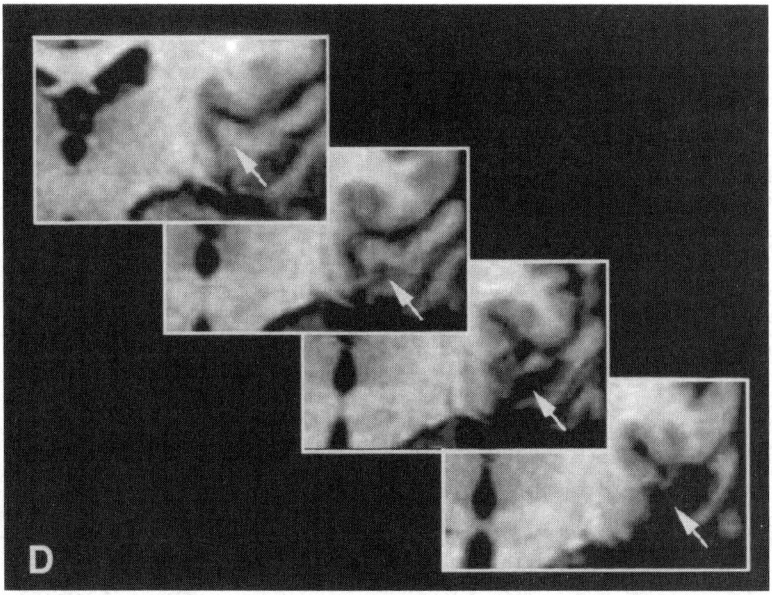


**FIGURE 2.** ***Continued*****. See Plate 4D in color section.**

More recently, functional imaging studies have begun to add important new evidence to our emerging view of the functional organization of the human brain. Among the many advantages of techniques, such as positron emission tomography (PET) and functional magnetic resonance imaging (fMRI), is that they provide a noninvasive means of obtaining anatomically accurate information about the cerebral areas that change their neural activity as a function of stimulus or task parameters. Thus, these techniques provide critical complementary evidence to the data derived from more traditional behavioral studies. However, whereas activation studies can identify the sites of neural activity associated with a particular function, they cannot determine the degree to which necessary computations are carried out. Conversely, only lesion techniques or stimulation techniques (such as transcranial magnetic stimulation, see Pascual-Leone, this volume) can provide information about essential areas, but they are constrained by various other limitations. Clearly, whenever converging evidence can be obtained from a variety of methods, the conclusions that can be drawn will be correspondingly stronger.

Several recent studies have examined pitch processing using functional imaging techniques. One relevant piece of evidence in favor of a special role for the right auditory cortex in pitch processing comes from PET studies of pitch production. Perry *et al.*[12] compared a condition in which subjects repeatedly produced a single vocal tone of constant pitch to a control in which they heard synthetic tones at the same rate. Increased cerebral blood flow (CBF) was noted in several motor areas, but within the superior temporal gyrus (STG), only right-sided activations were noted in both primary and secondary auditory cortices. The authors interpret this finding as a reflection of the specialization of the right auditory cortex for pitch perception. They

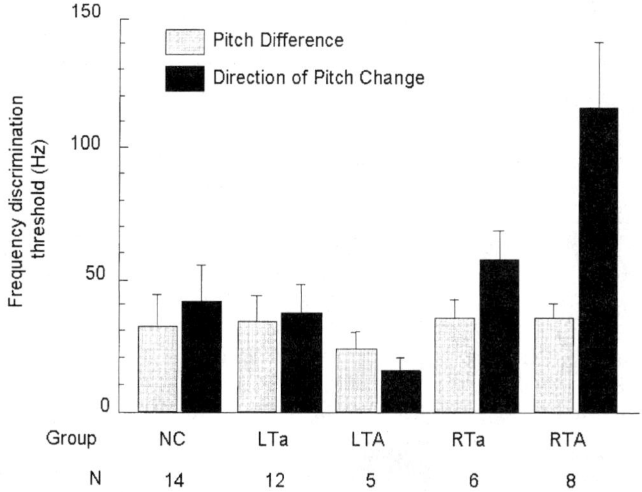


**FIGURE 3.** Mean frequency difference discrimination thresholds observed in two pitch discrimination tasks in patients with right or left (RT or LT) temporal-lobe excision, extending into the region of HG (**A**) or sparing this area (**a**). The simple discrimination thresholds were not different across the groups. Thresholds in the direction-of-pitch-change task were significantly higher in the RTA group than in any of the other groups. (Reproduced with permission from Johnsrude *et al.*[4])

argue that accurate perception of the pitch of one's own voice is required during singing in order to use feedback to maintain the desired pitch. Additional findings in favor of this conclusion come from regression analyses performed on these data, in which the accuracy of the subjects' singing was measured, and the degree of pitch excursion was covaried against CBF in the whole brain volume.[13] Strikingly, a region of covariation was identified in the right HG, such that increased CBF was observed with increasing pitch deviation. This result confirms the special role played by this region in pitch processing and suggests that a feedback mechanism may indeed be operative to maintain consistency of vocalized pitch.

## PROCESSING OF PITCH PATTERNS

The initial phase of pitch analysis set the stage for processing of pitch patterns. The relationships between individual tones form the basis for the most fundamental aspect of musical processing: the encoding and recognition of melodies. Melodic patterns, in turn, may be investigated at many different levels of analysis, from the basic interval relationships between tones to the semantic networks that are important for long-term representations of familiar tunes, and the output mechanisms necessary for singing or playing them. Here, we shall concentrate on the way in which auditory cortical systems are engaged in perception of melodies. One critical consideration in this respect is that all melodic perception (indeed, all auditory perception) involves working memory mechanisms. Since sounds unfold over time, in order to compute the relationships between successive events, an on-line retention system is evidently necessary. In the case of even a brief melody, this would mean that relationships between tones have to be computed and maintained over periods ranging from seconds to minutes in order to achieve a relatively stable and coherent internal representation.

Many studies of melodic perception with brain-lesioned individuals have demonstrated that damage to the right STG often leads to behavioral deficits in discrimination tasks,[7,14–16] although most of these studies have also noted that damage to the equivalent structures on the left may also lead to some degree of impairment. Indeed, Peretz *et al.*[17] (see also this volume) have noted that patients with bilateral lesions that involve portions of the STG and frontal cortical structures demonstrate much more severe disturbances in melodic tasks than most patients with unilateral damage. In fact, in each of the cases that Peretz and her colleagues have studied, the patients did not appear to show the specific musical deficit until after a second lesion had occurred, implying that unilateral damage had had at most only a mild effect not noted by the patients or their families. This point is important in that it suggests that the putative hemispheric specializations that are one theme of this chapter are likely to be relative rather than absolute.

The relative importance of right-hemisphere STG mechanisms has been demonstrated specifically for tonal working memory in two studies from our laboratory. In one,[18] we adapted a paradigm first developed by Deutsch[19] in which a target and a comparison tone are to be compared for pitch; the tones are either separated by a brief silent interval or by a series of interfering random tones. Deutsch had shown that memory for tones was relatively specific because it was not disrupted by other sounds, but only by other tones (see also Semal *et al.*[20]). When this tonal working-memory task was administered to temporal-lobectomy patients, performance was significantly worse following damage to the right than to the left temporal cortex for the interference task. On the other hand, unlike the studies of basic pitch processing discussed above, there was no difference in degree of behavioral deficit as a function of encroachment onto HG. Thus, areas of auditory cortex outside of the primary zone are most likely to be involved in this function. Finally, it was also noted that right frontal cortical excisions produced an impairment similar to that seen in the right temporal patients, a point that is taken up below.

To follow up on the question of tonal working memory function, Zatorre *et al.*[21] used PET to test 12 volunteer nonmusicians with unfamiliar melodies and noise bursts that had been constructed so as to approximate the acoustic characteristics of the melodies. Four conditions were run; subjects were asked to (1) listen to the noise bursts; (2) listen to the melodies without explicit instruction; (3) listen to the same melodies and compare the pitch of the first and second notes (two-note condition); and (4) listen to the melodies and compare the pitch of the first and last notes (first/last condition).

The comparison of listening to melodies minus noise permits examination of the cerebral regions specifically active during processing of novel melodies. The principal result indicated a significant CBF increase in the right superior temporal gyrus, anterior to the primary auditory cortex; a weaker CBF increase was also observed in the left STG. A reanalysis of these data with a smaller reconstruction filter and improved image registration[22] indicates that the right STG activation is in the ventrolateral aspect of the gyrus (FIG. 4). These findings point to differential activation of primary versus secondary auditory areas within the STG, depending on the stimulus features. Caution must still be exercised in interpreting the results, for the noise bursts clearly differ from the melodies on a number of dimensions (e.g., periodicity, frequency modulation, and spectral composition). It remains to be established, therefore, which specific features of the melodies may lead to the observed pattern of activation. The finding of an activation within the right STG while listening to melodies fits in well with the lesion data, however, and likely reflects the specialization of neuronal networks within the right secondary auditory cortices for perceptual analysis of tonal information.

Zatorre *et al.*[21] also examined working memory for pitch by subtracting the passive condition from the two-note condition, which resulted in significant activation within the right frontal lobe (both inferiorly and superiorly). The first/last minus passive melodies subtraction yielded a large number of cortical and subcortical activation sites in both hemispheres, including a focus in the right inferior frontal cortex identical to that observed in the two-note condition. A comparison not reported in the original study pits the first/last condition against the two-note condition. The advantage of this contrast is that both involve pitch comparisons, the main difference being the degree of working memory load. The interesting finding here (FIG. 4) is once again an area of increased blood flow in the right inferior frontal region, though not in the same location as in the other subtractions. The pattern of results from these conditions therefore implicates frontal-lobe mechanisms in effecting pitch comparisons, with a particularly important contribution from right-frontal regions. It should be recalled, however, that in the first/last minus passive melody comparison, a large number of separate foci of CBF change were seen over a wide swath of cortical and subcortical territory; this finding perhaps reflects the complexity and increased cognitive demands of the task, which was also manifested in increased error rate and slower reaction times.

One may speculate that the numerous frontal lobe sites observed might be associated with successful performance of distinct aspects of the task. The findings of Perry *et al.*[23] are directly relevant to this issue; they observed CBF increases with a rightward asymmetry in dorsolateral frontal areas while subjects were engaged in an auditory tonal working memory task that required them to monitor a sequence of tones. This finding suggests a functional dissociation between the cognitive act of monitoring information as opposed to holding sensory information in working memory. Perry *et al.*[23] suggest a hierarchical view, compatible with other evidence,[24]

---

**FIGURE 4. See Plate 5 in color section.** Selected PET CBF activation sites associated with processing of tonal melodies. The figures show averaged PET subtraction images superimposed on a corresponding section of an average anatomical MRI. (*Top*) Horizontal, coronal, and sagittal views through a focus located in the right superior temporal gyrus representing a significant blood flow increase while subjects listened to a series of short, unfamiliar tonal melodies, as compared with a baseline condition in which they listened to noise bursts that were acoustically equated to the melodies. Also visible in the horizontal section is a region of significant activity in the left temporal cortex. (*Bottom*) Horizontal, coronal, and sagittal views through a focus in the right frontal/opercular region showing activity associated with making judgments about the pitch of the first and last notes of the tonal sequences as compared to making judgments about the first two notes. Data reanalyzed from Zatorre *et al.*[21]

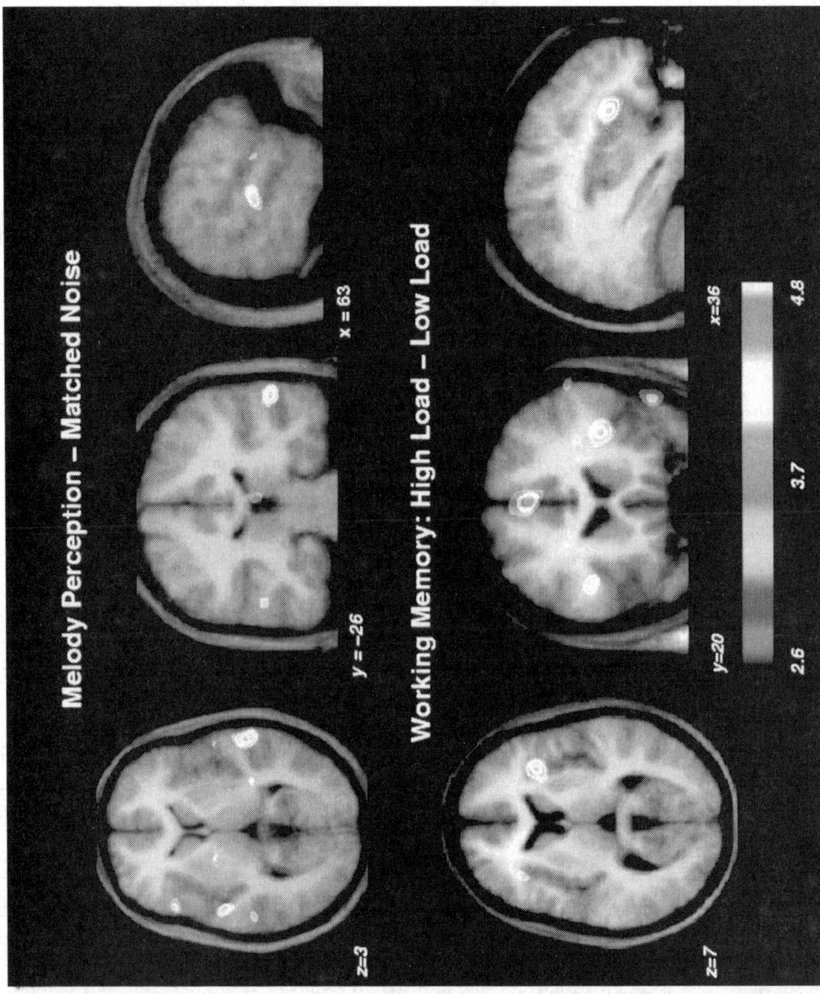


**FIGURE 4.** *See previous page for caption.*

such that right inferior lateral frontal areas are important for maintenance of tonal information, whereas dorsolateral frontal areas are required for higher-level functions such as monitoring the contents of working memory. This conclusion would be in accord with the lesion study described earlier,[18] as well as with PET studies[25] that also indicated increased activity in the right frontal cortex during pitch judgments on speech syllables.

Several other imaging studies over the past few years have supported the idea that the right secondary auditory cortex may be particularly important in processing tonal patterns (but see Griffiths, this volume, for evidence that the story is more complicated). The results of Démonet *et al.*[26,27] are relevant, since tone judgment tasks were used, in which subjects were required to detect pitch changes of tones within a sequence, and the difference in activity pattern between the tonal task and a language task was examined. The findings were consistent with data presented above, implicating the right auditory cortical areas in tonal processes, since foci were observed in posterior right STG and middle temporal areas. More recently, Binder *et al.*[28] reported whole-brain fMRI data using an active tone judgment task. This task, which required subjects to discriminate two pitches during a series of tones (which ranged from 3 to 7 tones in length) and then to make a judgment pertaining to the entire series of tones, emphasized not only pitch perception but also maintenance of simple pitch information in working memory. When this task was compared to an active task involving semantic categorization of words, stronger activation associated with the tone task was observed in the posterior STG bilaterally, the supramarginal region bilaterally (but much more on the right), the right posterior middle temporal gyrus, premotor cortex bilaterally (also much more on the right), and right superior parietal cortex.[28] These right temporoparietal foci are in agreement with the right posterior STG, middle temporal, and inferior parietal foci reported by Démonet in their tone-speech comparisons.[26, 27]

A further source of relevant evidence comes from studies of musical imagery (reviewed by Halpern, this volume). Those studies, including both behavioral-lesion experiments and functional imaging,[29–31] have indicated that a greater role is played by secondary auditory cortical areas on the right than on the left not only in perceptual tasks involving judgments of real melodies, but also in imaginal tasks in which subjects are invited to imagine a well-known tune in "the mind's ear." These studies, together with additional recent neuroimaging findings pointing in the same general direction,[32–34] provide additional evidence for a special role of several right-hemispheric cortical regions in various aspects of tonal pattern processing.

## HEMISPHERIC DIFFERENCES IN SPECTRAL PROCESSING

Much of the evidence presented above converges toward the conclusion that hemispheric specialization exists in the processing of pitch information, with right auditory cortical areas playing a more prominent role than left. Moving beyond the mere fact that differences do exist, it is critical to pose the question of why such lateralization might have evolved. One classic way of thinking about hemispheric differences, favored, for example, by investigators such as Teuber or Sperry, is in terms of complementary functional specialization—that is, that hemispheric processing advantages emerge because they confer some advantage over shared processing,

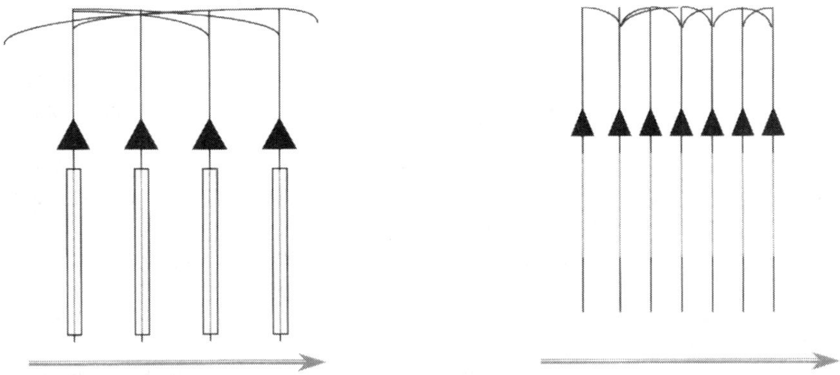

**FIGURE 5.** Cartoon depiction of putative differences in anatomical structure between left and right auditory cortical areas. *Triangles* and *lines* represent individual neurons with differing degrees of myelination (thicker on the left). Neurons on the left are larger than those on the right, have wider spacing across tonotopically organized cortex, and are more widely interconnected. These structural features would lead to relatively fast conduction but poor spectral resolution on the left, and to poorer temporal resolution but greater sensitivity to fine spectral differences on the right.

since each hemispheric system can specialize in one domain, thus leading to an overall enhancement for the organism as a whole.

If we think about the data presented earlier in this chapter within this context, it seems obvious that the two domains for which the auditory cortices in each hemisphere have become specialized are, roughly, speech and tonal patterns. A bit of additional reflection suggests a unifying hypothesis to explain these complementary functional specializations: speech and tonal stimuli differ in their acoustic structure, and hence in their processing requirements. Whereas the analysis of speech requires good temporal resolution to process rapidly changing energy peaks (formants) that are characteristic of many speech consonants (see, for example, the work of Tallal *et al.*[35]), it can be argued that tonal processes instead require good frequency resolution. In a truly linear system, temporal and spectral resolution are inversely related, so that improving temporal resolution can only come at the expense of degrading spectral resolution and vice versa. This tradeoff naturally arises from a fundamental physical constraint in acoustic processing: better resolution in the frequency domain can be obtained only at the expense of sampling within a longer time window, hence degrading temporal resolution; conversely, high resolution in the temporal domain entails a degraded spectral representation. The auditory nervous system is, of course, a highly nonlinear and distributed system; yet, it may also respect this fundamental computational constraint, such that in the left auditory cortex the high temporal resolution needed to process speech imposes an upper limit on the ability to resolve spectral information, and vice versa for the right auditory cortex. To put it more simply, the hypothesis is that there may be a tradeoff in processing in temporal and spectral domains, and that auditory cortical systems in the two hemispheres have evolved a complementary specialization, with the left having better temporal resolution, and the right better spectral resolution.

According to the foregoing idea, therefore, the hemispheric differences we see in the tonal perception literature reviewed above might reflect this more fundamental level of functional organization. If this hypothesis is correct, then we ought to be able to obtain evidence for differential response within left and right auditory cortices by manipulating temporal and spectral parameters of an auditory stimulus, even if it is not particularly perceived as speech or music. A recent functional imaging study from our lab[36] set out to do just that using a parametric approach. Rather than looking for differences in cerebral blood flow between a control and an activation condition, we examined the functional changes in the brain that correlated with a given input parameter. This approach can be particularly powerful since it helps to isolate brain activity that is specifically related to the parameter of interest. We first created nonverbal stimuli that varied independently and systematically along two dimensions, one temporal, the other spectral. The stimuli were merely a series of pure tones that varied in frequency and duration; in one set of conditions the frequency change was held constant and the temporal rate became faster across scans, while in the other set of conditions the rate was held constant and the frequency differences became finer across scans. We predicted that increasing the rate of temporal change would preferentially recruit left auditory cortical areas, while increasing the number of spectral elements would engage right auditory cortical regions.

The results of greatest relevance for the present discussion were that cerebral blood flow in a region of the left auditory cortex showed a greater response to increasing temporal than spectral variation, whereas a symmetrical area on the right showed the reverse pattern. A third area in the right superior temporal sulcus also showed a significant response to the spectral parameter but showed no change to the temporal parameter. Thus, the data supported the hypothesis that corresponding regions of the auditory cortex in the two hemispheres have different sensitivity to temporal and spectral information. Neurophysiological recordings in macaque monkeys have shown that auditory cortical neurons are highly sensitive to both spectral and temporal features of sounds simultaneously, and that they exhibit complex response functions.[37–39] It has also been shown that neurons in the belt region of the macaque auditory cortex are optimally responsive to stimuli of specific bandwidths, hence demonstrating sensitivity to a spectral parameter.[40] As well, Eggermont[41] noted that in two primary fields of the cat there was an inverse relation between bandwidth and temporal resolution, consistent with our proposal.

Additional evidence that left auditory cortical units have higher temporal resolution comes from recent depth-electrode recordings within Heschl's gyri[42] showing that the responses within the left auditory cortex encode the voice-onset time of a consonant, whereas the right auditory cortex did not show sensitivity to this temporal parameter. Even more striking consistency is offered by additional data from the same investigators (Liégeois-Chauvel *et al.*, this volume), who showed that intracortically recorded auditory evoked potentials were more sharply tuned to frequency in the right auditory cortex than in the left, as predicted by the hypothesis that resolution in the frequency and time domains are different in the two hemispheres.

This hypothesis has also received support from a PET study by Belin *et al.*,[43] in which the rate of formant transitions in a pseudospeech sound—not perceived as speech—was varied from 40 to 200 milliseconds. CBF changes in the left auditory region were found for both types of stimuli, indicating a capacity for processing spectral change over a wide range of durations; by contrast, regions in the right auditory cortex responded only to the slower rates, and not to the faster rate. Thus, the right auditory cortex seems unable to respond to fast formant transitions, whereas the homologous area on the left is better able to track rapidly changing acoustic information that would be relevant for speech processing. Finally, it is also relevant to recall the findings of Robin *et al.*,[9] who studied auditory discrimination in brain-damaged patients. They too found that damage to association cortices in the right hemisphere resulted in spectral but not temporal processing deficits, while the converse was observed after left-hemispheric damage.

## ANATOMICAL CONSIDERATIONS

One advantage of the hypothesis presented above is that it offers a unifying framework to understand some of the functional characteristics of the auditory nervous system that allow us to be able to process speech and tonal patterns. This model raises an important question, however: How are these functional differences instantiated in the brain? That is, can we find any eviidence in the structure of the auditory cortex that would not only support the existence of processing differences but would also suggest how these differences are implemented?

Some answers to these questions are provided by recent anatomical data from our laboratory[44] and from other investigators. Unlike the functional techniques described above, the aim of this research was to characterize the structural features (shape, volume, and position) of the human primary auditory cortical region *in vivo*. The approach taken was to use MRI scans from groups of normal right-handed volunteer subjects and to label the region of HG (FIG. 1) using interactive pixel-marking software that permits simultaneous viewing in all three planes of the section. The result of greatest relevance pertains to the estimates of the volume of HG, which was found to be significantly greater on the left than on the right in two independent samples of 20 subjects each.[44] Most interesting of all was that the differences in volume were found to be confined to the white matter underlying HG, and not to the volume of cortical tissue (grey matter) within the structure (this effect may be seen qualitatively in the MRI shown in FIG. 1). This finding suggests an anatomical asymmetry that arises from a difference in the volume of fibers that carry information to and from the primary auditory cortex and surrounding regions.

Although an MRI cannot reveal details of the underlying neuronal organization, the greater degree of white matter on the left could be consistent with the notion of greater speed of processing on the left, if the white-matter volume measures are related to degree of myelination. That is, a greater degree of left-sided myelination could lead to faster transmission of acoustically relevant information, because it is well established in neurophysiology that more heavily myelinated fibers transmit neuronal impulses more quickly. Confirmation that the left auditory cortical regions are indeed more heavily myelinated comes from a recent postmortem electron microscopy study showing that the myelin sheath is thicker in the left than in the right posterior temporal cortex.[45] As well, Hutsler and Gazzaniga[46] reported larger left-than right-layer IV pyramidal cells in the human auditory cortex, which would also lead to faster time constants.

In addition to the idea that left auditory cortical fibers are more heavily myelinated, there is also evidence for other hemispheric differences in the structural organization of the auditory cortex that are broadly consistent with the hypothesis presented above. For example, Seldon[47] reported that cortical columns in the left auditory cortex were more widely spaced than those on the right. Galuske *et al.*[48] recently found that clusters of neurons were spaced more widely in the left posterior temporal cortex, with more long-range intrinsic connections between these clusters. These features could be compatible with a greater degree of integration across tonotopically organized cortical areas on the left than on the right, hence leading to a relative degradation of spectral resolution on the left. Conversely, neurons in the right auditory cortex would have structural features that would enhance spectral resolution (much as shown by Liégeois-Chauvel, this volume) because they would be smaller, with fewer myelinated inputs, therefore perhaps more tightly packed together; and would integrate over narrower frequency regions. These characteristics would also result in poorer temporal resolution.

## CONCLUDING COMMENTS

We have seen how a variety of different methods and techniques point towards a general conclusion about the way in which the auditory nervous system processes information relevant for our ability to process speech and tonal sounds. These two domains of auditory processing are, of course, at the core of our uniquely human abilities to communicate. It is perhaps no coincidence that abilities to process both speech and music arise precociously in development and follow relatively fixed developmental sequences (see Trehub, this volume). It is perhaps also not irrelevant that both speech communication and tonal patterns appear to be ubiquitous across all human cultures (see both Huron and Cross, this volume).

The general approach we have defended, to view music within a biological framework, thus would seem to make sense of these disparate observations, if musical perceptual abilities—and by extension musical production abilities—are seen as emerging from functional and structural features of our nervous system. There is a final point that should perhaps be addressed, however, and this concerns the origins of music. Without entering into the controversies and speculations of this field of scholarship (see Huron, this volume), perhaps it is sufficient to point out that the approach offered here would provide a relatively parsimonious account. In order to understand the emergence of music, it might be sufficient to assume only one set of evolutionary pressures: the need to communicate verbally. If the specialization of the left auditory cortex discussed above reflects the nervous system's adaptation to this selective pressure, then all we need to assume is that the complementary structural and functional changes of the two hemispheres arose as a consequence of this specialization (see also Ref. 49 for further discussion).

The ideas presented in this chapter, or indeed in this entire volume, will ultimately be shown to be correct, or at least partially so, or in error by the usual slow and meandering path of science. Regardless of the details, however, the hope is that by viewing music as a phenomenon worthy of scientific study, we will enhance both our understanding of music within our species, as well as the workings of our nervous system that make music possible.

## ACKNOWLEDGMENTS

I wish to thank all my colleagues and students who contributed to the research described in this paper. Funding for the work has been provided by grants from the Canadian Institutes of Health Research, the McDonnell-Pew Cognitive Neuroscience Program, and the International Foundation for Music Research.

## REFERENCES

1. EVARTS, E.V. 1952. Effect of auditory cortex ablation on frequency discrimination in monkey. J. Neurophysiol. **15:** 443–448.
2. HEFFNER, H.E. & B. MASTERTON. 1978. Contribution of auditory cortex to hearing in the monkey (*Macaca mulatta*). Recent Advances in Primatology: Vol 1. Behaviour. D.J. Chivers & J. Herbert, Eds. Academic Press. New York.
3. JERISON, H.J. & W.D. NEFF. 1953. Effect of cortical ablation in the monkey on discrimination of auditory patterns. Fed. Proc. **12:** 237.
4. JOHNSRUDE, I.J., V.B. PENHUNE & R.J. ZATORRE. 2000. Functional specificity in right human auditory cortex for perceiving pitch direction. Brain **123:** 155–163.
5. PENHUNE, V., R. ZATORRE & W. FEINDEL. 1999. The role of auditory cortex in retention of rhythmic patterns in patients with temporal-lobe removals including Heschl's gyrus. Neuropsychologia **37:** 315–331.
6. ZATORRE, R.J. 1988. Pitch perception of complex tones and human temporal-lobe function. J. Acoust. Soc. Am. **84**(2): 566–572.
7. MILNER, B.A. 1962. Laterality effects in audition. *In* Interhemispheric Relations and Cerebral Dominance. V. Mountcastle, Ed.: 177–195. Johns Hopkins Press. Baltimore, MD.
8. DIVENYI, P. & A. ROBINSON. 1989. Nonlinguistic auditory capabilities in aphasia. Brain Lang. **37:** 290–326.
9. ROBIN, D.A., D. TRANEL & H. DAMASIO. 1990. Auditory perception of temporal and spectral events in patients with focal left and right cerebral lesions. Brain Lang. **39:** 539–555.
10. SAMSON, S. & R.J. ZATORRE. 1994. Contribution of the right temporal lobe to musical timbre discrimination. Neuropsychologia **32:** 231–240.
11. SIDTIS, J.J. & B.T. VOLPE. 1988. Selective loss of complex-pitch or speech discrimination after unilateral lesion. Brain Lang. **34:** 235–245.
12. PERRY, D.W., R.J. ZATORRE, M. PETRIDES, *et al.* 1999. Localization of cerebral activity during simple singing. NeuroReport **10:** 3979–3984.
13. PERRY, D., R. ZATORRE & A. EVANS. 1996. Co-variation of CBF during singing with vocal fundamental frequency. NeuroImage **3**(3): S315.
14. LIÉGEOIS-CHAUVEL, C., I. PERETZ, M. BABAÏ, *et al.* 1998. Contribution of different cortical areas in the temporal lobes to music processing. Brain **121:** 1853–1867.
15. SAMSON, S. & R.J. ZATORRE. 1988. Melodic and harmonic discrimination following unilateral cerebral excision. Brain Cogn. **7:** 348–360.
16. ZATORRE, R.J. 1985. Discrimination and recognition of tonal melodies after unilateral cerebral excisions. Neuropsychologia **23:** 31–41.
17. PERETZ, I., R. KOLINSKY, M. TRAMO, *et al.* 1994. Functional dissociations following bilateral lesions of auditory cortex. Brain **117:** 1283–1301.
18. ZATORRE, R.J. & S. SAMSON. 1991. Role of the right temporal neocortex in retention of pitch in auditory short-term memory. Brain **114:** 2403–2417.

19. DEUTSCH, D. 1970. Tones and numbers: specificity of interference in short-term memory. Science **168:** 1604–1605.
20. SEMAL, C., L. DEMANY, K. UEDA, *et al.* 1996. Speech versus nonspeech in pitch memory. J. Acoust. Soc. Am. **100:** 1132–1140.
21. ZATORRE, R.J., A.C. EVANS & E. MEYER. 1994. Neural mechanisms underlying melodic perception and memory for pitch. J. Neurosci. **14**(4): 1908–1919.
22. ZATORRE, R. & J. BINDER. 2000. Functional and structural imaging of the human auditory system. *In* Brain Mapping: The Systems. A. Toga & J. Mazziota, Eds.: 365–402. Academic Press. Los Angeles.
23. PERRY, D.W., M. PETRIDES, B. ALIVISATOS, *et al.* 1993. Functional activation of human frontal cortex during tonal working memory tasks. Soc. Neurosci. Abstr. **19:** 843.
24. PETRIDES, M. 1995. Functional organization of the human frontal cortex for mnemonic processing: evidence from neuroimaging studies. Ann. N.Y. Acad. Sci. **769:** 85–96.
25. ZATORRE, R.J., A.C. EVANS, E. MEYER, *et al.* 1992. Lateralization of phonetic and pitch processing in speech perception. Science **256:** 846–849.
26. DÉMONET, J.-F., F. CHOLLET, S. RAMSAY, *et al.* 1992. The anatomy of phonological and semantic processing in normal subjects. Brain **115:** 1753–1768.
27. DÉMONET, J.-F., C. PRICE, R. WISE & R.S.J. FRACKOWIACK. 1994. A PET study of cognitive strategies in normal subjects during language tasks. Brain **117:** 671–682.
28. BINDER, J., J. FROST, T. HAMMEKE, *et al.* 1997. Human brain language areas identified by functional magnetic resonance imaging. J. Neurosci. **17:** 353–362.
29. ZATORRE, R.J. & A.R. HALPERN. 1993. Effect of unilateral temporal-lobe excision on perception and imagery of songs. Neuropsychologia **31**(3): 221–232.
30. ZATORRE, R.J., A.R. HALPERN, D.W. PERRY, *et al.* 1996. Hearing in the mind's ear: a PET investigation of musical imagery and perception. J. Cognit. Neurosci. **8**(1): 29–46.
31. HALPERN, A.R. & R.J. ZATORRE. 1999. When that tune runs through your head: a PET investigation of auditory imagery for familiar melodies. Cereb. Cortex **9:** 697–704.
32. GRIFFITHS, T.D., I.S. JOHNSRUDE, J.L. DEAN, *et al.* 1999. A common neural substrate for the analysis of pitch and duration pattern in segmented sound? NeuroReport **10:** 3825–3830.
33. HUGDAHL, K., K. BRONNICK, S. KYLLINGSBAEK, *et al.* 1999. Brain activation during dichotic presentations of consonant-vowel and musical instrument stimuli: a 15O-PET study. Neuropsychologia **37:** 431–440.
34. TERVANIEMI, M., S. MEDVEDEV, K. ALHO, *et al.* 2000. Lateralized automatic auditory processing of phonetic versus musical information: a PET study. Hum. Brain Map **10:** 74–79.
35. TALLAL, P., S. MILLER & R. FITCH. 1993. Neurobiological basis of speech: a case for the preeminence of temporal processing. Ann. N.Y. Acad. Sci. **682:** 27–47.
36. ZATORRE, R. & P. BELIN. Spectral and temporal processing in human auditory cortex. Submitted.
37. DE CHARMS, R.C., D.T. BLAKE & M.M. MERZENICH. 1998. Optimizing sound features for cortical neurons. Science **280:** 1439–1443.
38. PHILLIPS, D.P. 1993. Representation of acoustic events in the primary auditory cortex. J. Exp. Psychol.: Hum. Percept. Perform. **19**(1): 203–216.
39. STEINSCHNEIDER, M., C. SCHROEDER, J. AREZZO, *et al.* 1995. Physiologic correlates of the voice onset time boundary in primary auditory cortex of the awake monkey: temporal response patterns. Brain Lang. **48:** 326–340.
40. RAUSCHECKER, J.P., B. TIAN & M. HAUSER. 1995. Processing of complex sounds in the *Macaque* nonprimary auditory cortex. Science **268:** 111–114.
41. EGGERMONT, J. 1998. Representation of spectral and temporal sound features in three cortical fields of the cat. Similarities outweigh differences. J. Neurophysiol. **80:** 2743–2764.
42. LIÉGEOIS-CHAUVEL, C., J. DE GRAAF, V. LAGUITTON, *et al.* 1999. Specialization of left auditory cortex for speech perception in man depends on temporal coding. Cereb. Cortex **9:** 484–496.
43. BELIN, P., M. ZILBOVICIUS, S. CROZIER, *et al.* 1998. Lateralization of speech and auditory temporal processing. J. Cognit. Neurosci. **10:** 536–540.

44. PENHUNE, V.B., R.J. ZATORRE, J.D. MACDONALD, *et al.* 1996. Interhemispheric anatomical differences in human primary auditory cortex: probabilistic mapping and volume measurement from magnetic resonance scans. Cereb. Cortex **6:** 661–672.
45. ANDERSON, B., B.D. SOUTHERN & R.E. POWERS. 1999. Anatomic asymmetries of the posterior superior temporal lobes: a postmortem study. Neuropsychiatry Neuropsychol. Behav. Neurol. **12:** 247–254.
46. HUTSLER, J. & M. GAZZANIGA. 1996. Acetylcholinesterase staining in human auditory and language cortices—regional variation of structural features. Cereb. Cortex **6:** 260–270.
47. SELDON, H. 1981. Structure of human auditory cortex. II: Axon distributions and morphological correlates of speech perception. Brain Res. **229:** 295–310.
48. GALUSKE, R., W. SCHLOTE, H. BRATZKE, *et al.* 2000. Interhemispheric asymmetries of the modular structure in human temporal cortex. Science **289:** 1946–1949.
49. JUSTUS, T. & J. BHARUCHA. Music perception and cognition. *In* Steven's Handbook of Experimental Psychology: Vol. 1. Sensation and Perception. H. Pashler & S. Yantis, Eds. John Wiley and Sons. New York. In press.

# Exploring the Functional Neuroanatomy of Music Performance, Perception, and Comprehension

LAWRENCE M. PARSONS

*University of Texas Health Science Center at San Antonio, San Antonio, Texas 78284-6240, USA*

**ABSTRACT: This chapter highlights findings by my colleagues and me in four neuroimaging and neurological studies of music performance, perception, and comprehension. These investigations elucidate the neural subsystems supporting musical pitch, melody, harmony, rhythm, tempo, meter, and duration. In a positron emission tomography (PET) study of pianists, a memorized performance of a musical piece was contrasted with that of scales to localize brain areas specifically supporting music. A second PET study assayed brain areas subserving selectively the comprehension of harmony, melody, and rhythm. Musicians sight-read a score while detecting specific melodic, harmonic, or rhythmic errors in its heard performance. In a third PET study, musicians and nonmusicians discriminated pairs of rhythms with respect to pattern, tempo, meter, or duration. Data in these studies implicated the cerebellum in nonmotor, nonsomatic, sensory, or cognitive processing. In a fourth study, neurological patients with degeneration of the cerebellum were found to be impaired in fine discrimination of pitch. Overall, these data suggest that the neural systems underlying music are distributed throughout the left and right cerebral and cerebellar hemispheres, with different aspects of music processed by distinct neural circuits. Also discussed are key issues for interpreting the role in music of brain areas implicated in neuroimaging studies.**

**KEYWORDS: Cerebellum; Neuroimaging studies; Musical performance; Harmony; Melody; Rhythm**

In the last decade researchers in the cognitive sciences, experimental psychology, behavioral neurology, ethology, and neuroimaging have significantly advanced the scientific understanding of music, after a long period of slower progress.[1–14] It is not possible to review these important and exciting advances here (see other chapters in this volume for a sampling of such work). This chapter complements these advances by reviewing findings by my colleagues and me in four recent neuroimaging and neurological studies of music performance, perception, and comprehension. These investigations were attempts to elucidate the neural subsystems supporting major

Address for correspondence: Lawrence M. Parsons, Research Imaging Center, University of Texas Health Science Center at San Antonio, 7703 Floyd Curl Drive, San Antonio, TX 78284-6240. Voice: 210-567-8189; fax: 210-567-8152.

parsons@uthscsa.edu.

components of music—specifically, pitch, melody, harmony, rhythm, tempo, meter, and duration.

## FUNCTIONAL NEUROANATOMY OF MUSICAL PERFORMANCE

Our first experiment focused on musical expression or performance. This experiment, conducted in collaboration with Peter Fox, Donald Hodges, and Justine Sergent, attempted to clarify and extend a pioneering functional neuroimaging study of right-handed piano performance of a sight-read musical score.[15] In this early study, positron emission tomography (PET) was used to measure regional cerebral blood flow, an index of neural activity, in performing pianists. Among the main findings of this study was the observation that cortical areas distinct from, but adjacent to, those underlying language operations were activated during sight-reading. The relationship between music and language has continued to be a target of fruitful investigations.[16–20]

In our study, which built upon the foregoing results, we used PET to study bimanual piano performance of memorized music.[21] By recording brain activity when both hands were equally and concurrently producing music, we examined neural systems in both cerebral hemispheres when the left and right sides were fully involved in performance. In the prior study, although the score was read with both eyes and piano output was heard with both ears, only the right hand performed the music. This feature of the study left unclear which particular right hemispheric areas may be involved in music performance.

In addition, by eliminating musical score reading from scanned task performance, we examined brain activation during a more purely musical performance. Sight-reading a score during performance adds an additional cognitive load not directly related to music per se. Indeed, there is a strong belief among musicians that a fully memorized piece, one performed without score reading, engenders a distinctly deeper understanding of the music and more satisfying realization of the piece in performance.

We furthermore contrasted brain activity during the piano performance of a musical piece by J. S. Bach with that during the two-handed performance of scales. Our intent was to reveal the outline of brain areas that are specifically involved in the cognitive, perceptual, and emotional representation and performance of effective, strongly engaging music per se. The Bach and scales performances here required movements of nearly comparable frequency and complexity from each hand. (However, the Bach required somewhat more complicated, finely controlled fingering than the scales, and we expected this to be reflected in different patterns of activity when Bach and scales brain states were directly contrasted.)

After giving informed consent, eight right-handed professional musicians participating in three PET conditions performed three times in pseudorandom order (FIG. 1). In the Bach condition, the third movement of the Italian Concerto (BMV 971) was performed from memory. Scales were executed synchronously (from memory) with both hands at a pace approximating that of the Bach performance. Subjects performed on a full-sized Yamaha P-132 electronic piano and heard the sounds they produced. In all conditions, subjects' eyes were closed and covered.

**FIGURE 1. See Plate 6 in color section.** A pianist in a positron emission tomography (PET) scanning session[21] performing the third movement of the Italian Concerto by J. S. Bach. Photograph by Dr. Stephan Elleringmann (Aurora/Bilderberg).

PET images, obtained by the $^{15}O$-bolus method, were coregistered on each participant's MRI image and spatially normalized.[22–27] The grand-mean image for each task was compared to control, forming grand-mean images of the task-induced changes in brain activity ($p < 0.001$). Grand-mean images were converted to z-score images using the population variance.[28] Activation loci were identified by 3-D center-of-mass address, peak-voxel intensity, and z-value, cluster size, and statistical significance.[29] Response coordinates were interpreted[30] in accordance with the Talairach and Tournoux stereotaxic atlas[23] and confirmed with coregistered group-mean MRI.

Only a brief summary of some of the principal findings can be presented here. I will focus on the auditory cortex (the temporal lobe). This is where we detected the primary difference in neural activation for Bach and for scales, as indicated in the analysis of the direct contrast between activity in the two conditions. As illustrated in FIGURE 2, many of the activations were in the anterior secondary auditory association areas, in Brodmann's areas (BA) 22, 21, and 20. Overall, the performance of the Bach activated the temporal lobe more strongly than did the performance of scales. Scales performance activated middle temporal areas bilaterally, with more on the left than on the right (FIG. 2). By contrast, the performance of the Bach activated superior, middle, and inferior temporal areas bilaterally, with more on the right than on the left.

The stronger activation in the right auditory cortex for the performance of the Bach is consistent with a variety of neuropsychological and neuroimaging results indicating that areas in right auditory association cortex are involved in the reception and expression of melody.[31–36] These findings suggest that some of the important

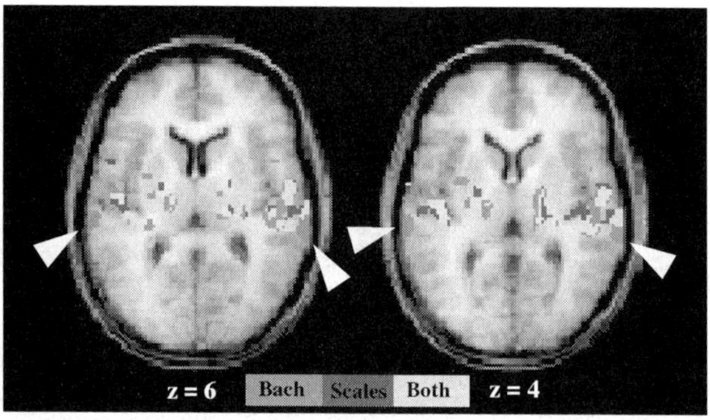


**FIGURE 2. See Plate 7 in color section.** Significant blood flow activations[21] in auditory (temporal) cortex specific to playing the Bach score (*red*), specific to playing scales (*blue*), and common to both performances (*cyan*). These are logical contrasts of Bach vs. rest and scales vs. rest shown in group-averaged PET images overlaid on anatomical magnetic resonance images (MRIs). The z-values indicate the axial height of the brain volume relative to the Talairach and Tournoux stereotactic atlas.[23] Throughout, left side of brain images shows the left side of the brain, and vice versa.

areas specifically representing the higher-order representations of musical meaning, particular in performance, lie in these regions of temporal cortex.

Unexpectedly, there was also strongly correlated activation during the performance of the Bach (FIG. 3) between right anterior temporal areas (BA 22, superior temporal gyrus) and left posterior lateral cerebellum ($r = + 0.77, p < 0.001$). Because left cerebellum has its primary connectivity with right cerebral cortex, we hypothesize that during the performance of the Bach, activation in right auditory cortex is related to that in left cerebellum. There was a corresponding, but less robust, correlation between left temporal areas and right posterior cerebellum. This contralateral coactivation of the auditory temporal cortex and lateral cerebellum suggests that they form a distributed circuit of processing. This cerebellar activity appears to be

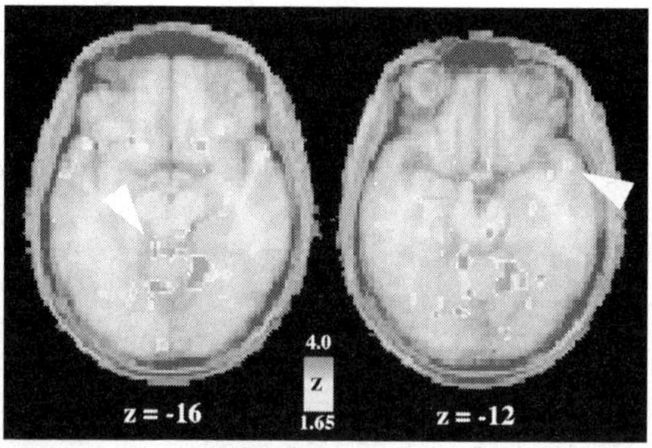


**FIGURE 3. See Plate 8 in color section.** Correlated blood flow activations in right anterior auditory cortex and left cerebellum (see *arrows*) as subjects played the Bach score.[21] These are group-averaged PET images for Bach (contrasted with rest) overlaid on anatomical MRIs. PET data are z-scores displayed on a color scale ranging from 1.65 (*yellow*; $p < 0.1$) to 4.0 (*red*; $p < 0.0001$).

effectively dissociated from motor components and thus appears to contradict the traditional assumption that the cerebellum possesses only motor functions. It is unclear from the present data what function this temporal-cerebellar circuit performs for higher-order representations of music. Interestingly, there is a rapidly growing set of neuroimaging and neurological findings indicating that the cerebellum possesses functions other than those classically attributed to it. These new data suggest that the cerebellum is involved in some way in a variety of sensory and cognitive tasks.[37–45] The present findings, and others discussed below, appear to extend this new view of cerebellar function to the perceptual representation of music.

In summary, these data assist in outlining some of the neural systems that support musical performance and understanding. These findings appear to confirm or extend other neuroimaging and neurological observations and pave the way for new, more focused, research.

## FUNCTIONAL NEUROANATOMY OF THE COMPREHENSION OF HARMONY, MELODY, AND RHYTHM

We pursued these results in our second functional neuroimaging study, in which we attempted to map the brain basis of the comprehension of individual principal components of music per se.[46] Using the same PET methodology as in our first study, my colleagues and I imaged eight right-handed, male Ph.D. faculty musicians (none of whom participated in our first study). These subjects are experts at monitoring musical scores while listening for accurate performance. Each subject was competent on several instruments, with primary specializations in either winds, strings, voice, or keyboard.

During the experiment, these subjects detected errors in a computer-generated instrumental performance of unfamiliar Bach chorales as they read the corresponding score. The subjects made no overt response during a trial; rather they acknowledged covertly the location and presence of specific errors. The errors, which we implanted on each trial, were exclusively either rhythmic, harmonic, or melodic. The errors occurred unpredictably every two to three beats. On rhythm trials, errors occurred only in the performed rhythm of a note in one of the four voices but were otherwise correct. The rhythm errors changed either the duration or onset of a written note only (leaving unchanged the other aspects of rhythm such as meter and tempo). On melody trials, errors perturbed a written note in the highest voice by raising or lowering it a chromatic half step; otherwise they were accurate. On harmony trials, the errors could occur in any one of the four voices but were otherwise correct. The harmony errors perturbed the written note by raising or lowering it a chromatic half step. Pilot studies were used to develop stimuli that equated for difficulty across the three conditions. In prescan and postscan performance, subjects were accurately detecting approximately 80% of the errors in each task.

After practicing the tasks, each subject was scanned performing each of the following five trials twice in pseudorandom order. On harmony, rhythm, and melody trials, subjects sight-read the score while listening to its errorful performance. On passive listening trials, subjects listened to a chorale performed without errors; they were not reading a score. On each scanned trial of the harmony, rhythm, melody, and

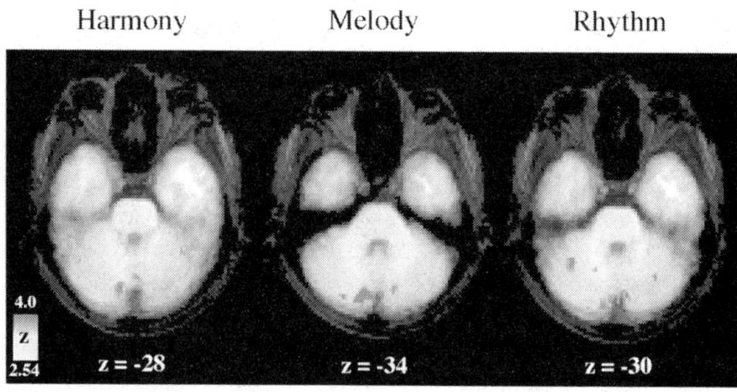


**FIGURE 4. See Plate 9 in color section.** Significant blood flow activations in the cerebellum as musicians sight- read a J. S. Bach chorale score and listen for performance errors in harmony, melody, or rhythm.[46] These are group-averaged PET images for each task (contrasted with passive listening control) overlaid on anatomical MRIs. PET data are z-scores displayed on a color scale ranging from 1.96 (*yellow*; $p < 0.05$) to 4.0 (*red*; $p < 0.0001$).

passive listening tasks, subjects heard a different Bach chorale. On rest baseline control trials, subjects fixated on a fixation point (neither reading a score nor listening to music). The image data during the primary three tasks were contrasted with those from listening to the errorless performance of a Bach chorale without reading a score, as well as to the rest baseline control. Analyses were conducted as described earlier for our preceding PET study.

To highlight some of our principal findings, the melody, harmony, and rhythm conditions each showed a distinct pattern of distributed, statistically significant brain activity compared to passive listening control ($p < 0.001$). Often, each condition activated different subareas of a particular major brain area. In broad anatomical terms, score reading and comprehension of melodic, rhythmic, and harmonic features of a piece were supported by processes in both cerebral hemispheres. More specifically, melodic comprehension activated each hemisphere equally, whereas harmonic and rhythmic comprehension activated more of the left hemisphere than the right one. In addition, the rhythm condition activated comparatively few brain areas outside the cerebellum.

All three tasks strongly activated bilateral cerebellum, primarily the lateral hemispheres, with little overlap in activated area across the three tasks (FIG. 4). Given the lack of explicit (and apparently implicit) movement, this cerebellar activity is likely related to sensory or cognitive processing, not to motoric processing (see below). Furthermore, activation of cerebellum in the rhythm condition was twice that in harmony and melody conditions. An involvement of the cerebellum in some way in processing rhythm is consistent with recent neurological and neuroimaging data.[42,47–49] Cerebellar activations during the processing of melody and harmony are relatively novel findings but are consistent with data above from our first study and with other recent studies showing lateral cerebellar activation for processing of tonal information (see later).

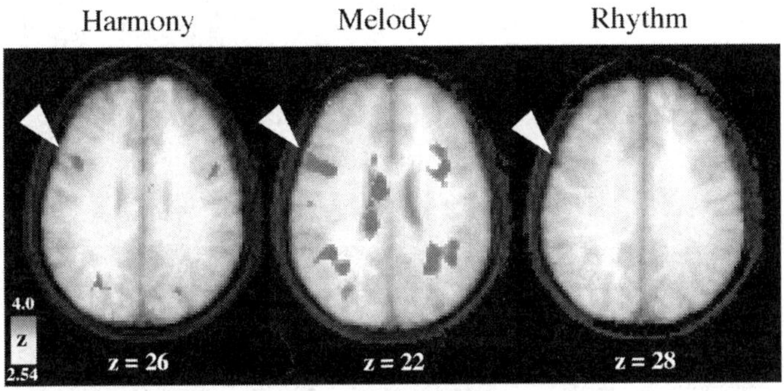


**FIGURE 5. See Plate 10 in color section.** Activations in different subareas of a major cytoarchitectonic area of frontal cortex (BA 44) as musicians sight-read a J. S. Bach chorale score and listen for performance errors in harmony, melody, or rhythm.[46] These are group mean PET images for each task (contrasted with passive listening control) overlaid on MRIs. PET data are z-scores displayed on a color scale ranging from 2.54 (*yellow*; $p < 0.01$) to 4.0 (*red*; $p < 0.0001$).

One area commonly activated by the harmony, melody, and rhythm tasks was in right fusiform gyrus. Because score reading was a central process common to all three conditions, the observed right fusiform activation may implicate that area in the visual processing of musical notes. Such a function would be analogous to visual word processing by left fusiform.[51–53] These implied parallel locations for comparable reading functions in music and language are consistent with other functional and structural correspondences.[16–20]

Each of the three conditions activated different subareas of two major areas in frontal cortex, a lateral region, BA 6, and an inferior lateral region, BA 44/45 (FIG. 5). These activations were bilateral, although stronger on the left side. In both of these major areas, the rhythm task activated superior subareas, the melody task activated inferior subareas, and the harmony task activated subareas between those two activations. These two major areas are currently associated with sequencing motor behavior and with recognizing actions.[54–59] However, there was no activation detected in supplementary motor area (SMA) medial BA 6, as has been found in neuroimaging studies of song imagery.[7] In addition, there was no activation in primary motor areas. Therefore, the activation of these areas in the present conditions does not seem related to motor preparation as such, whether for subvocal singing or for imagined playing of an instrument. It is possible that these activations are involved in translating the notes to abstract motor-auditory codes. Such a translation process may be a natural component of expert score reading.

The melody, harmony, and rhythm tasks each also produced distinct patterns of activity in auditory cortex (FIG. 6). During the melody condition, there were strong activations in secondary auditory association areas, specifically in bilateral superior (BA 22) and bilateral, strongly right, middle temporal (BA 21) cortex. The harmony task produced bilateral activation, more on the left side than the right, in middle (BA 21 and 20), posterior (BA 37), and superior (BA 22) temporal cortex. The rhythm condition produced only minor activation in left inferior (BA 20) and bilateral mid-

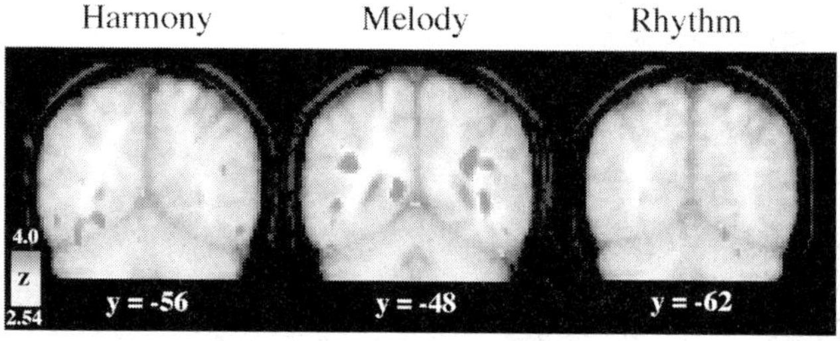


**FIGURE 6. See Plate 11 in color section.** Activations in different subareas of auditory cortex as musicians sight-read a J. S. Bach chorale score and listen for performance errors in melody, harmony, or rhythm.[46] These are group mean PET images for each task (contrasted with passive listening control) overlaid on MRIs. PET data are z-scores displayed on a color scale ranging from 2.54 (*yellow*; $p < 0.05$) to 4.0 (*red*; $p < 0.0001$).

dle (BA 21) temporal cortex. Superior temporal activation for melodic and harmony tasks, but not rhythmic processing, is consistent with neuropsychological data showing that superior temporal lesions spare rhythmic processing but cause impaired melodic perception.[6,11,60] The activation of such secondary auditory regions bilaterally during these comprehension tasks is also in accord with bilateral activation of those areas during the perception or performance of musical pieces and scales in our prior PET study.

The distributed pattern of activations observed in this investigation delineates some of the neural systems subserving musical comprehension. This study may provide an early glimpse of the functional neuroanatomy of expert musicians' comprehension of musical structure. These data confirm or extend prior understanding of the distributed system of brain areas supporting pitch and melody perception, auditory mental imagery, unimanual piano performance from a sight-read score, and bimanual piano performance of a memorized score. These findings are also consistent with and complement those reported in a contemporaneous neuroimaging study by Platel, Price, Baron, *et al.*[10] in which nonmusicians attended to either the timbre, melody, or rhythm properties of random-order, nonmusical tone sequences. Interestingly, the latter study reported that musical timbre, not studied here, was subserved by areas in right superior frontal, middle frontal, and precentral cortex.

The foregoing results suggest that music as a whole is represented in mechanisms widely distributed throughout the brain. However, different parts of the brain are activated when one pays close attention to different aspects of a piece of music. This observation is perhaps not surprising since music is a complex, multifaceted stimulus and activity. At the same time, the data illustrate the power of directed attention to change brain activity. The data also suggest that the specific subareas of major areas, such as cerebellum, and frontal and temporal cortex, perform different functions for particular subtasks of musical comprehension. More generally, the distributed neural system for musical comprehension revealed here can be compared in further studies to the better-known distributed neural systems for language tasks. Such studies can begin to identify independent and shared neural mechanisms for music and language and evaluate the scope of application for specific neural mechanisms.

## FUNCTIONAL NEUROANATOMY OF MUSICAL RHYTHM

The next study builds upon the findings of the preceding studies and aims to provide a more complete account of the functional neuroanatomy of the mental representation of musical rhythm. We examined the selective perception of individual principal components of musical rhythm—that is, pattern (phrasing), tempo, meter, and duration. In addition, in order to explore the effect of musical talent and training on the neural systems supporting musical rhythm perception, expert musicians were compared to nonmusicians.

Studies of brain-damaged patients have reported that the perception of rhythm can be impaired without affecting melodic processing, suggesting that there is a distinct neural system for the representation of rhythm.[61] Some studies have been inconclusive as to whether rhythm was a left hemispheric or bilateral neural process.[62,63] Interestingly, recent fMRI data suggest that memory for familiar, regular rhythms (with whole-number interval ratios) is supported by left frontal (BA 6) and parietal cortex and right anterior cerebellum.[48] On the other hand, these data suggested that memory for more unfamiliar rhythms (with fractionated interval ratios) was supported by right prefrontal, frontal (BA 6), and parietal areas, along with bilateral posterior cerebellum.

The involvement of the cerebellum in the processing of musical rhythm is consistent with data from our second PET study (see earlier) and those of two other studies.[10,42] This implication is congruent with neurological findings that cerebellar patients are impaired at the fine perception of temporal features of auditory stimuli.[47]

Involvement of parietal and prefrontal cortex in rhythm processing[48] is also consistent with data from our second PET study. In addition, this activity is congruent with recent neurological data implicating an inferior parietal-prefrontal circuit that supports time-dependent attention and working memory functions necessary for perception of duration of auditory stimuli.[64] Of particular note in this context is a study of unilateral temporal corticectomy patients in which evidence was found for a dissociation between the areas processing meter as opposed to rhythm pattern, with anterior superior temporal areas implicated for meter.[11]

In the present study, Michael Thaut and I explored the principal components of rhythm: duration, pattern, tempo (dynamically increasing or decreasing rate), and meter (e.g., the differing periodicities of 3/4, 4/4, 5/4, 5/8, 7/8, and 9/8). Using the same PET methods as in our preceding studies, we imaged nonmusicians and musicians, making covert discriminations of pairs of rhythmic auditory patterns. Each scanned set of trials required the subject to focus on a single one of four principal components of rhythm. We also included a pitch discrimination control task in which subjects compared pairs of auditory sequences. In the pitch discrimination task it was possible for a single note to vary in pitch in one member of the stimulus pair. The pitch discrimination stimuli possessed a metronomic rhythm (i.e., all notes had the same duration). As a control task, the pitch stimuli required comparable auditory perception, working memory, and comparison and decision processes as the rhythm tasks, without requiring the processing of rhythmic features of duration, meter, tempo, or pattern.

Participating in the study were five musicians, with (at least) an undergraduate university degree in music, and five nonmusicians, with no music training or performance experience beyond childhood. All 10 subjects were right-handed adult males. The stimuli (apart from the pitch control) were always 440-hertz computer-generated piano timbre sounds of 231- or 462-ms duration. The interval between tones in a stimulus sequence was a multiple of 231 milliseconds. The stimuli on the duration, pattern, and pitch control trials were modeled on the standardized Seashore Test.[65] The stimuli on the tempo and meter trials were modeled on the Gordon Musical Aptitude Profile.[66] These stimulus materials were adjusted (i.e., in tempo or accentuation) so that the stimuli on the tempo, phrasing, duration, meter, and pitch control trials were as similar as possible. Different conditions varied with respect to the exact nature of the task being performed, with stimuli being in almost all respects constant across conditions. Although tempo, pattern, and meter stimuli had more tones per sequence, between 9 and 12 events, approximately the same total number of tones per scanned interval were presented in the duration task as in the tempo, pattern, and meter tasks. A "different" trial contained the same number of events (and accents) and was of the same total duration as the "same" trials (by playing them at different tempi), except in the duration condition. In the pitch task, the tone alternated between 415, 440, and 466 hertz in sequences of 12 462-ms (quarter note) tones, without rests. Across all conditions, the intersequence interval was 1000 ms, and the intertrial interval was 1750 ms; the trial time was 14 s on average.

Stimuli were adjusted to produce comparable mean accuracy of musicians and of nonmusicians on each task. Each subject subsequently performed 11 PET trials: two trials each of the pattern task, duration task, meter task, tempo task, and pitch control task, as well as one rest trial. The subject's eyes were closed on all trials. During each 60 s of task there were four covert discrimination judgments. For half of the trials, the pair of stimuli was identical. After the PET session, subjects replicated the trials from the PET session, overtly indicating their responses. Their average performance across the latter trials confirmed that task difficulty was equated across conditions and across groups.

To highlight some of our principal findings, for both nonmusicians and musicians, distinct patterns of brain activity were detected for the discrimination of pitch and each musical rhythm. There were also distinct differences between nonmusicians and musicians during each discrimination.

For pitch (contrasted with rest), superior temporal areas (BA 21/22) were activated on the right in nonmusicians and on the left in musicians (FIG. 7A and D). However, middle and inferior temporal areas (BA 21/20) were activated on the right in both groups (FIG. 7B and E). These temporal activations confirm prior neurological and neuroimaging studies, including those for the comprehension of melody above.[6,7,11,34–36]

There were also strong cerebellar activations during pitch (FIG. 7C and F), again confirming our earlier PET data. The cerebellar activations are not likely related to motor activity since no overt motor activity occurred and there was little or no activity detected in neocortical motor areas. Instead, the cerebellar activity seems to support perceptual or cognitive processing of pitch (or melody). Interestingly, the nonmusicians showed strong predominantly left-sided activation in posterior lateral cerebellar hemispheres (FIG. 7C). The left cerebellum projects to the right cerebral cortex, so the activation in left cerebellum is likely supporting the right temporal activation for pitch. However, musicians showed bilateral cerebellar activation (FIG. 7F), apparently supporting bilateral temporal activation.

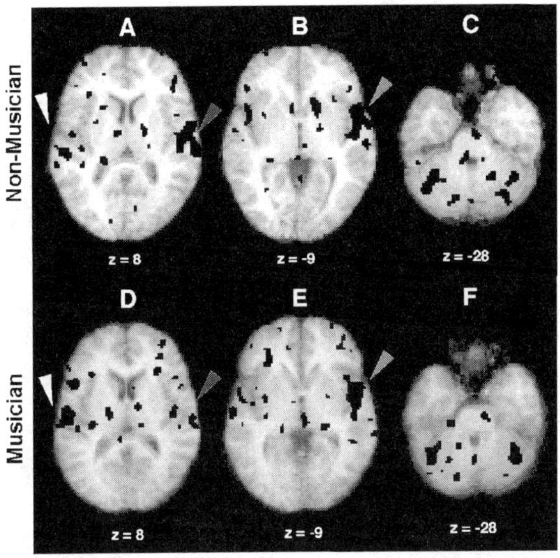


**FIGURE 7. See Plate 12 in color section.** Activations in superior and middle temporal cortex (**A**, **B**, **D**, and **E**) and in cerebellum (**C** and **F**) as nonmusicians and musicians discriminate pitch sequences.[49] These images show group mean PET data for each task (contrasted with rest) overlaid on MRIs. PET data are z-scores displayed on a color scale ranging from 1.96 (*yellow*; $p < 0.1$) to 4.0 (*red*; $p < 0.0001$).

The key rhythm-specific activations were as follows. First, certain activations were generally present for each rhythm task and for both nonmusicians and musicians, such as basal ganglia and cingulate cortex. The cingulate activations are likely involved in attention functions required in the task.[67–70]

For the tempo, pattern, and meter discriminations, there was a tendency for activation to be detected in nonmusicians that was not present for musicians. For tempo discrimination, nonmusicians showed activation in right medial frontal cortex (BA 9) and bilateral posterior lateral cerebellum, but this activation was very much weaker or absent in musicians (FIGS. 8 and 12). For pattern discrimination, nonmusicians showed activation in midbrain and bilateral posterior lateral cerebellum; however, this activation was absent and weaker in musicians (FIGS. 9 and 12). For meter discrimination, nonmusicians showed activation in right inferior frontal cortex (BA 47) and bilateral posterior lateral cerebellum (FIGS. 10 and 12), with little or no such activation in these areas for musicians. By contrast, during duration discrimination musicians showed bilateral medial frontal cortex (BA 9), bilateral inferior parietal cortex (BA 40), and strong activation in bilateral posterior lateral cerebellum (FIGS. 11 and 12). These activations were absent or very much weaker in nonmusicians.

As with pitch discrimination, the strong cerebellar activations during rhythm discriminations are not likely due to motoric processing per se but to a role in supporting nonmotor sensory or cognitive processing.[38–45] In addition, in the meter, tempo, and pattern discriminations, cerebellar activity is much stronger for nonmusicians than for musicians, but the reverse is true for the duration discrimination. Interest-

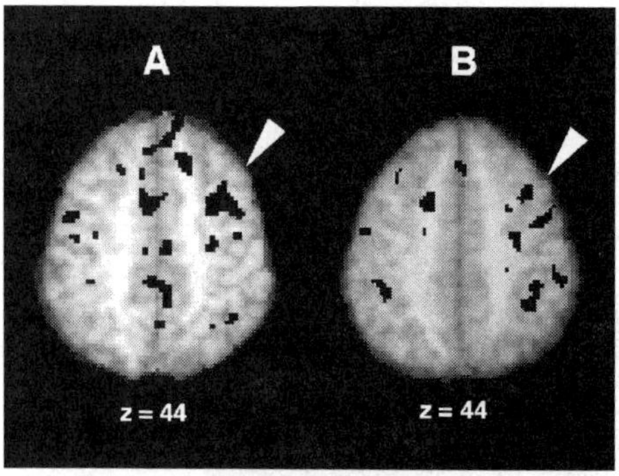


**FIGURE 8. See Plate 13 in color section.** Activations in left medial frontal cortex (BA 9) for nonmusicians (**A**) as compared to musicians (**B**) during the discrimination of musical tempi.[49] Shown are group mean PET images for each task (contrasted with rest) overlaid on MRIs.

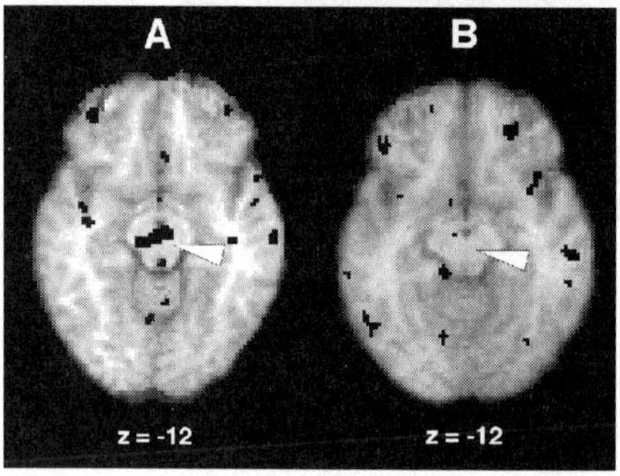


**FIGURE 9. See Plate 14 in color section.** Activations in midbrain for nonmusicians (**A**) as compared to musicians (**B**) during the discrimination of musical pattern.[49] These are group-averaged PET images for each task (contrasted with rest) overlaid on anatomical MRIs.

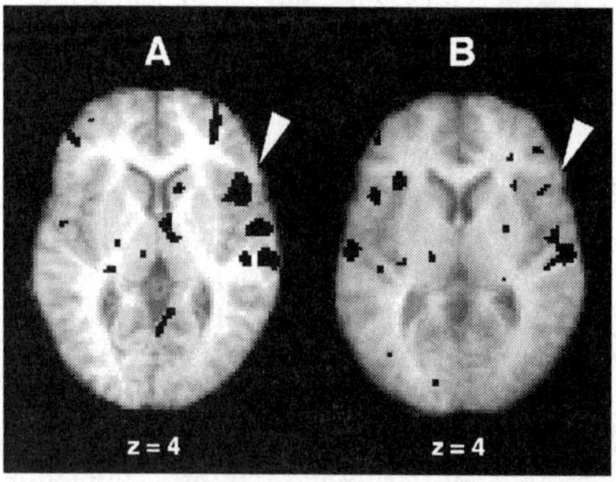


**FIGURE 10. See Plate 15 in color section.** Activations in right inferior frontal cortex (BA 47) for nonmusicians (**A**) as compared to musicians (**B**) during discrimination of musical meter.[49] Shown are group-averaged PET images for each task (contrasted with rest) overlaid on anatomical MRIs.

ingly, musicians found discriminating the duration of rhythmic sequences more novel than discriminating meter, tempo, and pattern. By contrast, the nonmusicians found discriminating the overall duration of rhythmic sequences less novel than discriminating other rhythm features. Thus, the increase in cerebellar activation appears to be associated with greater novelty of rhythm discrimination, consistent with other neuroimaging findings[71] and new hypotheses of cerebellar function.[37]

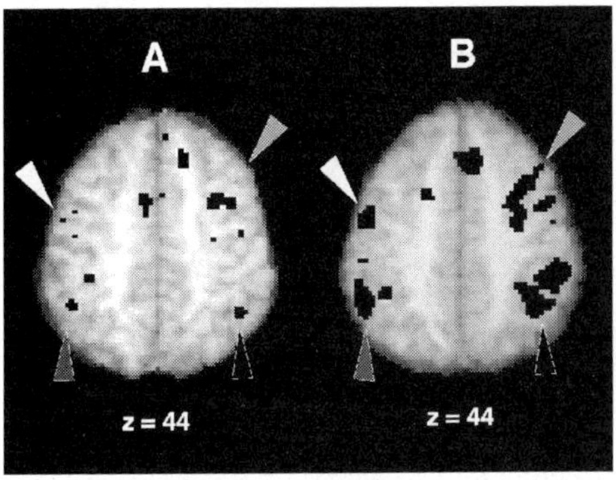


**FIGURE 11. See Plate 16 in color section.** Activations in bilateral medial frontal cortex (BA 9) and bilateral inferior parietal cortex (BA 40) for musicians (**B**) as compared to nonmusicians (**A**) during the discrimination of sequence durations.[49] Shown are group-averaged PET images for each task (contrasted with rest) overlaid on anatomical MRIs.

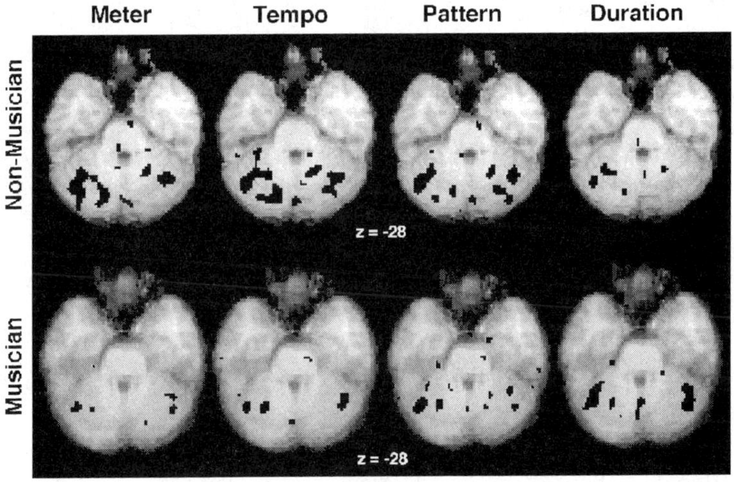


**FIGURE 12. See Plate 17 in color section.** Activations in posterior cerebellum as nonmusicians and musicians discriminate meter, tempo, pattern, and duration.[49] Shown are group-mean PET images for each task (contrasted with rest) overlaid on MRIs.

The activations in prefrontal cortical areas BAs 9, 10, and 47 are likely supporting working memory processes.[48] The activation in prefrontal and inferior parietal areas follows similar indications in the recent neurological study of tone duration discrimination.[64] We speculate that activations in temporal, frontal, and midbrain may be involved in processing the temporal grouping structure of rhythms.

These findings suggest some key brain areas in the distributed neural circuits underlying the perceptual/cognitive representation of central aspects of musical rhythm. These data suggest that distinct brain areas subserve musical meter, tempo, pattern, and duration. In addition, the results confirm that the neural subsystems underlying pitch discrimination are distinct from those subserving the elements of rhythm discrimination. The differences in activated brain areas for nonmusicians and musicians likely reflect differences in strategy, perceptual skill, and cognitive representation of the components of musical rhythm and pitch.

## NEUROLOGY OF PITCH DISCRIMINATION

In the foregoing review of neuroimaging and neurological studies of music, the cerebellum appears to be involved in some way in nonmotor, nonsomatic sensory or cognitive processing. This kind of conclusion is supported by many other recent neuroimaging and neurological findings.[37–45,73] Such findings challenge classical motor theories of cerebellar function and have created an environment in which researchers are increasingly questioning the extent to which the cerebellum is involved in motor control. Although in most cases, new alternative proposals are described as an extension of the role of the cerebellum in motor coordination itself, others, who have been driven primarily by new neuroimaging results, are proposing radically different functions for this structure.

James Bower and I, and our colleagues, have been using neuroimaging of tactile and cutaneous sensory and motor tasks to test the hypothesis that the cerebellum is involved in monitoring and optimizing the acquisition of information in various sensory modalities.[72–74] A direct prediction of this hypothesis is that damage to the cerebellum will impair performance on sensory tasks, especially those in which successful performance depends on sensory information processing at a very fine time scale. We evaluated this prediction for auditory perception by examining for the first time the effect of global cerebellar degeneration on auditory pitch discrimination in (nonmusician) patients.[75] This strictly auditory task does not depend on overt motor behavior, as do other sensory tasks.

Fifteen patients and fifteen healthy control subjects (matched in age and education) performed a pitch discrimination task measuring difference thresholds. The patients suffer from pancerebellar degeneration caused by hereditary, idiopathic, paraneoplastic, or postinfectious pancerebellitis. The degree of cerebellar degeneration in these patients, as confirmed by examinations of pancerebellar ataxia and anatomical imaging, ranged from minimal to severe, and included mild- and moderate-grade cases of ataxia. Patients and controls mostly had normal loudness thresholds; some had mild hearing loss in the 4000–8000-Hz range.

On each pitch discrimination trial, subjects decided whether a comparison tone was higher or lower than a preceding 500-Hz standard. All tones had suprathreshold loudness (80 dB) and 400 ms duration; the standard was followed 400 ms later by the comparison pitch. The tones were composed of the fundamental, three times the fundamental frequency at a 50% amplitude, and five times the fundamental at a 25% amplitude. The difference threshold estimation procedure used was PEST.[76,77] Subjects performed 60 trials, with threshold trials from above and below the standard pitch, mixed together in pseudorandom order. Each subject also performed two control tasks, the WAIS-III Digit Memory Span task, assessing auditory working memory capacity, and a simple auditory detection reaction time task, assessing simple motor reaction time for detecting audible sounds.

As shown in FIGURE 13, patients' pitch discrimination thresholds were on average 5.5 times that for controls ($p < 0.0001$). The degree of impairment was correlated

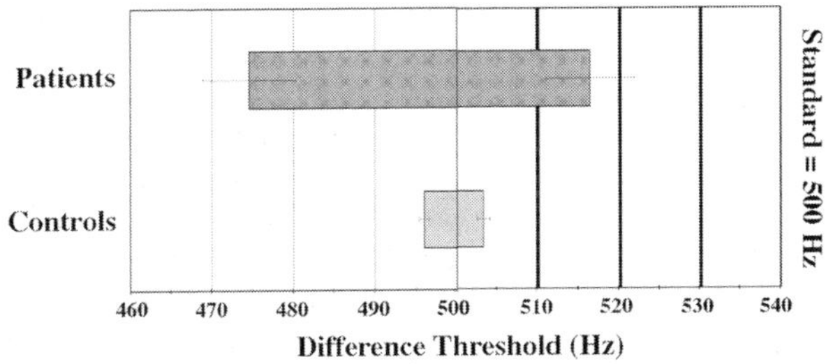


**FIGURE 13.** Average lower and upper difference discrimination thresholds for auditory pitch in pancerebellar degeneration patients and healthy controls.[75]

($r = +0.70$, $p < 0.003$) with the severity of patients' pancerebellar ataxia. Furthermore, the sizes of patients' pitch discrimination thresholds were uncorrelated with their loudness thresholds for the standard tone ($r = + 0.02, p < 0.79$) and for the wider range from 125 Hz to 8000 Hz ($r = + 0.06, p < 0.22$). Patient and control groups did not differ on digit span memory or auditory detection reaction time.

As a group, then, patients with pancerebellar degeneration, but no other neurological conditions, show strongly impaired pitch discrimination. However, the degree of impairment for individuals is proportional to the severity of their pancerebellar ataxia. The severity of the pancerebellar ataxia (e.g., with symptoms in gait, speech, limb, and eye behavior) is assumed to be indicative of the degree of pancerebellar atrophy. Although these data do not indicate the location of auditory processing regions in the cerebellum, it is assumed that atrophy in auditory regions of the cerebellum is proportional to the pancerebellar ataxia. Such atrophied auditory regions are assumed to be responsible for the pitch discrimination impairment. These cerebellar patients' auditory abilities are intact in a fundamental way, as indicated by reasonably normal auditory loudness thresholds. This observation is consistent with the idea that primary sensory processing is subserved in appropriate regions of cerebral cortex, which are spared in these patients. However, patients' loudness thresholds (ranging from normal hearing to mild hearing loss) are uncorrelated with the degree of their pitch discrimination impairment. Central to the idea of cerebellar involvement in sensory data acquisition is the suggestion that cerebellar effects would be most clearly evident in tasks that require highly optimized sensory data. In the current case, it seems reasonable to assume that the simple detection of the presence of a tone is unlikely to require the same quality sensory data as the fine discrimination of pitch.

Thus, these results suggest that the cerebellum has a fundamental supporting role in the information processing necessary for fine auditory discrimination. This conclusion is consistent with cerebellar activations reported in recent neuroimaging studies involving pitch or melody processing.[46,49] The pattern of coupled pitch and ataxia measures, and decoupled pitch and loudness measures, suggests that the role of the cerebellum is to monitor and optimize the acquisition of information in sensory modalities (audition). Patients and controls did not differ in digit span memory. This indicates that the differences between patients and controls in pitch discrimination were very likely not due to any working memory deficits that may possibly be associated with cerebellar damage.

An important independent corroboration of the implications of these data is provided by a recent study examining cerebellar degeneration patients' performance on a variety of visual tasks.[41] The authors report that the patients were impaired selectively for the detection of visual motion, speed, and direction, as dissociated from eye movement control and from any other attentional, memory, or cognitive task components. It should be noted that there are also neurological studies suggesting that the cerebellum may be instrumental in other nonmotor behaviors. These behaviors include judging the timing of events, solving perceptual and spatial reasoning problems, and generating words according to a semantic rule.[47,78–83] In terms of the sensory acquisition hypothesis we are testing, however, impairments in these specific tasks may be more related to the dependence of these tasks on well-controlled information processing (e.g., sensory acquisition) than to different specific roles for the cerebellum in each of those behaviors.

## GENERAL ISSUES IN INTERPRETING BRAIN ACTIVATIONS

Given the rapid emergence of promising new neuroimaging methods for the study of music performance, perception, and cognition, it may be useful in the present volume to discuss current and general methodological issues for interpreting the role of activated brain areas in the processing of music. Mapping elementary operations onto particular brain areas and mapping those operations/areas onto system-level models of specific musical and other cognitive tasks requires a convergence across paradigms (behavioral tasks) and independent experiments. No single task or paradigm used in neuroimaging or neurological investigations can isolate an elementary operation. Likewise, no single group of subjects (i.e., one study's sample) can fully capture the location and distribution of a brain area in a population. Thus, mapping localized operations entails sifting through multiple studies.

In functional neuroimaging, for example, the analysis of multiple studies, or metanalysis, can be rigorously and quantitatively performed[84] because the data are customarily reported in stereotaxic coordinates.[23,30] A growing number of rigorous metanalyses with important implications for mapping operations and systems, as well as for psychological models, are being conducted. For example, such analyses have been performed in motor processes and supplementary motor areas,[85] in oculomotor control and frontal eye fields,[86] in speech motor-sensory processes and the mouth area of primary motor-sensory cortex,[87,88] in visual information processing,[89–91] in visual attention,[91] and in memory.[92] Improved methods for such meta-analyses are currently being developed and validated.[84,87,88,93] Using such quantitative stereotactic-based metanalyses will yield important insights into the relationship between neural mechanisms supporting music and those supporting other cognitive processes.

As an aside, it should be noted that mapping neural mechanisms and systems will be advanced more efficiently if, whenever possible, the coordinate-based analyses just described replace analyses based only on lower resolution measures such as hemisphere, lobe, sulcus, gyrus, and, eventually, Brodmann area. Interestingly, close analysis suggests that much of the seemingly large variability in structure-function correlations remarked upon by some researchers using surface-based anatomical landmarks may vanish when the frame of analysis shifts to stereotactic coordinate–based analysis.[88]

Be that as it may, hypothesizing an operation (or computation) for an area requires a principled analysis of cognitive and behavioral processes and tasks. In spite of efforts in the cognitive sciences, there is no formal language in which to capture component operations that fractionate a task, including musical ones. A useful first step in that development would be for circumscribed communities of researchers studying a particular operation, area, or paradigm(s) in music processing to settle on an effective common vocabulary, with an eye toward establishing a working standard vocabulary with researchers studying other specific operations, areas, or paradigms. At the present time, this is rarely done, and apparently never in a concerted fashion. Standardized component models will be of great benefit for the process of mapping operations down to brain areas and up to system-level tasks in order to compare music and other cognitive processes.

Another significant point with regard to these goals is that whenever possible the operation hypothesized for a brain area should be consistent with its underlying microcircuitry. This is also rarely done in cognitive neuroscience. Take, for example, the case of hypothesizing operations for the cerebellum. The cerebellum has a uniform microcircuitry throughout, so until proved otherwise, we should assume that it performs a uniform operation, even if that operation may be performed on different kinds of information (as consistent with connectivity). Of course, even the most widely accepted facts about microcircuitry in an area can change from time to time. In the case of the cerebellum, David Marr's model of cerebellar function[94] was shown to be inconsistent with the underlying microcircuitry some time after its publication (although it continued to be influential).

Finally, there is a misguided tendency when interpreting neuroimaging data to assume that every activated area is directly involved in the cognitive or behavioral task eliciting it. A parallel tendency leads to the assumption that every cognitive or behavioral process affected by damage to a particular brain area necessarily implicates direct involvement of that area in those processes. As our sophistication in the computational and neural sciences grows, we will likely recognize that there are a range of indirect roles a mechanism might assume.

Consider again the cerebellum. The cerebellum is thought by various researchers to be directly involved in a variety of cognitive and behavioral processes. However, the consequences of injury to, or absence of, the cerebellum are often not fully appreciated. Although 70–80% of the neurons in the brain are packed tightly in the cerebellum, the symptoms are mild and transient in cases when nearly all of it is removed. Likewise, persons born without a cerebellum show few impairments. These observations imply that the function of the cerebellum is subtly supportive and not necessary. On the hypothesis of cerebellar function that my colleagues and I are testing,[37,73] a brain with a healthy cerebellum is assumed to be optimized for operations necessary for fine sensory acquisition. Furthermore, a brain without a cerebellum is viewed as subtly suboptimal, whereas a brain with a damaged cerebellum can range from mildly to very suboptimal, depending on degree and location of damage. In the more debilitating cases, the assumption is that a damaged cerebellum is causing other brain areas to operate on the basis of corrupted, noisy, poorly acquired and controlled sensory information. More generally, there may be a number of brain areas (such as basal ganglia or cingulate) with comparably indirect supporting roles in other cognitive or behavioral processes. Clearly, careful neurological studies (and, in some cases, transcranial magnetic stimulation studies) can assist in dissociating shades of involvement of a brain area in a particular psychological process.

## CONCLUSION

Musical experiences intricately and deeply engage the mind and the whole brain. Details of this process are being unveiled by new scientific advances well represented in this volume. Although many challenges lie ahead for fully understanding the biological foundations of music, the tools for meeting at least a fair share of those challenges appear to be in hand.

## ACKNOWLEDGMENTS

I am grateful to the following collaborators for their stimulating, intelligent assistance with the studies discussed in this chapter: Peter Fox, Donald Hodges, James Bower, Jeremy Schmahmann, Stephen Grill, Michael Thaut, Bruno Repp, Michael Martinez, and John Tennison. I also thank our subjects, musicians, and patients for their gracious participation in support of basic research. The work described in this article was supported by the EJLB Foundation, the National Association for Music Merchants (now the International Foundation for Music Research), the Texaco Foundation, a State of Colorado Center of Excellence Award, and NIH (NINDS) Grant NS3710901A1.

## REFERENCES

1. LERDAHL, F. & R. JACKENDOFF. 1983. A Generative Theory of Tonal Music. MIT Press. Cambridge, MA.
2. DOWLING, W. & D. HARWOOD. 1986. Music Cognition. Academic Press. Orlando, FL.
3. KRUMHANSL, C. 1990. Cognitive Foundations of Musical Pitch. Oxford University Press. Oxford.
4. REPP, B.H. 1992. Probing the cognitive representation of musical time: structural constraints on the perception of timing pertubations. Cognition **44:** 241–281.
5. SERGENT, J. 1993. Mapping the musician brain. Hum. Brain Map **1:** 20–38.
6. PERETZ, I., R. KOLINSKY, M. TRAMO, *et al.* 1994. Functional dissociations following bilateral lesions of auditory cortex. Brain **117:** 1283–1301.
7. ZATORRE R.J., A.R. HALPERN, D.W. PERRY, *et al.* 1996. Hearing in the mind's ear: a PET investigation of musical imagery and perception. J. Cognit. Neurosci. **8:** 29–46.
8. DELIEGE, I. & J. SLOBODA. 1997. Perception and Cognition of Music. Psychology Press. Wicyita, KS.
9. PALMER, C. 1997. Music performance. Annu. Rev. Psychol. **48:** 115–138.
10. PLATEL, H., C. PRICE, J.C. BARON, *et al.* 1997. The structural components of music perception. Brain **120:** 299–243.
11. LIEGEOIS-CHAUVEL, C., I. PERETZ, M. BABAI, *et al.* 1998. Contribution of different cortical areas in the temporal lobes to music processing. Brain **121:** 1853–1867.
12. PANTEV, C., R. OOSTENVELD, A. ENGELIEN, *et al.* 1998. Increased auditory cortical representation in musicians. Nature **392:** 811–814.
13. BLOOD, A.J., R.J. ZATORRE, P. BERMUDES, *et al.* 1999. Emotional responses to pleasant and unpleasant music correlate with activity in paralimbic brain regions. Nat. Neurosci. **2:** 382–387.

14. WALLIN, N.L., B. MERKER & S. BROWN, Eds. 1999. The Origins of Music. MIT Press. Cambridge, MA.
15. SERGENT, J., E. ZUCK, S. TERRIAH, *et al.* 1992. Distributed neural network underlying musical sight-reading and keyboard performance. Science **257:** 106–109.
16. BESSON, M. & F. FAITA. 1995. An event related potential (ERP) study of musical expectancy: comparison of musicians with non-musicians. J. Exp. Psychol.: Human Percept. Perform. **21:** 1278–1296.
17. PATEL, A.D. & I. PERETZ. 1997. Is music autonomous from language? A neuropsychological appraisal. *In* Perception and Cognition of Music. I. Deliege & J. Sloboda, Eds. Psychology Press. Wichita, KS.
18. BESSON, M., F. FAITA, I. PERETZ, *et al.* 1998. Singing in the brain: independence of lyrics and tunes. Psychol. Sci. **9:** 494–498.
19. PATEL, A.D., E. GIBSON, J. RATNER, *et al.* 1998. Processing syntactic relations in language and music: an event-related potential study. J. Cognit. Neurosci. **10:** 717–733.
20. PATEL, A.D., I. PERETZ, M. TRAMO, *et al.* 1998. Processing prosodic and musical patterns: a neuropsychological investigation. Brain Lang. **61:** 123–144.
21. FOX, P.T., L.M. PARSONS, D. HODGES, *et al.* Neural systems underlying musical performance. Submitted.
22. FOX, P.T., M.A. MINTUN, M.E. RAICHLE, *et al.* 1984. A noninvasive approach to quantitative functional brain mapping with $H_2{}^{15}0$ tissue activity. J. Nucl. Med. **30:** 141–149.
23. TALAIRACH, J. & P. TOURNOUX. 1988. Co-planar stereotaxic atlas of the human brain. Thieme Medical Publishers. New York.
24. FOX, P.T., M.A. MINTUN, E.M. REIMAN, *et al.* 1988. Enhanced detection of focal brain responses using inter-subject averaging and change-distribution analysis of subtracted PET images. J. Cereb. Blood Flow Metab. **8:** 642–653.
25. WOODS, R.P., S.R. CHERRY & J.R. MAZZIOTTA. 1992. Rapid automatic algorithm for aligning and reslicing PET images. J. Comp. Assist. Tomography **17:** 536–546.
26. WOODS, R.P., J.C. MAZZIOTTA & S.R. CHERRY. 1993. MRI-PET registration with automated algorithm. J. Comput. Assisted Tomogr. **17:** 536–546.
27. LANCASTER, J.L., T.G. GLASS, B.R. LANKIPLLI, *et al.* 1995. A modality-independent approach to spatial normalization of tomographic images of the human brain. Hum. Brain Map **3:** 209–223.
28. FOX, P.T. & M.A. MINTUN. 1989. Noninvasive brain mapping by change-distribution analysis of averaged PET images of $H_2{}^{15}O$ tissue activity. J. Nucl. Med. **30:** 141–149.
29. XIONG J., J-H. GAO, J.L. LANCASTER, *et al.* 1996. Clustered pixels analysis for functional MRI activation studies of the human brain. Hum. Brain Map. **3:** 287–301.
30. LANCASTER, J.L., M.G. WOLDORFF, L.M. PARSONS, *et al.* 2000. Automatic Talairach labels for functional brain mapping. Hum. Brain Map **10:** 120–131.
31. ZATORRE, R.J. 1988. Pitch perception of complex tones and human temporal-lobe function. J. Acoust. Soc. Am. **84:** 566–572.
32. ZATORRE, R.J. & S. SAMSON. 1991. Role of the right temporal neocortex in retention of pitch in auditory short-term memory. Brain **114:** 2403–2417.
33. ZATORRE, R.J., A.C. EVANS, E. MEYER, *et al.* 1992. Lateralization of phonetic and pitch discrimination in speech processing. Science **256:** 846–849.
34. ZATORRE, R.J., A.C. EVANS & E. MEYER. 1994. Neural mechanisms underlying melodic perception and memory for pitch. J. Neurosci. **14:** 1908–1919.
35. GRIFFITHS, T.D., C. BUCHEL, R.S.J. FRACKOWIAK, *et al.* 1998. Analysis of temporal structure in sound by the human brain. Nature Neurosci. **1:** 422–427.
36. PERRY, D.W., R.J. ZATORRE, M. PETRIDES, *et al.* 1999. Localization of cerebral activity during simple singing. NeuroReport **10:** 3979–3984.
37. BOWER, J.M. 1997. Is the cerebellum sensory for motor's sake, or motor for sensory's sake: the view from the whiskers of a rat? *In* Progress in Brain Research. C.I. de Zeeuw, P. Strata & J. Voogd, Eds.: 483–516. Elsevier. New York.
38. JUEPTNER, M., S. OTTINGER, S.J. FELLOWS, *et al.* 1997. The relevance of sensory input for the cerebellar control of movements. Neuroimage **5:** 41–51.
39. PARSONS, L.M. & P.T. FOX. 1997. Sensory and cognitive functions. *In* The Cerebellum and Cognition. J.D. Schmahmann, Ed.: 255–272. Academic Press. New York.
40. SCHMAHMANN, J.D., Ed. 1997. The Cerebellum and Cognition. Academic Press. New York.

41. DESMOND, J. & J. FIEZ. 1998. Neuroimaging studies of the cerebellum: language, learning, and memory. Trends Cognit. Sci. **2:** 355–362.
42. PENHUNE, V.B., R.J. ZATORRE & A.C. EVANS. 1998. Cerebellar contributions to motor timing: a PET study of auditory and visual rhythm reproduction. J. Cognit. Neurosci. **10:** 752–765.
43. THIER, P., T. HAARMEIER, S. TREUE, *et al.* 1999. Absence of a common functional denominator of visual disturbances in cerebellar disease. Brain **122:** 2133–2146.
44. PARSONS, L.M., D. DENTON, R. SHADE, *et al.* 2000. Neuroimaging evidence implicating cerebellum in support of sensory/cognitive processes associated with thirst. Proc. Natl. Acad. Sci. USA **97:** 2332–2336.
45. TESCHE, C.D. & J. KARHU. 2000. Anticipatory cerebellar responses during somatosensory omission in man. Hum. Brain Map. **9:** 119–142.
46. PARSONS, L.M., D.A. HODGES & P.T. FOX. The neural basis of the comprehension of musical harmony, rhythm, and melody. Submitted.
47. IVRY, R. & S. KEELE. 1989. Timing functions of the cerebellum. J. Cognit. Neurosci. **1:** 136–152.
48. SAKAI, K., O. HIKOSAKA, S. MIYAUCHI, *et al.* 1999. Neural representation of a rhythm depends on its interval ratio. J. Neurosci. **19:** 10074–10081.
49. PARSONS, L.M. & M. THAUT. Functional neuroanatomy of musical rhythm, meter, tempo, and duration. Submitted.
50. PETERSEN, S.E., P.T. FOX, M.I. POSNER, *et al.* 1989. Positron tomographic studies of the processing of single words. J. Cognit. Neurosci. **1:** 153–170.
51. MAZOYER, B.M., N. TZOURIO & V. FRAK, *et al.* 1993 The cortical representation of speech. J. Cognit. Neurosci. **5:** 467–479.
52. STROMSWOLD, K., D. CAPLAN, N. ALPERT, *et al.* 1996. Localization of syntactic comprehension by positron emission tomography. Brain Lang. **52:** 452–473.
53. PRICE, C. 1998. The functional anatomy of word comprehension and production. Trends Cognit. Sci. **2:** 281–288.
54. KURATA, K. 1993. Premotor cortex of monkeys—set-related and movement-related activity reflecting amplitude and direction of wrist movements. J. Neurophysiol. **73:** 836–854.
55. PASSINGHAM, R.E. 1993. The Frontal Lobes and Voluntary Action. Oxford University Press. New York.
56. KURATA, K. 1994. Information processing for motor control in primate premotor cortex. Behav. Brain Res. **61:** 135–142.
57. JACKSON, S.R. & M. HUSAIN. 1996. Visuomotor functions of the lateral pre-motor cortex. Curr. Opin. Neurobiol. **6:** 788–795.
58. RIZZOLATTI, G., L. FADIGA, M. MATELLII, *et al.* 1996. Localization of grasp representations in human by PET: I. Observations versus execution. Exp. Brain Res. **111:** 246–252.
59. CAREY, D.P., D.I. PERRETT & M.W. ORAM. 1997. Recognizing, understanding, and reproducing action. *In* Handbook of Neuropsychology, Vol. 11. F. Boller & J. Grafman, Eds.: 111–129. Elsevier. Amsterdam.
60. THOMPSON, W.F. 1994. Sensitivity to combinations of musical parameters: pitch with duration, and pitch pattern with durational pattern. Percept. Psychophys. **56:** 363–374.
61. PERETZ, I. & R. KOLINSKY. 1993. Boundaries of separability between melody and rhythm in music discrimination: a neuropsychological perspective. Q. J. Exp. Psychol. **46:** 1283–1301.
62. PERETZ, I. 1990. Processing of local and global musical information by unilateral brain-damaged patients. Brain **113:** 1185–1205.
63. POLK, M. & A. KERTESZ. 1993. Music and language in degenerative disease of the brain. Brain Cognit. **22:** 98–117.
64. HARRINGTON, D.L., K.Y. HAALAND & R.T. KNIGHT. 1998. Cortical networks underlying mechanisms of time perception. J. Neurosci. **18:** 1085–1095.
65. SEASHORE, C.E. 1938. The Psychology of Music. McGraw-Hill. New York.
66. GORDON, E.E. 1965. Musical Aptitude Profile. Houghton-Mifflin. Boston.
67. CORBETTA, M., F.M. MIEZIN, S. DOBMEYER, *et al.* 1991. Selective and divided attention during visual discriminations of shape, color, and speed: funtional anatomy by PET. J. Neurosci. **11:** 2383–2402.

# Color Plates

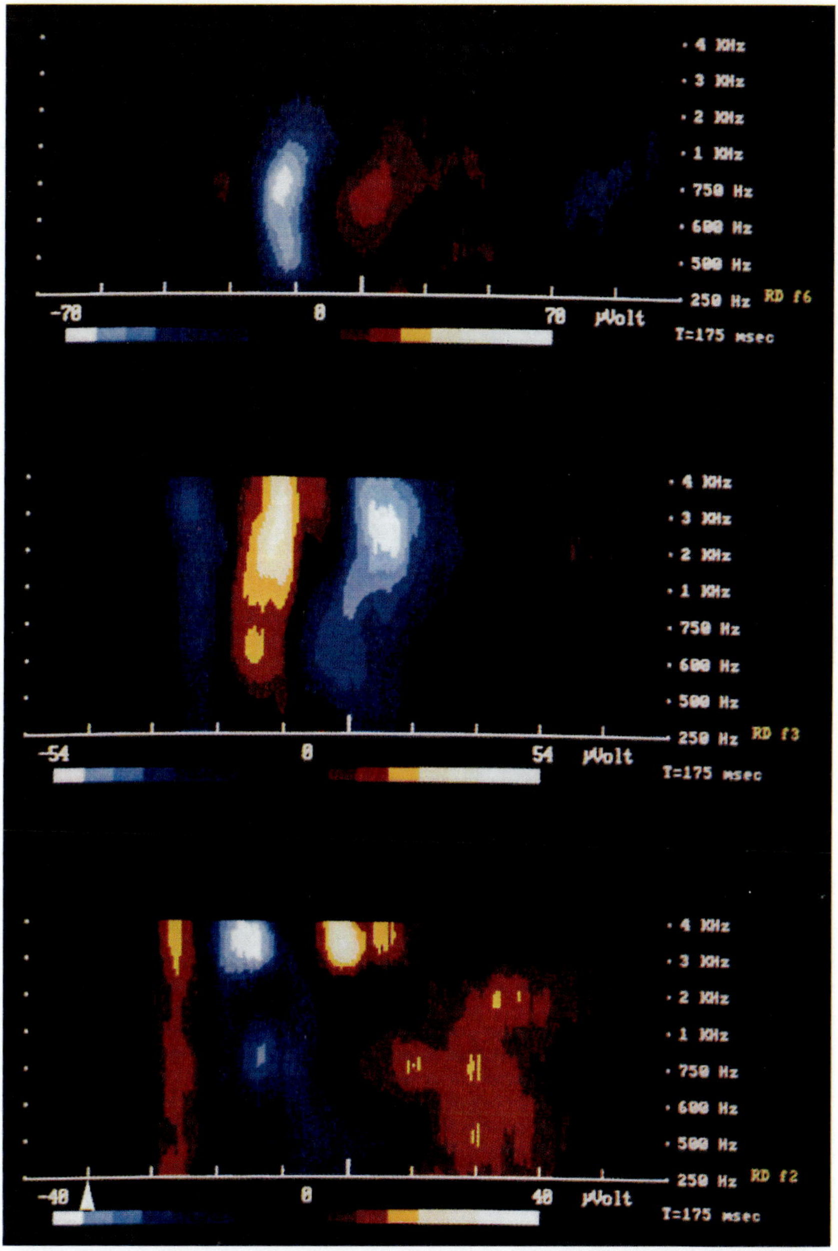

4 KHz
3 KHz
2 KHz
1 KHz
750 Hz
600 Hz
500 Hz
250 Hz
RD f6
-70
0
70
µVolt
T=175 msec
4 KHz
3 KHz
2 KHz
1 KHz
750 Hz
600 Hz
500 Hz
250 Hz
RD f3
-54
0
54
µVolt
T=175 msec
4 KHz
3 KHz
2 KHz
1 KHz
750 Hz
600 Hz
500 Hz
250 Hz
RD f2
-40
0
40
µVolt
T=175 msec


PLATE 1

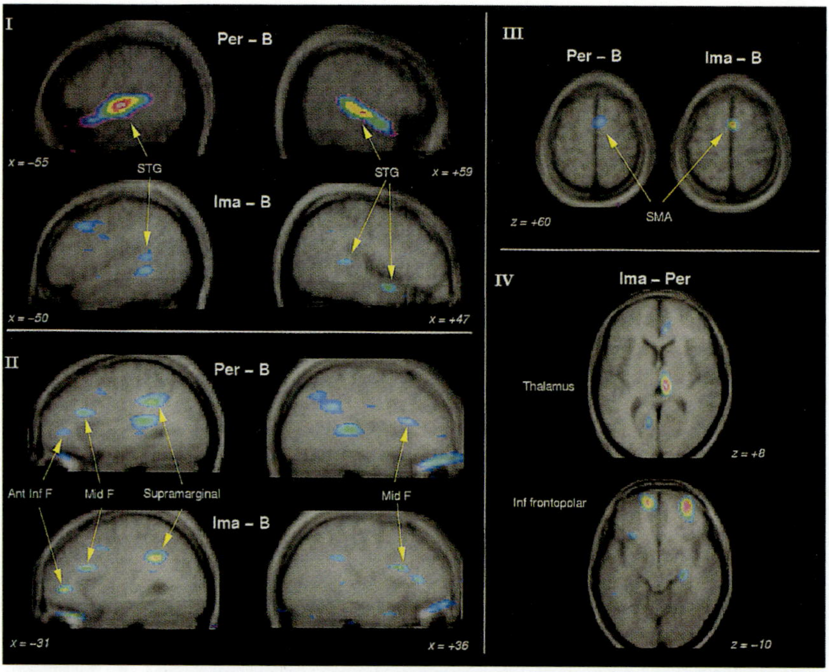


**PLATE 2**

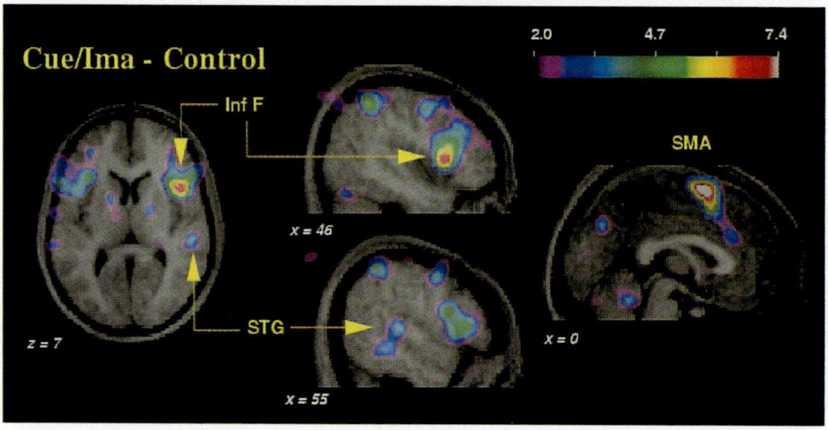


**PLATE 3**

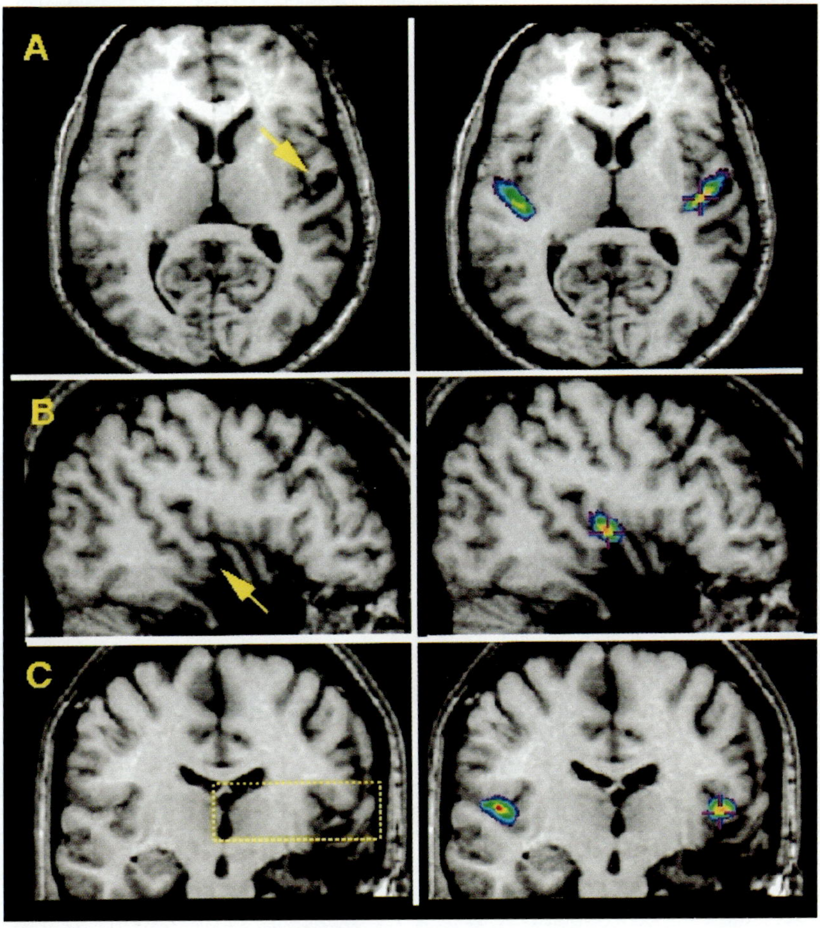


PLATE 4A–C

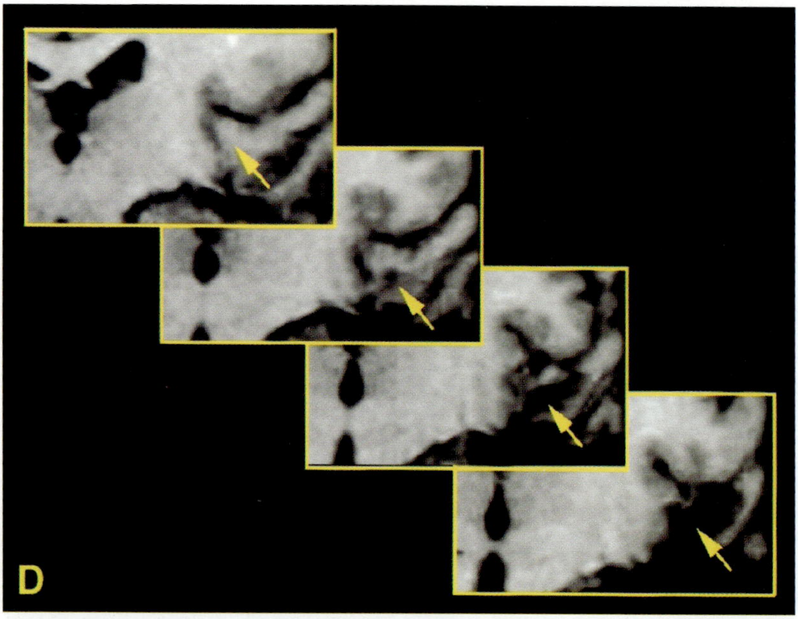


PLATE 4D

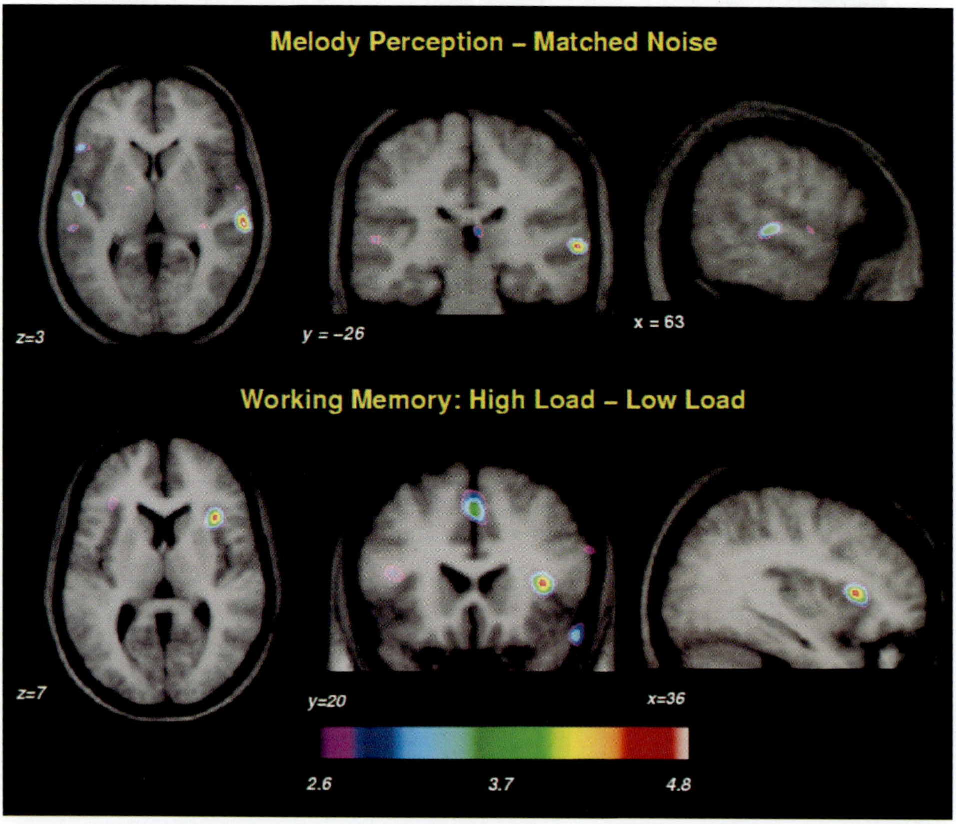


PLATE 5

PLATE 6

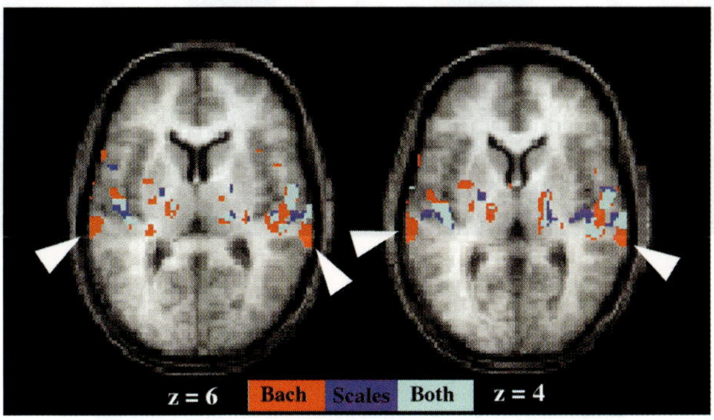


PLATE 7

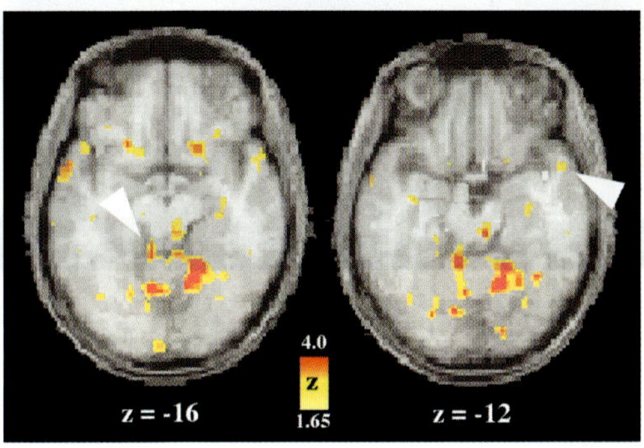


PLATE 8

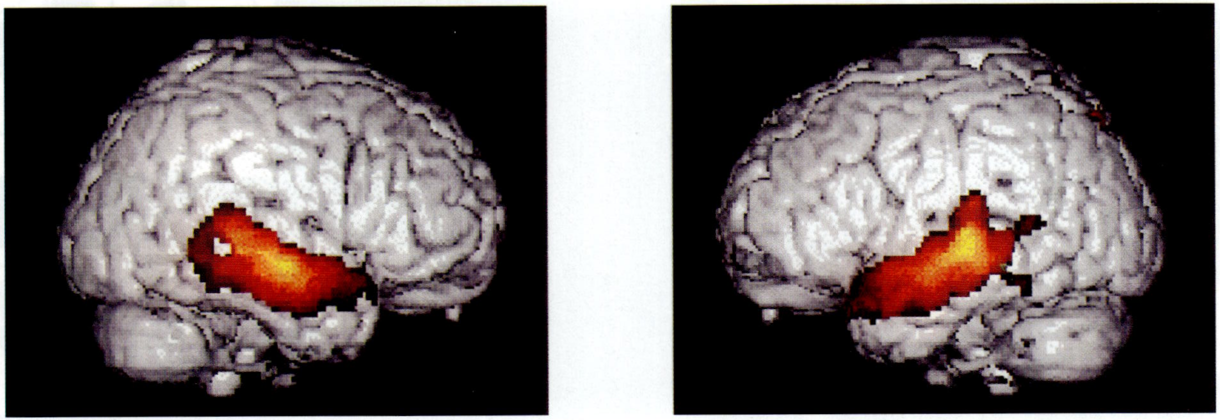

PLATE 23

68. GRASBY, P.M., C.D. FRITH, K.J. FRISTON, *et al.* 1993. Functional mapping of brain areas implicated in auditory/verbal memory function. Brain **116:** 1–20.
69. SHALLICE, T., P. FLETCHER, C.D. FRITH, *et al.* Brain regions associated with acquisition and retrieval of verbal episodic memory. Nature **368:** 633–635.
70. DOYON, J. 1997. Skill learning. *In* The Cerebellum and Cognition. J.D. Schmahmann, Ed.: 273–294. Academic Press. New York.
71. CORBETTA, M., G.L. SHULMAN, F.M. MIEZIN, *et al.* Superior parietal cortex activation during spatial attention shifts and visual feature conjunction. Science **270:** 802.
72. LIU, Y., Y. PU, J-H. GAO, *et al.* 2000. The human red nucleus and lateral cerebellum in cooperative roles supporting sensory discrimination. Hum. Brain Map **10:** 147–159.
73. PARSONS, L.M, J.M. BOWER, J-H. GAO, *et al.* 1997. Lateral cerebellar hemispheres actively support sensory acquisition and discrimination rather than motor control. Learn. Mem. **4:** 49–62.
74. GAO, J-H., L.M. PARSONS, J.M. BOWER, *et al.* 1996. Cerebellum implicated in sensory acquisition and discrimination rather than motor control. Science **272:** 545–547.
75. PARSONS, L.M., J.D. SCHMAHMANN, S. GRILL, *et al.* Neurological evidence implicating the cerebellum in fine auditory discriminations. Submitted.
76. TAYLOR, M. & C. CREELMAN. 1967. PEST: efficient estimates of probability functions. J. Acoust. Soc. Am. **41:** 782–787.
77. PENTLAND, A. 1980. Maximum likelihood estimation: the best PEST. Percept. Psychophys. **28:** 377–379.
78. BRACKE-TOLKMITT, R., A. LINDEN, G.M. CANAVAN, *et al.* 1989. The cerebellum contributes to mental skills. Behav. Neurosci. **103:** 442–446.
79. FIEZ, J.A., S.E. PETERSEN, M.K. CHENEY, *et al.* 1992. Impaired nonmotor learning and error-detection associated with cerebellar damage—a single case study. Brain **115:** 155–178.
80. GRAFMAN, J., I. LITVAN, S. MASSAQUOI, *et al.* 1992. Cognitive planning deficits in patients with cerebellar atrophy. Neurology **42:** 1493–1496.
81. COURCHESNE, E., J. TOWNSEND, N.A. AKSHOOMOFF, *et al.* 1994. Impairment in shifting attention in autistic and cerebellar patients. Behav. Neurosci. **108:** 848–865.
82. HORAK, F.B. & H.C. DIENER. 1994. Cerebellar control of postural scaling and central set in stance. J. Neurophysiol. **72:** 479–493.
83. DAUM, I., S. GRABER & H. ACKERMANN. 1998. Visuospatial processing in patients with cerebellar disorders. Cognit. Neurosci. Soc. Abstr.: 15.
84. FOX, P.T., L.M. PARSONS & J.L. LANCASTER. 1998. Beyond the single study: function/location metanalysis in cognitive neuroimaging. Curr. Opin. Neurobiol. **8:** 178–187.
85. PICARD, N. & P.L. STRICK. 1996. Motor areas of the medial wall: a review of their location and functional activation. Cereb. Cortex **6:** 342–353.
86. PAUS, T. 1996. Location and function of the human frontal eye-field: a selective review. Neuropsychologia **34:** 475–483.
87. FOX, P.T., J.L. LANCASTER, A. HUANG, *et al.* 1999. Functional volume modeling: scaling for group size in averaged images. Hum. Brain Map **8:** 143–150.
88. FOX, P.T., A. HUANG, L.M. PARSONS, *et al.* 2000. Location-probability profiles for the human primary-motor mouth representation. Neuroimage **13:** 196–201.
89. SHULMAN, G.L., M. CORBETTA & R.L. BUCKNER, *et al.* 1997. Top-down modulation of early sensory cortex. Cereb. Cortex **7:** 193–206.
90. SHULMAN, G.L., M. CORBETTA & R.L. BUCKNER, *et al.* 1997. Common blood flow changes across visual tasks: I. Increases in subcortical structures and cerebellum but not in nonvisual cortex. J. Cognit. Neurosci. **9:** 623–646.
91. SHULMAN, G.L., J.L. FIEZ, M. CORBETTA, *et al.* 1997. Common blood flow changes across visual tasks: II. Decreases in cerebral cortex. J. Cognit. Neurosci. **9:** 647–662.
92. BUCKNER, R.L. & S.E. PETERSEN. 1996. What does neuroimaging tell us about the role of prefrontal cortex in memory retrieval? Semin. Neurosci. **8:** 47–55.
93. FOX, P.T., J.L. LANCASTER, L.M. PARSONS, *et al.* 1997. Functional volume modeling. Hum. Brain Map. **5:** 306–311.
94. MARR, D. 1969. A theory of cerebellar cortex. J. Physiol. (Lond.) **202:** 437-470.

# Comparison between Language and Music

MIREILLE BESSON AND DANIELE SCHÖN

*Center for Research in Cognitive Neuroscience, CNRS-Marseille, France*

**ABSTRACT: Similarities and differences between language and music processing are examined from an evolutionary and a cognitive perspective. Language and music cannot be considered single entities; they need to be decomposed into different component operations or levels of processing. The central question concerns one of the most important claims of the generative grammar theory, that is, the specificity of language processing: do the computations performed to process language rely on specific linguistic processes or do they rely on general cognitive principles? Evidence from brain imaging results is reviewed, noting that this field is currently in need of metanalysis of the available results to precisely evaluate this claim. A series of experiments, mainly using the event-related brain potentials method, were conducted to compare different levels of processing in language and music. Overall, results favor language specificity when certain aspects of semantic processing in language are compared with certain aspects of melodic and harmonic processing in music. By contrast, results support the view that general cognitive principles are involved when aspects of syntactic processing in language are compared with aspects of harmonic processing in music. Moreover, analysis of the temporal structure led to similar effects in language and music. These tentative conclusions must be supported by other brain imaging results to shed further light on the spatiotemporal dynamics of the brain structure–function relationship.**

**KEYWORDS: Language and music; Music and language; Event-related brain potentials; Brain imaging**

## INTRODUCTION

> *Once, a long time ago [...] it so happened that people took to uncivilized ways, were ruled by lust and greed, behaved in angry and jealous ways with each other [...]. Seeing this plight, Indra and other gods approached god Brahma and requested him to give the people a toy, but one which could not only be seen, but heard, and this should turn out a diversion (so that people gave up their bad ways).*
>
> —A. Rangacharya, *Introduction to Bharata's Natya-Sastra*[1]

One main function of language is communication. Linguistic communication encompasses widely diverse uses of language, from the running of everyday life and basic interactions between individuals (*"Give me the salt, please."*) to the esthetics

Address for correspondence: Mireille Besson, Centre de Recherche en Neurosciences Cognitives, CNRS-CRNC, 31-Chemin Joseph Aiguier, 13402–Marseille Cedex 20, France. Voice: +33-491-164-305; fax: +33-491-774-969.

mbesson@lnf.cnrs-mrs.fr

of words, their combination in poetry (*"Come what come may/Hours and time run/ Through the roughest day,"* Shakespeare), and the telling of stories. Language is also necessary for expression of rational thought and organization of human societies. It may well have evolved from the need for social bonding between individuals belonging to the same group.[2] Language also permits projections into the past and into the future and is necessary for the transmission of knowledge.[3] While these characteristics, among others, make language specific to *Homo sapiens,* they also seem to contribute to the splendid isolation of the linguistic function among the other human cognitive activities. Largely because of the enormous impact in the cognitive sciences of the generative grammar theory, developed by Chomsky,[4] language is most often considered as relying on specific cognitive principles. Bickerton,[5] for instance, argues that the principles that govern language "seem to be specifically adapted for language and have little in common with general principles of thought or other apparatuses that might be attributable to the human mind" (p. 158). The specificity of the computations involved in language processing is the main question that we would like to address in this chapter. To try to delineate which computations are specific to language and which rely on more general cognitive principles, if any, we chose to compare language with another well-organized cognitive function that, although very different in many respects, nevertheless presents interesting similarities with language: music.

We start by reviewing some of the evidence in favor of the similarities and differences between language and music, based, first, on an evolutionary perspective, and second, on a cognitive perspective. We then report a series of experiments directly aimed at comparing different aspects of language and music processing.

## SIMILARITIES AND DIFFERENCES BETWEEN LANGUAGE AND MUSIC

### *Evolutionary Perspective*

While discussions on the origin of language were banished from the *Société de Linguistique* of Paris in 1866, and although the question of the anthropological foundations of language and music certainly remains difficult, there is renewed interest in evolutionary linguistics and musicology, as can be seen, for instance, from the recent and provocative book edited by Wallin *et al.*,[6] *The Origins of Music*. The question of a common or separate origin of language and music, which was at the center of hot debates between philosophers and scientists from the seventeenth to the nineteenth century,[7–10] is now being examined using new tools and technologies. In our opinion, one of the most promising avenues is functional brain imaging. By offering an excellent spatial and/or temporal resolution, brain imaging methods allow us to examine the brain regions that are activated by different aspects of information processing and how these processes unfold in time. Even if such methods may not help solve the problem of the common or separate origin of language and music, they provide invaluable information on the question of the similarities and differences between these two systems. Before reviewing some of the relevant brain imaging data, let us quickly consider how ideas on the origin of language and music have evolved throughout the centuries.

### *"The Old Masters"*

Rousseau[9] was a strong advocate of the view that music and language share some common ancestor and that language evolved out of music for the sake of a rational organization of human societies. Darwin[7] also argued for a common origin, but considered that music evolved out of the primate's reproductive calls and that language was first. Interestingly, most of the authors who addressed this issue in the book edited by Wallin *et al.*[6] also seem to share the opinion of a common ancestor to language and music. The concept that the first basic function of both language and music was to express emotive meaning through variations in the intonation of the voice (intonational prosody) and rhythm also seems to be an object of consensus.[10–12] In both language and music, emotional excitement is expressed through fast, accelerating, and high-registered sound patterns. In this again they join Rousseau,[9] who considered that the first languages were sung, not spoken; they were aimed at expressing emotions, love, hate, and anger. They were passionate before being rational.

### *Music and Culture*

In Western cultures, music has evolved to become more and more isolated from other expressive forms. Moreover, most studies in music cognition are concerned with music perception, and music performance has received much less attention (but see Sloboda[13]). By contrast, in other cultures, in which magical thought is still alive, the bounds among music, song, dance, poetry, and rite have not been lost.[14,15]

Furthermore, ethnomusicological studies have often emphasized music as a voluntary act: music is the acoustic result of action. Kubik,[16] for instance, pointed out that African music is not just sound; action is an intrinsic part of musical performance. Motor patterns are themselves sources of aesthetic pleasure, independent from the sound that they are associated with. This strong intertwining between music and action is even reflected in language, the same word being used in several African languages to refer to music and dance.

Blacking[14] defined music as "sound that is organized into socially accepted patterns." Moreover, he argues that every piece of music has its own inherent logic, as the creation of an individual reared in a particular cultural background. However, his claim that patterns of sound reflect patterns of social organization seems somewhat coarse and in need of further elaboration. Still, in much the same way that a context-sensitive grammar is a more powerful analytical tool than a context-free grammar, the cognitive systems underlying different styles of music shall be better understood if music is considered in context. Different musical styles should therefore be considered not as "sonic objects" but as humanly organized sound whose patterns are related to the social and cognitive processes of a particular society and culture.

## *Cognitive Perspective*

### *Structural Aspects*

Many definitions have been (and are still to be) proposed for language and music. The amazing point, however, is that the definition given for music will often apply to language as well, and vice versa. This is striking when we consider the comparison between language and music from both a structural and a functional perspective. Arom,[17] for instance, proposed two structural criteria to define music. One is rhythm and the temporal ratios that delineate a piece of music by the formalized segmentation of time. The other is that all cultures have divided the sound continuum into discrete pitches that form the musical scales. These two criteria may apply to language as well. Language is also composed of sequential events that unfold in time with a specific rhythm and specific segmental (phonemes) and suprasegmental (prosody) information. Moreover, the speech continuum is divided into discrete phonemes, the basic phonological unit. More generally, it is clear that both language and music are conveyed by sounds, are ubiquitous elements in all cultures, are specific to humans, and are cultural artifacts that do not correspond to natural objects.[12] They are rule-based systems composed of basic elements (phonemes, words, notes, and chords) that are combined into higher-order structures (musical phrases and sentences, themes and topics) through the rules of harmony and syntax. Therefore, there may be a musical grammar, and the experimental results to be described indeed point to the similarity of the brain's response to some specific violations of syntax in both language and music.

### *Functional Aspects*

From a functional perspective, several similarities also exist between language and music. In this respect, it is interesting to examine the following citation from Pinker[18] in his book, *The Language Instinct*: "Language is a complex, specialized skill, which develops in the child spontaneously, without effort or formal instruction, is deployed without awareness of its underlying logic, is qualitatively the same in every individual, and is distinct from more general abilities to process information or behave intelligently" (p. 18). Except for some constraints regarding music production, we could substitute "music" for "language" and the characteristics would apply as well. Both language and music rely on intentionality: All music implies "an act of creation that actualizes an intention,"[17] and this is true of language as well. In other words, both language and music require a theory of mind.[19] Both develop with specific learning according to more or less standardized procedures depending on the linguistic or musical culture. Even though perception precedes production in both domains, children acquire musical and linguistic rules in a similar, effortless way. Early on, children are able to create new musical and verbal sentences by applying a rule system that they have been able to abstract without conscious intentions. Both language and music involve memory; adults can recognize and reproduce learned melodies, words, poetry, and songs.

### *Similarities or Differences? A Question of Grain of Analysis?*

It should be noted, however, that the comparison between language and music might highlight either their similarities or their differences depending on the grain chosen for analysis. Thus, similarities at one level of processing may be interpreted as differences at another level. For example, while temporal structure and rhythmic organization play a fundamental role in both language and music, the metric structure is specific and consistent throughout a given musical piece, but the suprasegmental prosodic structure of language is less specific and more variable. Similarly, the segmentation of the sound continuum into discrete units (pitches or phonemes) is found in all music and languages. However, if we can eventually make an analogy between phonemes and intervals of a musical scale, we must also be aware of their differences. In fact, although the number of pitches by octave (degrees of the scale) is very similar across musical cultures (seven or fewer notes), the number of phonemes largely differs between languages (from 11 in Polynesian to 141 in the language of the Bushmen,[18] with 44 phonemes in English and 36 in French). Furthermore, some of the perceptual properties of the basic elements in music have no equivalent in language, as, for instance, the fact that octaves are perceived as equivalent in almost all cultures. This effect is linked with the finding that two notes separated by an octave are related by a simple frequency ratio of 2:1. Generally, the relationships between different pitches in a musical piece are much simpler than the relationships between different phonemes in a linguistic sentence.

As a last example, all languages are organized according to a syntactic structure that may be universal.[20,21] Indeed, verbs and nouns are always present. However, the order in which those elements are presented varies among languages: Subject–Verb–Object (French, English, etc.); Subject–Object–Verb (German, Japanese, etc.); Verb–Object–Subject (Malgache, etc.).[22] Even if it is common to refer to a musical syntax or a musical grammar, the extent to which this analogy extends beyond a simple metaphor remains to be determined. Music perception shares universal laws of auditory perception. For instance, the perception of a musical phrase is automatically influenced by factors such as the grouping of discrete notes into sequences (i.e., the melodic contour) and the feeling of closure that accompanies the playing of a cadence at the end of a phrase. Some of these factors are universally shared and others, just as verbal language, are culturally shared. However, even if there is such a thing as a musical grammar, the rules seem more flexible and ambiguous than the syntactic rules used in language. Ambiguity is a key element of the grammar and aesthetics of music.[23] There are always several ways to perceive and enjoy a musical piece. Finally, musical elements are most often played simultaneously, and each element may have its own "syntax." This vertical dimension of musical structure, commonly referred to as harmony, is not present in language. While different words sung at the same time may melt in a sublime combination of rhythm, melody, and harmony (as in the polyphonic madrigals of Monteverdi), different words produced at the same time by different speakers will only create an unpleasant cacophony, like that in a political debate.

### *Meaning and Expectancy*

Even if the similarities and differences between language and music depend on the level of details considered for the analysis, one fundamental difference nevertheless remains. Whereas the meaning of words is understood in relation to an extralinguistic designated space, music is considered mostly self-referential.[24–27] This does not mean that music is asymbolic. However, while the meaning of words is defined by an arbitrary convention relating sounds to meaning, notes or chords have no extramusical space in which they would acquire meaning. The internal sense of music may be conceived as something that goes beyond any objective reference structure and the possibilities of verbal language.[28] Much as Wittgenstein[29] who asked: "Describe the coffee aroma!", music is the kingdom of the ineffable. As stated by Meyer in his wonderful book *Emotion and Meaning in Music,*[25] "Music means itself. That is, one musical event (be it a tone, a phrase, or a whole section) has meaning because it points to and makes us expect another musical event" (p. 35). Interestingly, this statement not only highlights one of the most important differences between language and music, that is, the unsolved question of musical semantics, but also emphasizes their strongest similarity: in their own way, both systems generate strong expectancies. Just as a specific word is expected within a specific linguistic context, specific notes or chords are expected at a given moment within a musical phrase. Either these expectations are fulfilled, giving rise to resolution or satisfaction, or they are not fulfilled, giving rise to tension or surprise.

We should not believe, however, that expectation may "bear the entire burden of deriving affect."[30] Other factors such as tempo, volume, and nonmechanical interpretation of music certainly influence musical emotions. Still, the structure of music has intrinsic points of instability that tend to resolve, and the tension/resolution phenomenon results in affects. Moreover, tensions are perceived at different levels depending on the analysis performed. Jackendoff,[30] in analogy to language, points out that a modular and informational encapsulated parser might be at work independently from conscious memory. This independence from memory may explain why we keep enjoying a piece on repeated hearings "in spite of the consequences of an intuitive theory of affect based on expectation." In fact, an autonomous parser will keep analyzing and recreating whatever structure is retrieved from memory. Then "surprise will still occur within the parser."[30]

If rather than asking ourselves what is the meaning of music, we make a more fruitful reflection on "what can I do with sounds?" we may discover that music is, first of all, a set of choices. The flow of these choices might possibly become visible as a musical thought. Behind all this, the image of children playing appears. When the child plays with small wood blocks, we could say that the game is a way of answering the question: what can I do with my small wood blocks? Then, from the pleasure of playing we get directly into the aesthetic pleasure.

Concluding this brief and necessarily incomplete excursus, it is important to keep in mind that musical meaning is the sum of analytic approaches (musical parser), individual and/or cultural associations to the external/internal world (during some periods in the last centuries "music was conceived as conveying precise emotional and conceptual meanings, established by codes, or at least, *repertoires*"[31]), and aesthetic reaction. The importance of the aesthetic component of music becomes evident in considering that "the form of a work of art gains its aesthetics validity precisely in proportion to the number of different perspectives from which it can be viewed and understood."[32]

### *Levels of Processing*

From a cognitive perspective, language and music cannot be considered as single entities. To be analyzed and compared they need to be reduced to their constitutive elements. Within music, one classically differentiates the temporal (meter and rhythm), melodic (contour, pitch, and interval), and harmonic (chords) aspects. Each aspect most likely involves different types of processing, so that the processes called into play to process rhythm may differ from those involved in the processing of pitch and melodic intervals. Similarly, within language, at least four different levels of processing have been taken into consideration. The phonetic-phonological level, which comprises both segmental (phonemes) and suprasegmental (prosody) levels;

the morphosyntactic level, which encompasses the combination of phonemes into morphemes and of morphemes into words; the syntactic level, which governs the relations between words; and the lexicosemantic level, with access to the meaning of words and sentences. Finally, while often ignored in psycholinguistic and neurolinguistic experiments, the pragmatic level that comprises discourse organization and contextual influences represents an essential aspect of language organization.

Insofar as we agree with the concept that language and music cannot be considered as wholes but need to be subdivided into their component operations, it becomes unrealistic, for instance, to view the linguistic function as localized in the left hemisphere and music in the right. Rather, some aspects of language processing may preferentially involve left cerebral structures, whereas others require structures on the right. The same remark applies to music as well. With this view in mind, the task of the cognitive neuroscientist is to delineate the different computations performed within one level of processing, to understand the mechanisms that underlie these computations, and to localize where in the brain these mechanisms are implemented. This task is fraught with both philosophical and methodological problems,[19] but science is advancing rapidly, and new methods are now available to track these issues. In the second part of this presentation, we summarize the results of our research on the comparison between different levels of processing in language and music. Before going into the details of the experiments, we briefly review the position of linguistic theories on the question of the specificity of language processing.

## SPECIFICITY OF THE COMPUTATIONS INVOLVED IN LANGUAGE PROCESSING?

### *The Generative Grammar Theory*

One of the most important claims of the generative grammar (GG) theory is that language is autonomous from other cognitive functions.[4,18,21,33,34] Language is considered a computational module that entails its own functional and neural architecture.[12] Moreover, the linguistic module comprises different submodules, each responsible for different aspects of language processing, phonology, morphology, syntax, semantics, and pragmatics. Each submodule is encapsulated,[35] so that the processing of information in a module is performed independently of the processing of information in another submodule. Phonological processing, for instance, is realized without being influenced by the morphological, syntactic, semantic, or pragmatic aspects of language processing. Thus, the computations required to process language are specific to language, and the computations in one module are performed independently of those in the other modules.

Another basic claim of the GG theory is that languages are defined by their deep syntactic structure: syntax plays a dominant role in the structural organization of language. Moreover, from a functional perspective, syntax is first.[36] Logico-mathematic computations are first performed on symbols that have no intrinsic meaning; they only acquire meaning in a second step. Therefore, the chain of computations necessary to process language is considered to be serially and hierarchically organized.

### *Other Linguistic Theories*

It should be noted, however, that other linguistic theories have been developed in the last 20 to 30 years that advocate very different views of language structural and functional organization.[22,37] Although it is beyond the scope of this presentation to go into the details of these different linguistic theories, which differ from each other in many respects (functional grammar,[38,39] cognitive grammar,[40–42] and linguistic functional typology[43–45]), the important point is that these theories call into question the two basic claims of the GG theory just summarized. First, they reject the idea that language is an autonomous function, relying on its own structural and functional architecture. By contrast, they consider that languages rely on general cognitive principles, linked with perceptual and sensorimotor motor experiences.[46,47] Second, they reject the syntacticocentrism of the GG theory and the idea of the autonomy of syntax relative to phonology, morphology, semantics, and pragmatics.[22] Following Langacker,[41] for instance, semantics, morphology, and syntax form a continuum with specific meaning associated with lexicosemantic units and schematic meaning associated with grammatical units. Thus, in contrast to the GG view that grammatical units are semantically empty, all grammatical elements have meaning. Moreover, linguistic units are not static but constructed through a dynamic process influenced by the context of enunciation[48] and the interactions between individuals in a situation of communication.[49] Therefore, language should be studied not in isolation but in relation to other cognitive functions, specifically, attention and short-term and episodic memory.[45]

### *Success of Generative Grammar Theory in Cognitive Neurosciences*

Several reasons may explain the success of the GG theory in both linguistic and cognitive sciences, but two are of particular interest. First, the cognitive stakes of the GG theory have been clearly explained. It has therefore been possible to make predictions and design experiments to test these predictions.[21,47] Second, the modular organization of the functional aspects of language processing is clearly neurocompatible. The concept that language is organized in submodules, each responsible for one specific processing stage, finds strong support in the localizationist views of cerebral organization. The recent development of brain imaging methods, together with older data from the neuropsychological literature, largely contributes to the idea that specific functions are implemented in specific brain structures. A brief review of the literature shows that while this concept of the mapping of basic sensory functions into the organization of primary, sensory brain areas is probably correct, the story certainly becomes more complicated when trying to localize such higher-order cognitive abilities as language or music.

### *Evidence from Brain Imaging*

Brain imaging methods are aimed at understanding the functional activity of the brain either directly through measures of the electrical activity of single neurons (intracellular recordings), or of neuronal populations (electroencephalography, EEG), or through the magnetic activity that is coupled with the electrical activity (magnetoencephalography, MEG), or indirectly through the measures of brain metabolic activity (positon emission tomography, PET, and functional magnetic resonance imaging, fMRI). Overall, direct methods have excellent temporal resolution and relatively poor spatial resolution, whereas the reverse is true for indirect methods. Elegant works have been conducted using these different methods to demonstrate, for instance, the retinotopic organization of the visual cortex using fMRI[50] and the tonotopic organization of the auditory cortex using intracellular recordings (see Liégeois-Chauvel, this volume), MEG,[51] or fMRI.[52] Hence, there is strict mapping between the organization of the receptor fields at the periphery, in either the retina or the cochlea, and the functional organization of the primary visual and auditory cortex. Aside from extending to humans previous discoveries in animals, these findings validate the use of such complex methods as fMRI to study human perception and cognition.

To address the specificity of the brain structures involved in language processing, we would need metanalysis of the results obtained across the many experiments aimed at localizing the different aspects of language processing. We would then need to do the same for music or for any other cognitive function of interest and then compare the results of these metanalyses. Although such metanalyses are being performed for some aspects of language processing, such as language production[53] or prelexical and lexical processes in language comprehension,[54] data in the neuroimaging of music are still too scarce for such an enterprise. Moreover, assuming that such metanalyses are performed for music as well, it still remains extremely difficult to compare results of experiments that were not directly designed to compare language and music processing. Indeed, leaving aside the theoretical problem of which level of processing in language is best compared with which level of processing in music, the choice of the task to be performed on the stimuli, its difficulty, as well as experimental factors, such as the mode (blocked versus mixed) and rate of stimulus presentation, stimulus repetition, and data analysis (e.g., subtraction method, correlative analyses), have been shown to exert a predominant influence on the results obtained. With these remarks in mind, it is nevertheless interesting to mention some of the results found for language and music to determine the extent to which the brain structures that are activated are similar or different.

Few experiments have been designed to directly compare language and music using brain imaging methods. Binder *et al.*[55] compared tones and word processing in an fMRI study. Results showed that several brain structures, including the left superior temporal sulcus, middle temporal gyrus, angular gyrus, and lateral frontal lobe, showed stronger activation for words than tones. However, both types of stimuli activated the Heschl gyrus and the superior temporal plane, including the planum temporale. The investigators concluded that whereas the planum temporale is similarly involved in the auditory processing of words and tones, other broadly distributed areas are specifically involved in word processing. Gandour *et al.*[56] conducted a PET study in which both Thai and English participants were required to discriminate pitch patterns and Thai lexical tones derived from accurately filtered Thai words. Results of the tone minus pitch subtraction indicated that only native Thai speakers showed activation of the left frontal operculum (BA 44/45). This finding was taken as evidence that Thai lexical tones are meaningful for native Thai speakers but not for English speakers. However, for our purposes, it is also interesting that for both Thai and English speakers, several structures, including the left anterior cingulated gyrus (BA 32), the left and right superior temporal gyrus (BA 22), and the right cerebellum, were activated in both pitch and tone tasks.

More generally, results have shown that primary auditory regions (BA 41 and BA 42) respond in similar ways to speech and music.[57] Secondary auditory regions (BA 22) are activated by hearing and understanding words[58] as well as by listening to scales,[59] auditory imagery for sounds,[60] and access to melodic representations.[61] The supramarginal gyrus (BA 40) seems involved in understanding the symbolism of language[58] and the reading of musical scores.[59] Broca's area is known to be involved in motor activity related to language and was also shown to be active when playing music[59] and when musicians were engaged in a rhythmic task.[61] The supplementary motor areas (BA 6) and the right cerebellum are also active when playing and imaging playing music.[59,62] Although this list is far from exhaustive, it nevertheless suffices to show that some of the most important language areas are clearly involved in music processing as well. Some brain structures also seem to be specifically or preferentially involved in language processing,[63] and the converse is true for music.[64] Results of metanalysis of the type just mentioned should clearly help in drawing maps of the structure–function relationships known for language and music.

### *Semantics, Melody, and Harmony*

A starting point in the study of the neurophysiological basis of language processing was the discovery of the N400 component by Kutas and Hillyard.[65] This negative component of the event-related brain potentials (ERPs), peaking around 400 ms after word onset, is elicited by words that are semantically unexpected, incongruous, and within a linguistic context (e.g., "The pizza was too hot to cry"; FIG. 1). Further results have shown that N400 amplitude is modulated by semantic priming, so that an unexpected word related to the best sentence completion (e.g., "drink" when the expected word is "eat"; FIG. 1) elicits a smaller N400 than a completely unexpected word (e.g., "cry"[66]). These results, together with those issued from a large number of experiments, have led to the consensus that the N400 is a good index of the integration process of a word within its linguistic context.

The first experiments that we designed were aimed at finding out whether an N400 component would also be elicited when melodically and harmonically unexpected notes were presented within a melodic context.[67–69] We presented both familiar and unfamiliar monodic musical phrases to musicians and nonmusicians. The familiar melodies were chosen from the classical repertoire of Western occidental music from the eighteenth and nineteenth centuries, and the unfamiliar musical phrases were composed by a musician following the rules of tonal harmony (FIG. 2). These melodies were ended by the congruous or most expected note, by a note out of the tonality of the musical phrase (nondiatonic incongruities perceived as wrong notes), or by a note within the tonality but not the most expected ending (melodic or diatonic incongruities). Thus, we created a degree of musical incongruity from diatonic to nondiatonic.

Results clearly showed that both types of unexpected notes elicited the occurrence of late positive components, peaking around 600 ms (P600). As demonstrated for the N400 component, P600 amplitude was shown to depend on the degree of musical incongruity: it was larger for the most unexpected, nondiatonic wrong notes than for the less unexpected, diatonic incongruities. Moreover, the amplitude of the P600 was larger for familiar than for unfamiliar musical phrases and for musicians than for nonmusicians (FIG. 3). These findings clearly demonstrate not only that spe-

# *LANGUAGE*

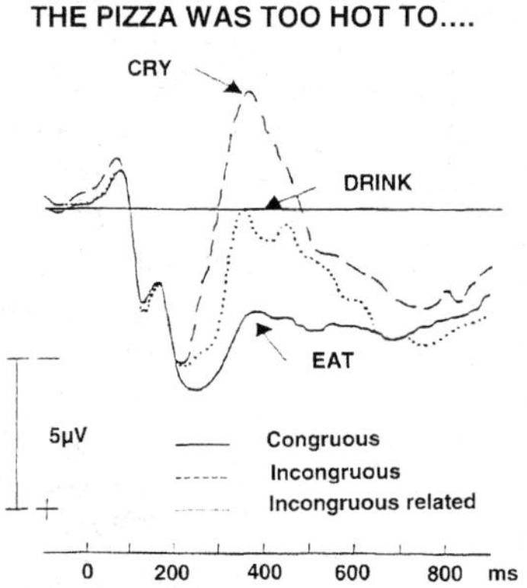


**FIGURE 1.** ERPs elicited by final sentence words at the central recording site (Cz) for congruous and incongruous words and for incongruous words that are semantically related to the best sentence completion. The amplitude of the negative component, peaking at 400 ms postfinal word onset (N400) is largest for incongruous words, intermediate for incongruous words related to the best sentence completion, and smallest for congruous words. In this and subsequent figures, amplitude (µV) is represented on the ordinate, with negative voltage up, and time (ms) on the abscissa. (Adapted from Kutas and Hillyard.[66])

cific notes are expected within a musical phrase, but also that such expectations depend on the familiarity of the musical excerpts and the expertise of the listener. Thus, an interesting similarity between language and music, just mentioned, their ability to generate strong expectancies, is supported by empirical evidence. However, our results also show that the processes that govern semantic expectancy and that are reflected by a negative component, peaking around 400 ms, the N400, are qualitatively different from those involved in musical expectancies and reflected by a positive component peaking around P600 ms, the P600. While, to our knowledge, the functional significance of positive versus negative polarities in the ERPs is not clearly established, our results, by demonstrating qualitative differences between language and music processing, nevertheless strongly argue for the specificity of the processes involved in computing the semantic aspects of language. Thus, one of the most important differences between language and music, outlined in the introduction, the fact that, in contrast to language, music has no intrinsic meaning and is a self-referential system, seems to find some support in these experimental findings.

### *Semantics and Harmony in Opera*

Opera is perhaps the most complete art form, as it calls upon music, language, drama, and choreography. It originated in Italy at the end of the sixteenth century with *Dafne*, set to music in 1594 by the Florentine composers Corsi and Peri. The first opera to survive intact is probably *Euridice,* set to music by Peri and Caccini

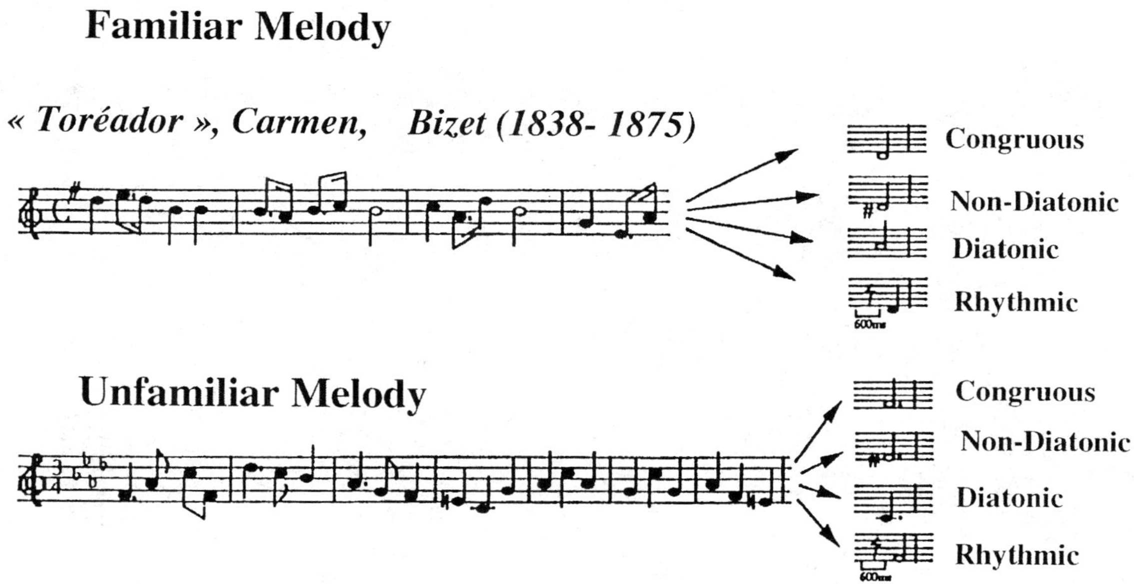


**FIGURE 2.** Examples of the stimuli used in the experiment. (Adapted from Besson and Faïta.[69])

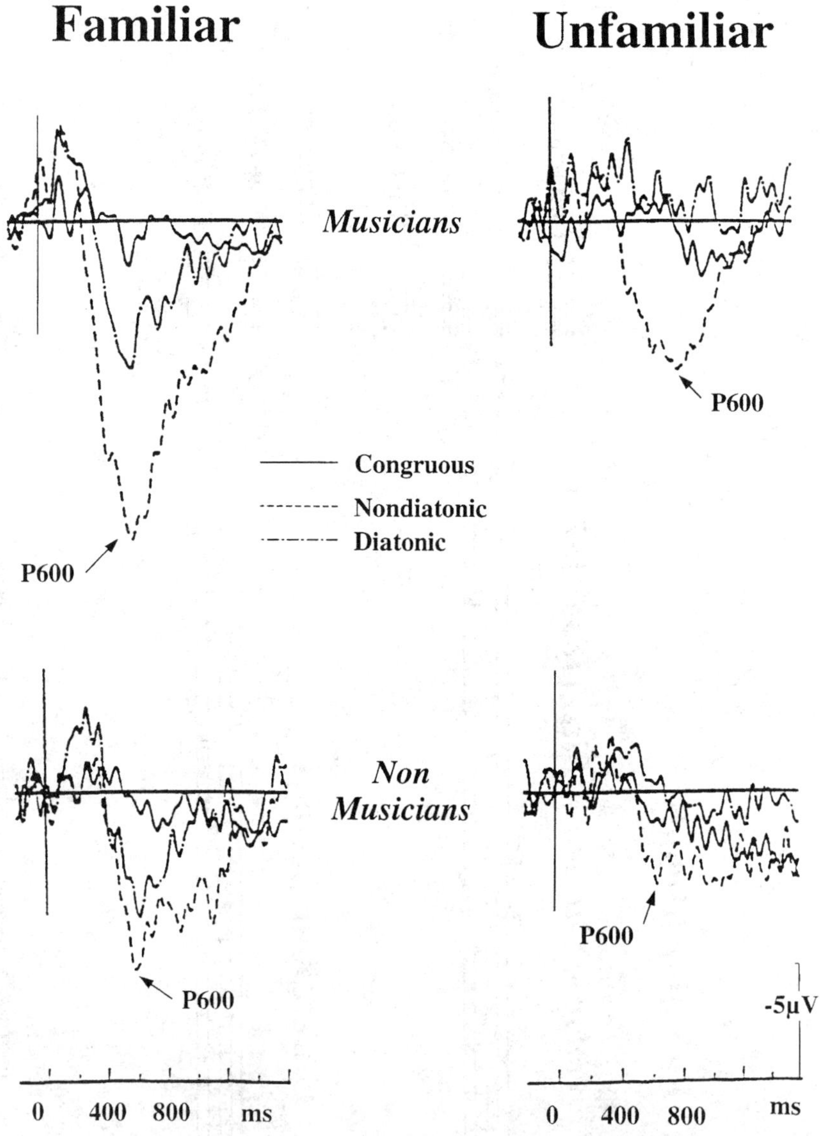


**FIGURE 3.** ERP results for musicians and nonmusicians are presented separately for familiar and unfamiliar musical phrases. The vertical lines mark the onset of the final note. Results are from one typical recording site, the parietal location (Pz). The amplitude of the positive component, P600, is larger for nondiatonic than for diatonic incongruity, for musicians than for nonmusicians and for familiar than for unfamiliar musical phrases. (Adapted from Besson and Faïta.[69]).

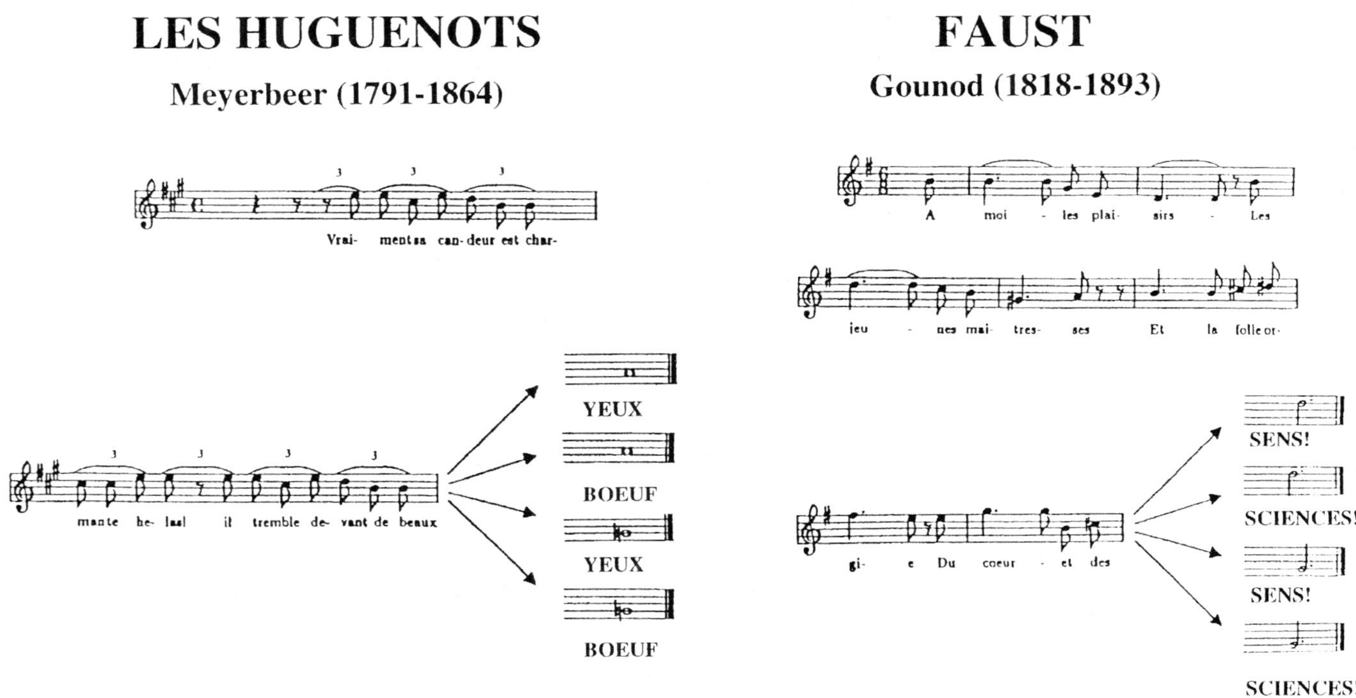


**FIGURE 4**. Example of the opera's excerpts used in the experiment. Approximate translation of the excerpts, from *Les Huguenots* (Meyerbeer): "Really, his naïvity is charming. However, he trembles in front of beautiful eyes," and from *Faust* (Gounod): "For me, the pleasures and young mistresses, the crazy orgy of the heart and the senses." Note that in French, the final incongruous words "boeufs" and "sciences" rhyme with the expected completions "yeux" and "sens." The final note of the excerpt is in or out of tune. (Adapted from Besson *et al.*[72])

and performed in 1600 as a wedding gift to Maria de Medici and Henri IV. Opera, as a new art form, then spread to other Italian courts with the better-known *Orfeo* of Monteverdi in 1604. Since that time, a question that has interested both music analysts and composers has been to determine which of the words or the music plays the most important role in opera. In his *Life of Rossini*, Stendhal (1783–1842) argued that music is most important: "its function is to animate the words." Later, ethnomusicologists, such as Levman,[70] pointed out that the lyrics are subordinate to the music in tribal songs and rituals. By contrast, Richard Wagner (1813–1883) considered that both aspects are intrinsically linked: "Words give rise to the music and music develops and reinforces the language," an opinion shared by Pierre Boulez[71]: "The text is the center and the absence of the musical piece." Richard Strauss (1864–1949) even composed an opera, *Capriccio* (1940), to illustrate the complementarity of words and music.

To determine, based on scientific grounds, if the words or the music is most important when we listen to opera, we selected 200 excerpts from French operas of the nineteenth and twentieth centuries.[72] Each excerpt lasted between 8 and 20 seconds and was sung *a capella* by a female singer under each of four experimental conditions, that is, the final word of the excerpt was (1) semantically congruous and sung in tune, (2) semantically incongruous and sung in tune, (3) semantically congruous and sung out of tune, and (4) both semantically incongruous and sung out of tune (FIG. 4).

Based on previous results,[65] it was of interest to determine whether semantically incongruous words will also elicit an N400 component when they are sung. Similarly, it was of interest to determine whether congruous words sung out of tune will also elicit a P600 component.[69] Of most interest was the double incongruity condition: will semantically incongruous words sung out of key elicit both an N400 and a P600 component? If language plays the most important role when we listen to opera, then results may show an N400 but not a P600. Conversely, if music is the cornerstone of opera, then results may show a P600 without an N400. Maybe both effects will be elicited, however; they may then be additive (i.e., equal to the sum of the effect associated with each type of incongruity alone) or interactive. To answer these questions, we recorded the ERPs associated with the final words of each excerpt, from 16 professional musicians from the opera company in Marseille.

To summarize, results demonstrated that sung incongruous words did elicit an N400 component, thus extending to songs results previously reported for written and spoken language[65,73] (FIG. 5A). Moreover, words sung out of tune did elicit a P600 component, thus extending to songs results previously reported for out-of-tune notes[69,74,76] (FIG. 5B). Most interesting are the results in the double incongruity condition. They show that incongruous words sung out of tune elicit both an N400 and a P600 component (FIG. 5C). Interestingly, the N400 occurred earlier than the P600, which is taken as evidence that the words were processed faster than the music. Finally, effects in the double incongruity condition were not significantly different from the sum of the effects observed in each condition of simple incongruity (see FIG. 6). This finding provides a strong argument in favor of the independence (i.e., the additivity) of the computations involved in processing the semantic aspects of language and the harmonic aspects of music. Therefore, when we listen to opera, we process both the lyrics and the tunes in an independent fashion, and language seems to be processed before music.

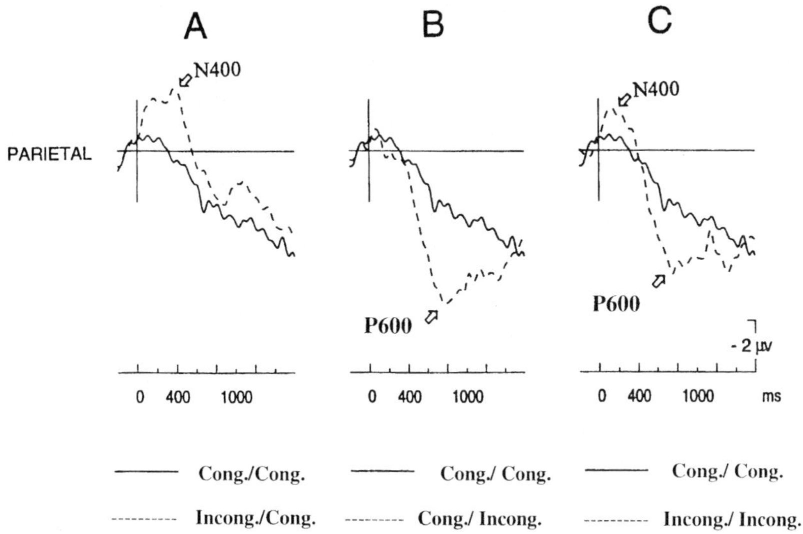


**FIGURE 5.** ERPs results averaged across 16 professional musicians and recorded from the parietal electrode (Pz). Terminal congruous words sung in key (Cong./Cong.) are compared to (**A**) semantically incongruous words sung in tune (Incong./Cong.), (**B**) semantically congruous words sung out of tune (Cong./Incong.), and (**C**) semantically incongruous words sung out of tune (Incong./Incong.). The *vertical lines* mark the onset of the final word of the excerpts. A large N400 component develops in the 50–600 ms that follow the presentation of semantically incongruous words (**A**). In marked contrast, a P600 develops in the 400–1200 ms that follow the presentation of words sung out of tune (**B**). Most importantly, both an N400 and a P600 develop in response to the double incongruity (**C**). (Adapted from Besson *et al.*[72]).

### *Influence of Attention*

We tracked these results further by conducting another series of experiments aimed at studying the effect of attention, again testing some professional musicians from the opera company in Marseille.[76] We hypothesized that if lyrics and tunes are processed independently, listeners should be able to focus their attention only on the lyrics or only on the tunes, depending on the instructions. Without going into the details of the results, an N400 component was elicited to sung incongruous words, and a P600 was associated with congruous words sung out of tune, thus replicating our previous findings.[72] Most interestingly, the N400 to incongruous words completely vanished when participants focused their attention on the music (FIG. 7). Thus, musicians were able not to process the meaning of words; they did not notice whether the terminal word made sense within the linguistic context when they only listened to the music. Conversely, P600 amplitude was significantly reduced when musicians focused attention on language, so that they did not hear that the final word was sung out of tune. Taken together these results again provide strong arguments in favor of the independence of lyrics and tunes. There is some limit to such processing independence, however. Results in the double incongruity condition showed that the

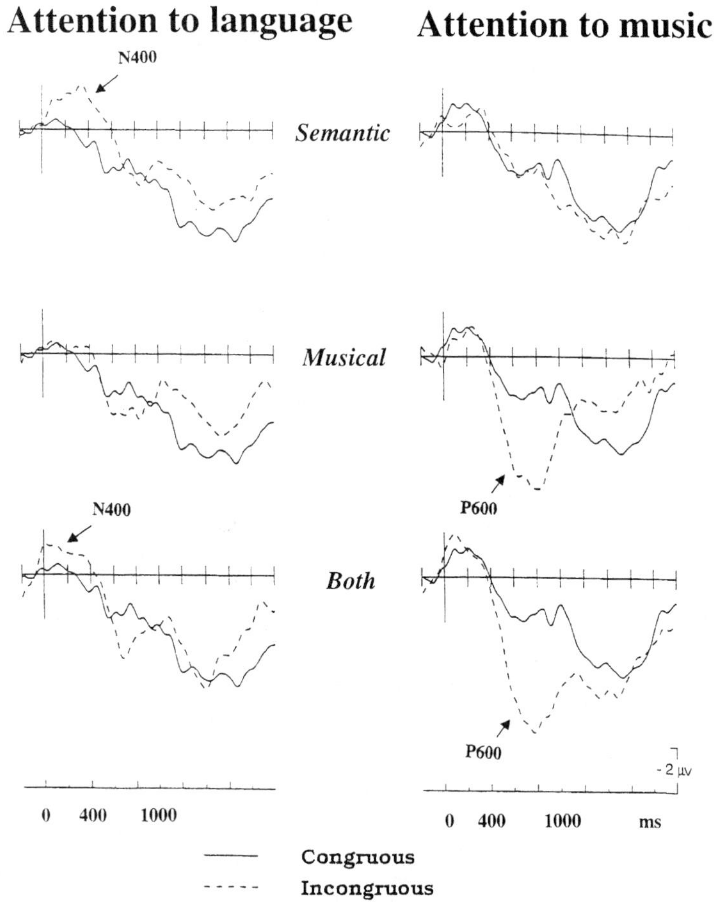


**FIGURE 6.** Overlapped are the ERPs to congruent and incongruent endings, recorded at the central recording site (Cz), when participants paid attention only to the language (*left column*) or only to the music (*right column*) of the opera's excerpts. A large N400 effect is generated when participants focus their attention on language. This effect completely vanishes when attention is focused on music (*top row*). Similarly, the P600 effect is much greater when participants paid attention to music than when they paid attention to language (*medium row*). Finally, when words are both semantically incongruous and sung out of tune, the N400 effect is greater when participants paid attention to the language, and the P600 effect is greater when they paid attention to the music (*bottom row*). (From Regnault and Besson, in preparation).

presence of one type of incongruity influenced the processing of the other type. When words were both semantically incongruous and sung out of tune, musicians could not help but hear the musical incongruity, even if they were asked to focus their attention on language.

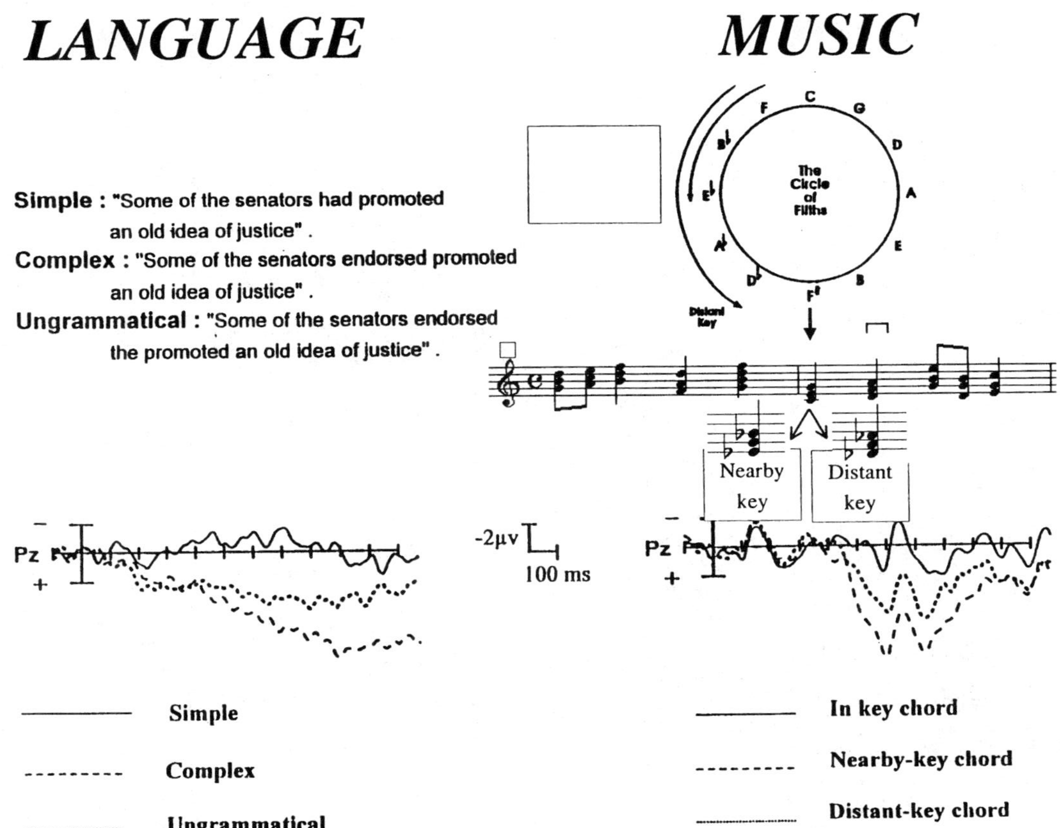


FIGURE 7. *See following page for caption.*

### *Syntax and Harmony*

The rules of harmony and counterpoint are often described as the grammar of tonal music. As syntax is used to extract the fundamental structure of an utterance by assigning different functions to different words, the rules of harmony allow us to specify the different elements, notes and chords, that fulfill a specific harmonic function. Results of experiments manipulating the harmonic function of target chords have shown that violations of harmonic expectancies are associated with P600 components.[77,76] Interestingly, research on syntax using ERPs has also shown that different types of syntactic violations, such as violations of gender, word order or noun-verb agreement, elicit a positive component, peaking around 600 ms.[78–80] Moreover, both components show a similar parietal distribution over the scalp, which, together with their similar polarity and latency, seems to indicate that they reflect qualitatively similar processes.

To further test this hypothesis, Patel and collaborators[81] conducted an experiment directly aimed at comparing the P600 components elicited by harmonic and syntactic violations. ERPs associated with a word within a grammatically simple, complex, or incorrect sentence were compared to those associated with the presentation of a chord that belonged to the tonality induced by the chord sequence, a nearby, or a distant tonality. Results showed that aside from early morphologic differences in the ERPs to words and chords, due to the differences in the acoustic characteristics of these two types of auditory signals, the effects associated with the violation of syntactic and harmonic expectancies were not significantly different (FIG. 8). Therefore, these results raise the interesting possibility that a general cognitive process is called into play when participants are asked to process the structural aspects of an organized sequence of sounds, be it language or music. Finally, an early right anterior negativity was found at around 300–400 ms in response to a chord belonging to a distant tonality. These results paralleled those obtained in language experiments showing that an early left anterior negativity is also associated with some syntactic violations.[82] Whereas these two negative components showed a different distribution over the scalp, with a left predominance for language and a right predominance for music, they may reflect functionally similar processes.

### *Temporal Structure*

Spoken language, as music, is composed of acoustic events that unfold in time. Because of the temporal structure inherent in both language and music, specific events are expected at specific times. The main question addressed in the next series of experiments was to determine whether the processes involved in analyzing tem-

**FIGURE 7.** (*Left*) Examples of the sentences presented in the auditory language experiment. Results showed increased positivity from simple to ungrammatical sentences. (*Right*) Representation of the circle of fifths. Examples of the stimuli used in the experiment. The target chord, shown by the downward-pointing *vertical arrow*, is the congruous chord. The two *arrows* below the musical notation point to moderately incongruous (nearby key) and highly incongruous (distant key) target chords. Results also showed increased positivity from the in-key chords to the distant-key chords. (Adapted from Patel *et al.*[81])

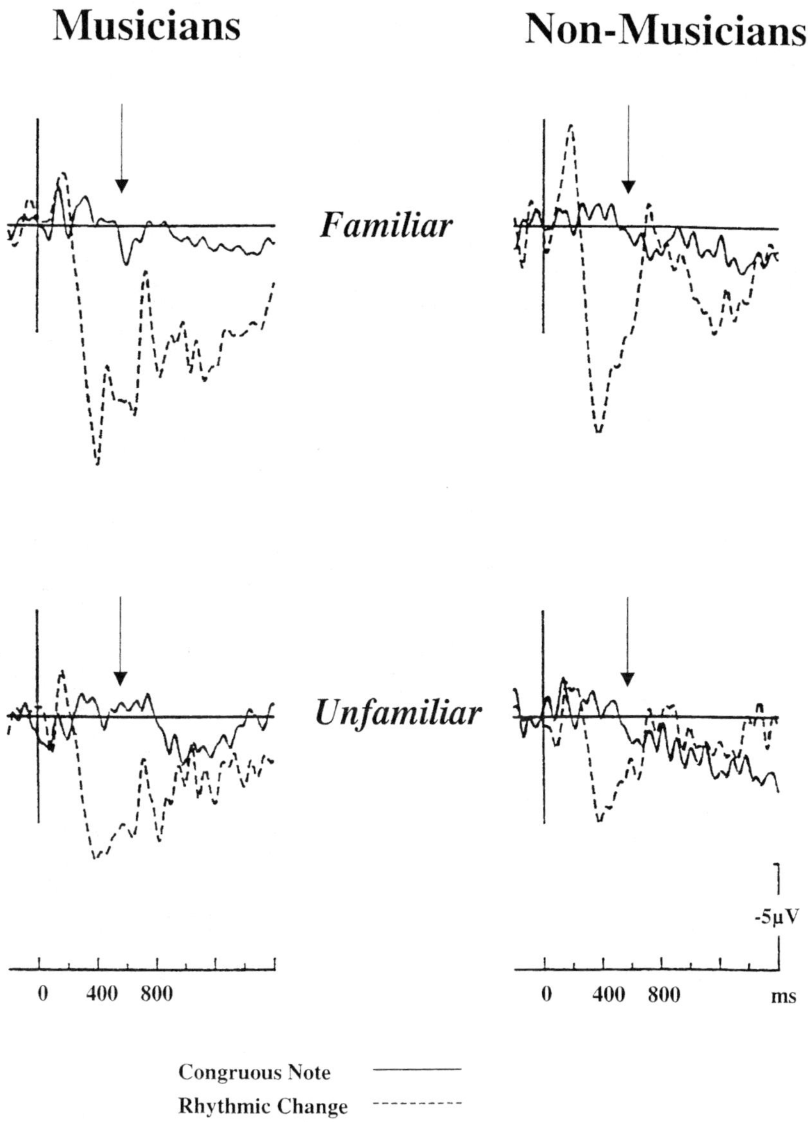


**FIGURE 8.** Overlapped are the ERPs to congruous notes and to the rhythmic incongruities ending familiar and unfamiliar musical phrases for musicians and nonmusicians. Recordings are from the parietal electrode (Pz). Large emitted potentials are elicited when the final note should have been presented (*vertical bar*) but was delayed by 600 ms. The *arrow* points to the moment in time when the final note was presented. (Adapted from Besson and Faïta.[69])

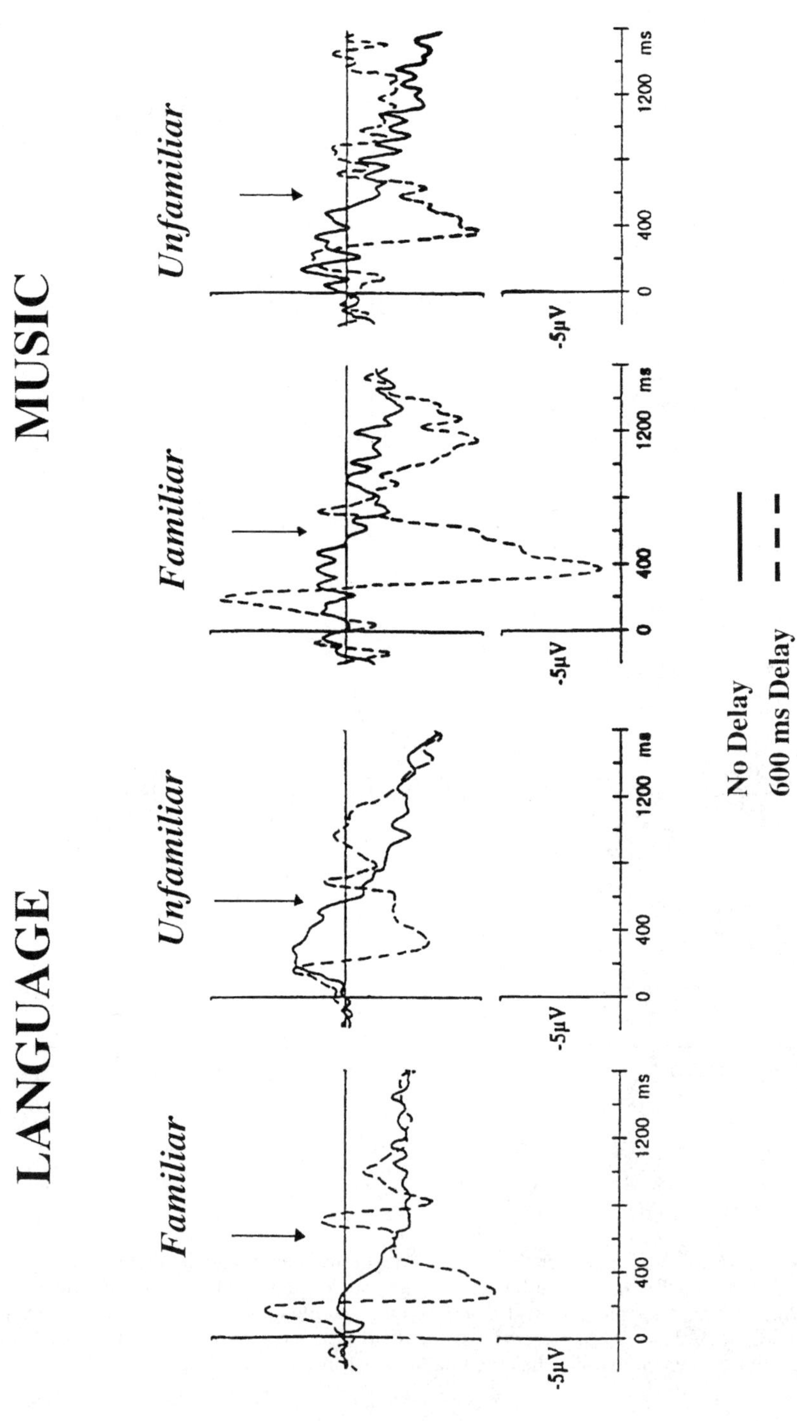


**FIGURE 9.** *See following page for caption.*

poral structures rely on general cognitive mechanisms or differ as a function of the specific characteristics of the materials to be processed. We used both the ERP and the MEG methods to analyze the time course of the effects of temporal structure violations in language and music, and fMRI to localize the cerebral structures activated by these violations. We hypothesized that if a general mechanism is responsible for processing the temporal structures in language and music, qualitatively similar effects should be revealed in the ERP and MEG recordings, and similar brain areas should be shown to be activated by temporal violations. By contrast, if processing temporal information in both systems relies on different mechanisms, qualitatively different effects and different brain areas should be found in language and music.

In previous experiments,[69] we introduced an unexpected silence between the penultimate note and the last note of a musical phrase (FIG. 2). Results showed that a large biphasic, negative then positive, potential, the emitted potential,[83] was elicited when the final note should have been presented but was not, because it was delayed by 600 ms. The amplitude of this effect was similar in musicians and nonmusicians, but it was larger for familiar than unfamiliar melodies (FIG. 9). These findings clearly indicate that both musicians and nonmusicians could anticipate the precise moment when the final note was to be presented and were surprised when it was not. Moreover, known melodies allowed participants to generate more precise expectancies than did unfamiliar melodies. Therefore, these results indicate that the occurrence of an emitted potential is a good index of temporal expectancy.

It was then of interest to determine whether similar results would be found for spoken language.[84] To this aim, we presented both familiar (e.g., proverbs) and unfamiliar auditory sentences to participants. In half of the sentences, final words occurred at their normal position, while in the other half, they were delayed by 600 ms. Results showed that an emitted potential, similar to the one described for temporal ruptures in music, developed when the final word should have been presented (FIG. 10). Therefore, these ERP results indicate that qualitatively similar processes seem to be responsible for temporal processing in language and music.

To strengthen this interpretation, it was important to determine whether the same brain structures are activated by the processing of temporal ruptures in language and music. As already mentioned, fMRI allows localization of brain activation with excellent spatial resolution. Moreover, the MEG permits localization of the generators of the effects observed on the scalp more precisely than the ERP method, while offering an excellent temporal resolution. Therefore, in collaboration with Heinze and his research team, we conducted three experiments in which we presented both auditory sentences and musical phrases.[85] These experiments used a blocked design in which only sentences or musical phrases without temporal ruptures were presented within a block of trials, and only sentences or musical phrases with temporal rup-

---

**FIGURE 9.** Comparison of the effects of temporal violations in language and music. Recordings are from the parietal electrode (Pz). On the *left*, overlapped are the ERPs to congruous words and to the temporal disruptions ending familiar and unfamiliar sentences. In both language and music, large emitted potentials are elicited when the final event should have been presented (*vertical bar*) but was delayed by 600 ms. Note that the amplitude of the emitted potential is greater in music than in language, but that, in both cases, its amplitude is greater for familiar than for unfamiliar materials. (Adapted from Besson *et al.*[84] [*left*] and Besson and Faïta[69] [*right*].)

tures at unpredictable positions were presented within another block of trials. The ERP method was used in the first experiment to replicate, within subjects, the results found previously with two different groups of subjects,[69,84] and the fMRI and the MEG methods were used, respectively, in the other two experiments, trying to localize the effects of interest.

Overall, the ERP results replicated, within subjects, those previously found in music and language separately (i.e., an emitted potential). However, comparison of the conditions with and without temporal violations revealed a different pattern of activation using the MEG and fMRI methods. Source localization based on MEG data revealed that the underlying generators of the biphasic potential recorded on the scalp were most likely located in the primary auditory cortex of both hemispheres. By contrast, fMRI results showed activation of the associative auditory cortex in both hemispheres as well as some parietal activation. Several factors may account for these differences,[85] but the main point is that similar brain areas were activated by temporal violations in both language and music. Therefore, taken together our results suggest that processing temporal information in both language and music relies on general cognitive mechanisms.

## CONCLUSION

We have addressed one of the central questions of human cognition, the specificity of language processing. Is language an autonomous system, independent from other human cognitive abilities or does language rely on general cognitive principles? To address this question, we have conducted several experiments aimed at comparing some aspects of language processing with some aspects of music processing. We mainly used the ERPs method, which offers excellent temporal resolution and therefore permits study of the time course of information processing and determination of whether the processes involved in language and music are qualitatively similar or different.

Taken together, results have shown that the semantic computations required to access the meaning of words, and their integration within a linguistic context, seem to be specific to language. Indeed, whereas unexpected words within a sentence context are associated with the occurrence of an N400 component, unexpected notes or chords within musical phrases elicit a P600 component. By contrast, words that are unexpected on the basis of the syntactic structure of the sentence, and chords that are unexpected as a function of the harmonic structure of the musical sequence, elicit similar effects in both cases, namely, a P600 component. Early negative effects, that is, left and right anterior negativity, which developed between 200 and 300 ms, have also been reported in experiments manipulating syntax and harmony, respectively. Although their different scalp distributions seem to reflect involvement of different brain structures, more research is needed to further track their functional significance. Finally, violations of temporal structure within language and music also elicit similar effects, a biphasic negative–positive complex, the emitted potential. The occurrence of the emitted potential shows that in both language and music, words and notes or chords are expected at specific moments in time. Therefore, when we listen to language and music, not only do we expect words or chords with specific meaning and function, but we also expect them to be presented on time!

The question of the specificity of language processing has broad implications in our understanding of the human cognitive architecture and, even more generally, for the fundamental problem of the relationship between structures (different brain regions) and functions (e.g., language, music). Although the research reported here sheds some light on certain aspects of language processing and highlights some similarities and differences with music processing, more research is clearly needed in this fascinating research domain. Of utmost interest is the use of brain imaging methods that offer complementary information about the spatiotemporal dynamics of brain activity in order to pinpoint the networks of cerebral structures that are involved in two of the most human cognitive abilities: language and music.

## REFERENCES


1. Rangacharya, A. 1966. Introduction to Bharata's Natya-Sastra. Popular Prakashan. Bombay, India.
2. Nadel, S. 1930. The Origin of Music. Musical Q. **16:** 531–546.
3. Leroi-Gourhan, A. 1988. Le geste et la parole. I: La mémoire et les rythmes. Sciences d'aujourd'hui. Albin-Michel. Paris.
4. Chomsky, N. 1957. Syntactic Structures. Mouton & Co. The Hague, the Netherlands.
5. Bickerton, D. 2000. Can biomusicology learn from language evolution studies? *In* The Origins of Music. N.L. Wallin, B. Merker & S. Brown, Eds.: 153–164. MIT Press. Cambridge, MA.
6. Wallin, N.L., B. Merker & S. Brown, Eds. 2000. The Origins of Music. MIT Press. Cambridge, MA.
7. Darwin, C. 1871. The Descent of Man, and Selection in Relation to Sex. Murray. London.
8. Descartes, R. 1618/1987. Abrégé de musique. Compendium musicae. Épiméthée. P.U.F. Paris.
9. Rousseau, J.J. 1781/1993. Essai sur l'origine des langues. Flammarion. Paris, France.
10. Spencer, H. 1857. The origin and function of music. Fraser's Mag. **56:** 396–408.
11. Brown, S. 2000. The "musilanguage" model of music evolution. *In* The Origins of Music. N.L. Wallin, B. Merker & S. Brown, Eds.: 271–300. MIT Press. Cambridge, MA.
12. Molino, J. 2000. Toward an evolutionary theory of music and language. *In* The Origins of Music. N.L. Wallin, B. Merker & S. Brown, Eds.: 165–176. MIT Press. Cambridge, MA.
13. Sloboda, J.A. 2000. Individual differences in music performance. Trends Cognit. Sci. **4:** 397–403.
14. Blacking, J. 1973. How Musical is Man? University of Washington Press. Seattle.
15. Schön, A. & D. Schön. 1999. Il potere del suono e della musica. Fuga a più voci. Psiche **1–2:** 159–165.
16. Kubik, G. 1969. Composition techniques in Kiganda xylophone music. Afr. Music J. **4.**
17. Arom, S. 2000. Prolegomena to a biomusicology. *In* The Origins of Music. N.L. Wallin, B. Merker & S. Brown, Eds.: 27–30. MIT Press. Cambridge, MA.
18. Pinker, S. 1994. The language instinct: how the mind creates language. Harper Perennial. New York.
19. Pacherie, E. 1999. Philosophie et sciences cognitives. *In* Encyclopédie Philosophique Universelle. J.F. Mattéi, Ed. PUF. Paris.
20. Chomsky, N. 1988. Language and Problems of Knowledge. The Managua Lectures. MIT Press. Cambridge, MA.
21. Chomsky, N. 1991. Linguistics and cognitive science: problems and mysteries. *In* The Chomskyan Turn. A. Kasher, Ed.: 26–53. Basil Blackwell. Cambridge, MA.
22. François, J. 1998. Grammaire fonctionnelle et dynamique des langues—de nouveaux modèles d'inspiration cognitive et biologique. Verbum **XX:** 233–256.
23. Aiello, R. 1994. Music and language: parallels and contrasts. *In* Musical Perceptions. R. Aiello & J. Sloboda, Eds.: 40–63. Oxford University Press. New York, NY.

24. JAKOBSON, R. 1973. Essais de linguistique générale. II. Rapports internes et externes du langage. Editions de Minuit, Arguments. Paris.
25. MEYER, L. 1956. Emotion and Meaning in Music. University of Chicago Press. Chicago, IL.
26. KIVY, P. 1991. Music Alone. Philosophical Reflection on the Purely Musical Experience. Cornell Paperbacks. New York.
27. BOUCOURECHLIEV, A. 1993. Le langage musical. Collections les chemins de la musique. Fayard. Paris.
28. PIANA, G. 1991. Filosofia della Musica. Guerini e associati. Milan.
29. WITTGENSTEIN, L. 1953. Philosophical investigations. G.E.M. Anscombe & R. Rhees, Eds.: 209. Blackwell. Oxford.
30. JACKENDOFF, R. 1991. Musical parsing and musical affect. Music Percept. **9:** 199–230.
31. ECO, U. 1979. Trattato di Semiotica Generale. Bompiani. Milano.
32. ECO, U. 1962. Opera Aperta. Bompiani. Milan.
33. JACKENDOFF, R. 1997. The Architecture of the Language Faculty. MIT Press. Cambridge, MA.
34. POLLOCK, J.Y. 1997. Langage et cognition. Introduction au programme minimaliste de la grammaire générative. PUF. Paris.
35. FODOR, J. 1983. Modularity of Mind. MIT Press. Cambridge, MA.
36. FRAZIER, L. 1987. Sentence processing: a tutorial review. *In* Attention and Performance. XII. M. Coltheart, Ed.: 559–586. Erlbaum. Hillsdale, NJ.
37. VICTORRI, B. 1999. Le sens grammatical. Languages **136:** 85–105.
38. DIK, S. 1997. The Theory of Functional Grammar. Mouton De Gruyter. Berlin.
39. VAN VALIN, R.D. & R.J. LAPOLLA. 1997. Syntax Structure, Meaning and Function. Cambridge University Press. Cambridge.
40. LAKOFF, G. 1987. Women, Fire and Dangerous Things. University of Chicago Press. Chicago, IL.
41. LANGACKER, R.W. 1987. Foundations of Cognitive Grammar. Vol. I: Theoretical Prerequisites. Stanford University Press. Stanford, CA.
42. TALMY, L. 1988. The relation of grammar to cognition. *In* Topics in Cognitive Linguistics. B. Rudzka-Ostyn, Ed. Benjamins. Amsterdam.
43. CROFT, W. 1995. Autonomy and functionalist linguistics. Language **71:** 490–532.
44. GREENBERG, J. 1995. The diachronic typological approach to language. *In* Approaches to Language Typology. M. Shibatani &. Bynon T., Eds.: 145–166. Clarenton. Oxford.
45. GIVÓN, T. 1995. Functionalism and Grammar. Benjamins. Amsterdam.
46. FUCHS, C. 1997. Diversité des représentations linguistiques: Quels enjeux pour la cognition? *In* Diversité des langues et représentations cognitives. C. Fuchs & S. Robert, Eds.: 5-24. Ophrys. Paris.
47. ROBERT, S. 1997. Variation des représentations linguistiques: des unités à l'énoncé. *In* Diversité des langues et représentations cognitives. C. Fuchs & S. Robert, Eds.: 25–39. Ophrys. Paris.
48. CULIOLI, A. 1999. Pour une linguistique de l'énonciation. Ophrys. Paris.
49. FAUCONNIER, G. 1997. Mappings in Thought and Language. Cambridge University Press. Cambridge, MA.
50. TOOTELL, R.B.H., N.K. HADJKHANI, J.D. MANDOLA, *et al.* 1998. From retinotopy to recognition: fMRI in human visual cortex. Trends Cognit. Sci. **2:** 174–183.
51. PANTEV, C., M. HOKE, B. LUETKENHOENER, *et al.* 1989. Tonotopic organization of the auditory cortex: pitch versus frequency representation. Science **246:** 486–488.
52. STRAINER, J.C., J.L. ULMER, F.Z. YETKIN, *et al.* 1997. Functional magnetic resonance imaging of the primary auditory cortex: analysis of pure tone activation and tone discrimination. Am. J. Neuroradiol. **18:** 601–610.
53. INDEFREY, P. & W.J.M LEVELT. 2000. The neural correlates of language production. *In* The New Cognitive Neurosciences. M.S. Gazzaniga, Ed.: 845–865. MIT Press. Cambridge, MA.
54. NORRIS, D. & R. WISE. 2000. The study of prelexical and lexical processes in comprehension: psycholinguistics and functional neuroimaging. *In* The New Cognitive Neurosciences. M.S. Gazzaniga, Ed.: 867–880. MIT Press. Cambridge, MA.

55. Binder, J.R, J.A. Frost, T.A. Hammeke, *et al.* 1996. Function of the left planum temporale in auditory and linguistic processing. Brain **119:** 1239–1247.
56. Gandour, J., D. Wong & G. Hutchins. 1998. Pitch processing in the human brain is influenced by language experience. Neuroreport **9:** 2215–2119.
57. Zatorre, R.A. C. Evans, E. Meyer, *et al.* 1992. Lateralization of phonetic pitch discrimination in speech processing. Science **256:** 846–849.
58. Falk, D. 2000. Hominid brain evolution and the origin of music. *In* The Origins of Music. N.L. Wallin, B. Merker & S. Brown, Eds.: 197–216. MIT Press. Cambridge, MA.
59. Sergent, J., E. Zuck, S. Terriah, *et al.* 1992. Distributed neural network undelying musical sight-reading and keyboard performance. Science **257:** 106–109.
60. Zatorre, R.J., A. Halpern, D.W. Perry, *et al.* 1996. Hearing in the mind's ear: a PET investigation of musical imagery and perception. J. Cognit. Neurosci. **8:** 29–46.
61. Platel, H., C. Price, J.C. Wise, *et al.* 1997. The structural components of music perception. Brain **120:** 229–243.
62. Chen, W., T. Kato, X.H. Zhu, *et al.* 1996. Functionnal mapping of human brain during music imagery processing. NeuroImage **3:** S205.
63. Grabowski, T.J. & A.R. Damasio. 2000. Investigating language with functional neuroimaging. *In* Brain Mapping: The Systems. A.W. Toga & J.C. Mazziotta, Eds.: 425–458. Academic Press. London.
64. Zatorre, J. & J.R. Binder. 2000. Functional and structural imaging of the human auditory system. *In* Brain Mapping: The Systems. A.W. Toga & J.C. Mazziotta, Eds.: 365–402. Academic Press. London.
65. Kutas, M. & S.A. Hillyard. 1980. Reading sens..ess sentences: brain potentials reflect semantic incongruity. Science **207:** 203–205.
66. Kutas, M. & S.A. Hillyard. 1984. Event-related brain potentials (ERPs) elicited by novel stimuli during sentence processing. Ann. N.Y. Acad. Sci. **425:** 236–241.
67. Besson, M. & F. Macar. 1987. An event-related potential analysis of incongruity in music and other non-linguistic contexts. Psychophysiology **24:** 14–25.
68. Besson, M., F. Faïta & J. Requin. 1994. Brain waves associated with musical incongruity differ for musicians and non-musicians. Neurosci. Lett. **168:** 101–105.
69. Besson, M. & F. Faïta. 1995. An event-related potential (ERP) study of musical expectancy: comparison of musicians with non-musicians. J. Exp. Psychol. Human Percept. Perform. **21:** 1278–1296.
70. Levman, B.G. 1992. The genesis of music and language. Ethnomusicology **36:** 147–170.
71. Boulez, P. 1966. Relevés d'apprenti. Editions du Seuil. Paris.
72. Besson, M., F. Faita, I. Peretz, *et al.* 1998. Singing in the brain: independence of lyrics and tunes. Psychol. Sci. **9:** 494–498.
73. MacCallum, W.C., S.F. Farmer & P.V. Pocock. 1984. The effects of physical and semantic incongruities on auditory event-related potentials. Electroencephal. Clin. Neurophysiol. **59:** 477–488.
74. Paller, K.A., G. McCarthy & C.C. Wood. 1992. Event-related potentials elicited by deviant endings to melodies. Psychophysiology **29:** 202–206.
75. Verleger, R. 1990. P3-evoking wrong notes: unexpected, awaited or arousing? Int. J. Neurosci. **55:** 171–179.
76. Regnault, P., E. Bigand & M. Besson. 2001. Different brain mechanisms mediate sensitivity to sensory consonance and harmonic context: evidence from auditory event related brain potentials. J. Cognit. Neurosci. **13:** 1–17.
77. Janata, P. 1995. ERP measures assay the degree of expectancy violation of harmonic contexts in music. J. Cognit. Neurosci. **7:** 153–164.
78. Hagoort, P., C. Brown & J. Groothusen. 1993. The syntactic positive shift as an ERP-measure of syntactic processing. Lang. Cognit. Processes **8:** 439–483.
79. Osterhout, L. & P.J. Holcomb. 1992. Event-related brain potentials elicited by syntactic anomaly. J. Mem. Language **31:** 785–804.
80. Friederici, A.D. 1998. The neurobiology of language comprehension. *In* Language Comprehension: A Biological Approach. A.D. Friederici, Ed.: 263–301. Springer. New York.

81. PATEL, A.D., E. GIBSON & J. RATNER. 1998. Processing syntactic relations in language and music: an event-related potential study. J. Cognit. Neurosci. **10:** 717–733.
82. FRIEDERICI, A.D., E. PFEIFER & A. HAHNE. 1993. Event-related brain potentials during natural speech processing: effects of semantic, morphological and syntactic violations. Cognit. Brain Res. **1:** 182–192.
83. SUTTON, S., M. BRAREN, J. ZUBIN & E.R. JOHN. 1967. Evoked potential correlates of stimulus uncertainty. Science **150:** 1187–1188.
84. BESSON, M., F. FAÏTA, C. CZTERNASTY & M. KUTAS. 1997. What's in a pause: event-related potential analysis of temporal disruptions in written and spoken sentences. Biol. Psychol. **46:** 3–23.
85. WEYERT, H., M. BESSON, C. TEMPELMANN, *et al.* 2001. An analysis of temporal structure in language and music using ERPs, MEG and fMRI techniques. In preparation.

# Musical Sound Processing in the Human Brain

## Evidence from Electric and Magnetic Recordings

MARI TERVANIEMI

*Cognitive Brain Research Unit, Department of Psychology, University of Helsinki, Helsinki, Finland, and Department of Psychology, University of Jyväskylä, Jyväskylä, Finland*

**Abstract: Recently, our knowledge regarding the brain's ability to represent invariant features of musical information even during the performance of a simultaneous task (unrelated to the sounds) has accumulated rapidly. Recordings of the change-specific mismatch negativity component of event-related brain potentials have shown that temporally and spectrally complex sounds as well as their relations are automatically processed by human auditory cortex. Furthermore, recent magnetoencephalographic and positron emission topographic investigations indicate that this processing differs between phonetic and musical sounds within and between the cerebral hemispheres. These data thus suggest that despite the focus of listeners' conscious attention, relatively complex sound phenomena can be encoded by neural mechanisms that are specialized for musical sounds.**

**Keywords: Music sound perception; Preattentive processing; Event-related potentials; EEG; MEG**

## INTRODUCTION

Neural sound processing can be probed with a millisecond accuracy by recording the event-related potential (ERP). The first ERP components of interest in neuromusicology were the P3[1–5] and late positive component (LPC, also termed P600).[6–9] These late auditory ERP components reflect the degree of encountered expectancy violation in musical context. For their elicitation, subjects' attentional focus needs to be directed to the sounds.

However, the importance of focused attention in music perception was questioned shortly after the first P3 and P600 investigations. These studies on automatic neural encoding of musical material employed the mismatch negativity (MMN) paradigm.[10,11] The MMN is evoked by an infrequently presented sound ("deviant") differing from the frequently occurring stimuli ("standard") in one or several physical or abstract parameters. Its presence implies that the invariant parameters of the standard sound were neurally encoded and differed from the parameters of the deviant sound.

Address for correspondence: Dr. Mari Tervaniemi, Cognitive Brain Research Unit, Department of Psychology, P.O. Box 13, FIN-00014 University of Helsinki, Finland. Voice: +358-9-1912-3408; fax: +358-9-1912-2924.

tervanie@helsinki.fi

The MMN can be recorded even when the subject is performing a task unrelated to the stimulation under interest, such as reading a book or playing a computer game. Thus it offers a direct measure for the similarity of neural sound representations, without being contaminated by differences, for instance, in attentional or motivational involvement of the subject. Yet, several data indicated that the MMN parameters closely correlate to the indicators of the subjects' perceptual accuracy as determined in a separate behavioral session. For instance, the MMN amplitude and latency reflect perceptual accuracy, as determined by tests of musical aptitude (for a review, see Ref. 12) and by hit rates and reaction times.[13–15] In addition, the degree of perceptual similarity between different musical instrument timbres highly correlates with the MMN amplitude.[16] This correlation between the MMN parameters and behavioral responses implies that preattentive neural functions determine the accuracy of the subsequent attentive processes.[10,17]

Since the MMN is elicited without attention, it received particular interest during past years among clinicians who need to evaluate the integrity of neurocognitive functions in patients unable or unwilling to participate in neuropsychological (behavioral) testing (e.g., the MMN elicitation predicts the recovery of consciousness in comatose patients[18,19]) (for reviews, see Refs. 20 and 21). In addition, the deficits in MMN elicitation might help target perceptual training or rehabilitation in milder perceptual disorders. For instance, recently it was revealed that automatic discriminative functions of dyslexic adults are worse than those of control subjects.[22–25]

In the paper, an overview of studies using musical sounds and sound successions to investigate the automatic neural sound processing will be given. In addition, recent studies comparing musicians and nonmusicians will be reviewed.

## NEURAL GENERATORS OF THE MMN

MMN is generated mainly in the primary auditory cortex or in its immediate vicinity. This has been indicated recently by several brain research methods, such as magnetoencephalography (MEG),[26–28] intracranial electrode recordings,[29–30] positron emission tomography (PET),[31] and functional magnetic resonance imaging (fMRI).[32,33] In addition, this is reflected in electric recordings by polarity reversal above the sylvian fissure from the frontocentral negative maximum to positivity in the mastoids leads (when the nose reference is used) (for a review, see Ref. 34 [FIG. 1]). Additional generators have been found in the frontal[35,36] and parietal[37] lobes. Within the auditory cortex, MMN may have multiple generators because its scalp topography in electric recordings differs as a function of the parameter in which the deviance occurs (such as intensity, duration, and frequency).[38] This suggests that the neural representations for separate sound parameters are spatially separate.

The possible existence of spatially distinct areas underlying MMN generation for musical sounds when compared with either pure sinusoidal tones or speech sounds was addressed by three consecutive studies. The first was conducted with a whole-head 122-channel MEG.[39] This study used sounds with an occasional frequency change of identical magnitude, namely, that of one whole-tone step (G4–A4, 392–440 Hz), but in three different contexts: among single tones, parallel chords, and sequential chord patterns. It was found that frequency change elicited the MMN with all three sounds without latency or amplitude differences between the sound types.

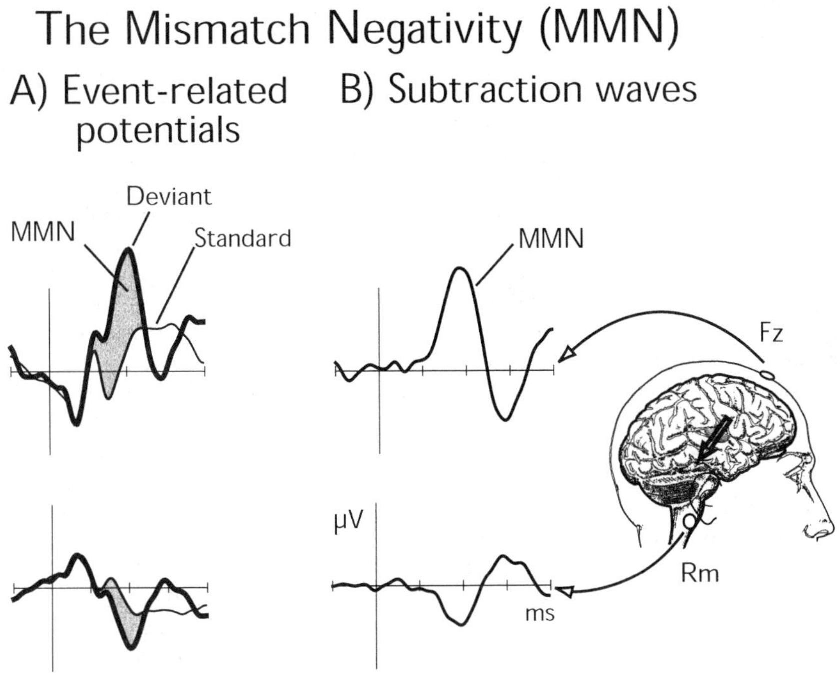


**FIGURE 1.** A typical mismatch negativity response to a 2.5% pitch change (average across 10 subjects). (**A**) The thin line represents an ERP to a frequently presented "standard" tone. The thick line represents an ERP to an infrequently presented "deviant" tone. The sounds were delivered binaurally via headphones to healthy adult subjects who concentrated on reading a self-chosen book during the recordings. The *x* axis represents the time after the sound onset in milliseconds and the *y* axis the strength of activity in microvolts. (**B**) Prior to numerical MMN analysis, the subtraction (difference) waveform is calculated by subtracting the ERP to standard tone from the ERP to deviant tone. The resulting waveform is considered to reflect the differential neuronal processing of a deviant tone compared to that of a standard tone. Thereafter, the MMN latency and amplitude are quantified for subsequent statistical analysis. The electrode locations in which the recordings were made (Fz, Rm) are illustrated in the schematic brain on the right.

However, comparisons of the frequency MMN generators revealed that the MMN source was different for complex than for simple sounds. More precisely, when the recorded MEG signal was modeled as equivalent current dipoles (a commonly used method for localizing MEG activity[40]), the MMN generator was, on the average, 1 cm more medial in the auditory cortex with sequential and parallel chords than with the single tones. This suggests that even within the auditory cortex, there are spatially separable areas specialized in encoding spectrally/temporally complex information.

The second MEG experiment was conducted to determine whether the auditory cortex has spatially distinct areas for encoding musical versus phonetic sounds that did not differ in their spectral complexity.[41] The subjects, while watching a silent movie, were presented with frequent and infrequent phonemes (/e/ versus /o/) or chords (A major versus A minor). These phonetic and musical stimuli were matched in complexity as well as in the magnitude of the frequency change embedded in them. It was found that in both hemispheres the source of the magnetic MMN (MMNm) elicited by the deviant phonemes or chords was located posterior to the source of the earlier exogenous P1m component.[a] In addition, the MMNm source for a phoneme change was located superior to that of the chord change. However, P1m sources did not differ between phonemes and chords. These data thus indicate that there are distinct cortical areas specialized in representing phonetic and musical sounds in both hemispheres. In other words, the functional specialization does not only cover stimulus complexity (single tone versus chord)[39] but also the informational content (phonetic versus musical).[41] However, this specialization is not present prior to memory-related auditory processing as suggested by the dissociation between the present P1m and MMNm data (FIG. 2).

Additionally, the data reviewed above indicated that in the right hemisphere the MMNm was larger in amplitude for changes among chords than among phonemes. However, in the left hemisphere, no corresponding dominance for phoneme changes was found when compared with other types of sound changes. This result was clearly discrepant from the previous results.[42,43] To solve this discrepancy, a further study was conducted with PET technique.[b]

This PET study[31] used the phonetic and musical sounds developed for the MEG study described above.[41] However, the phoneme duration was prolonged from 200 ms to 400 ms, because the presence of deviant phonemes elicited no changes in brain metabolism with phonemes of 200 ms duration. The subjects were concentrating on classifying the gender of visually presented words as they were presented with sound sequences consisting of (1) both deviant and standard sounds or (2) standard sounds only (phonemes and chords in separate sequences). The data, obtained after subtracting the activity elicited by the standard sounds only from those elicited by both deviant and standard sounds when intermixed, showed that the change from vowel /e/ to /o/ was processed in the left auditory cortex, more specifically, in the middle and supratemporal gyri. In a mirror-like manner, the change from A major to A minor chord was processed in the right auditory cortex, in the supratemporal gyrus. These data thus indicate that hemispheric specialization for phonetic versus musical processing, previously seen in dichotic-listening studies and brain-imaging studies using an active task, may be present even during a performance of a parallel task, unrelated to the sound stimulation. However, this phenomenon is very vulnerable to changes in stimulus parameters (see also Refs. 46–48 for recent evidence about the effects of acoustic noise on lateralization of phonetic processing).

[a]Early exogenous ERP components and their MEG equivalents are mainly determined by the physical sound characteristics, for instance, the intensity and rise time, while endogenous ERP/MEG components with longer onset latencies are mainly modulated by cognitive and attentional factors.[10]

[b]Localizing the active neural population with MEG (and EEG) necessarily involves inverse modeling of the active neural source. Accuracy of this operation depends on the adequacy of the head model available.[40] This source of uncertainty can be avoided by using brain-imaging techniques such as PET and fMRI. Those methods detect directly the locus of activation change in brain metabolism, presumably caused by increased neural activity.[44,45]

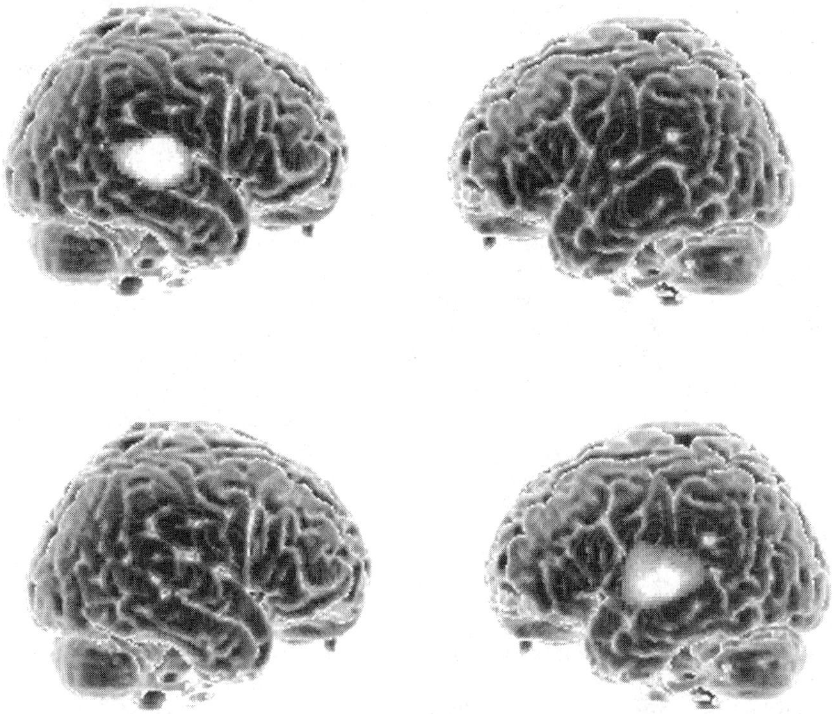

**FIGURE 2. See Plate 18 in color section.** The functional lateralization of the mismatch negativity (MMN) as evidenced with positron emission tomography (PET) (average across 15 subjects in each data plot). During the PET scanning, the native Russian subjects were engaged in a gender classification task of visually displayed Russian words. (*Top*) Difference in brain metabolism between a condition in which only A-major chords were presented and a condition in which A-major chords were frequently and A-minor chords were infrequently presented. The differential metabolic activity reflecting the neural MMN response is seen in the right superior temporal gyrus of the temporal lobe. This indicates that the change detection for musical material is right lateralized even during performance of a parallel visual task. (*Bottom*) Difference in brain metabolism between a condition in which only /e/ phonemes were presented and a condition in which /e/ phonemes were frequently and /o/ phonemes were infrequently presented. The differential metabolic activity reflecting the neural MMN response is seen in the left superior and medial temporal gyri of the temporal lobe. This indicates that the auditory change detection for phonetic material during performance of a parallel visual task is left lateralized. Adapted from figure in Ref. 31.

## REGULARITIES OF THE SOUND SEQUENCE

Originally, the MMN studies used two types of tones that differed from each other in their basic physical sound parameters between standard and deviant tones, such as frequency, duration, intensity, or spatial locus of origin (for a review, see Ref. 10). However, soon after these fundamental findings were established, several investigations were targeted to reveal whether more cognitive operations with regard to successive sounds also could be preattentively made.

In the late 1980s it had already been investigated whether succession of two different frequencies is automatically encoded when sounds are presented with constant stimulus onset asynchrony (SOA) of 800 milliseconds.[49] The subjects were presented with sounds of two frequencies in alternating order. Infrequently either lower or higher frequency was repeated (schematically denoted A-B-A-B-B-A). During the recordings, the subjects were reading a book of their own choice. The authors found that the sound repetitions, violating the "rule" of regular frequency alternation, elicited the MMN. This original finding indicates that automatically activated auditory mechanisms can form perceptual entities of sequentially presented sounds, any interruption in the regularity being encoded as a new event. This perceptual entity may thus be more variable in its physical features than originally proposed. More recently, the effects of sound presentation rate and frequency difference in this automatic regularity extraction were parametrically investigated.[50] It was found that the MMN elicited by a sound repetition is strongest (largest in amplitude and shortest in latency) when the sounds are presented with a relatively rapid stimulation rate and when the frequency difference between them is large. This implies that the neural sound representations for such a continuous sound stream are more readily formed if the sounds clearly differ from each other in frequency and if they fall into the same temporal window of integration (see below).

Subsequent studies clarified the encoding of tone pairs whose members differed from each other in frequency by one whole-tone step (E–F#)[51] (for related MEG evidence, see also Refs. 52 and 53). Sinusoidal tones of 60 ms in duration were separated by 40-ms silent within-pair interstimulus intervals. It was found that the order reversals of the tone-pair members (AB, AB, **BA**) elicited the MMN as did the tone repetitions (AB, AB, **AA**; AB, AB, **BB**) or tone omissions (AB, AB, A_). This was taken as evidence that two closely spaced tones may be neurally represented as an integrated unit. The second part of the experiment wished to probe the time window of such integration by determining the time behavior of the omission-evoked MMN. The silent within-pair interstimulus interval (ISI) of E–F# tone pairs in different conditions was parametrically prolonged by 100 steps from 40 to 340 milliseconds. In parallel, the between-pair ISI was shortened from 640 ms to 340 milliseconds. The sound omissions elicited the MMN only when the within-pair interval was 40 or 140 milliseconds. This result was postulated to confirm the existence of a temporal window of integration, previously evidenced in loudness enhancement with sounds shorter than 200 ms and by backward masking within the same time span (for a review, see Ref. 54). More recently, this conclusion was replicated by presenting the subjects with equispaced tones, sound omission serving as the deviant event.[55] It was found that when SOA exceeded 175 ms, sound omission no longer elicited an MMN, whereas with shorter SOAs an MMN with auditory cortex origin was elicited.

To investigate how any other sound parameter in addition to frequency is encoded by the auditory cortex during a performance of a parallel task, intensity changes were presented in paired tones[56] and in regularly spaced tones.[57] In the first study, the standard sound pairs, presented with 0, 125, and 250 ms with within-pair intervals (in different conditions) and 900 ms between-pair intervals, consisted of loud (70 dB) and soft (58 dB) sounds. In the deviant sound pair, the order of these loud and soft sounds was reversed. In the second study, the sounds were presented with a regular 200 ms onset-to-onset interval. In the standard event, loud and soft sounds were alternatively presented, and in the deviant event, either a loud or soft sound was repeated (L-S-L-S-S; S-L-S-L-L). In both studies it was found that the deviant events, pairs with reversed intensity relation[56] and intensity repetitions,[57] evoked the MMN, supporting the hypothesis that sound relations can be processed despite the attentional focus on sounds in sound parameters other than frequency.

Taken together, the data reviewed above suggest that, in addition to physical sound parameters, automatically formed cortical sound representations may encode regularities in sound relations (successions).

## ABSTRACTIONS AND EXPECTATIONS

Despite the emphasis given in the above-described ERP studies on sound processing accuracy at the sensory level, auditory perception, especially music perception and cognition, cannot be reduced merely to a successful sensory-level performance. For instance, despite transpositions and modulations within a musical piece, the "core" of the melody can be easily remembered and recognized; in other words, abstract sound features and rules are perceived. Consequently, it was questioned whether abstract sound relations and primitive concepts such as "ascending" or "descending" interval, pitch continuum, or chord cadences could also be encoded during a performance of a parallel task.

### *Abstractions*

The first study investigating the interval perception used tone pairs similar to those described above with a whole-tone frequency change within a tone pair.[51] However, here the frequency level of the standard tone pairs was given randomly and varied at five different frequency levels (AB, CD, BC,…EF, DE, AB,…).[58] Randomly presented deviant pairs had either reversed frequency change (BA,...FE) or a frequency repetition (AA, BB,…EE), also at five different frequency levels. It was found that despite this complexity in stimulation and parallel reading task given to the subjects, the deviant sound pairs elicited MMN, indicating that the change in the direction of the interval was discriminated in the auditory cortex. This line of research has successfully been continued by Paavilainen and his colleagues.[59] They have found that in addition to the direction of the tone interval, the interval size is also automatically encoded in the auditory cortex. This was evidenced by the MMN elicitation by smaller or larger intervals among minor seventh standard intervals (all intervals were randomly presented at 10 frequency levels). Furthermore, the cortical interval representation may also be formed on the basis of converging information from both ears as indicated by the MMN elicitation when the two successive sounds of an interval are delivered to separate ears.[60]

### *Expectations*

In the two studies summarized below, the regularity of the sound change itself created a rule about the sound continuum: in other words, an expectation for the next sound of the sequence either in terms of pitch or sound location. An MMN elicitation by abrupt changes in such sound continuums would strengthen the above-

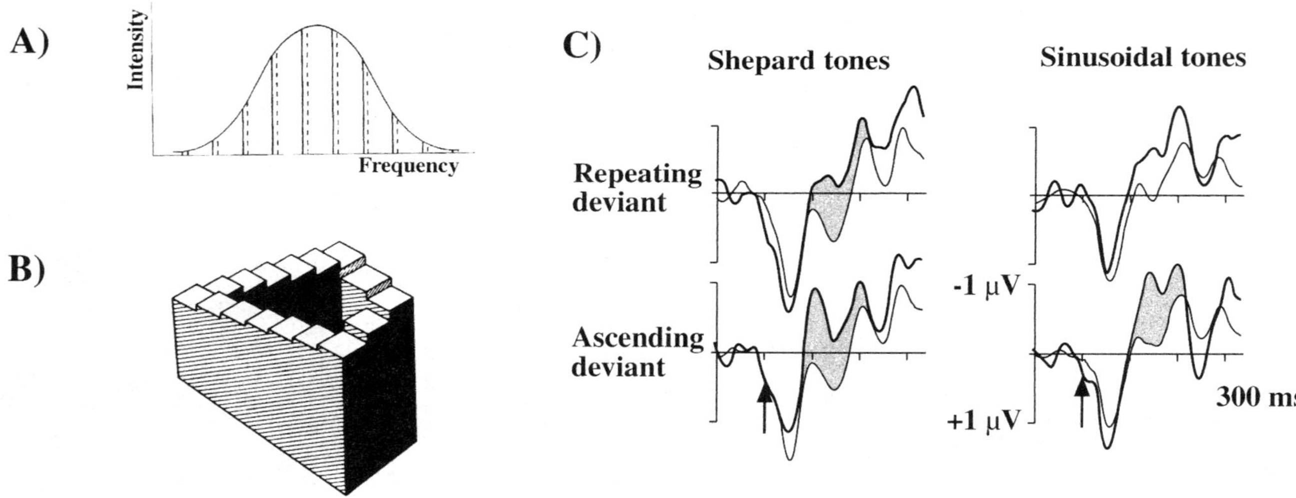


**FIGURE 3.** ERP evidence for neural mechanisms automatically interpolating incoming auditory input (average across 8 subjects). (**A**) A spectrum of a Shepard sound. When Shepard sounds are presented in ascending/descending sequences in one semitone steps, perception of tone height (the sense of octave range) disappears and, thus, the pitch is perceived to ascend/descend in an endless manner. (**B**) A visual analogy of the Shepard illusion. (**C**) The ERPs recorded at Fz electrode to Shepard and sinusoidal tones that were binaurally delivered via headphones (thin line: ERP to standard stimulus; thick line: ERP to deviant stimulus). During the recording, the subjects concentrated on reading a book of their own choice. *Left*: Regularly descending Shepard sound sequence randomly replaced by repetitive (*top*) or ascending (*bottom*) deviant sounds. *Right*: Regularly descending pure sinusoidal sound sequence randomly replaced by repetitive (*top*) or ascending (*bottom*) deviant sounds. The arrow indicates the onset of the stimulus and the shadowed area, the MMN, which was significant in statistical analysis. The data indicate that, especially with Shepard sounds creating strong expectation for a constantly descending pitch, a sound with a repeating or ascending pitch evokes the MMN, although the subjects concentrated on a reading task during the recordings. This suggests that the human auditory cortex automatically detects deviations from the regularity of the acoustic environment, despite the physically changing nature of the environment. **C** is adapted from figure in Ref. 61.

documented conclusions about the "intelligence" of automatically formed neural sound representations in extracting invariant sound features from continuously variable acoustical information.

Tervaniemi *et al.* presented reading subjects with regularly spaced sound sequences consisting of a regular pitch decrease of one semitone as the standard event.[61] In two conditions, the sounds were Shepard sounds and in two conditions there were pure sinusoidal tones. Randomly, a pitch repetition or an ascending pitch interrupted this line of regularly descending pitch. The MMN was elicited by ascending sounds, be they Shepard or sinusoidal sounds, and by pitch repetition when they were Shepard sounds. These data were suggested to reflect the existence of primitive expectation formation even without conscious attention being directed to the sounds (FIG. 3).

To confirm and expand the previous data obtained using sound sequences based on regular pitch change, a further experiment was conducted in which the regularity was created by virtual sound movement.[62] In this experiment, sounds were presented via nine loudspeakers symmetrically extending in a horizontal arc in front of the subject. Short noise bursts were presented about three times a second, continuously moving back and forth between the two extreme loudspeakers. Occasionally the regular virtual sound movement was broken by the sound skipping two loudspeakers, which is experienced as a jump in the movement. ERPs recorded from reading subjects showed that jumps elicited a large centrally negative response. This suggests that expectation about the source location of the next sound was automatically formed (for related data about automatic sound source discrimination, see Refs. 63–65).

### *Chord Cadences*

The studies of the most recent approach used cadences formed by five chords (first 4 chords 600 ms, fifth chord 1200 ms in duration) that were continuously presented at several randomly varying frequency levels.[66] The majority of the chords were presented with synthesized piano timbre, the task of the subject being to detect the chords presented with a different timbre (e.g., celesta, marimba). The study was conducted to reveal whether chords not belonging to the tonal context established by the previous chords would elicit differential ERPs from those belonging to that tonal context despite the subject's attention being directed away from these tonality-violating chords—that is, to the timbre-deviant chords.

The ERPs, recorded from subjects without any formal training in music, were found to differ at two latencies between tonal-context chords and (Neapolitan) chords violating the tonal context established by the previous chords. The first difference was labeled as the early right-anterior negativity (ERAN) according to its relatively early peak latency at around 200 ms, right preponderant distribution, and negative polarity (its linguistic counterpart being labeled as the left anterior negativity, ELAN[67]). (See also Ref. 9 for related data in music cognition.) The second difference was labeled as the N5 (negative polarity and peak latency at around 500 ms, without hemispheric dominance in distribution). The authors concluded that musically untrained subjects may have several neural mechanisms activated for extrapolating musical sequences, and, further, that this may occur despite the direction of their conscious attention.

Taken together, the auditory system appears to be able to encode automatically regularity of the sound environment. For this regularity extraction, acoustic repetitiveness of the sound material is not necessary; instead, more abstract sound relations (direction of the sound change, interval size, chord progression) are also encoded.

## MUSICAL EXPERTISE

Musical expertise is reflected in early auditory ERPs even 100 ms after the sound onset, at least when the attention of the subject is not directed away from the sounds.[68] Likewise, it is reflected in the cognitive late-latency ERPs such as P3 and LPC when the subjects are required to attend to the sound stimulation.[1,2,8] Recently it was also investigated whether musical expertise may be reflected in the auditory ERPs during a performance of a parallel task in another modality. These studies addressed the accuracy of violin players' and, more generally, musicians' auditory systems in processing slight pitch changes.

In the first of the studies, three-part major chords with a perfect fifth were used as the standard stimulus.[69] The deviant stimulus was the same chord as the standard stimulus, except that the middle tone of the chord was marginally mistuned (<1%), leading the chord toward a minor chord. This stimulation was presented to subjects while they were reading a book (the first and third blocks of the experiment; reading condition) and while they were asked to detect the deviant chords (the second block; discrimination condition). In the reading condition, the deviants elicited the MMN only in musicians. In the discrimination condition, nonmusicians detected about 10%, and musicians about 80% of the deviant chords. In the discrimination condition, a significant MMN was followed by N2b and P3b deflections in musicians. This N2b-P3 complex reflects higher cognitive processes concerned with the conscious detection and evaluation of deviants. Nonmusicians had a small, but statistically significant, MMN without subsequently elicited N2b or P3. The third block was presented to see whether the intermediate attentive task facilitated subjects' automatic pitch processing. This, however, was not the case. Musicians showed a MMN that did not differ from their MMN of the first block, and nonmusicians showed no MMN. These results demonstrate that professional violin players automatically detected tiny pitch changes in auditory information, which were undetectable for nonmusicians and, further, that these automatic functions could not be modified by attentional manipulations. Interestingly, the MMN evoked by a small (<1%) or large (10%) pitch change in pure sinusoidal tones did not differentiate violinists from nonmusicians. This suggests that the superior pitch processing accuracy of the violinists was activated only when the pitch change was presented among musical sounds.

In the subsequent experiment, the superiority of musicians in automatic and attentive pitch discrimination, originally observed in violin players,[69] was further investigated.[70] Ten professional musicians and 10 nonmusicians were presented with harmonically rich sounds of 300 ms in duration. Standard sounds were of 528 Hz in fundamental frequency while deviant sounds were 0.8, 2, or 4% higher in fundamental frequency. It was found that in the discrimination condition, the musicians detected the pitch changes faster and more accurately than the nonmusicians. This was reflected also in their N2b and P3b components, which were significant earlier and larger in musicians than in nonmusicians. However, their event-related responses recorded during the reading condition did not differentiate the subject groups from each other—the MMN amplitude or latency was not different between the musicians and nonmusicians.

In sum, the data reviewed above on facilitated accuracy in pitch discrimination in violinists suggest that musical expertise is reflected at preattentive cortical auditory processing, especially when the sound structure is musically relevant and when the stimulus change is of musically relevant magnitude. However, in musicians with other instrumental training (guitar, flute, piano, wind instruments), despite their superiority in pitch discrimination during attentional listening, preattentive neural mechanisms were not enhanced when compared with those of nonmusicians. This implies that automatized neural functions in pitch discrimination are most sensitive in musicians who most need such ability.

## CONCLUSIONS AND FUTURE PERSPECTIVES

The data reviewed above, obtained by EEG, MEG, and PET investigations on healthy human subjects, suggest that the cortical networks automatically represent temporally and spectrally complex musical sounds as well as their abstract relations. The loci of these neural representations seem to differ between musical and phonetic sounds within and between the cerebral hemispheres. In addition, musical expertise is reflected in automatically elicited responses to pitch changes.

However, musical experience cannot be reduced to sensory-level discrimination of tiny sound changes. Therefore, in the future, it will be highly important to expand the experimental sound material, for instance, to include transpositions and longer musical passages. Results thus obtained would significantly contribute to the present knowledge about the brain's ability to represent natural, spectrally and temporally complex, sounds. Furthermore, by comparing musicians with training in musical genres or instruments that differ in their emphasis given to musical parameters (pitch, harmony, rhythm), the specificity or generality of neural mechanisms behind musical expertise could be further elucidated.

## REFERENCES

1. Verleger, R. 1990. P3-evoking wrong notes: unexpected, awaited, or arousing? Int. J. Neurosci. **55:** 171–179.
2. Crummer, G.C., J.P. Walton, J.W. Wayman, *et al.* 1994. Neural processing of musical timbre by musicians, nonmusicians, and musicians possessing absolute pitch. J. Acoust. Soc. Am. **95:** 2720–2727.
3. Beisteiner, R., M. Erdler, D. Mayer, *et al.* 1999. A marker for differentiation of capabilities for processing of musical harmonies as detected by magnetoencephalography in musicians. Neurosci. Lett. **277:** 37–40.
4. Janata, P. 1995. ERP measures assay the degree of expectancy violation of harmonic contexts in music. J. Cognit. Neurosci. **7:** 153–164.
5. Nittono, H., T. Bito, M. Hayashi, *et al.* 2000. Event-related potentials elicited by wrong terminal notes: effects of temporal disruption. Biol. Psychol. **52:** 1–16.

6. BESSON, M. & F. FAÏTA. 1996. An event-related potential (ERP) study of musical expectancy: comparison between musicians and non–musicians. J. Exp. Psychol.: Human Percept. Perf. **21:** 1278–1296.
7. BESSON, M., F. FAÏTA, I. PERETZ, *et al.* 1998. Singing in the brain: independence of lyrics and tunes. Psychol. Sci. **9:** 494–498.
8. BESSON, M., F. FAÏTA & J. REQUIN. 1994. Brain waves associated with musical incongruities differ for musicians and non-musicians. Neurosci. Lett. **168:** 101–105.
9. PATEL, A.D., E. GIBSON, J. RATNER, *et al.* 1998. Processing syntactic relations in language and music: an event-related potential study. J. Cognit. Neurosci. **10:** 717–733.
10. NÄÄTÄNEN, R. 1992. Attention and Brain Function. Lawrence Erlbaum Associates. Hillsdale, NJ.
11. NÄÄTÄNEN, R. & I. WINKLER. 1999. The concept of auditory stimulus representation in cognitive neuroscience. Psychol. Bull. **125:** 826–859.
12. TERVANIEMI, M. 1999. Pre-attentive processing of musical information in the human brain. J. New Music Res. **28:** 237–245.
13. TIITINEN, H., P. MAY, K. REINIKAINEN, *et al.* 1994. Attentive novelty detection in humans is governed by pre-attentive sensory memory. Nature **372:** 90–92.
14. KRAUS, N., T. MCGEE, T.D. CARRELL, *et al.* 1996. Auditory neurophysiologic responses and discrimination deficits in children with learning problems. Science **273:** 971–973.
15. TREMBLAY, K., N. KRAUS & T. MCGEE. 1998. The time course of auditory perceptual learning: neurophysiological changes during speech-sound training. Neuroreport **9:** 3557–3560.
16. TOIVIAINEN, P., M. TERVANIEMI, J. LOUHIVUORI, *et al.* 1998. Timbre similarity: convergence of neural, behavioral, and computational approaches. Music Percept. **16:** 223–241.
17. NOVAK, G.P., W. RITTER, H.G. VAUGHAN, JR., *et al.* 1990. Differentiation of negative event-related potentials in an auditory discrimination task. Electroencephalogr. Clin. Neurophysiol. **75:** 255–275.
18. KANE, N.M., S.H. CURRY, C.A. ROWLANDS, *et al.* 1996. Event-related potentials—neurophysiological tools for predicting emergence and early outcome from traumatic coma. Intensive Care Med. **22:** 39–46.
19. FISCHER, C., D. MORLET, P. BOUCHET, *et al.* 1999. Mismatch negativity and late auditory evoked potentials in comatose patients. Clin. Neurophysiol. **110:** 1601–1610.
20. GENE-COS, N., H.A. RING, R.C. POTTINGER, *et al.* 1999. Possible roles for mismatch negativity in neuropsychiatry. Neuropsychiatry Neuropsychol. Behav. Neurol. **12:** 17–27.
21. NÄÄTÄNEN, R. & M. HOKE, Eds. 2000. Special issue on mismatch negativity and its clinical applications. Audiol. Neuro-Otol. **5:** 101–246.
22. SCHULTE-KORNE, G., W. DEIMEL, J. BARTLING, *et al.* 1998. Auditory processing and dyslexia: evidence for a specific speech processing deficit. Neuroreport **9:** 337–340.
23. SCHULTE-KORNE, G., W. DEIMEL, J. BARTLING, *et al.* 1999. Pre-attentive processing of auditory patterns in dyslexic human subjects. Neurosci. Lett. **276:** 41–44.
24. BALDEWEG, T., A. RICHARDSON, S. WATKINS, *et al.* 1999. Impaired auditory frequency discrimination in dyslexia detected with mismatch evoked potentials. Ann. Neurol. **45:** 495–503
25. KUJALA, T., K. MYLLYVIITA, M. TERVANIEMI, *et al.* 2000. Basic auditory dysfunction in dyslexia as pinpointed by brain-activity measurements. Psychophysiology **37:** 262–266.
26. HARI, R., M. HÄMÄLÄINEN, R. ILMONIEMI, *et al.* 1984. Responses of the primary auditory cortex to pitch changes in a sequence of tone pips: neuromagnetic recordings in man. Neurosci. Lett. **50:** 127–132.
27. TIITINEN, H., K. ALHO, M. HUOTILAINEN, *et al.* 1993. Tonotopic auditory cortex and the magneto encephalographic (MEG) equivalent of the mismatch negativity. Psychophysiology **30:** 537–540.
28. WINKLER, I., M. TERVANIEMI, M. HUOTILAINEN, *et al.* 1995. From objective to subjective: sound representation in the human auditory cortex. Neuroreport **6:** 2317–2320.
29. KROPOTOV, J.D., R. NÄÄTÄNEN, A.V. SEVOSTIANOV, *et al.* 1995. Mismatch negativity to auditory stimulus change recorded directly from the human temporal cortex. Psychophysiology **32:** 418–422.

30. HALGREN, E., K. MARINKOVIC & P. CHAUVEL. 1997. Generators of the late cognitive potentials in auditory and visual oddball tasks. Electroencephalogr. Clin. Neurophysiol. **106:** 156–164.
31. TERVANIEMI, M., S.V. MEDVEDEV, K. ALHO, *et al.* 2000. Lateralized automatic auditory processing of phonetic versus musical information: a PET study. Hum. Brain Map. **10:** 74–79.
32. OPITZ, B., A. MECKLINGER, D.Y. VON CRAMON, *et al.* 1999. Combining electrophysiological and hemodynamic measures of the auditory oddball. Psychophysiology **36:** 142–147.
33. OPITZ, B., A. MECKLINGER, A.D. FRIEDERICI, *et al.* 1999. The functional neuroanatomy of novelty processing: integrating ERP and fMRI results. Cerebr. Cortex **9:** 379–391.
34. ALHO, K. 1995. Cerebral generators of mismatch negativity (MMN) and its magnetic counterpart (MMNm) elicited by sound changes. Ear Hear. **16:** 3–50.
35. GIARD, M.H., F. PERRIN, J. PERNIER, *et al.* 1990. Brain generators implicated in the processing of auditory stimulus deviance: a topographic event-related potential study. Psychophysiology **27:** 627–640.
36. RINNE, T., K. ALHO, R.J. ILMONIEMI, *et al.* 2000. Separate time behaviors of the temporal and frontal mismatch negativity sources. Neuroimage **12:** 14–19.
37. LEVÄNEN, S., A. AHONEN, R. HARI, *et al.* 1996. Deviant auditory stimuli activate human left and right auditory cortex differently. Cerebr. Cortex **6:** 288–296.
38. GIARD, M.H., J. LAVIKAINEN, K. REINIKAINEN, *et al.* 1995. Separate representation of stimulus frequency, intensity and duration in auditory sensory memory: an event-related potential and dipole-model analysis. J. Cognit. Neurosci. **7:** 133–143.
39. ALHO, K., M. TERVANIEMI, M. HUOTILAINEN, *et al.* 1996. Processing of complex sounds in the human auditory cortex as revealed by magnetic brain responses. Psychophysiology **33:** 369–375.
40. HÄMÄLÄINEN, M., R. HARI, R.J. ILMONIEMI, *et al.* 1993. Magnetoencephalography–theory, instrumentation, and applications to noninvasive studies of the working human brain. Rev. Mod. Phys. **65:** 413–497.
41. TERVANIEMI, M., A. KUJALA, K. ALHO, *et al.* 1999. Functional specialization of the human auditory cortex in processing phonetic and musical sounds: a magnetoencephalographic study. Neuroimage **9:** 330–336.
42. NÄÄTÄNEN, R., A. LEHTOKOSKI, M. LENNES, *et al.* 1997. Language-specific phoneme representations revealed by the electric and magnetic brain responses. Nature **385:** 432–434.
43. RINNE, T., K. ALHO, P. ALKU, *et al.* 1999. Hemispheric asymmetry of cortical activation as reflected by the mismatch negativity reveals when a sound is processed as speech. Neuroreport **10:** 1113–1117.
44. TERVANIEMI, M. & T.L. VAN ZUIJEN. 1999. Methodologies of brain research in cognitive musicology. J. New Music Res. **28:** 200–208.
45. RUGG, M.D. 1999. Functional neuroimaging in cognitive neuroscience. *In* The Neurocognition of Language. C.M. Brown & P. Hagoort, Eds.: 15–36. Oxford University Press. Oxford.
46. SHTYROV, Y., T. KUJALA, J. AHVENINEN, *et al.* 1998. Noise-induced shift in hemispheric lateralization of speech processing. Neurosci. Lett. **251:** 141–144.
47. SHTYROV, Y., T. KUJALA, R.J. ILMONIEMI, *et al.* 1999. Noise affects speech-signal processing differently in the cerebral hemispheres. Neuroreport **10:** 2189–2192.
48. HERRMANN, C.S., U. OERTEL, Y. WANG, *et al.* 2000. Noise affects auditory and linguistic processing differently: an MEG study. Neuroreport **11:** 227–229.
49. NORDBY, H., W.T. ROTH & A. PFEFFERBAUM. 1988. Event–related potentials to breaks in sequences of alternating pitches or interstimulus intervals. Psychophysiology **25:** 249–261.
50. ALAIN, C., D.L. WOODS & K.H. OGAWA. 1994. Brain indices of automatic pattern processing. Neuroreport **6:** 140–144.
51. TERVANIEMI, M., J. SAARINEN, P. PAAVILAINEN, *et al.* 1994. Temporal integration of auditory information in sensory memory as reflected by the mismatch negativity. Biol. Psychol. **38:** 157–167.
52. HARI, R., J. RIF, J. TIIHONEN, *et al.* 1992. Neuromagnetic mismatch fields to single and paired tones. Electroencephalogr. Clin. Neurophysiol. **82:** 152–154.

53. CSEPE, V., C. PANTEV, M. HOKE, *et al.* 1997. Mismatch field to tone pairs: neuromagnetic evidence for temporal integration at the sensory level. Electroencephalogr. Clin. Neurophysiol. **104:** 1–9.
54. SCHRÖGER, E. 2000. Top-down effects on auditory sensory memory processing. *In* Contributions to Psychological Acoustics, Results of the Eight Oldenburg Symposiums on Psychological Acoustics. A. Schick, M. Meis & C. Reckhardt, Eds.: 337–353. Bibliotheks und Informationssystem der Universität Oldenburg. Oldenburg, Germany.
55. YABE, H., M. TERVANIEMI, J. SINKKONEN, *et al.* 1998. The temporal window of integration of auditory information in the human brain. Psychophysiology **35:** 615–619.
56. SCHRÖGER, E., M. TERVANIEMI & R. NÄÄTÄNEN. 1995. Time-course of loudness in tone patterns is automatically represented by the human brain. Neurosci. Lett. **202:** 117–120.
57. SCHRÖGER, E., M. TERVANIEMI, C. WOLFF, *et al.* 1996. Preattentive periodicity detection in auditory patterns as governed by time and intensity information. Cognit. Brain Res. **4:** 145–148.
58. SAARINEN, J., P. PAAVILAINEN, E. SCHRÖGER, *et al.* 1992. Representation of abstract stimulus attributes in human brain. Neuroreport **3:** 1149–1151.
59. PAAVILAINEN, P., M. JARAMILLO, R. NÄÄTÄNEN, *et al.* 1999. Neuronal populations in the human brain extracting invariant relationships from acoustics variance. Neurosci. Lett. **265:** 179–182.
60. PAAVILAINEN, P., M. JARAMILLO & R. NÄÄTÄNEN. 1998. Binaural information can converge in abstract memory traces. Psychophysiology **35:** 483–487.
61. TERVANIEMI, M., S. MAURY & R. NÄÄTÄNEN. 1994. Neural representations of abstract features in the human brain as reflected by the mismatch negativity. Neuroreport **5:** 844–846.
62. WINKLER, I., W. TEDER, M. TERVANIEMI, *et al.* 2001. Event-related potential (ERP) correlates of unexpected jumps of a continuously moving auditory stimulus. In preparation.
63. PAAVILAINEN, P., M.L. KARLSSON, K. REINIKAINEN, *et al.* 1989. Mismatch negativity to changes in the spatial location of an auditory stimulus. Int. J. Psychophysiol. **7:** 342–343.
64. SCHRÖGER, E. & C. WOLFF. 1996. Mismatch response of the human brain to changes in sound location. Neuroreport **7:** 3005–3008.
65. WINKLER, I., M. TERVANIEMI, E. SCHRÖGER, *et al.* 1998. Processing of spatial cues used for auditory lateralization as revealed by the mismatch negativity. Neurosci. Lett. **242:** 49–52.
66. KOELSCH, S., T.C. GUNTER, A.D. FRIEDERICI, *et al.* 2000. Brain indices of music processing: "non-musicians" are musical. J. Cognit. Neurosci. **12:** 520–541.
67. HAHNE, A. & A.D. FRIEDERICI. 1999. Electrophysiological evidence for two steps in syntactic analysis: early automatic and late controlled processes. J. Cognit. Neurosci. **11:** 194–205.
68. PANTEV, C., R. OOSTENVELD, A. ENGELIEN, *et al.* 1998. Increased auditory cortical representation in musicians. Nature **392:** 811–814.
69. KOELSCH, S., E. SCHRÖGER & M. TERVANIEMI. 1999. Superior attentive and pre-attentive auditory processing in musicians. Neuroreport **10:** 1309–1313.
70. TERVANIEMI, M., V. JUST, S. KOELSCH, *et al.* 2001. The specificity of the facilitated pitch-discrimination accuracy evidenced in musicians. In preparation.

# How Many Music Centers Are in the Brain?

ECKART O. ALTENMÜLLER

*Institut für Musikphysiologie und Musiker-Medizin der Hochschule für Musik und Theater Hannover, 30175 Hannover, Germany*

**ABSTRACT: When reviewing the literature on brain substrates of music processing, a puzzling variety of findings can be stated. The traditional view of a left–right dichotomy of brain organization—assuming that in contrast to language, music is primarily processed in the right hemisphere—was challenged 20 years ago, when the influence of music education on brain lateralization was demonstrated. Modern concepts emphasize the modular organization of music cognition. According to this viewpoint, different aspects of music are processed in different, although partly overlapping neuronal networks of both hemispheres. However, even when isolating a single "module," such as, for example, the perception of contours, the interindividual variance of brain substrates is enormous. To clarify the factors contributing to this variability, we conducted a longitudinal experiment comparing the effects of procedural versus explicit music teaching on brain networks. We demonstrated that cortical activation during music processing reflects the auditory "learning biography," the personal experiences accumulated over time. Listening to music, learning to play an instrument, formal instruction, and professional training result in multiple, in many instances multisensory, representations of music, which seem to be partly interchangeable and rapidly adaptive. In summary, as soon as we consider "real music" apart from laboratory experiments, we have to expect individually formed and quickly adaptive brain substrates, including widely distributed neuronal networks in both hemispheres.**

KEY WORDS: **Music; Brain; Neuromusicology; Music centers**

## CHANGING CONCEPTS IN NEUROMUSICOLOGY

During the past two decades, the concepts of brain substrates underlying music processing have changed. Although never unequivocally supported by classical lesion studies,[1,2] traditional theories proposed a simple right-versus-left–hemisphere dichotomy, with music being processed in the right brain, language in the left. This simple viewpoint—still represented in many textbooks—could not be held any longer, when in 1974 Bever and Chiarello[3] were able to demonstrate the influence of professional training on hemispheric lateralization during music processing, nonmusicians exhibiting right, professionals left hemispheric preponderance. In the following years, results of several brain imaging studies supported this idea. In our

Address for correspondence: Prof. Dr. med. Eckart Altenmüller, MD, MA, Institut für Musikphysiologie und Musiker-Medizin der Hochschule für Musik und Theater Hannover, Plathnerstr. 35, 30175 Hannover, Germany. Voice: +49 (0) 511 3100 552; fax: +49 (0) 511 3100 557.
altenmueller@hmt-hannover.de

laboratory, for example, brain activation patterns were investigated during demanding harmonic and melodic discrimination tasks in a large group of nonmusicians, amateurs, and professional musicians. Professional musicians processed these tasks primarily in the left frontotemporal lobes, whereas amateurs as well as nonmusicians bilaterally activated the frontal lobes and the right temporal lobe.[4] It was assumed that professional musicians had access to different cognitive strategies ascompared to amateurs and nonmusicians. The left hemisphere activation in professionals was attributed to covert inner speech, since trained musicians—as a consequence of hundreds of lessons of ear-training and solfège—reported that they named the intervals and harmonies more or less automatically during processing of the task. In other words and speaking more generally, access to "auxiliary" representations of music acquired during training was discussed as accounting in part for the variability in brain activation patterns during music listening. Consequently, brain substrates of music processing were supposed to reflect the *way of listening and processing* rather than more-or-less fixed "music centers."

## LISTENING TO MUSIC: CONCEPTS OF PERCEPTIVE MODULES AND HIERARCHIES

Before considering brain substrates of music processing, we have to clarify what we term "music" in this context. To our understanding, music is not a mere acoustic structure in time, or a stimulus created in a laboratory to fit a well-controlled experimental design, but a phenomenon of subjective human experience. Such an experience is not based on a uniform mental capacity but on a complex set of perceptive and cognitive operations represented in the central nervous system. These operations act interdependently in some parts, independently in others. They are integrated in time and linked to previous experiences with the aid of memory systems, thus enabling us to perceive, or better, to "feel" a sort of meaning while listening. Neuromusicology has been profoundly influenced by the idea of the *modularity of musical functions*.[5,6] According to Fodor,[5] a module corresponds to a specialized computational device that is devoted to the execution of some biologically important function. Applied to music, this concept has been put forward by the groups of Isabelle Peretz and Robert Zatorre, demonstrating convincingly the neuropsychological fractionation of different musical subfunctions in patients following brain lesions.[7–12] Taking together the results of these studies, we see that a complex pattern of distinct dissociation syndromes with isolated loss of cognitive subunits of music processing following a lesion emerges. For example, there is evidence that separate modules are processing time or pitch structures of complex musical stimuli. Generally speaking, time structures seem to be processed to a greater extent in the left temporal lobe, whereas pitch structures may be processed primarily in right temporal lobe networks. According to recent results, a predominance of the posterior portions of the right supratemporal lobe for processing of pitch structures may exist.[13]

Such a modular organization concerning processing of segregated physical (temporal or pitch) properties of musical structures could account for the involvement of separated, in part, overlapping, neuronal substrates. However, the situation becomes more complex when one considers that perception of music may occur on different hierarchical levels. With respect to temporal structures, for example, two levels of organization may be distinguished: meter and rhythm. Rhythm is defined as the serial relation of durations between different acoustical events in a train of sounds—that is, rhythm represents a serial durational pattern—whereas meter involves a temporal invariance in terms of the regular recurrence of pulses or beats marking off equal durational units, which can be organized as measures. Meter therefore represents a more complex acoustical "gestalt," since its perception and production require information on sound intensity (accented and unaccented events) and on periodicity of rhythmical events, the latter based on integration of information over longer time periods. Again, by means of neuropsychological fractionation of perceptive subfunctions in patients with brain lesions, it was possible to isolate neuronal networks processing meter and rhythm, demonstrating spared metric judgment but disrupted rhythmic discrimination.[7] It is beyond the scope of the present contribution to discuss the complex issue of local versus global gestalt processing in the auditory domain. Whereas in the visual modality, a dissociation of neuronal substrates concerned with local (left parietal) or global (right parietal) processing can clearly be demonstrated,[14] in the auditory modality such a clear-cut neuroanatomical distinction is still under debate. This may be related to the methodological problem of fractionating time structures in appropriate perceptive units allowing for a clear and interindividually consistent discrimination between local or global processing units. In fact, most experimental designs have not been very convincing in excluding variable time fractionation based on rapidly changing individual auditory habits.

## "MUSIC CENTERS" IN THE BRAIN REFLECTING THE AUDITORY BIOGRAPHY

Although lesion studies conducted in larger groups of patients suffering from well-defined lesions support the idea of modularity and hierarchy and yield relatively clear results, it still remains a scientific challenge to further clarify the nature and function of these hypothesized modules and hierarchies and to delineate their degree of autonomy and specificity to music. Another, in our opinion, even more urgent problem arises when taking a look at the individual data. In a recently published study,[15] where procedures were performed on patients suffering from small unilateral ischemic lesions of the temporal, parietal, or frontal lobe, respectively, a surprising heterogeneity of the patterns of impairment with respect to modular subfunctions of auditory processing emerged. Whereas none of the patients had a deficit in basic auditory perception as, for example, in pitch discrimination, in some individuals right hemispheric lesions produced deficits in more complex temporal or pitch organization tasks, including in all instances, however, a combined deficit of local as well as global processing stages. Following left hemispheric stroke, surprisingly dissociated impairments of rhythm, meter, interval, or contour processing occurred, irrespective of whether the lesion was localized anterior or posterior to the central sulcus. In summary, it emerged that individually variable varieties of brain regions were necessary to ensure complex auditory functions, including parts of the posterior parietal lobes and the frontal lobes. It therefore becomes evident that not only the temporal lobes, as suggested in the comprehensive study of Liégois-Chauvel and coworkers,[13] but widespread and individually developed neuronal networks may underlie music processing. Without going into methodological details, it should be mentioned in this context that our patients were investigated relatively early—7 to 14 days following the ischemic stroke—before profound plastic changes compensating for impaired functions had reached their full extent. The discrepancies with earlier studies applying similar test batteries can in part be explained by the early timing of the investigation.

When considering individual factors as possible sources of variability in brain substrates of music processing, the next step is to further delineate the nature of these factors. In order to clarify whether the *way of learning music* plays an important role, we investigated the impact of music education on brain activation patterns in a group of 13- to 15-year-old students[16] in close cooperation with the music educator Wilfried Gruhn. The hypotheses were that (1) learning music and acquiring a new mental representation of music changes brain activation patterns while listening to music, and that (2) different ways of music learning may cause various mental representations that are reflected in different cortical activation patterns.

The task was to judge formal aspects of symmetrically structured phrases, so-called musical periods that consist of corresponding parts, "antecedent" and "consequent." Students had to distinguish between correct and incorrect (balanced or unbalanced) phrases. Whereas the antecedent phrase ends in a weak cadence on the dominant, suspending the expected tonic (half cadence), the consequent phrase leads to a stable ending on the tonic (perfect cadence). This different quality of cadences and the balance of the two melodic parts can easily be recognized merely by an internal feeling of musical balance and the tension of the cadence. For training, subjects were divided into three subgroups: (1) a "declarative" learner group that received traditional instructions about the antecedent and consequent and their tonal relationship with respect to the closing on a complete or incomplete cadence (the instructions included verbal explanations, visual aids, notations, verbal rules, and some musical examples that were played for the subjects, but never sung or performed); (2) a "procedural" learner group that participated in musical experiences for establishing genuine musical representations by singing and playing, improvising with corresponding rhythmic and tonal elements, or performing examples from the music literature; and (3) a control group of nonlearners who did not receive any instruction about or in music. Low-frequency DC shifts of the electroencephalogram (EEG) were measured prior to learning and after a five-week training period.

In FIGURE 1, the main results of the study are summarized. After learning, in the verbally trained "declarative" group, music processing produced an increased activation of the left frontotemporal brain regions, which probably reflects inner speech and analytical, step-by-step processing. By contrast, the musically trained "procedural" group showed increased activation of the right frontal and of bilateral parietooccipital lobes, which may be ascribed to a more global way of processing and to visuospatial associations. These results demonstrate for the first time directly that musical expertise influences auditory brain activation patterns and that changes in these activation patterns depend on the teaching strategies applied. In other words, *brain substrates of music processing reflect the auditory learning "biography," the personal experiences accumulated over time.* Listening to music, learning to play an instrument, formal instruction, and professional training result in multiple, in many instances multisensory, representations of music, which seem to be partly interchangeable and rapidly adaptive.

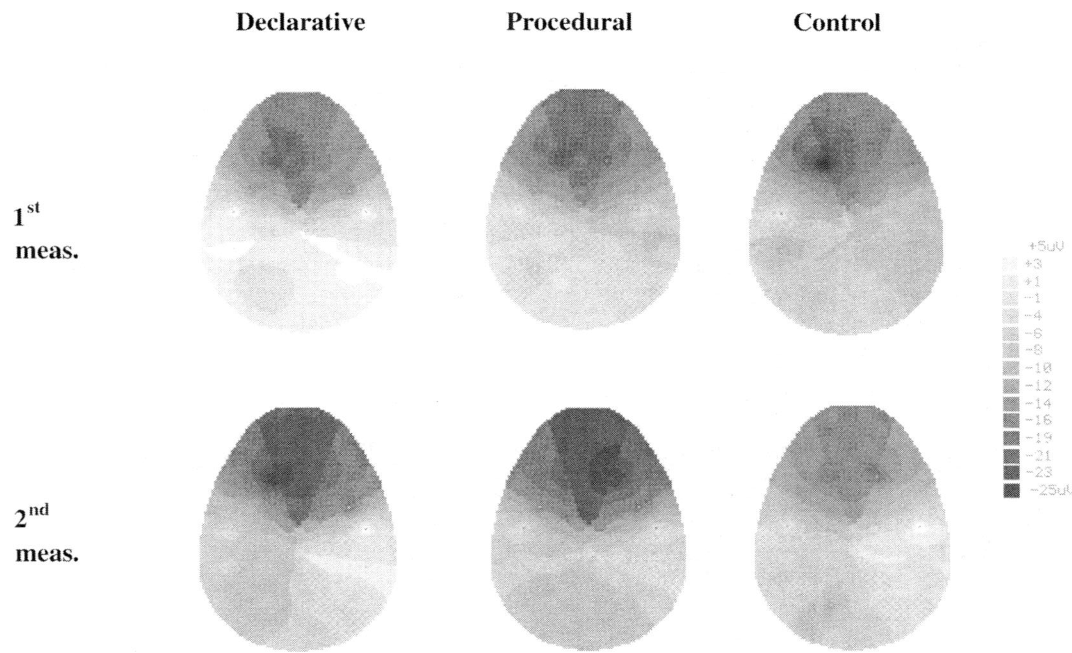


**FIGURE 1.** Brain maps demonstrating cortical activation patterns before (*upper row*) and after (*lower row*) learning in the "Declarative" learning group, in the "Procedural" learning group, and in the "Control" group. Group averages are displayed. Activation is *dark*, inactivation is *white* (see microvolt scale on the *right*). The brain diagrams are displayed as *top views*, frontal regions up, left hemisphere on the *left*, right hemisphere on the *right*. As can be recognized, declarative, mainly verbally mediated training leads to an increase in brain activity over the left frontal areas, whereas procedural, genuinely musical training produces an increase in activity over right frontal and bilateral parietooccipital regions. In controls, overall activity decreased slightly. (Modified from Altenmüller *et al.*[16])

## A TENTATIVE MODEL ON BRAIN SUBSTRATES OF MUSIC PROCESSING

It is not a particularly fruitful approach to look at brain substrates of music processing from such a solipsistic viewpoint, stating that presently around six billion different auditory biographies may produce the same number of different "music centers" in the brain. Many aspects of auditory processing in general and music processing in particular are necessarily bound to fixed neuronal substrates common to all humans and mainly located in the superior temporal gyrus. As a sort of division of labor, hemispheric specialization may take place at a very early stage of auditory processing: the left superior temporal gyrus seems to be specialized for very rapid processes requiring high temporal resolution such as the identification of different phonemes, whereas the right superior temporal lobe is specialized for spectral analyses of sound.[17] Even at such an early stage, however, the brain substrates underlying auditory processing remain adaptive and subject to plastic changes as a consequence of conditioning or training.[18,19]

In FIGURE 2, a tentative model illustrating the interrelationship between the complexity of neuronal networks involved in music processing ($y$-axis) and the complexity of auditory processing ($x$-axis) is outlined. An additional dimension can be added on the $z$-axis, accounting for effects of acculturation. The complexity of neuronal networks increases with the complexity of processing demands. Training and practice add additional mental representations of music that rely on different brain substrates. Therefore, professionals presumably use larger and more complex neuronal networks during music processing than do nonprofessionals. The actual network engaged in a defined processing task is not fixed, but is subject to short- and long-term plastic changes, allowing at early processing stages for adaptations and compensations in case of damage to certain brain substrates. For example, a bilateral lesion of the primary auditory cortices with complete cortical deafness may be compensated for.[20] The vertical axis of the small cross "*P.*" (for *p*lasticity) on the left symbolizes such a reciprocal replacement of brain functions by other brain structures. The horizontal axis relates to the fact that listeners may add or reduce complexity of auditory processing by adapting their listening strategies. When entering more advanced processing stages, including many neuropsychological laboratory experiments, the learning biography, determining the multiplicity of different auditory representations may influence the actual network used for music processing. According to actual demands or as a reaction to brain lesions, processing strategies may lead to a simplified or to a more complex way to listen to the stimuli or music, respectively. Learning biography (*L.B.*) and auditory strategies are symbolized by the larger cross on the right in the diagram. For the sake of clarity, other variables influencing brain activation patterns during music processing have been omitted: guidance of attention, emotions, working memory, procedural and explicit short- and long-term memory systems, and effects of acculturation play an important role, rendering the brain substrates of music processing still more complex.

In summary, as soon as we consider "real music" apart from laboratory experiments, we have to expect individually formed and quickly adaptive brain substrates, including widely distributed neuronal networks in both hemispheres. In our laboratories, we are just beginning to face the enormous challenges linked to the clarification of rules determining this puzzling variety of findings, determining the

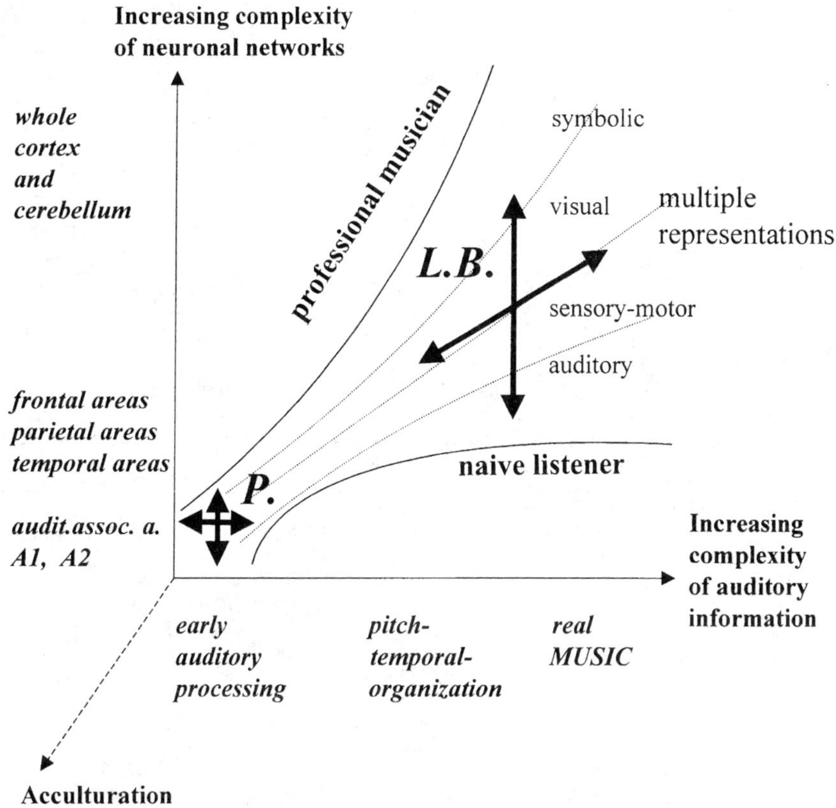


**FIGURE 2.** A simplified model demonstrating the interdependency between increasing complexity of auditory information processing ($x$-axis) and increasing complexity of neuronal networks ($y$-axis). *P.* = plasticity, *L.B.* = learning biography. *A1*, *A2*, and *audit. assoc. a* stand for primary auditory cortex, secondary auditory cortex, and auditory association cortices. For further explanation, see text.

complexity and transitoriness of neuronal interactions during music processing. Two hundred years ago, the German poet Johann Wolfgang von Goethe recognized and expressed this problem in a masterful way in the "scene in the laboratory:"

> *Homunculus*: That is the way that things are apt to take:
> The cosmos scarce will compass nature's kind
> But man's creations need to be confined.
>
> (Johann Wolfgang von Goethe, *Faust*, Part II, Act II. Scene in the Laboratory. Translated by Philip Wayne.)

## REFERENCES

1. Henschen, S.E. 1920. Über Amusie. *In* Klinische und anatomische Beiträge zur Pathologie des Gehirns. Vol. 5: 137–213. Nordiska Bokhandeln. Stockholm.

2. UVSTEDT, H.J. 1937. The method of examination in amusia. Acta Psychiatr. Neurol. **12:** 447–455.
3. BEVER, T.G. & R.I. CHIARELLO. 1974. Cerebral dominance in musicians and non-musicians. Science **185:** 537–540.
4. ALTENMÜLLER, E. 1989. Cortical DC-potentials as electrophysiological correlates of hemispheric dominance of higher cognitive functions. Int. J. Neurosci. **47:** 1–14.
5. FODOR, G.A. 1983. The Modularity of Mind. MIT Press. Cambridge, MA.
6. GARDNER, H. 1983. Frames of Mind. The Theory of Multiple Intelligences. Basic Books. New York.
7. PERETZ, I. 1990. Processing of local and global musical information by unilateral brain-damaged patients. Brain **113:** 1185–1205.
8. PERETZ, I. 1993. Auditory agnosia: a functional analysis. *In* Thinking in Sound: The Cognitive Psychology of Human Audition. S. McAdams & E. Bigand, Eds.: 199–230. Clarendon Press. Oxford.
9. PERETZ, I., R. KOLINSKY, M. TRAMO, *et al.* 1994. Functional dissociations following bilateral lesions of auditory cortex. Brain **117:** 1283–1301.
10. PERETZ, I. & L. GAGNON. 1999. Dissociation between recognition and emotional judgements for melodies. Neurocase **5:** 21–30.
11. ZATORRE, R.J. & S. SAMSON. 1991. Role of right temporal neocortex in retention of pitch in auditory short-term memory. Brain **114:** 2403–2417.
12. ZATORRE, R.J., A.C. EVANS & E. MEYER. 1994. Neural mechanisms underlying melodic perception and memory for pitch. J. Neurosci. **14:** 1908–1919.
13. LIÉGEOIS-CHAUVEL, C., I. PERETZ, M. BABAÏ, *et al.* 1998. Contribution of different cortical areas in the temporal lobes to music processing. Brain **121:** 1853–1867.
14. HEINZE, H.J., G.R. MANGUN, W. BURCHERT, *et al.* 1994. Combined spatial and temporal imaging of brain activity during visual selective attention in humans. Nature **372:** 543–546.
15. SCHUPPERT, M., T.F. MÜNTE, B.M. WIERINGA, *et al.* 2000. Receptive amusia: evidence for cross-hemispheric neural networks underlying music processing strategies. Brain **123:** 546–559.
16. ALTENMÜLLER, E., W. GRUHN, D. PARLITZ, *et al.* 1997. Music learning produces changes in brain activation patterns: a longitudinal DC-EEG-study. Int. J. Arts Med. **5:** 28–34.
17. ZATORRE, R.J. 2000. Neural specialization for tonal processing. Abstract of this conference.
18. PANTEV, C., A. WOLLBRINK, L.E. ROBERTS, *et al.* 1999. Short-term plasticity of the human auditory cortex. Brain Res. **842:** 192–199.
19. WEINBERGER, N.M. & J.S. BAKIN. 1998. Learning-induced physiological memory in adult primary auditory cortex: receptive field plasticity, model and mechanisms. Audiol. Neuro-Otol. **3:** 145–167.
20. TRAMO, M.J., J.J. BHARUCHA & F. MUSIEK. 1989. Music perception and cognition following bilateral lesions of auditory cortex. J. Cognit. Neurosci. **2:** 195–212.

# The Brain of Musicians

## A Model for Functional and Structural Adaptation

GOTTFRIED SCHLAUG

*Department of Neurology, Beth Israel Deaconess Medical Center and Harvard Medical School, Boston, Massachusetts 02215, USA*

**ABSTRACT: Musicians form an ideal subject pool in which one can investigate possible cerebral adaptations to unique requirements of skilled performance as well as cerebral correlates of unique musical abilities such as absolute pitch and others. There are several reasons for this. First, the commencement of musical training usually occurs when the brain and its components may still be able to adapt. Second, musicians undergo long-term motor training and continued practice of complicated bimanual motor activity. Third, imaging studies from our group as well as other groups have shown that motor learning and the acquisition of skills can lead to changes in the representation of motor maps and possibly also to microstructural changes. Whether the unique musical abilities and structural differences that musicians' brains show are due to learning, perhaps during critical periods of brain development and maturation, or whether they reflect innate abilities and capacities that might be fostered by early exposure to music is largely unknown. We will report studies that indicate that certain regions in the brain (corpus callosum, motor cortex, cerebellum) may show some form of adaptation to extraordinary challenges and requirements of performance. These challenges may eventually lead to functional and structural cerebral changes to accommodate the requirements for musical performance. Furthermore, we will also show the neural correlates of one unique musical ability, absolute pitch. This ability may be linked to one structure in the human brain (planum temporale), which is preferentially activated in musicians who have absolute pitch during tone tasks. This structure may undergo some form of functional plasticity that is possible only during a critical period of brain development.**

**KEYWORDS: Brain; Music; Development; Motor learning; Absolute pitch**

### INTRODUCTION

Musicians form an ideal subject pool in which one can investigate adaptations to unique requirements of skilled performance as well as cerebral correlates of unique musical abilities such as absolute pitch or musical sight-reading. The most important aspect of using musicians as a model for functional and structural adaptation of the brain due to extraordinary challenges is that the exposure to music and the commencement of musical training usually occurs when the brain and its components

Address for correspondence: Gottfried Schlaug, M.D., Ph.D., Department of Neurology, Beth Israel Deaconess Medical Center and Harvard Medical School, 330 Brookline Ave., Boston, MA 02215. Voice: 617-667-8202; fax: 617-667-8695.
gschlaug@caregroup.harvard.edu

may still be able to adapt to these challenges. This allows one to examine various aspects of skill learning and acquisition as well as functional and structural changes in the brains of musicians as a result of their unique training and motor experiences. Musicians exhibit some unique abilities such as the ability to memorize long and complex bimanual finger sequences, to translate musical symbols into motor sequences during sight-reading, and to perceive and identify tones absolutely in the absence of a reference tone (absolute pitch). Whether these abilities can be learned, perhaps during critical periods of brain development and maturation, or whether they reflect innate abilities and capacities that might be fostered by early exposure to music is largely unknown. We are just beginning to understand what the cerebral correlates of these abilities are.

In this paper, we will report studies that indicate that certain regions in the brain may show some form of adaptation to extraordinary challenges. These challenges may eventually lead to functional and structural changes to accommodate the requirements for musical performance. However, we will also show that there are other regions that, although they are the neural substrate for some musical abilities, may undergo functional adaptations during critical periods of development but not necessarily plastic structural changes. These regions could be the mediators of certain functions or abilities due to their preexisting size or hemispheric asymmetry.

## THE PLASTIC BRAIN

The unique training and motor experiences of musicians provides an ideal experimental design to investigate whether previously shown functional cerebral adaptations in response to skill learning or sensory stimulation are correlated with long-term micro- and macrostructural cerebral changes. These changes may reflect adaptations among others due to the continual practice of complicated bimanual motor skills. Functional reorganization of adult mammalian sensory and motor cortical representations after peripheral or central stimulation or as an adjustment after injury has been found in many different experimental animal models of brain plasticity.[1–5] Similar adaptive changes in the cortical organization after skill learning and as an adjustment after injury have been found in humans using electrophysiological or neuroimaging methods.[6–17] These findings have advanced the understanding that functional properties of central nervous system neurons as well as the neural circuitry either within the same or different brain areas are malleable and retain a significant degree of functional plasticity, which could lead to microstructural changes. The term *plasticity* is broad and can mean an adjustment or adaptation of a sensory or motor system to environmental stimuli or performance requirements or a compensation of some cerebral structures for others that are impaired due to injury or deafferentiation.[18–20]

Three particular human studies are relevant to those we carried out on structural adaptation that show differences in the sensorimotor representation maps as a response to skill learning and acquisition of new skills. Pascual-Leone *et al.*[10] showed that as subjects learned a five-finger exercise on the piano over the course of five days, the cortical representation area targeting the long finger flexor and extensor muscles enlarged. Karni *et al.*[14] showed that a few minutes of daily practice of a sequential finger opposition task induced large, incremental performance gains over a few weeks of training, which was associated with changes in cortical movement representation within the primary motor cortex. These authors argued that the changes in the primary motor cortex reflected the setting up of a task-specific motor processing routine, while the subsequent changes indicated the consolidation of a motor program. Although several reports have stressed the rapid reversal of representational changes, other studies have found persistent representational changes in response to the early acquisition of fine sensorimotor skills such as having a larger sensory finger representation in the left hand of string players.[21]

The underlying neurophysiological changes of such functional adaptations are still a matter of intense research. Adaptations could include a strengthening of existing synapses, the formation of new synapses, or the recruitment of cortical tissue into the activated cortex that was previously not recruited by changing the local balance of excitation and inhibition. On the other hand, it may well be a combination of all of these mechanisms.[1–4,9–11,13,19,20,22]

In a series of experiments conducted over the last 10 years, several groups have shown that significant microstructural changes can be induced in motor-related brain regions as a consequence of intense and prolonged motor activity. Most of these studies have focused on the cerebellum and have used adult rats as experimental animal models.[23–25] Results from these studies indicate that acrobatic training or complex motor learning result in an increase in the number of synapses per Purkinje cell as well as an increase in the number of glial cells per neuron in the cerebellar cortex. In conditions of long-term vigorous physical activity without motor skill learning (running on a treadmill) the main finding is the formation of new blood capillaries and an insignificant difference in the number of synapses per neuron.[23,24] Similar results have been obtained with the motor cortex. An increased number of synapses per neuron are found after motor skill learning[26] as well as after long-term stimulation of thalamic afferents to the motor cortex.[27] Of interest is also that the hippocampus can contain an increased number of neurons in adult mice living in an enriched environment compared to a control group.[28]

One of our main interests has been to determine whether the sum of these use-dependent microstructural changes, including changes in the number of synapses, microglia, and capillaries, can lead to volumetric changes detectable on a macrostructural level using magnetic resonance imaging (MRI) methods.

## THE MUSICIAN BRAIN: A PROTOTYPICAL MODEL FOR REGIONAL STRUCTURAL BRAIN PLASTICITY

### *Subjects*

All our study participants were either right-handed, classically trained professional musicians or nonmusicians. A professional musician was defined as someone who was either enrolled as a student in a full-time music program in music school or music conservatory, or someone who was a graduate of a music program and had his or her main income derived from a professional career in music. The majority of musicians in our studies were keyboard players. We also had a subgroup of string players, who were mostly keyboard players as well. For our studies, we contrasted different groups of musicians with groups of nonmusicians who were matched for age, gender, and handedness. A nonmusician was defined as someone who did not have any formal training in music and never played a musical instrument for any reasonable period of time. Hand preferences were typically assessed with the 12-item Annett questionnaire.[29] Consistent right-handedness corresponded to performance of all 12 tasks with the right hand with up to two "either hand" preferences being acceptable.[30–33] In addition to hand preference, we also assessed distal hand motor skills using an index finger tapping test.[34] A laterality coefficient was calculated according to the following formula (R–L)/(R+L), where L (left) and R (right) were the number of finger taps with either hand within 20 seconds. A subgroup of our subjects also underwent a battery of behavioral and cognitive tests, including subsets taken from the Wechsler Adult Intelligence Scale tests (WAIS-R[35]), as well as other tests of verbal intelligence (Shipley-Hartford Scale) and visual-spatial tests such as a mental rotation task and the block design test from theWAIS-R[35] inventory, as well as other memory tests and tests of attention.

### *Neuroimaging Studies and Image Analysis*

Neuroimaging studies were performed using a whole-body Siemens 1.5 T MR machine acquiring volumetric T1-weighted brain images with a typical voxel resolution of 1.0 × 1.0 × 1.0 millimeters. Functional MR imaging (fMRI) was done using a T2*-weighted MR sequence that was sensitive to changes in the local concentrations of oxy- and deoxyhemoglobin. Changes in the relative concentrations of oxy-/deoxyhemoglobin are indirect markers of regional blood flow changes that again are indirect markers of changes in regional neuronal synaptic activity. Analysis of fMRI data was done using SPM99, AFNI2.1, and custom-made software. All morphometric analyses were done using custom-made software running on HP workstations and mostly implemented in the Advanced Visual System (AVS) image analysis package. Morphometric studies were typically done by two independent investigators who were blinded to the identity as well as to the hemisphere of each brain. Interobserver correlations for the anatomical measurements were usually higher than 0.9.

### *The Corpus Callosum*

The morphometry of the corpus callosum is of particular interest for studies examining brain asymmetry and interhemispheric exchange for several reasons. First, the corpus callosum is the main interhemispheric fiber tract and plays an important role in interhemispheric integration and communication. Group differences in callosal size or shape observed in morphometric studies are generally regarded as neuroanatomical substrates of differences in cerebral asymmetry and interhemispheric connectivity.[36–38] Second, there is evidence that the functional maturation and possibly structural maturation of the corpus callosum extends into late childhood and early adolescence and coincides with the termination of its myelination cycle.[39] This is supported by *in vivo* imaging, which revealed that increases in the midsagittal callosal size can be seen even beyond the first decade with a maximum change in size during the first decade of human life (TABLE 1).[40,41] Third, there is a general consensus that movement control and motor coordination as well as intermanual transfer of sensorimotor information improves gradually from ages 4 to 11 years, an age span coinciding with callosal maturation.[42–45] Fourth, a positive correlation between the

**TABLE 1. Midsagittal area measurements of the corpus callosum (CC) in mm$^2$ (mean ± SD)**

| | Total CC Area | Anterior CC Area[a] | Posterior CC Area |
|---|---|---|---|
| All musicians ($n$ = 30) | 687 ± 85 | 371 ± 46 | 314 ± 43 |
| Musicians with commencement of musical training ≤7 years of age | 709 ± 81 | 384 ± 42 | 321 ± 44 |
| Musicians with commencement of mussical training >7 years of age | 637 ± 77 | 340 ± 43 | 297 ± 38 |
| Nonmusician controls ($n$ = 30) | 649 ± 88 | 344 ± 48 | 305 ± 43 |

[a]Significant differences are those between controls and all musicians, between controls and musicians with early commencement of musical training, and between the two subgroups of musicians with or without early commencement of musical training.

midsagittal callosal area and the number of fibers crossing through the corpus callosum has been established.[46] Our main hypothesis was that early and intensive training in keyboard and string players and the requirement for increased and faster interhemispheric exchange in order to perform bimanual complex motor sequences might lead to structural changes in the callosal anatomy. In examining 30 professional musicians and 30 age-, sex-, and handedness-matched nonmusician controls, we found that the anterior half of the corpus callosum was significantly larger in musicians ($p$ = 0.031). A MANOVA comparing subgroups of musicians with (≤7 years) and without early commencement of musical training and controls indicated group differences ($p$ = 0.009). Post hoc tests revealed a significantly larger anterior corpus callosum in musicians with early commencement compared to musicians starting later and compared to controls (FIG. 1).

The differences in callosal size might be due either to more fibers crossing through the corpus callosum, to a larger proportion of thicker myelinated fibers with faster interhemispheric transfer, or to fibers with thicker axons or more axon collateral. Previous anatomical studies found a positive correlation between midsagittal callosal size and the number of fibers crossing through the corpus callosum. The anterior part of the corpus callosum contains mainly fibers from frontal motor-related regions and prefrontal regions,[47] and the anterior corpus callosum matures the latest of all callosal subregions. Therefore, this anatomical difference in the midsagittal area of the corpus callosum has to be seen in the context for a requirement for increased interhemispheric communication subserving complex bimanual motor sequences in musicians. It could have been triggered by performing and continuously practicing complicated and independent bimanual finger movements. Our results indicate that environmental factors might play a role in the determination of callosal fiber composition and fiber size.

### *The Motor Cortex*

As reviewed above, there is much evidence supporting the notion that plastic changes can be induced in the functional organization of the human sensorimotor cortex following sensory stimulation or following the acquisition of new motor skills. These functional changes after skill acquisition may be related to microstructural

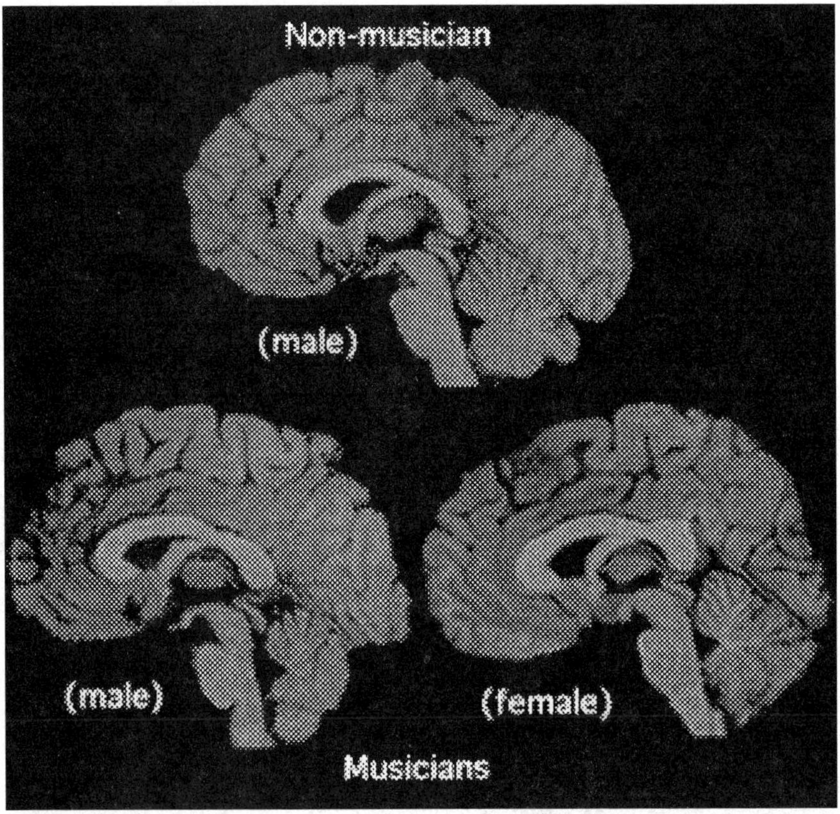


**FIGURE 1.** Post hoc tests revealing a significantly larger anterior corpus callosum in musicians with early commencement compared to musicians starting later and compared to controls.

changes such as increased numbers of synapses per neuron, increased numbers of glial cells per neuron, and/or more capillaries as has been shown in animal experiments.[26] Therefore, we asked a fundamental question whether consistent and in most cases daily practice of complicated bimanual finger sequences led to macrostructural changes in the human motor cortex if this training occurred during a critical period of brain development. Previously, we and others described macrostructural correlates of handedness in the primary motor cortex.[48,49] Since the exact histological extents of the primary motor cortex on MR images could not be defined, we used gross anatomical markers that corresponded to anatomical boundaries of the primary motor cortex. We used the intrasulcal length of the posterior bank of the precentral gyrus (ILPG) as a gross anatomical marker of the human primary motor cortex. All MR data sets were coded and, in a random manner, half of the MR data sets were mirrored in the horizontal plane, so that neither subject identification nor side-of-hemisphere was known to the examiner during the measurements. All brains were spatially oriented and normalized to the coordinate system of Talairach and

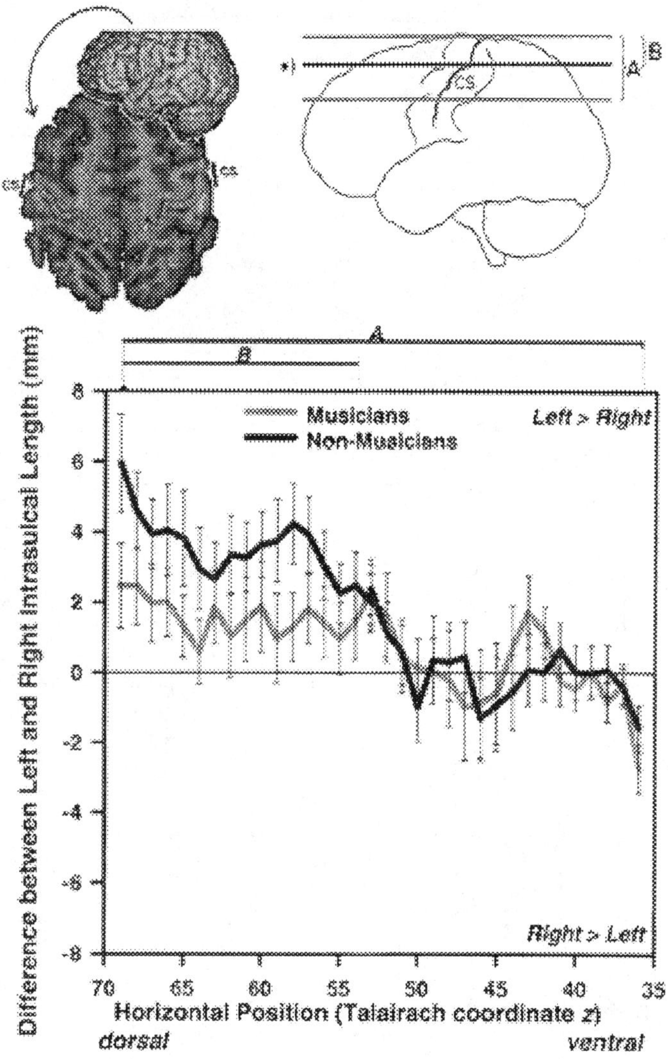


**FIGURE 2.** The intrasulcal length of the posterior bank of the precentral gyrus (ILPG) measured in horizontal slices (parallel to the AC–PC plane) from the deepest point of the central sulcus following the contour line of the posterior bank of the precentral gyrus to a lateral surface tangent that connected the crests of the pre- and postcentral gyrus.

Tournoux.[50] The ILPG was measured in horizontal slices (parallel to the AC–PC plane) from the deepest point of the central sulcus following the contour line of the posterior bank of the precentral gyrus to a lateral surface tangent that connected the crests of the pre- and postcentral gyrus. All horizontal slices from Talairach coordinates z = 69 to z = 35 were included in the analysis (FIG. 2).

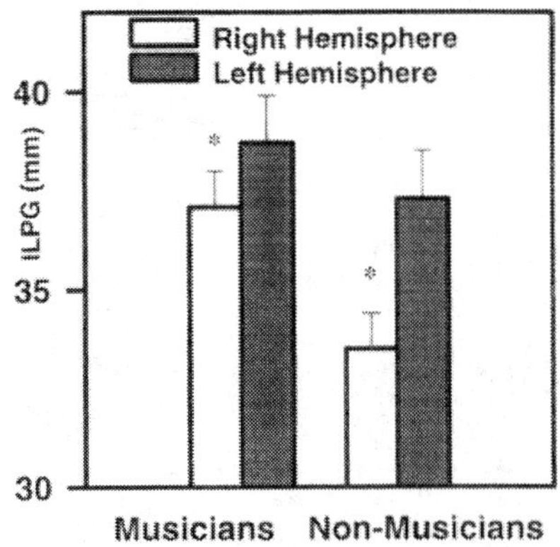


**FIGURE 3.** Pairwise multiple tests showing a significantly greater right intrasulcal length in musicians than in controls, while there were no significant between-group differences in this measure for the left hemisphere.

In each horizontal section, the total intrasulcal length of each hemisphere as well as an asymmetry score of both hemispheric measures were determined. In order to strengthen the statistical analysis, two mean asymmetry scores were computed, one comprising the entire set of horizontal sections and one only a dorsal subregion. The extent of this dorsal subregion was determined based on prior fMRI data reporting the strongest activity changes for hand/finger movements between Talairach coordinates $z = 69$ to $z = 55$.[51] For the asymmetry scores of regions A and B, significant leftward asymmetries were evident, both for musicians and controls. Leftward asymmetry was most pronounced in the dorsal subregion (region B). The absolute means of the left and right ILPG of this region (region B) for each subject were subjected to a two-way analysis of variance with musicians and controls as factors and right and left hemisphere measurements as repeated measure. The test revealed a significant interaction ($F = 5.02$, $p < 0.05$) as well as main effects ($F = 28.5$, $p < 0.0001$ for repeated measurements factor). Subsequent pairwise multiple tests showed a significant greater right intrasulcal length in musicians than in controls, while there were no significant between-group differences in this measure for the left hemisphere (FIG. 3). Thus, a greater symmetry of the ILPG in musicians was mainly due to a larger size of the ILPG in the right hemisphere controlling the nondominant left hand. These results could be explained by a simple scaling effect, since absolute brain volumes did not show a significant different for musicians versus controls in this study. The reduced asymmetry of the ILPG was paralleled by a reduced hand skill asymmetry using the tapping test in musicians.

In order to examine whether the mean individual ILPG was related to the age of commencement of musical training, correlation analyses were performed. We found strong correlations between the time at which musical training had begun and right and left mean ILPG (Spearman rank order correlation; $r = -0.63$ for the right hemisphere, and $r = -0.60$ for the left hemisphere, both $p < 0.01$; FIG. 4).

These data serve as evidence for macrostructural differences in a gross morphological marker of the human primary motor cortex between musicians and nonmusi-

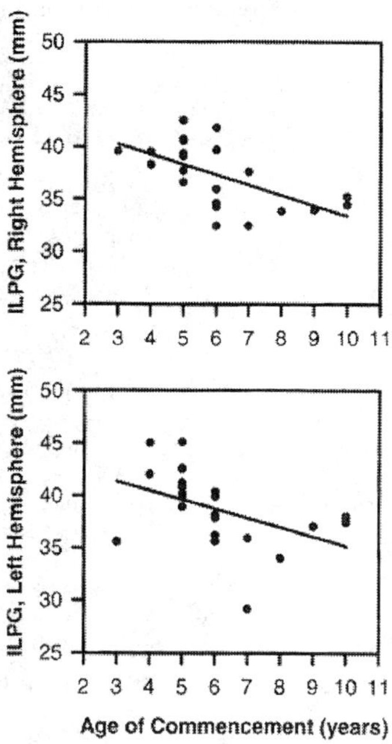


**FIGURE 4.** Correlations between the time at which musical training had begun and right and left mean ILPG. (See text for details.)

cians. Although we do not have proof that musical training induced these changes, the correlation between early commencement of musical training and pronounced structural differences points in the direction that these differences evolved over time. Similarly, we found evidence that the primary motor cortex exhibited macrostructural interhemispheric differences that were related to handedness.[52] Handedness is regarded as the most obvious functional asymmetry that develops between the second and third year of life. The alternative explanation is of course that all of this is hardwired, and musicians—in particular keyboard players—are preselected by having a more symmetrical anatomical organization in their hand representation. These individuals with larger motor cortices and less interhemispheric asymmetry of their motor cortex could excel in motor skill performance and surpass competitors "suffering" from a more asymmetric or smaller motor cortex, or both. However, the finding of an association between early commencement of musical training and ILPG lends support to the hypothesis of a training-induced anatomical plasticity. This explanation concurs with several functional studies showing experience-related reorganization of representational motor maps.[10–14]

### *The Cerebellum*

Although the cerebellum comprises only a tenth of the whole brain volume, the number of cells in the human cerebellar cortex exceeds the total number of cells in the cerebral cortex by four times.[53] The only outputs from the cerebellar cortex are

**TABLE 2. Morphometric data on cerebellar volume (means ± SD) in musicians and nonmusicians**

| | %CV | aCV (in cc) |
|---|---|---|
| Male musicians ($n$ = 32) | 10.30 (0.64) | 145.3 (9.7) |
| Male nonmusicians ($n$ = 24) | 9.85 (0.68)* | 139.6 (15.4) |
| Female musicians ($n$ = 24) | 10.43 (0.65) | 134.7 (12.1) |
| Female nonmusicians ($n$ = 15) | 10.43 (0.82) | 131.8 (12.9) |
| All males ($n$ = 56) | 10.11 (0.69) | 142.8 (12.6) |
| All females ($n$ = 34) | 10.43 (0.72) | 133.3 (12.3) |

ABBREVIATIONS: %CV = % cerebellar volume of total brain volume; aCV = absolute cerebellar volume in cubic centimeters.

[a]Significant differences between the groups of male nonmusicians and musicians in relative cerebellar volume.

Purkinje cells, which project through inhibitory connections to the deep cerebellar nuclei. There are two major inputs to the cerebellum in the form of climbing fibers from the inferior olivary nuclei and mossy fibers from different brain nuclei and the spinal cord. Due to these unique connections and the abundance of neurons, the cerebellum has attracted widespread attention for its role in motor learning as well as in other cognitive processes. As reviewed above, several experimental studies have already provided much evidence that microstructural changes (e.g., increased synaptic density, glial cells, and capillaries) can be observed after intense and prolonged motor activity. These findings as well as the role that the cerebellum plays in movement coordination, timing of sequential movements, and possibly other cognitive functions, prompted us to examine whether the cerebellum shows structural differences between adult musicians and nonmusicians.

In order to take statistical advantage of an increased number of subjects, we conducted this study retrospectively using all available MRI data from musicians and nonmusicians having participated in previous anatomical and functional MRI studies. A total of 51 classically trained, professional musicians (32 male, 19 female) and a total of 39 nonmusicians (24 male, 15 female) were available for analysis. All subjects were right-handed. The male group was matched relatively well according to hand performance data and body height, while the female subgroup was much smaller and not matched as well, exhibiting a 3 cm group mean difference in body height. All MR images were segmented into brain and nonbrain tissue using an interactively controlled procedure. The cerebellum was segmented from the brain stem and cerebellar peduncles using a priori established anatomical criteria. The interobserver reliability for cerebellar volume was 0.92 for a randomly selected set of 15 cases. Only one observer analyzed all 90 cases blinded to the identity of each case.

In order to compensate for the high intersubject variability in cerebellar volume, absolute cerebellar volume was normalized as a percentage of the total brain volume. Results of this study are summarized in TABLE 2. One of the main findings of the study was that there was a significant gender effect, with female subjects having a higher relative cerebellar volume (10.47% vs. 10.11%, $p$ <0.02), although males

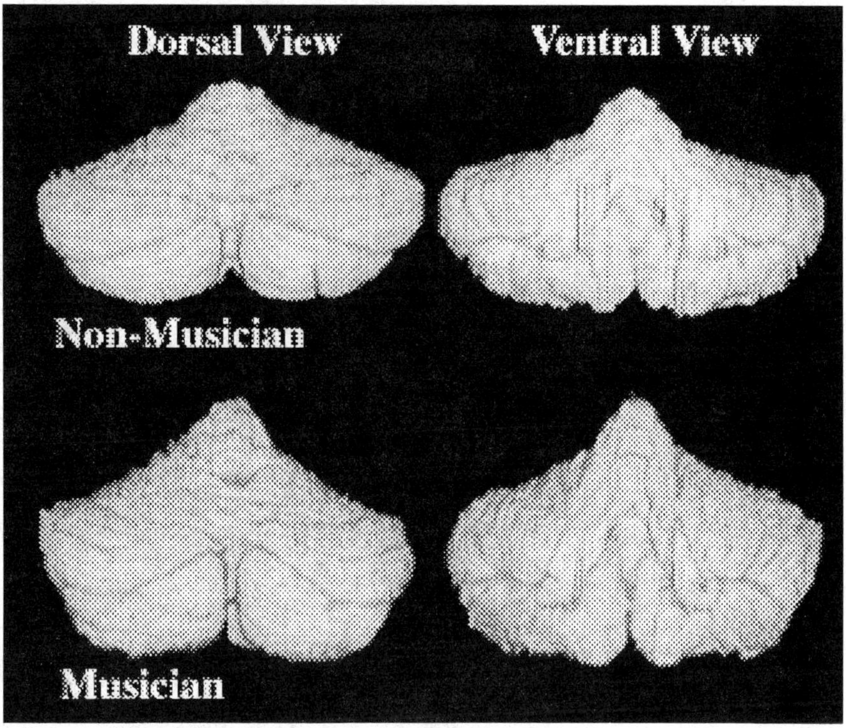


**FIGURE 5.** Relative volume difference of about 5% in the cerebellum comparing male musicians with nonmusicians.

showed a trend for a higher mean absolute cerebellar volume and total brain volume. The second main finding was that male musicians had a significantly higher mean relative cerebellar volume compared to male nonmusicians ($F(1,54) = 6.51$, $p = 0.014$), which was not due to a difference in total brain volume, since male musicians and male nonmusicians did not show a significant difference in total brain volume in this study ($F(1,54) = 0.0308$, $p = 0.86$). No significant difference in relative cerebellar size was found when female musicians ($n = 19$) were compared with their nonmusician counterparts ($n = 15$) ($F(1,32) = 0.14$, $p = 0.71$). The third finding of this study was that there was a positive trend between intensity of musical training (practice time per day and across lifetime) and relative cerebellar volume.[54]

The results in the male subgroup are interpreted as evidence for microstructural adaptations in the human cerebellum in response to early commencement and continual practice of complicated bimanual finger sequences. This interpretation is based mainly on animal experiments that showed evidence for microstructural changes (i.e., increased synaptic density, increased number of glial cells and capillaries) in rats exposed to continuous motor activity and complex motor learning tasks.[23,25,26] These microstructural changes may amount to structural differences detectable at a macroscopic level. The absence of an observable effect in the female subgroup may be due to a smaller group size and a less well-matched sample; in ad-

**TABLE 3. Means (± SD) for degree of anatomical planum temporale asymmetry (δPT), and size of left and right PT in right-handed AP musicians, non-AP musicians, and nonmusicians**

| | | PT size (mm$^2$) | |
|---|---|---|---|
| Subject | δPT | Left | Right |
| AP musicians ($n$ = 27) | –0.50 (0.27)[a] | 1381 (449) | 822 (236) |
| Non-AP musicians ($n$ = 24) | –0.24 (0.14) | 1350 (340) | 1062 (267) |
| Nonmusicians ($n$ = 27) | –0.28 (0.24) | 1341 (306) | 008 (285) |

[a]Significant differences between AP musicians and non-AP musicians as well as between AP-musicians and nonmusicians.

dition, many of the female nonmusicians were athletes. Also, there seems to be synaptic up- and downregulation during the female menstrual cycle, which was not taken into account in this study.[55] Nevertheless, the relative volume difference between male musicians and nonmusicians was about 5%, which is a remarkable difference in size and could be further evidence for the plastic properties of the central nervous system, particularly of the cerebellum (FIG. 5).

## EVIDENCE FOR FUNCTIONAL ADAPTATION IN THE MUSICIAN BRAIN: THE PLANUM TEMPORALE AND ABSOLUTE PITCH

The planum temporale (PT) has long been used as a marker of hemispheric asymmetry and cerebral dominance (TABLE 3).[56,57] It has been found that the majority of right-handers have a leftward planum temporale asymmetry, whereas the majority of left-handers have either a symmetric planum temporale or show a rightward asymmetry.[57] The PT has been taken as a structural marker for left hemispheric language dominance in right-handers. However, the planum temporale is not only a structural marker of left-hemispheric dominance for language, it is also involved in auditory processing and as such is of great interest for studies investigating laterality of auditory processing. Recent studies have found associations between PT surface area and functional dominance during story listening.[58] Neural systems within the superior part of the temporal lobe are involved in the perceptual analysis of musical stimuli.[59,60] Anatomical and functional studies have suggested an anterior–posterior intrahemispheric gradient of specialization in the superior temporal lobe.[60]

In an anatomical MRI study designed to test the hypothesis that musicians might have a different degree of hemispheric dominance than nonmusicians, we found that musicians differed significantly from nonmusicians by having an increased left-sided asymmetry of the planum temporale. The surprising finding was that a subgroup of musicians, those with absolute pitch, explained all the difference between the musician and nonmusician group. The absolute pitch musician subgroup showed an increased left-sided asymmetry of the planum temporale.[31] The difference in planum temporale asymmetry between musicians with and without absolute pitch has been replicated in the meantime using a slightly different morphometric technique.[61]

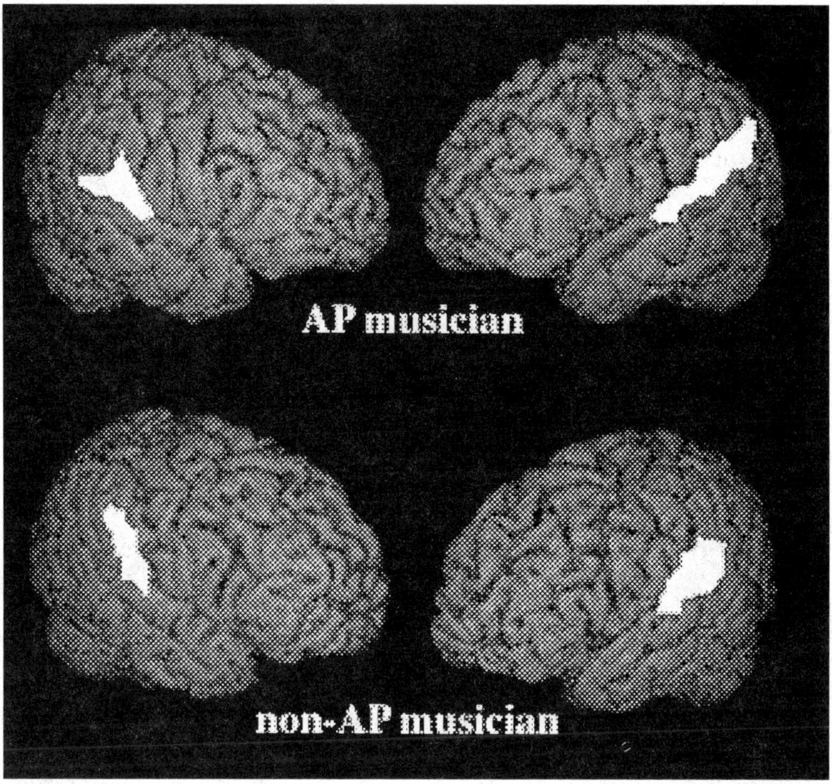


**FIGURE 6.** Significantly greater leftward asymmetry of the planum temporale of AP musicians when compared to non-AP musicians.

More recently, we have replicated our own finding and have significantly increased our sample size.[62] This allowed us also to examine whether there was a difference between the PT asymmetry comparing two groups of musicians with each other that differed in their absolute pitch ability, but that had a similar early commencement of musical training. We confirmed our previous finding of an increased left-sided PT asymmetry in musicians who have absolute pitch (AP musicians) that was not seen in the control group of non-AP, early beginning musicians (FIG. 6).

As an advance on our previous studies,[31] we estimated the surface of the PT using splines instead of polygons to segment this region in sagittal sections, and we did a surface mesh interpolation between sections to estimate the true surface area instead of just adding up the distance and multiplying it by slice thickness as we had done previously. This explains the larger absolute left and right PT size compared to our previous publications.[31] The PT asymmetry index (δPT) in this new group of AP musicians was almost identical to what we reported previously.

From our own sample of AP musicians and other published reports,[63,64] it appears that two factors are important in determining the acquisition of AP. The first is age: It is extremely seldom that someone develops absolute pitch if they do not start musical training or are not exposed to music before the age of seven. In our own sample of now more than 50 AP musicians (right- and left-handers), only one AP musician started after the age of 7. The second factor is the increased left-sided PT asymmetry. Although the functional significance of the increased left-sided PT asymmetry in AP musicians is not clear, it should be seen in the context of their ability to assign any pitch to a verbally labeled pitch class in the absence of a reference tone.[65] Siegel[66] demonstrated the influence of possession of verbal labels on recognition memory for pitch. In her study, AP subjects were able to assign different verbal labels to tones that belonged to different pitch classes, resulting in a better performance than non-AP subjects. Comparing different tones that belonged to the same pitch class did not result in a performance difference between AP and non-AP subjects, since both groups were supposedly using a sensory coding strategy. These results could be taken as evidence for the categorical nature of absolute pitch.[67]

With reference to Siegel's[66] observation, we designed an fMRI experiment that aimed to investigate the functional significance of the increased left-sided PT asymmetry in AP by contrasting two conditions with each other, a tone task and a phoneme task. In the tone condition, subjects heard a string of sine wave tones (500 msec long with an attack and decay period of 30 msec). A two-alternative forced-choice paradigm was used requiring the subjects to indicate whether each tone was the "same" or "different" compared to two tones before. The second task was a phonological task using five different consonant-vowel syllables presented in the same way as the tone task. Again, a two-alternative forced-choice paradigm was used requiring subjects to make a decision whether each of the currently presented phonemes were the "same" or "different" than two phonemes before. We had two control conditions. In the first control condition, subjects listened to the regular scanner noise and alternated button press responses between the right and left hand after each MR sound at about the same speed as during the tone and phoneme presentation. In the second control condition, subjects listened to white noise bursts (500 msec long with an attack and decay period of 30 msec) and alternated button press responses between the right and left hand after each noise burst. In addition, we had a "label" condition in which AP musicians covertly categorized each tone according to pitch class.

We hypothesized that AP musicians would perceive tones as well as phonemes categorically. The left superior temporal lobe might play a role in this and might actually be the mediator of categorical perception. Non-AP control subjects would only perceive phonemes categorically, but not tones.

The selected statistical procedure was the two-factor 3D ANOVA implemented in the AFNI 2.1 image analysis software package with the first factor representing the three groups of subjects (AP musicians, non-AP musicians, nonmusicians) and the second factor the three types of stimulus pairs (noise vs. rest, tones vs. rest, phonemes vs. rest). This procedure analyzed the variances of the coefficients of general linear models of our data. These coefficients represented degrees of activations with respect to a boxcar stimulus presentation. This analysis generated a systematic and exhaustive list of results from $F$-tests, $t$-tests on group means, and $t$-tests on contrasts.

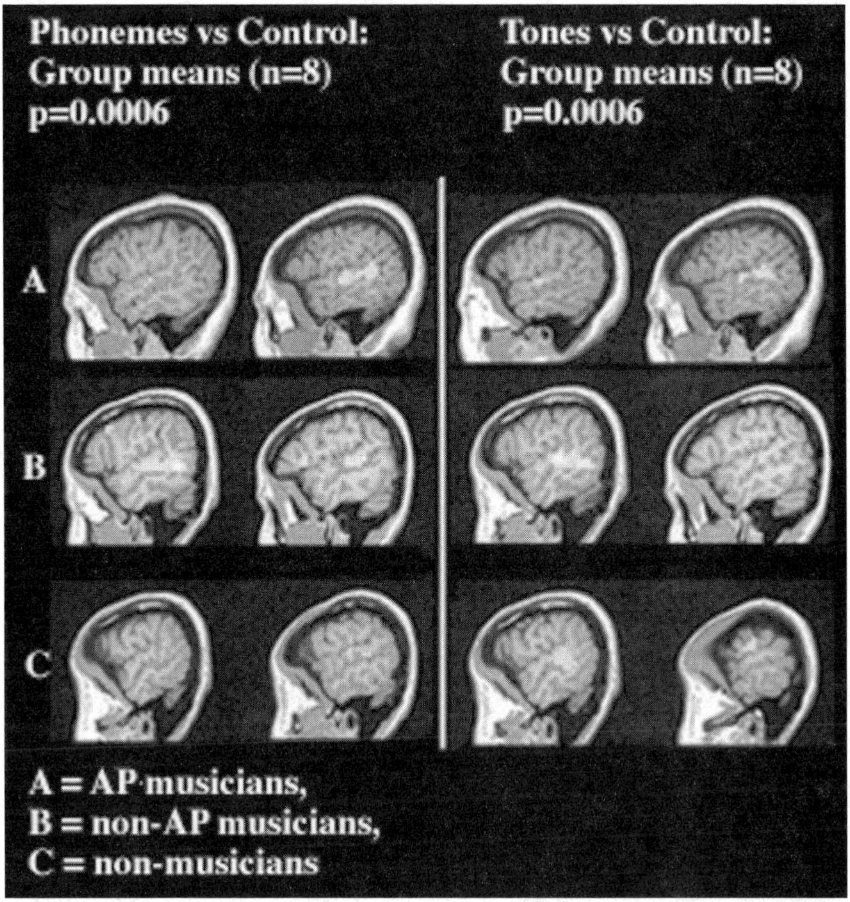


**FIGURE 7. See Plate 19 in color section.** Representative results of an fMRI experiment aimed to investigate the functional significance of the increased leftsided PT asymmetry in AP musicians by contrasting two conditions with each other, a tone task and a phoneme task. The left column of each block is a representative slice taken from the right hemisphere. The right column of each block is a representative slice taken from the left hemisphere.

Representative results of this analysis taken from the group mean data are presented in FIGURE 7. AP musicians, non-AP musicians, and nonmusicians showed pronounced left-sided superior temporal lobe activation in the phonemes vs. control task comparison. Similar activations were seen in the tones condition for the AP musicians. However, the non-AP musicians as well as the nonmusicians showed either symmetric or right-sided activation of the superior temporal lobe in the tones task. We argued that this is evidence that in AP musicians, the perception of phonemes and tones is mediated by the same left-hemispheric superior temporal lobe structures.

More recently Pantev and colleagues[68] found evidence for increased auditory cortical representation of piano tones in highly skilled musicians that was not seen for pure tones. There was a correlation between the enhanced representation of tones in the auditory cortex and age of commencement of musical training. No significant difference was seen between musicians with and without absolute pitch. This may have been due to the fact that mainly the N1, which occurs up to 100 ms after tone onset, was examined. In a subsequent report, a functional difference was found between musicians with and without absolute pitch. AP musicians showed a distinct activity more posterior in the left posterior superior temporal lobe than in the nonmusicians.[69] This was interpreted as a result of cortical plasticity in AP musicians, either induced by training and/or inherent due to structural differences.

## SUMMARY

It is our hope that the material presented in this paper provides some evidence for the plastic nature of the human brain. The musician is apparently an ideal model to investigate functional and structural adaptation of the motor and auditory system. Experimental animal studies strongly support the existence of microstructural changes in the number of synapses per neuron, glial cells counts, and capillary density. We have further provided evidence that certain musical abilities have neural correlates and that the asymmetry or dominance of certain homologue structures can have functional implications, for example, the left planum temporale or, more generally, the left superior temporal lobe might be the mediator for a very specific kind of perception, categorical perception, which may underly absolute pitch perception.

## ACKNOWLEDGMENTS

We would like to thank all the subjects, musicians, and nonmusicians, who participated in these various studies and allowed us and others to learn more about the brain that makes music. We also acknowledge the enormous contributions that have been made during various parts of these ongoing studies by my former mentor Helmuth Steinmetz as well as by my friends, colleagues, and co-workers: Lutz Jaencke, Katrin Amunts, Ben Martin Bly, Andrea Halpern, Michael Charness, and Ven Thangaraj. Furthermore, this work would have not been possible without the invaluable help, dedication, and efforts of my students Leslie Lee, Maria Picone, Katrin Schulze, and Eileen Lüders. I am also very thankful to my wife, Lisa Jones, for her constant support and her many invaluable contributions to this work.

## REFERENCES

1. MERZENICH, M.M., G. RECANZONE, W.M. JENKINS, *et al.* 1988. Cortical representational plasticity. *In* Neurobiology of Neocortex. P. Rakic & W. Singer, Eds.: 41–47. John Wiley & Sons Limited. New York.
2. RECANZONE, G.H., T.T. ALLARD, W.M. JENKINS, *et al.* 1990. Receptive-field changes induced by peripheral nerve stimulation in S1 of adult cats. J. Neurophysiol. **63:** 1213–1225.
3. JACOBS, K.M. & J.P. DONOGHUE. 1991. Reshaping the cortical motor map by unmasking latent intracortical connections. Science **251:** 944–947.

4. NUDO, R.J., W.M. JENKINS, M.M. MERZENICH, *et al.* 1992. Neurophysiological correlates of hand preference in primary motor cortex of adult squirrel monkeys. J. Neurosci. **12:** 2918–2947.
5. WANG, X., M.M. MERZENICH, K. SAMESHIMA, *et al.* 1995. Remodeling of hand representation in adult cortex determined by timing of tactile stimulation. Nature **378:** 71–75.
6. COHEN, L.G., S. BANDINELLI, T.W. FINDELEY, *et al.* 1991. Motor reorganization after upper limb amputation in man. Brain **114:** 615–627.
7. SEITZ, R.J., P.E. ROLAND, C. BOHM, *et al.* 1990. Motor learning in man: a positron emission tomographic study. Neuroreport **1:** 57–60.
8. JENKINS, I.H., D.J. BROOKS, P.D. NIXON, *et al.* 1994. Functional reorganization of primary somatosensory cortex in adult owl monkeys after behaviorally controlled tactile stimulation. J. Neurosci. **14:** 3775–3790.
9. PASCUAL-LEONE, A., J. GRAFMAN & M. HALLETT. 1994. Modulation of cortical motor output maps during development of implicit and explicit knowledge. Science **263:** 1287–1289.
10. PASCUAL-LEONE, A., N. DANG, L.G. COHEN, *et al.* 1995. Modulation of muscle responses evoked by transcranial magnetic stimulation during the acquisition of new fine motor skills. J. Neurophysiol. **74:** 1037–1044.
11. PASCUAL-LEONE, A., E.M. WASSERMANN, N. SADARO, *et al.* 1995. The role of reading activity on the modulation of motor cortical outputs to the reading hand in Braille readers. Ann. Neurol. **38:** 910–915.
12. SCHLAUG, G., U. KNORR & R. SEITZ. 1994. Inter-subject variability of cerebral activations in acquiring a motor skill: a study with positron emission tomography. Exp. Brain Res. **98:** 523–534.
13. KARNI, A., G. MEYER, P. JEZZARD, *et al.* 1995. Functional MRI evidence for adult motor cortex plasticity during motor skill learning. Nature **377:** 155–158.
14. KARNI, A., G. MEYER, C. REY-HIPOLITO, *et al.* 1998. The acquisition of skilled motor performance: fast and slow experience-driven changes in primary motor cortex. Proc. Natl. Acad. Sci. USA **95:** 861–868.
15. SEITZ, R.J., P. HOFLICH, F. BINKOFSKI, *et al.* 1998. Role of premotor cortex in recovery from middle cerebral artery infarction. Arch. Neurol. **55:** 1081–1088.
16. HUND-GEORGIADIS, M. & Y. VON CRAMON. 1999. Motor-learning-related changes in piano players and non-musicians revealed by functional magnetic-resonance signals. Exp. Brain Res. **125:** 417–425.
17. CHARNESS, M. & G. SCHLAUG. 2000. Cortical activation during finger movements in concert pianists, dystonic pianists, and non-musicians. Neurology **54:** A221.
18. ZILLES, K. 1992. Neuronal plasticity as an adaptive property of the central nervous system. Ann. Anat. **174:** 383–391.
19. HALLETT, M. 1995. The plastic brain. Ann. Neurol. **38:** 4–5.
20. RAUSCHECKER, J.P. 1995. Compensatory plasticity and sensory substitution in the cerebral cortex. Trends Neurosci. **18:** 36–43.
21. ELBERT, T., C. PANTEV, C. WIENBRUCH, *et al.* 1995. Increased cortical representation of the fingers of the left hand in string players. Science **270:** 305–307.
22. NUDO, R.J., G.W. MILLIKEN, W.M. JENKINS, *et al.* 1996. Use-dependent alterations of movement representations in primary motor cortex of adult squirrel monkeys. J. Neurosci. **16:** 785–807.
23. BLACK, J.E., K.R. ISAACS, B.J. ANDERSON, *et al.* 1990. Learning causes synaptogenesis, whereas motor activity causes angiogenesis, in cerebellar cortex of adult rats. Proc. Natl. Acad. Sci. USA **87:** 5568–5572.
24. ISAACS, K.R., B.J. ANDERSON, A.A. ALCANTARA, *et al.* 1992. Exercise and the brain: angiogenesis in the adult rat cerebellum after vigorous physical activity and motor skill learning. J. Cereb. Blood Flow Metab. **12:** 110–119.
25. ANDERSON, B.J., X. LI, A.A. ALCANTARA, *et al.* 1994. Glial hypertrophy is associated with synaptogenesis following motor-skill learning, but not with angiogenesis following exercise. Glia **11:** 73–80.
26. KLEIM, J.A., E. LUSSNIG, E.R. SCHWARZ, *et al.* 1996. Synaptogenesis and FOS expression in the motor cortex of the adult rat after motor skill learning. J. Neurosci. **16:** 4529–4535.

27. KELLER, A., A. KOSTADINK & H. ASANUMA. 1992. Synaptic proliferation in the motor cortex of adult cats after long-term thalamic stimulation. J. Neurophysiol. **68:** 295–308.
28. KEMPERMANN, G., H.G. KUHN & F.H. GAGE. 1997. More hippocampal neurons in adult mice living in an enriched environment. Nature **386:** 493–495.
29. ANNETT, M. 1970. A classification of hand preference by association analysis. Br. J. Psychol. **61:** 303–321.
30. SCHLAUG, G., L. JÄNCKE, Y. HUANG, *et al.* 1995. Increased corpus callosum size in musicians. Neuropsychologia **33:** 1047–1054.
31. SCHLAUG, G., L. JÄNCKE, Y. HUANG, *et al.* 1995. *In vivo* evidence of structural brain asymmetry in musicians. Science **267:** 699–701.
32. JAENCKE, L. 1996. The hand performance test with a modified time limit instruction enables the examination of hand performance asymmetries in adults. Percept. Mot. Skills **82:** 735–738.
33. JAENCKE, L., G. SCHLAUG & H. STEINMETZ. 1997. Hand skill asymmetry in professional musicians. Brain Cognit. **34:** 424–432.
34. PETERS, M. & B.M. DURDING. 1978. Handedness measured by finger tapping: a continuos variable. Can. J. Psychol. **32:** 257–261.
35. WECHSLER, D. 1981. Wechsler Adult Intelligence Scale—Revised. The Psychological Corporation. New York.
36. O'KUSKY, J., E. STRAUSS, B. KOSAKA, *et al.* 1988. The corpus callosum is larger with right-hemisphere cerebral speech dominance. Ann. Neurol. **24:** 379–383.
37. WITELSON, S.F. & D.L. KIGAR. 1992. Sylvian fissure morphology and asymmetry in men and women: bilateral differences in relation to handedness in men. J. Comp. Neurol. **323:** 326–340.
38. STEINMETZ, H., L. JAENCKE, A. KLEINSCHMIDT, *et al.* 1992. Sex but no hand difference in the isthmus of the corpus callosum. Neurology **42:** 749–752.
39. YAKOVLEV, P.I. & A.R. LECOURS. 1967. The myelogenetic cycles of regional maturation of the brain. *In* Regional Development of the Brain in Early Life. A. Minkowski, Ed.: 3–70. Blackwell. Oxford.
40. COWELL, P.E., L.S. ALLEN, N.S. ZLATIMO, *et al.* 1992. A developmental study of sex and age interactions in the human corpus callosum. Dev. Brain Res. **66:** 187–192.
41. PUJOL, J., P. VENDRELL, C. JUNQUÉ, *et al.* 1993. When does human brain development end? Evidence of corpus callosum growth up to adulthood. Ann. Neurol. **34:** 71–75.
42. DENCKLA, M. 1973. Development of motor co-ordination in normal children. Dev. Med. Child Neurol. **16:** 729–741.
43. HICKS, J.A. 1930. The acquisition of motor skills in young children. Child Dev. **1:** 90–103.
44. KERR, R. 1975. Movement control and maturation in elementary-grade children. Percept. Mot. Skills **41:** 151–154.
45. MÜLLER, K. & V. HÖMBERG. 1992. Development of speed of repetitive movements in children is determined by structural changes in corticospinal efferents. Neurosci. Lett. **144:** 57–60.
46. ABOITIZ, F., A.B. SCHEIBEL, R.S. FISHER, *et al.* 1992. Fiber composition of the human corpus callosum. Brain Res. **598:** 143–153.
47. PANDYA, D.N. & B. SELTZER. 1986. The topography of commissural fibers. *In* Two Hemispheres—One Brain: Functions of the Corpus Callosum. F. Lepore, M. Ptito & H.H. Jasper, Eds.: 47–73. Alan R. Liss. New York.
48. WHITE, L.E., G. LUCAS, A. RICHARDS, *et al.* 1994. Cerebral asymmetry and handedness. Nature **368:** 197–198.
49. AMUNTS, K., G. SCHLAUG, A. SCHLEICHER, *et al.* 1996. Asymmetry in the human motor cortex and handedness. Neuroimage **4:** 216–222.
50. TALAIRACH, J. & P. TOURNOUX. 1988. Coplanar Stereotaxic Atlas of the Human Brain. Thieme. Stuttgart.
51. SCHLAUG, G., J.N. SANES, V. THANGARAJ, *et al.* 1996. Cerebral activation covaries with movement rate. Neuroreport **7:** 879–883.
52. AMUNTS, K., G. SCHLAUG, L. JÄNCKE, *et al.* 1997. Motor cortex and hand motor skills: structural compliance in the human brain. Hum. Brain Mapping **5:** 206–215.

53. ANDERSEN, B.B., L. KORBO & B. PAKKENBERG. 1992. A quantitative study of the human cerebellum with unbiased stereological techniques. J. Comp. Neurol. **326:** 519–560.
54. SCHLAUG, G., L.H.L. LEE, V. THANGARAJ, *et al.* 1998. Macrostructural adaptation of the cerebellum in musicians. Soc. Neurosci. **24:** 842.7.
55. WOOLLEY, C.S. & B.S. MCEWEN. 1992. Estradiol mediates fluctuation in hippocampal synapse density during the estrous cycle in the adult rat. J. Neurosci. **12:** 2549–2554.
56. GESCHWIND, N. & W. LEVITZKY. 1968. Human brain: left–right asymmetries in temporal speech region. Science **161:** 186–187.
57. STEINMETZ, H. & R.J. SEITZ. 1991. Functional anatomy of language processing: neuroimaging and the problem of individual variability. Neuropsychologia **29:** 1149–1161.
58. TZOURIO, N., B. NKANGA-NGILA & B. MAZOYER. 1998. Left planum temporale surface correlates with functional dominance during story listening. Neuroreport **9:** 829–833.
59. ZATORRE, R.J., A.C. EVANS & E. MEYER. 1994. Neural mechanisms underlying melodic perception and memory for pitch. J. Neurosci. **14:** 1908–1919.
60. RAUSCHECKER, J.P., B. TIAN & M. HAUSER. 1995. Processing of complex sounds in the macaque non-primary auditory cortex. Science **268:** 111–114.
61. ZATORRE, R.J., D.W. PERRY, C.A. BECKETT, *et al.* 1998. Functional anatomy of musical processing in listeners with absolute pitch and relative pitch. Proc. Natl. Acad. Sci. USA **95:** 3172–3177.
62. CHEN, C., A. HALPERN, B.M. BLY, *et al.* 2000. Planum temporale asymmetry and absolute pitch. Neuroimage **11:** S114.
63. SERGEANT, D. 1969. Experimental investigation of absolute pitch. J. Res. Music Ed. **17:** 135–143.
64. BAHARLOO, S., P.A. JOHNSTON, S.K. SERVICE, *et al.* 1998. Absolute pitch: an approach for identification of genetic and nongenetic components. Am. J. Hum. Genet. **62:** 224–231.
65. TAKEUCHI, A.H. & S.H. HULSE. 1993. Absolute pitch. Psychol. Bull. **113:** 345–361.
66. SIEGEL, J.A. 1974. Sensory and verbal coding strategies in subjects with absolute pitch. J. Exp. Psychol. **103:** 37–44.
67. RAKOWSKI, A. 1993. Categorical perception in absolute pitch. Arch. Acoust. **18:** 515–523.
68. PANTEV, C., R. OOSTENVELD, A. ENGELIEN, *et al.* 1998. Increased auditory cortical representation in musicians. Nature **392:** 811–814.
69. HIRATA, Y., S. KURIKI & C. PANTEV. 1999. Musicians with absolute pitch show distinct neural activities in the auditory cortex. Neuroreport **10:** 999–1002.

# Representational Cortex in Musicians

## Plastic Alterations in Response to Musical Practice

C. PANTEV,[a] A. ENGELIEN,[a,b] V. CANDIA,[c] AND T. ELBERT[c]

[a]*Institute for Experimental Audiology, University of Münster, Münster, Germany*

[b]*Functional Neuroimaging Laboratory, Weill Medical College of Cornell University, New York, New York USA*

[c]*Department of Psychology, University of Konstanz, Konstanz, Germany*

**ABSTRACT: The lifelong ability to adapt to environmental needs is based on the capacity of the central nervous system for plastic alterations. In a series of neurophysiological experiments, we studied the impact of music and musical training in musicians on the specific functional organization in auditory and somatosensory representational cortex. In one such study, subjects listened to music from which one specific spectral frequency was removed. This led to rapid and reversible adaptation of neuronal responses in auditory cortex. Further experimental evidence demonstrated that long years of practice and training by professional musicians to enable them to reach their capacity is associated with enlarged cortical representations in the somatosensory and auditory domains. This tuning of neuronal representations was specifically observed for musical tones and was absent when pure sinusoidal tones were used as stimuli. In the somatosensory cortex, plastic changes proved to be specific for the fingers frequently used and stimulated. These changes were not detected in the fingers of the hand that were not involved in playing the particular instrument. Neuroplastic alterations also may be driven into a domain where they may become maladaptive. The clinical syndrome of focal hand dystonia that may occur in musicians who engage in forceful practice may be one such consequence. We will discuss the possibilities of reversing maladaptive responses leading to the successful treatment of focal hand dystonia, which relies on basic research about cortical reorganization. This example elucidates how neuroscientific progress can guide the development of practice guidelines and therapeutic measures for the benefit of professional musicians.**

## INTRODUCTION

The structural and functional organization of the human brain becomes increasingly differentiated during child development. In higher mammals, including humans, neurons are formed prenatally as are some of their interconnections into neural networks. For many years the prevailing opinion suggested that network connections between neurons are built primarily during cerebral maturation processes in childhood, with the exception of only those structures that were directly involved

Address for correspondence: Cristo Pantev, Ph.D., D.Sc., Professor, Head MEG Unit, Rotman Research Institute for Neuroscience, Baycrest Centre for Geriatric Care, 3560 Bathurst Street, Toronto, Ontario, Canada M6A 2E1. Voice: 416-785-2500 ext. 2690; fax: 416-785-2862.

antev@uni-muenster.de

in memory. It was thought that this network pattern, almost like a connection diagram, would not change later. However, humans respond with considerable flexibility to new challenges throughout their entire life. Since the early 1980s, increasing experimental evidence demonstrated that the connectivity of the adult brain is in fact only partially determined by genetics and early development, and may be substantially modified through sensory experiences, even during adulthood.

Practicing and performing music entails intense auditory and somatosensory peripheral sensory excitations, which are transmitted via specific receptors and fiber systems to the corresponding specialized regions of the cerebral cortex. Because the functional organization of representational cortex has been intensely studied, the examination of sensory cortical areas in the somatosensory and auditory systems provides an excellent model for studying the plastic changes that are associated with being a musician. The sensory cortex of both systems has a known topographical order of neuronal representations: a homuncular mapping of the body surface on somatosensory cortex[1–3] and a tonotopic mapping of acoustic frequency in auditory cortex.[4–7] Thus, modifications of representations can be specifically assessed. For example, different fingers of a given hand may not be equally used (and thus not equally stimulated) when playing certain musical instruments.

Based on W. James's suggestion at the end of the nineteenth century that learning may alter synaptic connectivity, the prominent Canadian neuropsychologist Donald Hebb formulated an important and innovative theory to view the brain not as a static but as a dynamic system. Hebb's rule asserts that effective connections between neurons are formed depending on synchronous activation: "Cells that fire together, wire together."[8] The development of new scientific methods for recording neuronal activity has made it possible to prove the hypothesis of plasticity in functional neuronal networks. The electrical activity of single neurons can only be recorded invasively in animals, but derived potentials that reflect the activity of a group of neurons can be recorded noninvasively on the scalp surface by means of electroencephalography (EEG). The magnetic counterpart of the EEG is the magnetoencephalography (MEG), which has become an established method for noninvasive study of the activity of the human cortex.[9] The main sources of cortical evoked magnetic fields are the pyramidal cells, which produce currents flowing tangentially to the surface of the head. Though MEG measurements provide only a macroscopic view of the function of the brain, the spatial resolution achieved with this technique is sufficient to give indications of functional organization and reorganizational plasticity of the human cortex by localizing the sources of evoked magnetic fields, which are elicited by defined and standardized peripheral excitation.

The first studies to clearly demonstrate reorganizational plasticity of the adult cortex were performed in the 1980s in "deafferentation" experiments. Typically, in these studies the afferent information influx specific to certain cortical areas was eliminated.[10–15] One mechanism of change observed was that neurons that had lost their regular input were recruited by neighboring regions. For example, neurons that were specialized for processing information from the fifth digit (little finger) may process information coming in on the second digit after the fifth digit was amputated. Similarly, neurons that were specific to a certain acoustic frequency may respond to neighboring spared frequencies after the part of the cochlea transmitting their originally preferred input was destroyed.

Experimental studies in monkeys have demonstrated that intensive sensory stimulation may lead to an expansion of the corresponding cortical area.[16,17] Here we briefly review the experimental paradigm for one such study in the somatosensory modality concerning finger representation in adult monkeys. The animals were trained to touch a rotating disk with the tip of their index and middle fingers for 15 seconds. The surface of that disk was an irregular grid, so that it caused a sensation (and somatosensory excitation) specifically for those two fingertips. Anytime the monkeys touched the disk for 15 s, they received a reward (chips with banana taste). After about 600 stimulation periods, the representation of the hand in the somatosensory cortex was measured. The representation for the two stimulated fingers had become enlarged. Recanzone *et al.* trained owl monkeys for 60–80 daily sessions to make fine-pitch discriminations in selected regions of the auditory frequency spectrum (these regions differing among animals).[17] Tonotopic mapping carried out invasively afterward showed that the cortical area tuned to the trained frequencies was enlarged by a factor of 2 to 3 compared to untrained monkeys or to animals that experienced the same acoustic stimuli passively while being trained on a somatosensory discrimination task. Thus, the organization of the brain seems capable of significant change to adapt to the changing demands of the organism's environment. Synchronized sensory stimulation may be of particular importance for such changes. The common observation captured in the German proverb *Übung macht den Meister* [practice makes perfect] could therefore have its neurobiological correlate in an augmented simultaneous stimulation of neurons, which entrains a reorganization of the functional neuronal network.

Musicians practice and train for many years before they achieve their professional skills. Coordination and synchronization of somatosensory and motor control on the one hand, and audition on the other hand, are crucial when playing a musical instrument. We therefore hypothesized that musical aptitude and the training to fulfill it may be associated with plastic changes of neuronal representations in the cortical organization. We will describe below a series of experiments that support this hypothesis for the somastosensory and auditory domains.[18–20]

Neuronal representations in the given cortical field of interest were determined by measuring evoked magnetic fields contralateral to the stimulated side with a 37-channel BTi Magnes system. For the somatosensory modality, a brief pneumatic stimulation was applied to the fingertips; for the auditory modality, a variety of different auditory stimuli were delivered to the right ear. The following sections describe some experimental examples in more detail.

## SHORT-TERM PLASTICITY EFFECT OF THE AUDITORY CORTEX INDUCED BY NOTCHED MUSIC

Most studies of deafferentation-induced cortical reorganization have investigated cortical reorganization on a timescale of days to weeks, or longer. However, other more recent studies have documented rapid changes in cortical dynamics following deafferentation. These studies have shown that neurons broaden and shift their receptive fields to sensory surfaces near or beyond the edge of the lesioned zone within minutes of deafferentation in the somatosensory[21] and visual systems,[22] and within hours in the auditory one.[23] Rapid retuning of sensory neurons has also been observed following reversible "functional" deafferentation in which sensory input from the environment is altered by procedures such as artificial scotomota[24] or digit ligation[25] rather than by permanent lesions of the receptor organs.

A study of "functional deafferentation" was carried out to determine whether plastic changes of frequency representation occur on a short timescale of a few hours when the adult human auditory cortex is deprived of sensory input. Ten normal test subjects were asked to provide three favorite CDs from their CD collection. The music was manipulated in such a way that a notch between 0.7 and 1.3 kHz, centered around 1 kHz, was produced using a band rejection filter (Bessel, 96 dB/oct) in the broad-band spectrum of the music (cf. FIG. 1a). Although this manipulation initially produced a clearly noticeable change in perception, subjects reported that they quickly adapted to the modified sound and that their appreciation of the music during the listening time was unchanged. The subjects were asked to listen attentively to the music for a continuous period of three hours. Due to the notch filtering during this period, any afferent input to cortical neurons tuned to frequencies around 1 kHz was abolished. In order to measure the effect of notching of the music on the neuronal representation of 1 kHz, MEG recordings to the test stimulus (band-passed noise bursts centered at 1 kHz) were compared to MEG recordings to the control stimulus (band-passed noise bursts centered at 0.5 kHz; cf. FIG. 1a). The experiment was repeated three times in each subject on consecutive days in order to address the time course and the reversibility of cortical remodeling induced by this procedure. These repetitions served to investigate the dynamics of cortical reorganization over a period of 24 h. The auditory evoked fields (AEF, channels 12 and 35, depicting the maximum and minimum of the AEF, respectively) obtained for test and control stimuli before and after listening to notched music are shown for one representative subject in FIGURE 1b. Whereas the AEF amplitudes measured before and after listening to notched music are almost the same for the control stimulus, the AEF amplitude for the test stimulus is about 10% smaller after listening to notched music than before. FIGURE 1c demonstrates the differences of the strength of the cortical sources before and after listening to the notched music for the test and for the control stimulus, averaged over all subjects and days. Whereas the strength of the cortical source decreased significantly after the listening to the notched music for the test stimulus ($p < 0.01$), this value did not change appreciably for the control stimulus. In order to provide information on the reversibility of the notching effect within the time period of 24 h, the measurements taken before listening to notched music on each day were compared. Between days, no significant differences between the strength of the cortical sources were observed for these measures, a result that proves the reversibility of the short-term plasticity effect provided by the "functional deafferentation" with respect to listening to the notched music.

Taken together, our results suggest that reorganization of cortical representations can occur within time periods as short as a few hours following functional deafferentation of the adult human auditory cortex. The temporal properties of the notching effect are consistent with animal studies that have shown that cortical neurons deafferented by cochlear lesions display elevated response thresholds initially and then shift their tuning preferences away from the lesioned area to frequencies near the edge of the deafferented region over a period of 1–3 h or longer.[23] Several interre-

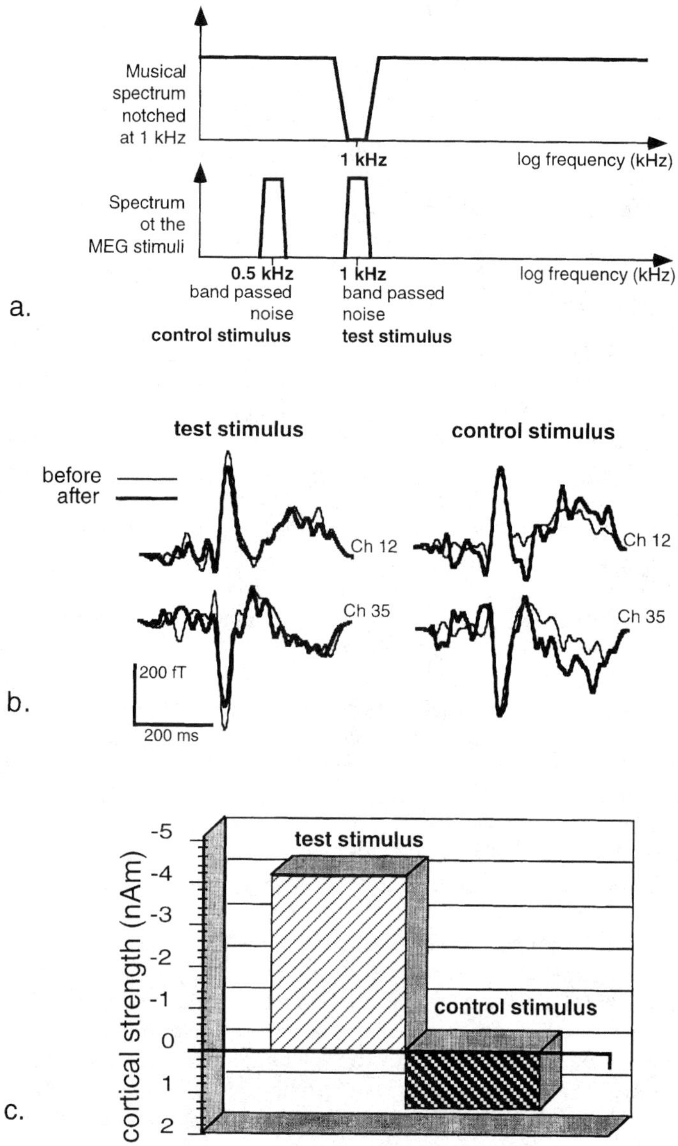


**FIGURE 1.** Short-term plasticity effect of the auditory cortex induced by notched music. (a) Music spectrum notched at 1 kHz (*top*) and spectral characteristics of the test and control stimuli for MEG recording (*bottom*). (b) Auditory evoked fields (AEF) of a single subject recorded for test stimulus (1 kHz, *left panel*) and control stimulus (0.5 kHz, *right panel*) before and after listening to notched music. (c) Change in cortical representation or respectively in the strength of the cortical source produced by listening to the notched music and calculated as the moment (Q, nAm) of the equivalent current dipole fitted to the N1m component of the AEF for test and control stimuli averaged over subjects and days (Modified from Pantev *et al.*[20])

lated mechanisms appear to contribute to cortical remodeling induced by deafferentation, including (1) changes in the efficacy of existing excitatory synapses unmasked by lesioning,[26] (2) modification of synaptic efficacy by transcription of immediate early genes,[27] and (3) the sprouting of new connections.[28] Of these mechanisms the first and second appear to be probable candidates to account for our findings. Synaptogenesis may occur within hours,[29] but axonal sprouting and dendritic growth may require more time and hence would be less likely to be involved.

## CORTICAL PLASTICITY AND MUSICAL TRAINING

### *Somatosensory Representations*

Two groups of subjects were examined. The first group consisted of nine musicians who played string instruments: six violinists, two violoncellists, and one guitarist. The subjects' average age was 24 ± 3 years. They had started to play their instruments 12 years earlier on average, ranging between 7 and 17 years. They practiced on their instruments for an average of nine to ten hours per week. The second group consisted of six control subjects who had never played a musical instrument and did not frequently carry out tasks that involved a systematic and rhythmic finger stimulation such as typing (on a typewriter or computer). All subjects were right-handed. The fingers D1 (the thumb) and D5 (the little finger) of both hands were consecutively excited with a brief, nonpainful standardized pneumatic pressure.

After excitation of the left-hand fingers D1 and D5, the strength of the cortical sources as determined with MEG were stronger in musicians than the corresponding sources in control subjects (FIG. 2). Thus, the cerebral representation was increased representing the excitation of those fingers that are intensively used in string instrument musicians. This effect was particularly pronounced for the fifth digit (D5). The cortical representation of the left thumb was also enhanced, but not as strongly as the one for left D5. Cortical representations obtained for right-hand stimulation did not differ between control subjects and musicians.

The amount of increase in somatosensory cortical representations of left-hand fingers in musicians depended on the age at which the musicians had started to play their instrument (FIG. 3). The cortical response for stimulation of D5 was greater in those musicians who had begun to play their instrument earlier. Among the musicians who started to play violin or violoncello later (after the age of 13 years), the cortical representations for D5 were somewhat smaller, but still larger than those observed in the control subjects who had never played an instrument.

### *Auditory Representation*

Three groups of subjects were examined. The first group consisted of musicians with absolute pitch ($n = 9$), the second group of musicians with relative pitch ($n = 11$), and the third group of subjects were controls who had never played a musical instrument ($n = 13$). In the first group, the mean number of years the musicians had played their instruments was 21 ± 6, in the second group 15 ± 3 years. The musicians were recruited from the conservatory in Münster. The musicians with absolute pitch reported practicing their instruments for 27 ± 14 h per week, and the musicians with

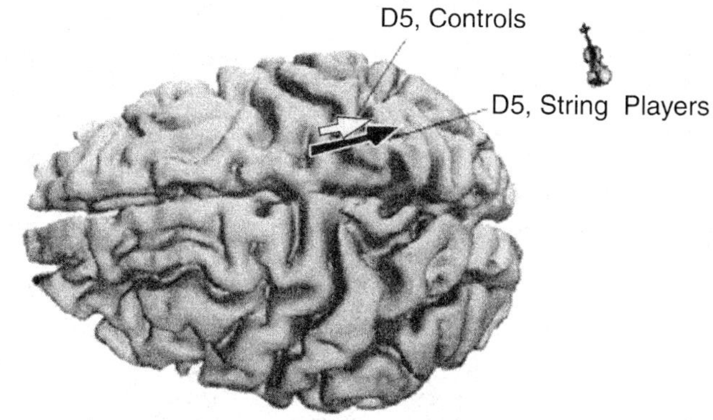


**FIGURE 2.** Equivalent current dipoles (ECD) elicited by stimulation of the digit 5 (D5) of the left hand of control subjects and string players, superimposed onto an MRI-reconstruction of the cerebral cortex of a control subject, selected to provide anatomical landmarks for the interpretation of the MEG-based localization. The arrows represent the location and orientation of the ECD vector for D5 averaged across musicians (*black arrow*) and control subjects (*white arrow*). The length of the arrows represents the mean magnitude of the dipole moment for D5 in each group. The dipole moment is larger for the musicians' D5 as indicated by the greater magnitude of the black arrow. (Modified from Elbert *et al.*[39])

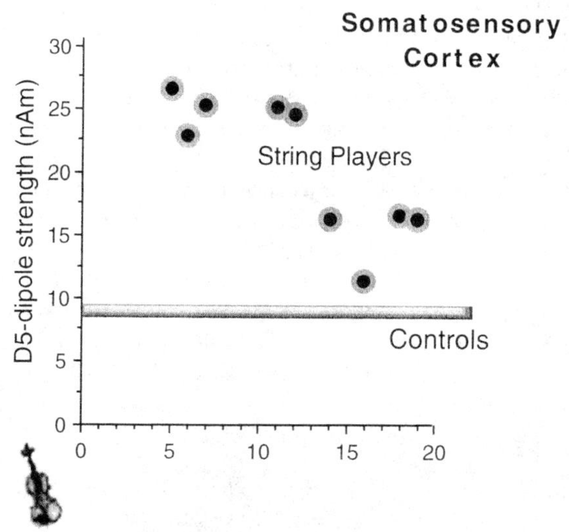


**FIGURE 3.** The magnitude of the D5 cortical strength as a function of the age of inception of musical practice: string players (*circles*), mean value for control subjects (*line*). Note the larger dipole moment for individuals beginning musical practice before the age of 12. (Modified from Elbert *et al.*[39])

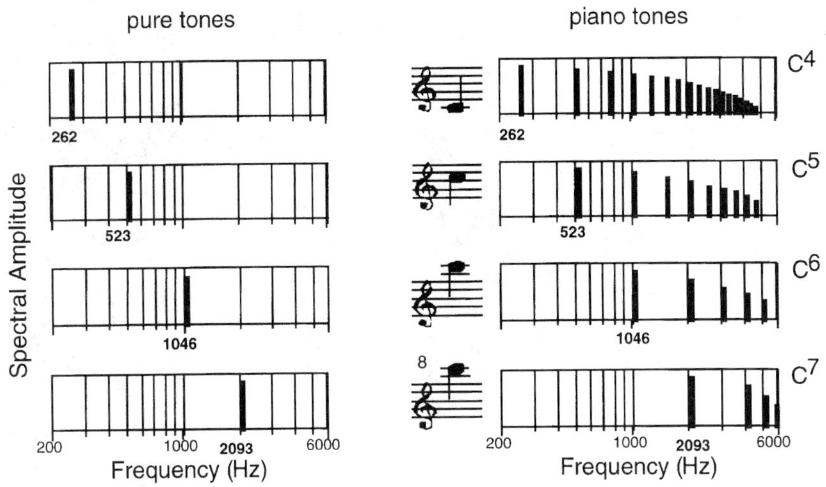


**FIGURE 4.** Spectra of piano tones C4, C5, C6, and C7 (American notation, having fundamental frequency at 262, 523, 1046, and 2093 Hz, respectively) and four pure tones of the same fundamental frequencies. (Modified from Pantev *et al.*[19])

relative pitch 23 ± 12 h per week, during the five years preceding participation in the study. The musicians who claimed to have absolute pitch were subjected to a testing procedure that was developed according to criteria established in the musical psychological literature.[30–34] A random sequence of 35 piano tones from H2 to C7 was presented to the musicians. For our study, in order to be considered a musician with absolute pitch, correct recognition of at least 90% of the tones was required. Musicians with relative pitch were either self-identified ($n$ = 9) or were subjects who failed the test for absolute pitch ($n$ = 2). The mean age in the three groups was 29 ± 6 years for musicians with absolute pitch, 26 ± 5 years for musicians with relative pitch, and 26 ± 4 years for control subjects. All participants were right-handed, as established by the Edinburgh handedness questionnaire[35] and did not have recent or past audiology, otology, or neurology medical history. Audiological status was normal at the time the study was performed for each subject.

The acoustic stimulation consisted of a pseudorandom sequence of four piano tones (C4, C5, C6, and C7; with the fundamental frequencies of 262, 523, 1046, and 2093 Hz), and four pure (sinusoidal) tones matching the fundamental frequencies (FIG. 4). Each stimulus was presented 128 times to determine the location and strength of the electrical source in the auditory cortex.

In all musicians larger neuronal excitation was noted for piano tones as compared to pure tones. Our measure for the cortical neuronal representation, the strength of the electrical dipole moment in both musician groups, was significantly 25% stronger for the piano tones as compared to the one for the pure tones ($p < 0.001$). By contrast, in the control subjects, who never played an instrument, no such difference was noted (FIG. 5).

This finding demonstrates an increase in neuronal representation specific for the processing of the tones of the musical scale in musicians. Similar to the described

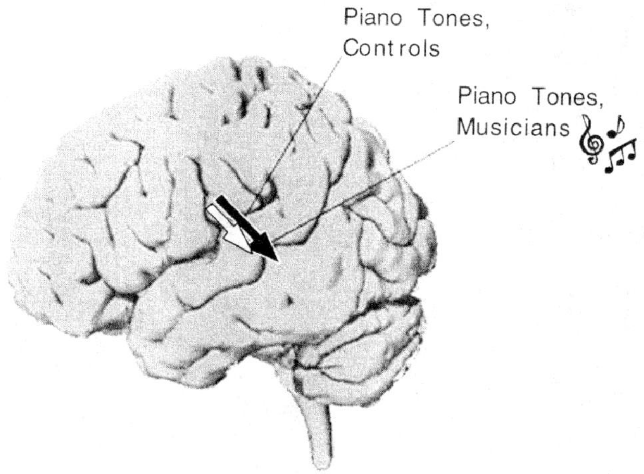


**FIGURE 5.** Mean values of the strength of cortical activation during the auditory N1 evoked field for piano tones as compared to pure tones in control subjects and musicians with absolute or relative pitch. (Modified from Pantev *et al.*[19])

enhancement of somatosensory representation, the degree of this increase was also dependent on the age at which the musicians had started to play their instruments. The earlier the initiation of musical practice, the stronger the neuronal response to the piano tones. In this study a marked difference was found between those who had begun practicing music before and after nine years of age.

The cortical representations for processing information from the fingers and the ears of musicians give an impressive demonstration that not only deafferentation, but also intensive training, can trigger a functional adaptation of the cortical organization. In general, this suggests that training can induce plastic changes of the adult human brain. In the musicians proficient in playing stringed instruments, the representation of the frequently used left-hand fingers is enlarged relative to control subjects. The more a given finger is stimulated, the larger the expansion of the cortical response. The representation for the thumb, which is not as frequently used as the other fingers, is not increased as much. However, the amount of practice is not the only factor that influences the organization of the somatosensory cortex. The plastic change of cortical representation is not directly correlated to the number of hours played per week, but rather to the age at which the training was started in childhood.

For the auditory system, the representation of pure tones, which are not part of our natural acoustic environment and also not encountered in musical training and practice, does not differ between musicians and control subjects. By contrast, the representation of piano tones, which in control subjects is of the same strength as the one for pure tones, is specifically enhanced in musicians. The analysis of reports of these musicians revealed that neither the passive exposure to music as a child nor the amount of time spent listening to music passively as an adult significantly influences functional cortical organization. It seems that active practice is necessary to induce plastic alterations. This reorganization of the auditory system also exhibits a trend

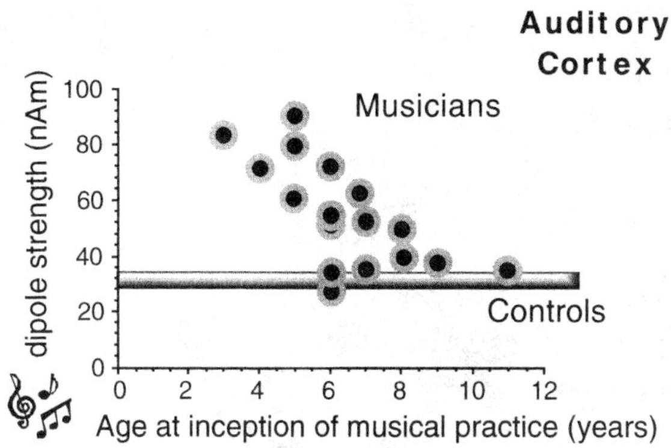


**FIGURE 6.** Mean strength of cortical activation for the different piano tones as a function of the age at which musical training began in musicians with absolute or relative pitch. The line denotes mean dipole moment in control subjects who never played a musical instrument. (Modified from Pantev *et al.*[19])

toward more pronounced brain plasticity with a younger age of starting to play the musical instrument (FIG. 6). The musicians who started playing before they were nine years old demonstrated the largest cortical representations.

The fact that the degree of expansion of the neuronal representations depends on the age training is started suggests that the enormous capacity of plasticity in the adult brain is, however, more limited than the plasticity of the infant brain. This corresponds to the prescientific observation that it is more difficult for adults to learn to play a musical instrument. To achieve an equivalent musical aptitude, adults can still adapt their cortical organization but they have to work harder to do so.

## THE DARK SIDE OF NEUROPLASTICITY IN MUSICIANS

As mentioned above, neuronal networks are thought to be particularly plastic during "sensitive periods" in the development of cortical structures. However, they maintain the ability to alter their architecture and function to afferent input throughout life. Perceptual and behavioral correlates of this reorganization indicate that neuroplasticity can be adaptive, as illustrated previously. Another type of reorganization includes the smearing of representational zones. An example has been described in blind multifinger Braille readers where overlap of digit representation may aid the ability to process different types of digital input simultaneously, thus enabling more rapid reading.[36] However, cortical reorganization can also be maladaptive, as is indicated by its association with phantom limb pain,[37] tinnitus,[38] and focal hand dystonia.[39,40] Intense training, as found in professional musicians, employs mechanisms of cortical plasticity with the inherent danger that maladaptive processes may also occur. Evidence from several laboratories suggests that one such danger is focal hand dystonia. Focal dystonia of the hand falls into a class often termed oc-

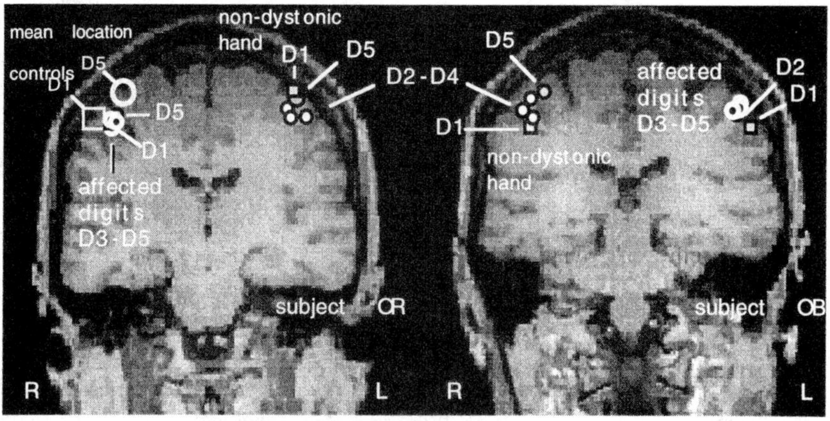


**FIGURE 7.** A coronal MRI section is shown through the somatosensory cortices of an organ player (*left*) and an oboist (*right*), both suffering from focal hand dystonia. The symbols indicate cortical representations (dipole locations) of digits 1–5 (D1 to D5) resulting from contralateral stimulation of the finger tips. The large *open symbols* in the right hemisphere of subject O.R. indicate the mean location of dipoles for D1 and D5 in normal control subjects. (Modified from Elbert *et al.*[39] )

cupational hand cramp.[41,42] It involves manual discoordination occurring in individuals engaging in extensive or forceful use of the fingers. The disorder involves a disorganization of control of the digits such that the movements of some fingers on a hand become involuntarily linked to those of others, something particularly incapacitating to professional instrumental musicians.

Repetitive and behaviorally relevant stimulation to several digits performed with healthy subjects results in a fusion and in a disordered arrangement of the representation of individual digits.[36,43] At the same time subjects in these experiments consistently mislocalize light pressure stimuli applied to the fingertips. Consistent with earlier work in animals, source imaging in humans demonstrates that synchronous stimulation of the digits may create a fusion of cortical representational zones, whereas asynchronous stimulation leads to separation. Could such a process of cortical reorganization become maladaptive and thus contribute to focal dystonia? By means of magnetic and neuroelectric source imaging, an overlap or smearing of the homuncular organization of the representation of the digits in the primary somatosensory and the motor cortex was indeed observed (FIG. 7).[39,44,45]

Work by N. Byl, M. Merzenich, and colleagues in New World monkeys has indicated that lack of digital motor coordination resulting from digital overuse is associated with an induced disorder in the representation of the digits in the somatosensory cortex.[42,46] These findings suggest that overuse-dependent central nervous system (CNS) plasticity is the basis of the focal hand dystonia. Therefore, because evidence suggests that behavioral usage and a CNS plastic response to this usage could be the cause of both the cortical disorder and the involuntary lack of movement coordination, it was thought that a behavioral intervention focusing on movement could be of value in reducing or eliminating these conditions. Any such treatment approach should include practice in a behaviorally relevant setting. A number of studies have shown that the behavioral relevance of the stimulation is the critical factor for reorganization to occur, whereas simple passive stimulation seems to be ineffective in altering cortical connectivity.[17] For example, Kilgard *et al.* suggested that the cholinergic nucleus basalis might play an important role in labeling stimuli as behaviorally relevant because nucleus basalis activation facilitated cortical plasticity induced by correlated stimulation in the periphery.[47] Further evidence points to the additional involvement of dopaminergic pathways. It seems possible that blockade of dopamine, released by a potentially rewarding condition, for example, through neuroleptics, will not allow cortical reorganization to occur.

On the basis of these considerations, Candia *et al.* have developed a successful therapy for focal hand dystonia.[40] Professional musicians (pianists, guitarists, and wind players), with long-standing symptoms, who had previously received a variety of treatments, practiced their instrument according to rules derived from insights into the mechanisms of cortical reorganization. The therapy involved immobilization by splints of one or more of the digits other than the focal dystonic finger. The focal dystonic finger was required to carry out repetitive exercises in coordination with one or more of the other digits for 1.5 to 2.5 h a day over an intended period of eight consecutive days under supervision of a therapist. The patients were instructed to continue the exercises using the splint at home for one hour every day for one year posttreatment. The wind players, who, in effect, constituted placebo controls, did not improve substantially. However, each of the pianists and guitarists showed marked and significant improvement in spontaneous repertoire performance without the splint at the end of treatment, up into the normal range. Neuroimaging results obtained in successfully treated subjects indicated normalization of the cortical representational maps, highlighting again the continued role of plasticity in the adult human brain for successful adaptation. The outcome demonstrates that learning-induced alterations in the functional architecture of the brain can be maladaptive, and that the resulting pathology can be treated using behavioral techniques that arc based on learning principles that take into account recent research on neuroplasticity.

## CONCLUSIONS

By use of MEG, it was shown that it is possible to noninvasively study mechanisms of cerebral organization and reorganization in humans. Listening to music attentively for as little as three hours can have a temporary influence on auditory cortical representation, as was shown in our study with music notched at a specific frequency. It was further shown that musical education and training is reflected in the organization of auditory and somatosensory representational cortex in musicians. From the results obtained, the following questions arise: Can we draw conclusions on the behavioral plasticity after knowing about the neuronal one? How do alterations in functional cortical organization affect musical accomplishment? Do the best virtuosos have the most atypical cortical organization?

The results of the studies presented here do not allow final conclusions. However, as shown in the case of focal dystonia, current research can lead to recommendations about musical practice and help design methods of treatment. The observed correlation between the age of initiation of musical practice and the amount of cortical change should not be overinterpreted from a music-pedagogical perspective. The complex relationships between genetic and familial, as well as educational pedagogical factors, would have to be studied by an interdisciplinary team of scientists. Such a team should comprise neuroscientists, as well as musicians and experts in music psychology, because it would be necessary to evaluate subtle differences in musical achievement to address these questions and establish causal links.

## ACKNOWLEDGMENTS

This research was supported by the Deutsche Forschungsgemeinschaft. The work presented here was made possible through multiple collaborations with Drs. Larry Roberts, Brigitte Rockstroh, Bernhard Ross, Edward Taub, and Christian Wienbruch. We wish to thank Deborah Urdang and Dr. Harvey Salomon for comments on the manuscript.

## REFERENCES

1. Clark, S.A., *et al.* 1988. Receptive fields in the body-surface map in adult cortex defined by temporally correlated inputs. Nature **332:** 444–445.
2. Narici, L., *et al.* 1991. Neuromagnetic somatosensory homunculus: a non-invasive approach in humans. Neurosci. Lett. **121:** 51–54.
3. Elbert, T., *et al.* 1995. The separation of overlapping neuromagnetic sources in first and second somatosensory cortices. Brain Topogr. **7:** 275–282.
4. Merzenich, M.M., P.L. Knight & G.L. Roth. 1975. Representation of cochlea within primary auditory cortex in the cat. J. Neurophysiol. **38**: 231–249.
5. Pantev, C., *et al.* 1988. Tonotopic organization of the human auditory cortex revealed by transient auditory evoked magnetic fields. EEG Clin. Neurophysiol. **69**: 160–170.
6. Pantev, C., *et al.* 1989. Tonotopic organization of the auditory cortex: pitch versus frequency representation. Science **246:** 486–488.
7. Pantev, C., *et al.* 1995. Specific tonotopic organizations of different areas of the human auditory cortex revealed by simultaneous magnetic and electric recordings. EEG Clin. Neurophysiol. **94:** 26–40.
8. Hebb, D.O. 1949. The organization of behavior. John Wiley & Sons. Chichester, UK.
9. Hari, R. 1990. Magnetic evoked fields of the human brain: basic principles and applications. EEG Clin. Neurophysiol. Suppl. **41:** 3–12.
10. Merzenich, M.M., *et al.* 1984. Somatosensory cortical map changes following digit amputation in adult monkeys. J. Comp. Neurol. **224:** 591–605.
11. Merzenich, M.M., *et al.* 1983. Topographic reorganization or somatosensory cortical areas 3b and 1 in adult monkeys following restricted deafferentation. Neuroscience **8:** 33–55.
12. Elbert, T., *et al.* 1994. Extensive reorganization of the somatosensory cortex in adult humans after nervous system injury. Neuroreport **5:** 2593—2597.
13. Elbert, T., *et al.* 1997. Input-increase and input-decrease types of cortical reorganization after upper extremity amputation in humans. Exp. Brain Res. **117:** 161—164.
14. Rajan, R., *et al.* 1993. Effect of unilateral partial cochlear lesions in adult cats on the representation of lesioned and unlesioned cochleas in primary auditory cortex. J. Comp. Neurol. **338:** 17—49.
15. Irvine, D.R.F. & R. Rajan. 1995. *In* Advances in Hearing Research. G.A. Manley *et al.*, Eds.: 3—23. World Scientific Publishing Co. Singapore.
16. Jenkins, W.M., *et al.* 1990. Functional reorganization of primary somatosensory cortex in adult owl monkeys after behaviorally controlled tactile stimulation. J. Neurophysiol. **63:** 82—104.

17. RECANZONE, G.H., C.E. SCHREINER & M.M. MERZENICH. 1993. Plasticity in the frequency representation of primary auditory cortex following discrimination training in adult owl monkeys. J. Neurosci. **13:** 87–103.
18. ELBERT, T., *et al.* 1995. Increased cortical representation of the fingers of the left hand in string players. Science **270:** 305–307.
19. PANTEV, C., *et al.* 1998. Increased auditory cortical representation in musicians. Nature **392:** 811–814.
20. PANTEV, C., *et al.* 1999. Short-term plasticity of the human auditory cortex. Brain Res. **842:** 192–199.
21. DOETSCH, G.S., *et al.* 1996. Short-term plasticity in primary somatosensory cortex of the rat: rapid changes in magnitudes and latencies of neuronal responses following digit denervation. Exp. Brain Res. **112:** 505–512.
22. GILBERT, C.D. 1997. Cortical dynamics. Acta Paediatr. Suppl. **422:** 34–37.
23. ROBERTSON, D. & D.R. IRVINE. 1989. Plasticity of frequency organization in auditory cortex of guinea pigs with partial unilateral deafness. J. Comp. Neurol. **282:** 456–471.
24. GILBERT, C.D., *et al.* 1996. Spatial integration and cortical dynamics. Proc. Natl. Acad. Sci. USA **93:** 615–622.
25. ROSSINI, P.M., *et al.* 1994. Short-term brain "plasticity" in humans: transient finger representation changes in sensory cortex somatotopy following ischemic anesthesia. Brain Res. **642:** 169–177.
26. ZIEMANN, U., M. HALLET & L.G. COHEN. 1998. Mechanisms of deafferentation-induced plasticity in human motor cortex. J. Neurosci. **18:** 7000–7007.
27. KACZMAREK, L. & A. CHAUDHURI. 1997. Sensory regulation of immediate-early gene expression in mammalian visual cortex: implications for functional mapping and neural plasticity. Brain Res. Brain Res. Rev. **23:** 237–256.
28. FLORENCE, S.L., H.B. TAUB & J.H. KAAS. 1998. Large-scale sprouting of cortical connections after peripheral injury in adult macaque monkeys. Science **282:** 1117–1121.
29. ENGERT, F. & T. BONHOEFFER. 1999. Dendritic spine changes associated with hippocampal long-term synaptic plasticity. Nature **399:** 66–70.
30. BACHEM, A. 1937. Various types of absolute pitch. J. Acoust. Soc. Am. **9:** 146–151.
31. MIYAZAKI, K. 1988. Musical pitch identification by absolute pitch possessors. Percept. Psychophys. **44:** 501–512.
32. NEU, D.M. 1947. A critical review on the literature on "absolute pitch." Psychol. Bull. **44:** 249–266.
33. TAKEUCHI, A. & S.H. HULSE. 1993. Absolute pitch. Psychol. Bull. **113:** 345–361.
34. WARD, W.D. 1998. *In* The Psychology of Music. D. Deutsch, Ed. Academic Press. San Diego, CA.
35. OLDFIELD, R.C. 1971. The assessment and analysis of handedness: the Edinburgh inventory. Neurophysiologia **9:** 97–113.
36. STERR, A., *et al.* 1998. Perceptual correlates of changes in cortical representation of fingers in blind multifinger Braille readers. J. Neurosci. **18:** 4417–4423.
37. FLOR, H., *et al.* 1995. Phantom limb pain as a perceptual correlate of massive cortical reorganization in upper limp amputees. Nature **375:** 482–484.
38. MUHLNICKEL, W., *et al.* 1998. Reorganization of auditory cortex in tinnitus. Proc. Natl. Acad. Sci. USA **95:** 10340–10343.
39. ELBERT, T., *et al.* 1998. Alteration of digital representations in somatosensory cortex in focal hand dystonia. Neuroreport **9:** 3571–3575.
40. CANDIA, V., *et al.* 1999. Constraint-induced movement therapy for focal hand dystonia in musicians. Lancet **353:** 42.
41. LEDERMAN, R.G. 1988. Occupational cramp in instrumental musicians. Med. Probl. Perform. Art **3:** 45–51.
42. BYL, N.N., M.M. MERZENICH & W.M. JENKINS. 1996. A primate genesis model of focal dystonia and repetitive strain injury: I. Learning-induced dedifferentiation of the representation of the hand in the primary somatosensory cortex in adult monkeys. Neurology **47:** 508–520.
43. BRAUN, C., *et al.* 2000. Differential activation in somatosensory cortex for different discrimination tasks. J. Neurosci. **20:** 446–450.

44. ELBERT, T. & A. KEIL. 2000. Imaging in the fourth dimension. Nature **404:** 29–31.
45. BARA-JIMENEZ, W., *et al.* 1998. Abnormal somatosensory homunculus in dystonia of the hand. Ann. Neurol. **44:** 828–831.
46. BYL, N.N., *et al.* 1997. A primate model for studying focal dystonia and repetitive strain injury: effects on the primary somatosensory cortex. Phys. Ther. **77:** 269–284.
47. KILGARD, M.P. & M.M. MERZENICH. 1998. Cortical map reorganization enabled by nucleus basalis activity. Science **279:** 1714–1718.

# The Brain That Plays Music and Is Changed by It

ALVARO PASCUAL-LEONE

*Behavioral Neurology Unit, Beth Israel Deaconess Medical Center, Harvard Medical School, Boston, Massachusetts 02215, USA*

**ABSTRACT: Playing a musical instrument demands extensive procedural and motor learning that results in plastic reorganization of the human brain. These plastic changes seem to include the rapid unmasking of existing connections and the establishment of new ones. Therefore, both functional and structural changes take place in the brain of instrumentalists as they learn to cope with the demands of their activity. Neuroimaging techniques allow documentation of these plastic changes in the human brain. These plastic changes are fundamental to the accomplishment of skillful playing, but they pose a risk for the development of motor control dysfunctions that may give rise to overuse syndromes and focal, task-specific dystonia.**

**KEYWORDS: Brain plasticity; Musical training**

## INTRODUCTION

> *The most intricately and perfectly coordinated of all voluntary movements in the animal kingdom are those of the human hand and fingers, and perhaps in no other human activity do memory, complex integration, and muscular coordination surpass the achievements of the skilled pianist.*
>
> — Homer W. Smith, *From Fish to Philosopher*

Playing a musical instrument requires more than factual knowledge about the musical instrument and the mechanics of how it is played. For example, given complete information about hand position, finger motions, and sequence of keys to push for how long and with what force, I would still be unable to play even the simplest piano sonata. The central nervous system has to acquire and implement a "translation mechanism" to convert knowledge into action. These translation capabilities constitute the skill that enables the pianist to act on memory systems, select the relevant facts, choose the proper response goals, activate the necessary sensorimotor structures, and execute the sonata successfully. We generally think of such a skill as being acquired with practice. The pianist confronted with a new composition, after understanding the task and its demands, develops a cognitive representation of it and ini-

Address for correspondence: Alvaro Pascual-Leone, M.D., Ph.D., Department of Neurology, Behavioral Neurology Unit, Beth Israel Deaconess Medical Center, 330 Brookline Avenue, Kirstein Building KS 454, Boston, MA 02215. Voice: 617-667-0203; fax: 617-975-5322.
apleone@caregroup.harvard.edu

tiates a first, centrally guided response that results in sensorimotor feedback and movement correction. It seems certain that both sensory and motor aspects have to be exquisitely coordinated. At the beginning, the limbs move slowly, with fluctuating accuracy and speed, and success requires visual, proprioceptive, and auditory feedback. Eventually, each single movement is refined, the different movements chained into the proper sequence with the desired timing, a high probability of stability in the ordered sequence attained, and a fluency of all movement developed. Only then can the pianist shift his or her attentional focus away from the mechanical details of the performance towards the emotional content of the task. We can think of the acquisition of such a skill as the conversion of declarative knowledge (facts) into procedural knowledge (actions, skills).[1–3]

Learning and memory might be considered integral parts of all the operations of any neural circuit, a concept for which Fuster[4] recently made an eloquent and convincing argument. In this view, "perception and action are phenomena of memory and, conversely, memory is an integral part of perceptual and motor processing" (Ref. 4, p. 21). The nervous system comes to be viewed as a dynamic, dialectic organization in which plasticity is an intrinsic property that relates to the acquisition of new memories and skills as an obligatory consequence of perceptions and motor actions. To play an instrument, the nervous system is modified as a consequence of practice to yield the necessary changes in ability. We refer to this experience-dependent modification in neural structure as plasticity. These changes take place both in sensory and motor systems as well as in their interface. The consequence of this notion is that these changes do not necessarily represent behavioral benefits to the subject but might in fact be misguided and functionally deleterious. The development of noninvasive imaging and neurophysiologic techniques enables us to pursue the study of such changes in humans.

## LEARNING TO PLAY THE PIANO CHANGES YOUR BRAIN

> *...the work of a pianist...is inaccessible for the untrained human, as the acquisition of new abilities requires many years of mental and physical practice. In order to fully understand this complicated phenomenon it is necessary to admit, in addition to the strengthening of pre-established organic pathways, the establishment of new ones, through ramification and progressive growth of dendritic arborizations and nervous terminals. ...Such a development takes place in response to exercise, while it stops and may be reversed in brain spheres that are not cultivated.*
>
> —Santiago Ramón y Cajal, *Textura del Sistema Nervioso del Hombre y de los Vertebrados*

Normal subjects were taught to perform with one hand a five-finger exercise on a piano keyboard connected by a MIDI interface to a computer.[5] Subjects did not play any musical instrument, did not know how to typewrite using all fingers, and held jobs not demanding skillful hand and finger activities. The exercise required pressing a piano key sequentially with thumb (C), index finger (D), middle finger

(E), ring finger (F), little finger (G), ring finger (F), middle finger (E), index finger (D), thumb (C), index finger (D), and so forth. The subjects were instructed to attempt to perform the sequence of finger movements fluently, without pauses and without skipping any key, while paying particular attention to keep the interval between the individual key presses constant and the duration of each key press the same. A metronome gave a tempo of 60 beats per minute, which the subjects were asked to aim for. Subjects performed the exercise under auditory feedback. They were studied on five consecutive days, and each day they had a two-hour practice session followed by a test. The test consisted of the execution of 20 repetitions of the five-finger exercise and was eventually analyzed with the computer for the exact sequence of key presses, the interval between keys, and the duration and velocity of each key press. Following each test the subjects were given feedback about their performance and tips about how to improve. FIGURE 1 illustrates the great improvement in the subjects' playing skill over the course of the five study days. The number of sequence errors and the duration and variability of the intervals between key presses (as marked by the metronome beats) all decreased significantly.

The aim of the experiment was to correlate skill acquisition with changes in motor cortical output. Therefore, before the first practice session on the first day of the experiment, we used focal transcranial magnetic stimulation (TMS) to map the motor cortical areas targeting long finger flexor and extensor muscles bilaterally. Thereafter, this mapping was repeated daily. The baseline maps can be compared with those obtained during the five days. As the subjects' performance improved, the threshold for TMS activation of the finger flexor and extensor muscles decreased steadily; even taking into account this change in threshold, the size of the cortical representation for both muscle groups increased significantly (FIG. 1).[5] However, this increase could be demonstrated only when the cortical mapping studies were conducted following a 20- to 30-minute rest after the practice (and test) session. No such modulation in the cortical output maps was noted when maps were obtained before each daily practice session (FIG. 1).[6]

When a near-perfect level of performance was reached at the end of a week of daily practice, subjects were randomized into two groups (FIG. 2). Group 1 continued daily practice of the same piano exercise during the following four weeks. Group 2 stopped practicing. During the four weeks of follow-up, cortical output maps for finger flexor and extensor muscles were obtained in all subjects on Mondays (before the first practice session of that week in group 1) and on Fridays (at the end of the last practice session for the week in group 1). In the group that continued practicing (group 1), the cortical output maps obtained on Fridays showed an initial plateau and eventually a slow decrease in size despite continued performance improvement (FIG. 2). On the other hand, maps obtained on Mondays before the practice session and after the weekend rest showed a small change from baseline with a tendency to increase in size over the course of the study (FIG. 2). In group 2, who stopped practicing after one week, the maps returned to baseline after the first week of follow-up and remained stable thereafter.

This experiment reveals that acquisition of the necessary motor skills to perform a five-finger movement exercise correctly is associated with reorganization in the cortical motor outputs to the muscles involved in the task. There are two main mechanisms to explain this reorganization: establishment of new connections, or sprouting, and unmasking of previously existing connections. The rapid time course in the

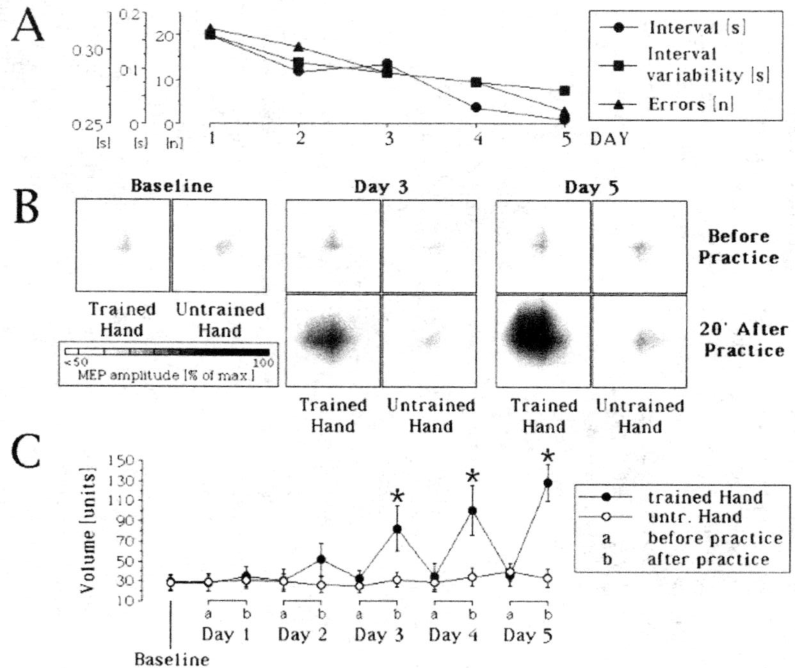


**FIGURE 1.** (**A**) Behavioral results of the week-long practice of the five-finger exercise on the piano. Interval between key presses, variability of these intervals, and number of errors in 20 repetitions of the exercise all show a highly significant decrease. This documents learning for all subjects studied. (**B**) Cortical output maps for the finger flexors of the trained and untrained hands of a representative subject. (See text and Pascual-Leone *et al.*[5] for details on mapping method.) Note the marked changes of the output maps for the trained hand and the lack of changes for the untrained hand and the maps obtained for either hand before the daily practice sessions. (**C**) Graphic display of mean (± SD) volumes of the cortical output maps for all subjects studied. Note the significant (*star*) changes in cortical output maps for the trained hands after practice sessions on days 3 to 5.

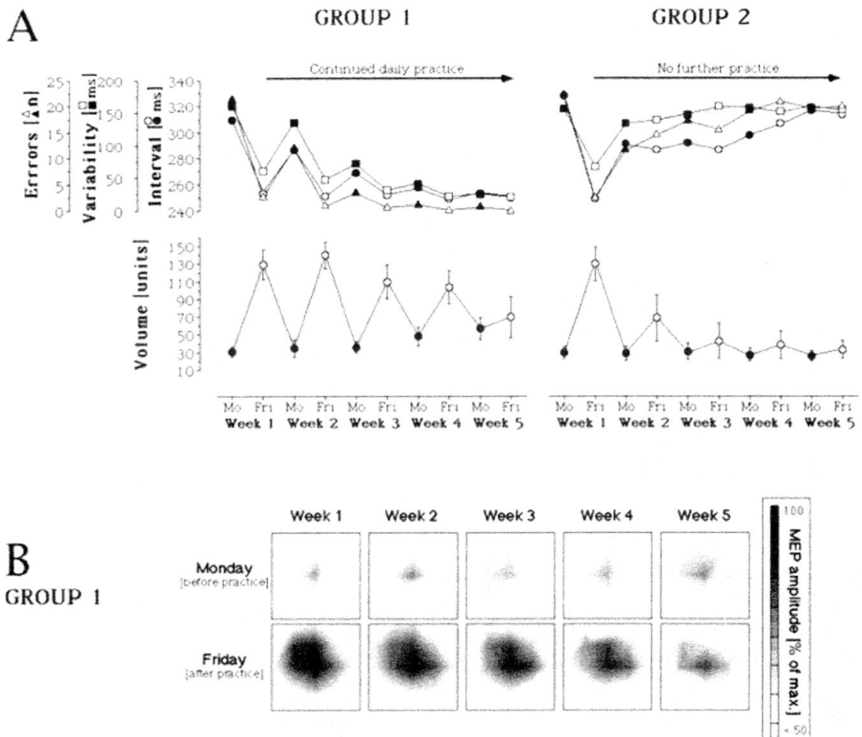


**FIGURE 2.** (**A**) Behavioral results for subjects in groups 1 (continued practice after 1 week) and 2 (practice limited to 1 week) learning the five-finger exercise on the piano. Interval between key presses, variability of these intervals, and number of errors in 20 repetitions of the exercise are shown. Data from Monday and Friday of each week are shown. (**B**) Cortical output maps for the finger flexors of the trained hand of a representative subject. (See text and Pascual-Leone *et al.*[5] for details on mapping method.) Note the marked changes in the output maps. There appear to be two parallel processes in action, one accounting for the rapid modulation of the maps from Monday to Friday and the other responsible for the slow and more discrete changes in Monday maps over time.[5,6]

initial modulation of the motor outputs, by which a certain region of motor cortex can reversibly increase its influence on a motoneuron pool, is most compatible with the unmasking of previously existing connections.[7–10] Supporting this notion, the initial changes are transient, demonstrable after practice, but return to baseline after a weekend of rest. We suggest that such flexible, short-term modulation represents a first and necessary step, leading to longer-term structural changes in the intracortical and subcortical networks as the skill becomes overlearned and automatic.

It is important to realize that our TMS mapping technique demonstrates a trace or memory of the activation of the motor cortical outputs that took place during the performance of the task rather than the activation during the task itself, as would be the case with neuroimaging studies. Long-term potentiation has been demonstrated in the motor cortex,[11,12] and our results might reveal a similar phenomenon. During the learning of the task, the cortical output maps obtained after task performance show a progressive increase in size, suggesting that skill acquisition is associated with a change in the pattern of activation of the executive structures. These changes are not demonstrable before the task performance, and we might hypothesize that "learning" to activate the cortical outputs appropriately (in this case, activating a progressively larger cortical output map) constitutes the neurophysiological correlate of performance improvement.

As the task becomes overlearned over the course of five weeks, the pattern of cortical activation for optimal task performance might change as other neural structures take a more leading role in task performance. This might result in the decreasing size of the cortical output maps after practice and the increasing size before practice. For example, if basal ganglia play a more important role in driving task performance, the changed activity in the basal ganglia motor circuit might enhance thalamocortical connections. This may account for changes in the cortical motor representation, which would be more stable than that observed during the initial acquisition of a skill. Greenough *et al.*[13] showed that motor training is associated with changes in the dendritic branching patterns of motor and sensory cortical cells involved in the performance of a task. Sprouting may account for plastic changes in such situations, as likely occurred in monkeys deafferented for 10 years,[14] and represent the correlate of long-standing "memories." Neuroimaging studies suggest a similar phenomenon.[15–22] Related changes in the pattern of cortical activation and the resulting modulation of cortical outputs might be induced by changes in the subject's strategy as explicit learning mechanisms are engaged.[23]

Our findings stress the role of the primary motor cortex (M1) in skill acquisition. It is not unreasonable to expect plastic changes in M1 during motor skill learning, because M1 is clearly involved in movement, and its cells have complex patterns of connectivity, including variable influences on multiple muscles within a body part.[24,25] Recent animal studies also illustrate the importance of the M1 in skill learning. For example, increases in excitability of primary motor cortex neurons have been found during conditioning,[26,27] and repeated activation of somatosensory inputs into the motor cortex results in long-term potentiation of motor neurons.[11,12] Even so, at the low stimulus intensities used in our mapping studies, TMS activates cortical cells largely transsynaptically.[28,29] Therefore, the demonstrated modulation in motor cortical outputs might be conditioned by changes in premotor areas projecting to M1, rather than by changes in M1 itself.

## IF YOU CANNOT DO IT, AT LEAST THINK ABOUT IT: MENTAL PRACTICE

In the quote reproduced above, Cajal talks about rapid and slower plastic changes in the brain in the context of practice. Our results confirm his intuitions. Furthermore, Cajal writes about physical and *mental* practice. Might the latter result in plastic brain changes similar to those induced by the former?

Mental practice is the imagined rehearsal of a motor act with the specific intent of learning or improving it, without overt movement output. Mental practice can be viewed as a virtual simulation of behavior by which the subject develops and "internally" rehearses a cognitive representation of the motor act. When confronted with a new motor task, the subject must develop a cognitive representation of it and initiate a centrally guided response, which secondarily can be improved using sensorimotor feedback. Mental practice may accelerate the acquisition of a new motor skill by providing a well-suited cognitive model of the demanded motor act in advance of any physical practice.[30, 31]

Mental practice has found wide acceptance in the training of athletes.[32] Musicians also have long recognized the benefit of mental rehearsal. Harold Schoenberg, in his fascinating books on virtuoso intrumentalists,[33,34] provides ample information on this topic. For example, Horowitz is supposed to have practiced mentally before concerts to avoid disturbing his motor skills by the feedback of pianos other than his own Steinway. Rubinstein, eager to enjoy life to its fullest and dedicate as little time to practice as possible, found mental rehearsal the best way to minimize the number of hours spent sitting in front of the piano while maintaining his skill.

Using the same experimental design as described above for the five-finger exercise, we studied subjects who, instead of practicing at the keyboard two hours daily for five days spent time at the keyboard visualizing, rather than executing, the movements.[5] They were told to repeat the movement mentally, as if they were playing, but without moving their fingers. They could rest their fingers on the piano keyboard, but the lack of voluntary movements was monitored using electromyography and video. Such mental practice resulted in a reorganization of the motor outputs to finger flexor and extensor muscles similar to the one observed in the group of subjects who physically practiced the movements and led to similar improvement in their ability to perform the five-finger movement exercise (FIG. 3).

Studies of regional cerebral blood flow (rCBF) suggest that the prefrontal and supplementary motor areas, basal ganglia, and cerebellum are part of the network involved in the mental simulation of motor acts.[35–38] Therefore, mental simulation of movements activates some of the same central neural structures required for the performance of the actual movements. In so doing, mental practice alone seems to be sufficient to promote the modulation of neural circuits involved in the early stages of motor skill learning. This modulation not only results in marked improvement in performance, but also seems to place the subjects at an advantage for further skill learning with minimal physical practice.[5,6] The combination of mental and physical practice leads to greater performance improvement than does physical practice alone, a phenomenon for which our findings provide a physiological explanation.

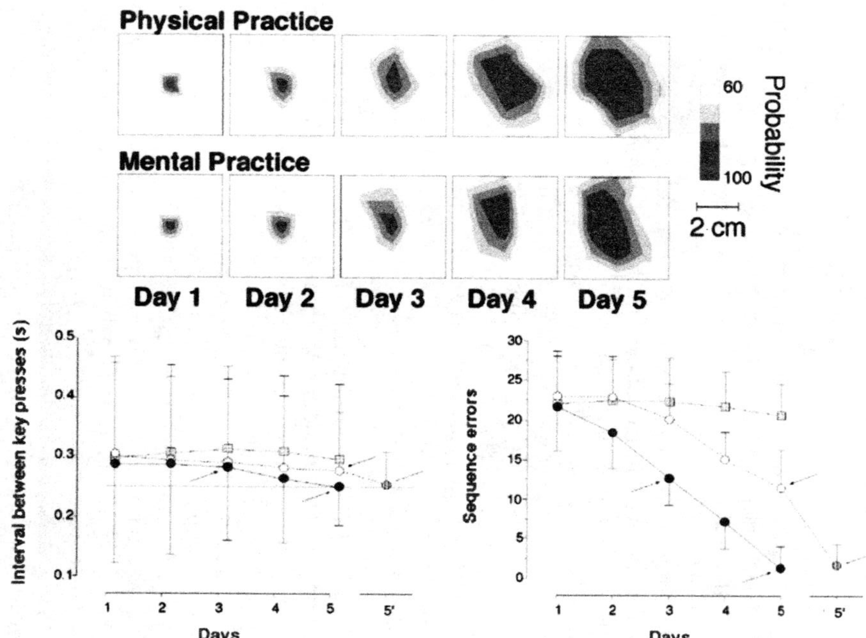


**FIGURE 3.** (*Top*) Cortical output maps for the finger flexors of the trained hand in a representative subject with physical versus mental practice alone. Modified from Pascual-Leone *et al.*;[5] see also text for details. Note the parity of changes with either form of practice. (*Bottom*) Graphic representation of behavioral data in a group of control subjects (no practice, *squares*) and a group of test subjects performing mental practice (*open circles*) versus physical practice (*filled circles*). Note the behavioral advantage of the physical practice group, but the significant improvement of the mental practice group as compared with controls. The last data point (5′) represents the performance achieved by the mental practice group after 5 minutes of physical practice at the end of day 5. Note the apparent advantage of mental practice, because subjects go from a performance equivalent to day 3 of the physical practice group to matching their day 5 performance with only 5 minutes of practice. Modified from Pascual-Leone *et al.*[5]

## THE RISK OF THE CHANGE: SENSORIMOTOR MISMATCH AND TASK-INDUCED DYSTONIA

As we have seen, skill acquisition requires plastic changes in the brain. This plastic reorganization is driven by efferent demand and afferent input. However, a system capable of such flexible reorganization harbors the risk of unwanted change. Increased demand of sensorimotor integration poses such a risk. We can postulate that faulty practice may result in unwanted cortical rearrangement and set the stage for motor control problems such as overuse syndrome and focal, task-specific dystonias. The style of piano playing—for example, the Russian versus the German school—seems to play a critical role in the risk of development of motor control problems. Forceful playing with the fingers bent and executing hammer-like movements is more frequently associated with overuse syndrome and dystonia than is softer playing with extended fingers "caressing" the keys. This stresses the importance of proper, well-guided practice and illustrates the need for greater understanding of the neurobiology underlying music playing to define what proper practice actually is.

Focal hand dystonia in musicians is a strongly task-related movement disorder that can end an instrumentalist's career. Typically, symptoms become manifest only when players execute specific overpracticed skilled exercises on their instrument. Suddenly, a finger moves involuntarily, voluntary motor control is lost, the muscles tense up excessively, and pain develops. Playing is disturbed. For years, focal, task-specific dystonias were thought to be psychiatric in nature. It seemed too bizarre that involuntary muscle contraction might occur when playing a certain passage but not with any other activity or when playing on a certain instrument (organ) but not another (pianoforte). We now know that dystonias are neurologic involuntary movements due to disturbances in motor programs.[39] What, however, is their underlying pathophysiology?

We examined five guitarists with functional magnetic resonance imaging (fMRI) during dystonic symptom provocation by means of an adapted guitar inside the magnet.[40] As reference, we used the activation pattern obtained in the same subjects during other hand movements and in matched guitar players without dystonia during execution of the same guitar playing exercises. A 1.5-Tesla system equipped with echo-speed gradients and single-shot echoplanar imaging (EPI) software was used. Data acquisition was centered on the cortical motor system encompassed in eight contiguous slices.

Dystonic musicians compared in both control situations showed significantly greater activation of the contralateral primary sensorimotor cortex, which contrasted with conspicuous bilateral underactivation of premotor areas (FIG. 4). Our results agree with studies of other types of dystonia in that they show abnormal recruitment of cortical areas involved in the control of voluntary movement. They do suggest, however, that rather than being hypoactive in idiopathic dystonic patients, the primary sensorimotor cortex may be overactive when tested during full expression of the task-induced movement disorder.

Although the primary manifestation of dystonia is abnormal motor function, evidence is increasing for a dysfunction of sensory processing that may be an associated or contributing factor.[39,41–43,46,47] In fact, dedifferentiation of the normally independent sensory representations of multiple digits may be a causative element

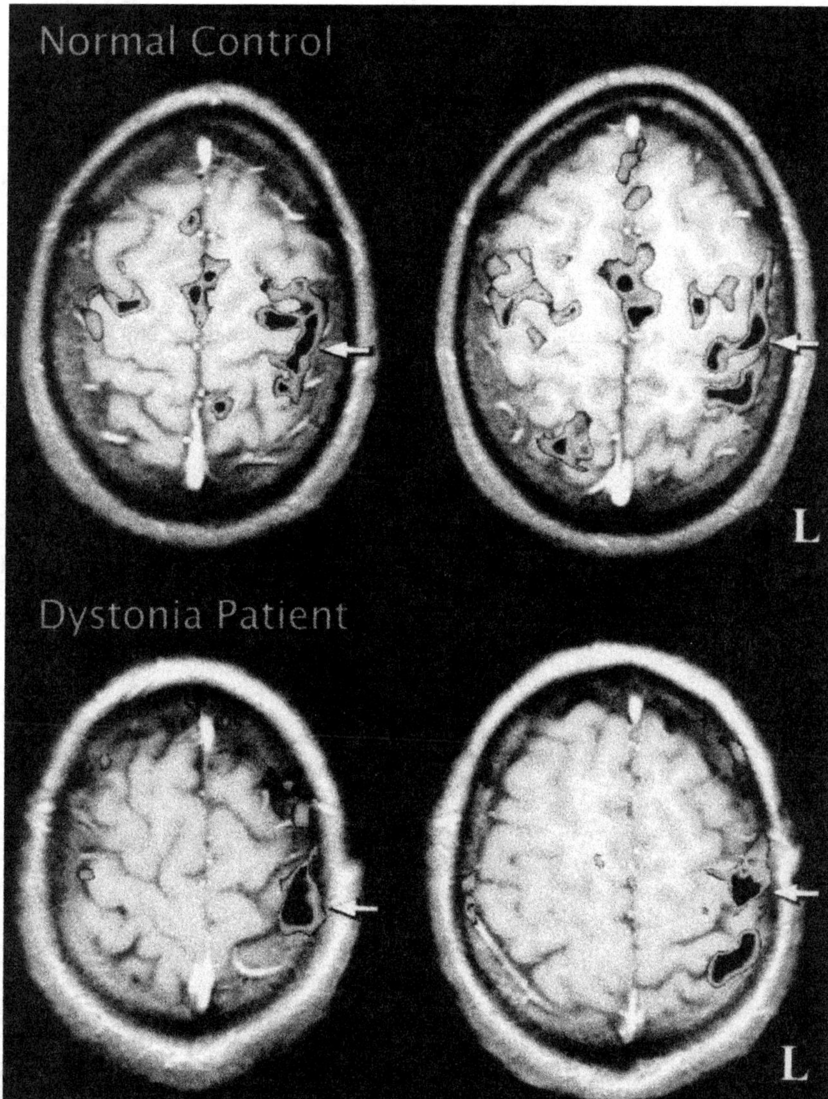


**FIGURE 4. See Plate 20 in color section.** The BOLD fMRI images of a normal and dystonic guitar player executing right hand arpeggios in the scanner are displayed. Note the greater activation of the sensorimotor cortex (*arrows*) and the lack of activation of premotor and supplementary motor cortices in the dystonic patient. Modified from Pujol *et al.*[40]

in the etiology of dystonia.[44,45] For example, in musicians, extensive practice of co-ordinated hand postures in which various digits function as a unit, such as arpeggios, could eventually induce changes in the sensory representation of the hand with blurring of the segregation of different digits. This might be the case particularly when

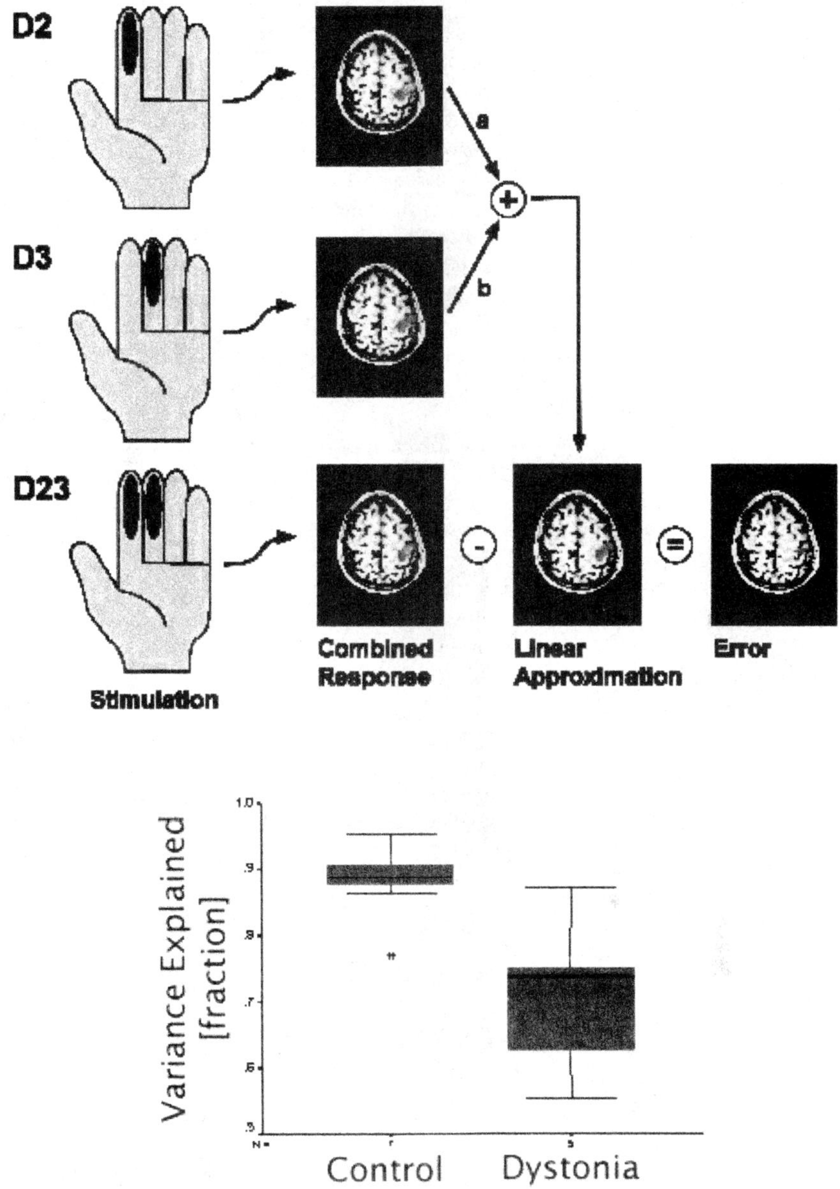


**FIGURE 5. See Plate 21 in color section.** (*Top*) Schematic representation of the experimental design and representative fMRI data in one subject. Sensory stimulation was applied to the index (D2) or middle (D3) finger alone or to both fingers at the same time (D23) while the fMRI BOLD signal was recorded. A linear approximation of the fMRI signal associated with combined stimulation of D2 and D3 was calculated and substracted from the measured signal, giving rise to the "error" of "variance." (*Bottom*) Variance results in graphic form for all control subjects and patients with dystonia. Modified from Sanger *et al.*[46,47]

small repeated traumas are added, as in forceful, "hammer-finger" piano playing. Disorganization and consequent confusion of sensory inputs could potentially lead to poorly differentiated control of motor representations and be the mechanisms underlying the risk of faulty motor control in some instrumentalists.

Sanger *et al.*[46,47] investigated whether evidence from fMRI supports abnormal receptive fields in primary somatosensory cortex that span more than a single finger surface. We developed a new technique for investigating overlap of sensory cortical receptive fields (FIG. 5) and hypothesized that the combined metabolic demands of two spatially separated populations of cells, when activated simultaneously, would be approximated by the sum of the metabolic demands of each population individually. Therefore, if the populations of cells activated by the second and third digits in normal subjects are distinct, we expect that the task-related component of the blood oxygenation level–dependent (BOLD) signal obtained by simultaneous activation of the two digits will be a linear combination of the task-related BOLD signals when each finger is activated individually. Conversely, if there is overlap of the receptive fields so that the same sensory populations respond to two different fingers, we do not necessarily expect linearity, because the metabolic demands of a single population stimulated by two different fingers may be determined by a complex function of the stimulus strength.

BOLD contrast was measured with fMRI during tactile stimulation of the index finger, the middle finger, or both simultaneously in five patients with focal dystonia and seven control subjects. In the control subjects, a linear combination of activation patterns for individual finger stimulation predicts an average of 88% of the variance in the pattern of activity for combined stimulation. In patients with dystonia, the linear combination predicted only 70% of the combined stimulation pattern ($p = 0.008$; FIG. 6). Therefore, our results suggest that in patients with dystonia, the same region of sensory cortex may respond to tactile stimuli on more than one finger. Disorganization of sensory representations appears to be part of the pathophysiology of focal dystonia and may contribute to motor abnormalities. Hence, emphasis needs to be placed in the sensory as well as on the motor aspects of skill acquisition and practice in musicians. Perhaps all instrumentalists should learn Braille to enhance the functional segregation of individual fingers and minimize the risk of dystonia.

## CONCLUSION

Tools are now available to study the neurophysiological correlates of skill learning in humans. The motor cortex plays an important role in motor skill learning, but so does the sensory cortex. The sensorimotor cortex changes as a consequence of skill acquisition. These plastic changes, which probably include functional and structural components, place the subjects at great advantage for skillful task performance, but harbor the risk of the development of motor control disorders. The results of our studies may be useful in understanding not only the physiology of skill acquisitions, but also the pathophysiology of movement disorders in skilled performers. Further work along these lines may lead to helpful insight into the appropriate teaching/learning technique for fine motor skills.

## ACKNOWLEDGMENTS

Part of the work summarized in this article was done in collaboration with Mark Hallett, Nguyet Dang, Leonardo Cohen, José María Tormos, Josep Valls-Solé, Terrance Sanger, Gottfried Schlaug, Daniel Tarsy, Jesus Pujol, and Jaume Roset-Llobet. This work was supported in part by the General Clinical Research Center at Beth Israel Deaconess Medical Center (National Center for Research Resources MO1 RR01032) and grants from the National Institute of Mental Health (MH57980 and MH60734) and the National Eye Institute (EY12091).

## REFERENCES

1. CRITCHLEY, M. & R.A.E. HENSON. 1977. Music and the Brain: Studies in the Neurology of Music. The Camelot Press. Southampton.
2. HOLDING, D.H.E. 1989. Human Skills, 2nd Ed. John Wiley & Sons. Chichester. New York.
3. WILSON, F.R. 1989. Acquisition and loss of skilled movement in musicians. Sem. Neurol. **9:**146–151.
4. FUSTER, J.M. 1995. Memory in the Cerebral Cortex. An Empirical Approach to Neural Networks in the Human and Nonhuman Primate. The MIT Press. Cambridge, MA.
5. PASCUAL-LEONE, A., D. NGUYET, L.G. COHEN, *et al.* 1995. Modulation of muscle responses evoked by transcranial magnetic stimulation during the acquisition of new fine motor skills. J. Neurophysiol. **74:**1037–1045.
6. PASCUAL-LEONE, A., F. TARAZONA & M.D. CATALA. 1999. Applications of transcranial magnetic stimulation in studies on motor learning. Electroencephalogr. Clin. Neurophysiol. Suppl. **51:** 157–161.
7. JACOBS, K.M. & J.P. DONOGHUE. 1991. Reshaping the cortical motor map by unmasking latent intracortical connections. Science **251:** 944–947.
8. SANES, J.N., J. WANG & J.P. DONOGHUE. 1992. Immediate and delayed changes of rat motor cortical output representation with new forelimb configurations. Cerebr. Cortex **2:** 141–152.
9. BRASIL-NETO, J.P., A. PASCUAL-LEONE, J. VALLS-SOLÉ, *et al.* 1992. Focal transcranial magnetic stimulation and response bias in a forced-choice task. J. Neurol. Neurosurg. Psychiatry **55:** 964–966.
10. BRASIL-NETO, J.P., J. VALLS-SOLÉ, A. PASCUAL-LEONE, *et al.* 1993. Rapid modulation of human cortical motor outputs following ischaemic nerve block. Brain **116:** 511–525.
11. IRIKI, A., C. PAVLIDES A. KELLER & H. ASANUMA. 1989. Long-term potentiation in the motor cortex. Science **245:** 1385–1387.
12. IRIKI, A., C. PAVLIDES, A. KELLER & H. ASANUMA. 1991. Long-term potentiation of thalamic input to the motor cortex induced by coactivation of thalamocortical and corticocortical afferents. J. Neurophysiol. **65:** 1435–1441.
13. GREENOUGH, W.T. 1984. Structural correlates of information storage in the mammalian brain: a review and hypothesis. Trends Neurosci. **7:** 229–233.
14. PONS, T.P., P.E. GARRAGHTY, A.K. OMMAYA & J.H. KAAS. 1991. Massive cortical reorganization after sensory deafferentation in adult macaques. Science **252:** 1857–1860.
15. SEITZ, R.J., E. ROLAND, C. BOHM, *et al.* 1990. Motor learning in man: a positron emission tomographic study. Neuroreport **1:** 57–60.
16. GRAFTON, S.T., J.C. MAZZIOTTA, S. PRESTY, *et al.* 1992. Functional anatomy of human procedural learning determined with regional cerebral blood flow and PET. J. Neurosci. **12:** 2542–2548.
17. GRAFTON, S.T., R.P. WOODS & J.M. TYSZKA. 1994. Functional imaging of procedural motor learning: relating cerebral blood flow with individual subject performance. Human Brain Mapping **1:** 221–234.

18. GRAFTON, S.T., E. HAZELTINE & R. IVRY. 1995. Functional mapping of sequence learning in normal humans. J. Cognit. Neurosci. **7:** 497–510.
19. JENKINS, I.H., D.J. BROOKS, P.D. NIXON, *et al.* 1994. Motor sequence learning: a study with positron emission tomography. J. Neurosci. **14:** 3775–3790.
20. KARNI, A., G. MEYER, P. JEZZARD, *et al.* 1994. Where practice makes perfect: an fMRI study of long-term motor cortex plasticity associated with the acquisition of a motor skill (abstr). Soc. Neurosci. Abstr. **20:** 1291.
21. KARNI, A., G. MEYER, P. JEZZARD, *et al.* 1995. Functional MRI evidence for adult motor cortex plasticity during motor skill learning. Nature **377:** 155–158.
22. KARNI, A., G. MEYER, C. REY-HIPOLITO, *et al.* 1998. The acquisition of skilled motor performance: fast and slow experience-driven changes in primary motor cortex. Proc. Natl. Acad. Sci. USA **95:** 861–868.
23. PASCUAL-LEONE, A., J. GRAFMAN & M. HALLETT. 1994. Explicit and implicit learning and maps of cortical motor output. Science **265:** 1600–1601.
24. SCHIEBER, M.H. 1990. How might the motor cortex individuate movements. Trends Neurosci. **13:** 440–445.
25. SCHIEBER, M.H. & L.S. HIBBARD. 1993. How somatotopic is the motor cortex hand area? Science **261:** 489–492.
26. AOU, S.A., C.D. WOODY & D. BIRT. 1992. Increases in excitability of neurons of the motor cortex of cats after rapid acquisition of eye blink conditioning. J. Neurophysiol. **12:** 560–569.
27. WOODY, C.D. 1986. Understanding the cellular basis for learning and memory. Annu. Rev. Psychol. **37:** 433–493.
28. PASCUAL-LEONE, A., J.M. TORMOS, J. KEENAN, *et al.* 1998. Study and modulation of human cortical excitability with transcranial magnetic stimulation. J. Clin. Neurophysiol. **15:** 333–343.
29. PASCUAL-LEONE, A., D. BARTRES-FAZ & J.P. KEENAN. 1999. Transcranial magnetic stimulation: studying the brain-behaviour relationship by induction of 'virtual lesions'. Philos. Trans. R. Soc. Lond. B Biol. Sci. **354:** 1229–1238.
30. MENDOZA, D.W. & H. WICHMAN. 1978. 'Inner' darts: effects of mental practice on performance of dart throwing. Percept. Motor Skills **47:** 1195–1199.
31. MCBRIDE, E.R. & A.L. ROTHSTEIN. 1979. Mental and physical practice and the learning and retention of open and closed skills. Percept. Motor Skills **49:** 359–365.
32. DENNIS, M. Visual imagery and the use of mental practice in the development of motor skills. Can. J. Appl. Sport Sci. **10:** 4S–16S.
33. SCHONBERG, H. 1987. Great Pianists. Fireside Books. St. Louis, MO.
34. SCHONBERG, H. 1988. The Virtuosi. Vintage-Random House. New York.
35. INGVAR, D.H. & L. PHILIPSON. 1977. Distribution of the cerebral blood flow in the dominant hemisphere during motor ideation and motor performance. Ann. Neurol. **2:** 230–237.
36. ROLAND, P.E., L. ERICKSSON, S. STONE-ELANDER & L. WIDEN. 1987. Does mental activity change the oxidative metabolism of the brain? J. Neurosci. **7:** 2373–2389.
37. DECETY, J. & D.H. INGVAR. 1990. Brain structures participating in mental simulation of motor behavior: a neuropsychological interpretation. Acta Psychol. (Amst.) **73:** 13–34.
38. DECETY, J., D. PERANI, M. JEANNEROD, *et al.* 1994. Mapping motor representations with positron emission tomography. Nature **371:** 600–602.
39. HALLETT, M. 1998. The neurophysiology of dystonia. Arch. Neurol. **55:** 601–603.
40. PUJOL, J., J. ROSET-LLOBET, D. ROSINES-CUBELLS, *et al.* 2000. Brain cortical activation during guitar-induced hand dystonia studied by functional MRI. Neuroimage **12:** 257–267.
41. HALLETT, M. 1995. Is dystonia a sensory disorder? Ann. Neurol. **38:** 139–140.
42. BARA-JIMENEZ, W., M.J. CATALAN, M. HALLETT & C. GERLOFF. 1998. Abnormal somatosensory homunculus in dystonia of the hand. Ann. Neurol. **44:** 828–831.
43. ELBERT, T., V. CANDIA, E. ALTENMULLER, *et al.* 1998. Alteration of digital representations in somatosensory cortex in focal hand dystonia. Neuroreport **9:** 3571–3575.
44. BYL, N.N. & M. MERZENICH. 1997. The neural consequences of repetition: clinical implications of a learning hypothesis. J. Hand Ther. **10:** 160–174.

45. BYL, N.N., M.M. MERZENICH & W.M. JENKINS. 1996. A primate genesis model of focal dystonia and repetitive strain injury. I. Learning-induced dedifferentiation of the representation of the hand in the primary somatosensory cortex in adult monkeys. Neurology **47:** 508–520.
46. SANGER, T., A. PASCUAL-LEONE, D. TARSY & G. SCHLAUG. 2001. Non-linear sensory cortex response to simultaneous tactile stimuli in writer's cramp. Movement Dis. In press.
47. SANGER, T., D. TARSY & A. PASCUAL-LEONE. 2001. Abnormalities of spatial and temporal sensory discrimination in writer's cramp. Movement Dis. In press.

# Cortical Plasticity and Music

JOSEF P. RAUSCHECKER

*Georgetown Institute for Cognitive and Computational Sciences, Georgetown University Medical Center, Washington, DC 20007, USA*

**ABSTRACT: Auditory experience changes the make-up of areas in the cerebral cortex that are involved in the processing of complex sounds, including music. Evidence for this comes from various lines of research. Early blindness results in an expansion of auditory-responsive areas in the parietal cortex and a refinement in the selectivity of neurons in the auditory cortex. Occipital areas normally used only for vision are activated by auditory stimuli in the early blind. This lends credibility to the claim that blind individuals have greater musical abilities. The cross-modal changes in auditory cortical representations are based on activity-dependent modifications of synaptic circuitry. Imagery and anticipation of music also lead to activation of the auditory (and frontal) cortex. It is conceivable, therefore, that even with mental practice alone we can sharpen our musical representations in the cerebral cortex.**

**KEYWORDS: Brain plasticity; Blindness and enhanced sound localization; Conductors, orchestral, and enhanced sound localization**

Remarkable as our cerebral cortex is in many respects, its ability to self-organize in response to extrinsic stimuli is perhaps the most astonishing property of all. The three chapters in this section are all impressive testimony to this ability. The effects of musical experience can be found on various levels, such as the gross-anatomical differences among professional musicians with absolute pitch (Ref. 1 and Schlaug, this volume) and the subtle functional differences after musical training found with magnetoencephalography (Ref. 2 and Pantev, this volume) or transcranial magnetic stimulation (Ref. 3 and Pascual-Leone, this volume). Of course, even the ability to learn and memorize a simple tune is an expression of the brain's ability to change with musical experience. In this case, the changes would have to be sought in even finer modifications of synaptic strength in distributed cortical networks.

## BASIC PRINCIPLES OF CORTICAL PLASTICITY

The tenet that learning and memory are based on changes of synaptic efficacy was propagated, among others, by Hebb.[4] More specifically, he postulated that the conjunction of pre- and postsynaptic activity contributes to a strengthening of synaptic connections. Hebb's ideas were applied to many different areas of neuroplasticity,

Address for correspondence: Josef P. Rauschecker, Georgetown Institute for Cognitive and Computational Sciences, Georgetown University Medical Center, New Research Building, Room WP15, 3970 Reservoir Road NW, Washington, DC 20007-2197. Voice: 202-687-1580; fax: 202-687-0617.

rauschej@georgetown.edu

but perhaps most prominently in the field of visual cortical plasticity. Development of the visual cortex became a model system for the study of neural plasticity when Wiesel and Hubel established many of the fundamental facts, such as ocular dominance plasticity after monocular lid closure.[5] The development of orientation selectivity was also proposed early on as an example of experience-dependent plasticity at the single-neuron level[6] and was later shown to follow Hebbian rules.[7]

## IMPROVED AUDITORY ABILITIES IN THE BLIND

Deprivation in one sensory modality, such as vision, can also lead to dramatic reorganization in other modalities, such as hearing or touch.[8] The mechanisms of cross-modal plasticity of the auditory cortex appear to be the same as those of visual cortical plasticity.[9] Compensatory improvement of auditory abilities in the blind had long been postulated, claiming, among other things, improved musical abilities.

While it is difficult to quantitatively assess musical abilities, other auditory perceptual capacities can be tested more easily. An ability of blind people to localize sounds with greater precision has been found in a number of recent studies.[10,11] These studies followed on the footsteps of work in visually deprived cats[12] and ferrets[13] that had also demonstrated improved sound localization abilities in these animals. Animal studies can additionally be used to find the neural bases of these behavioral improvements. In blind cats, visually deprived by binocular lid suture from birth, an area in the parietal cortex (the anterior ectosylvian visual area [AEV]) that is normally innervated by visual fibers is completely taken over by auditory and somatosensory inputs.[14] At the single-unit level, auditory cortical neurons show greater selectivity for the position of sounds in space.[15] Taken together, these findings suggest a greater sampling density of auditory space in the cerebral cortex as the basis for the greater precision in sound localization of blind individuals.[8] Furthermore, the expansion of auditory territory in the cerebral cortex would appear to enable other refinements of auditory perception as well.

Modern neuroimaging techniques, such as positron emission tomography (PET), can be used to look for such cross-modal map changes directly in blind humans. During localization of sounds in virtual auditory space via headphones, sighted humans normally activate parts of their posterior parietal cortex in an area separate from known visual areas involved in spatial analysis.[16] Blind humans, in addition to activating a much more extended area in the same region, also activate large parts of the occipital lobe, which is normally used for seeing[17] (FIG. 1). The results of cross-correlating activities in different brain regions suggest that the auditory input to the occipital cortex is relayed via posterior parietal cortex. Similar results have been found with a visual-to-auditory substitution device activating occipital cortex during pattern recognition.[18] These surprising results warrant further testing in animals with neurophysiological and neuroanatomical techniques.

## "WHAT" AND "WHERE" IN AUDITORY CORTEX

Another fundamental tenet of Hebbian models of long-term memory is that information gets stored in the same places where it is processed. We are far from under-

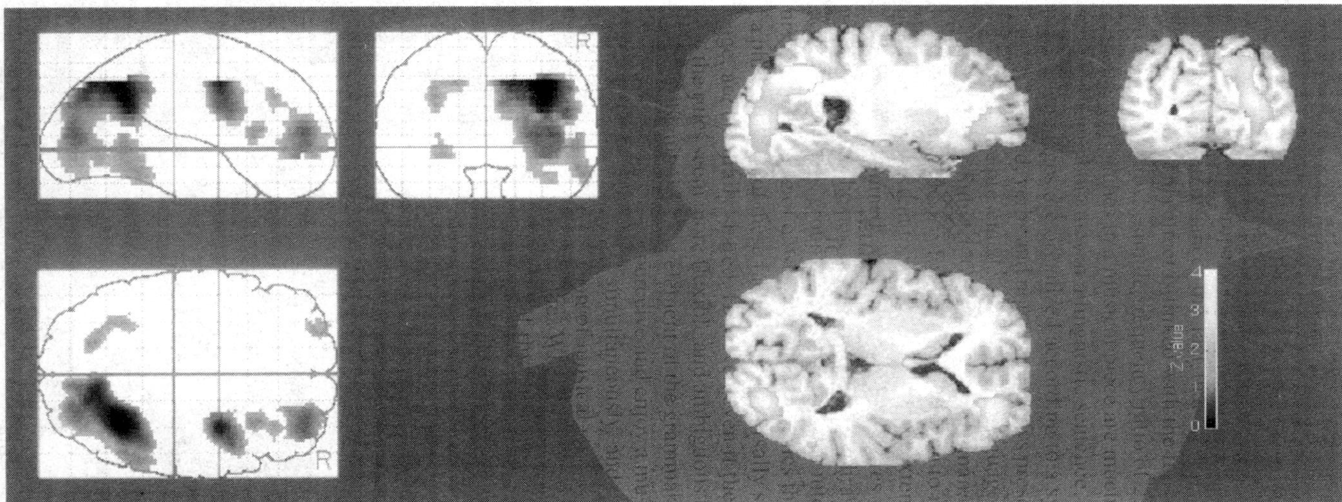


**FIGURE 1. See Plate 22 in color section.** Results of PET scanning in congenitally blind patients during localization of sounds presented via headphones in virtual auditory space (data from Ref. 17). Areas in the occipital cortex that are driven only by visual stimuli in sighted individuals are now activated during sound localization. The "rewiring" seems to happen via the parietal cortex, consistent with studies on visually deprived animals.[8]

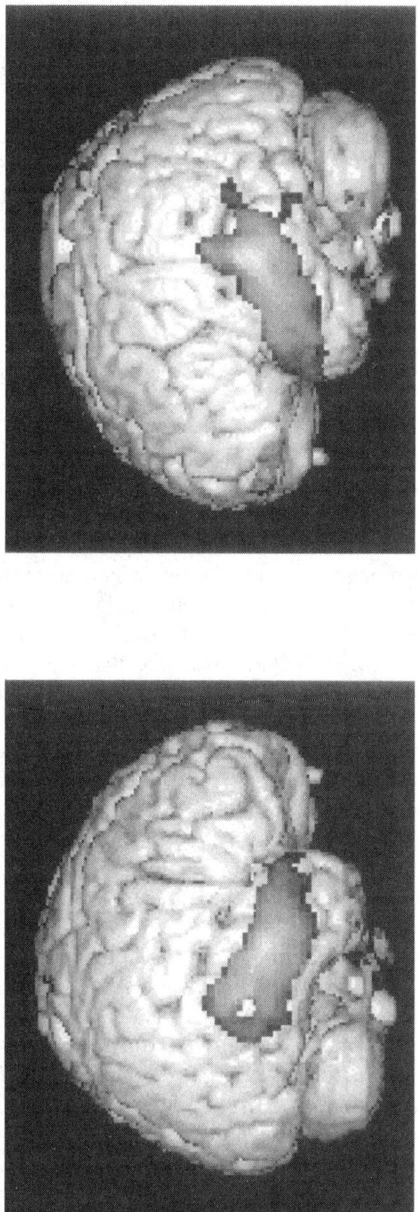

**FIGURE 2. See Plate 23 in color section.** Auditory cortical activation while listening to music. Averaged group results of functional magnetic resonance imaging are shown.

standing how lengthy strings of musical melodies or even more complex pieces of music are stored in our brains. What we are beginning to understand, however, is that complex sounds are first broken down into their frequency components in the periphery of the auditory system. At later stages of cortical processing this tonal information is integrated into more complex response properties,[19] and single units respond to more and more complex contents. The hierarchical processing of such "what" information seems to take place in more anterolateral portions of the superior temporal gyrus (STG), all the way to the temporal pole, whereas the location of sounds is determined in more posterior aspects of the STG, which project to the posterior parietal cortex.[20,21] The anterior STG projects to the orbitofrontal cortex, whereas the caudal STG projects to the dorsolateral prefrontal cortex, directly and via the parietal cortex, again consistent with its role in spatial processing.[22]

## MUSICAL LEARNING AND AUDITORY IMAGERY

If the anterior STG and orbitofrontal cortex are responsible for the processing of complex auditory patterns, they should also be the storage houses of musical memories. Indeed, when we try to imagine music, we activate parts of the auditory cortex that are anterior of Heschl's region plus areas in the frontal cortex.[23] Even when we simply anticipate familiar music, the inferior frontal cortex lights up.[24] This may correspond to the involvement of left frontal regions in the retrieval of complex memories, as predicted by Tulving's HERA model.[25] Other brain regions, such as the cerebellum and the anterior cingulate cortex, are also active during anticipatory imagery of music. This justifies the idea that mental rehearsal, by activating some of the same brain regions as during a real performance, helps to practice and memorize music, even though we are not actually playing the instrument at that time. It is also well known that experienced conductors not only are able to conduct hundreds of pieces by heart but also rehearse them mentally, which stresses again the partial equivalence of brain activation through attention and mental exercise.

"What" and "where" information in music come together for conductors of large orchestras. In these situations, the content of what is being played is of utmost importance, but the location of a particular instrument on the stage helps the conductor to pinpoint the source of errors. Training this ability obviously helps, as has been found in a recent study with event-related brain potentials.[26] Conductors show greater differentiation between positions in peripheral auditory space than nonmusicians or pianists. As attention has been shown to increase brain activation in nonprimary auditory cortex,[27] attentional modulation during rehearsal would be expected to sharpen conductors' skills by modulating the activity of brain regions encoding auditory space. This situation is not unlike that of blind individuals focusing completely on nonvisual events, although in the latter situation attention provides only the initial vehicle that leads to permanent change. At the single-neuron level, cells have been found in area CL of the monkey that are selective to both the content and the location of complex sounds.[21] Such cells could well provide the neural substrate for the above effects.

## SENSITIVE PERIODS OF AUDITORY CORTICAL PLASTICITY

Auditory cortical plasticity can be demonstrated in everyday processes, such as musical learning, and it becomes even more overtly obvious in the reorganization of the auditory cortex across areal borders after early blindness. Age certainly plays an important role in these plastic changes, as the age of onset of musical training as well as the age of onset of blindness have both been shown to be critical for the extent of reorganization. Learning a musical instrument is easier during childhood not only because our motor abilities are more malleable during these early phases, but also because our auditory cortex has sensitive periods for self-organization. This does not mean that we cannot acquire these abilities any more later in life, but an equivalent outcome will require greater effort. Similar sensitive periods apply to language learning: learning a second language after the age of eight is definitely still possible, but there will usually be an accent.[28]

## CONCLUSIONS

The connection between brain activation and musical perception is becoming clearer as more and more refined techniques become available to image neural activity. Brain imaging can be performed while subjects are listening to music or imagining it (FIG. 2). One can measure the effects of music on brain activation and cerebral blood flow as well as its effects on brain development and later brain organization. Although animals do not have music (just as, in a strict sense, they do not speak), one can still get a grasp of the exact neural mechanisms of music perception by measuring the responses of single neurons to various types of complex sounds in animals.

## ACKNOWLEDGMENTS

This work was supported in part by a grant from the National Institute of Deafness and Other Communication Disorders (R01-DC03489).

## REFERENCES

1. SCHLAUG, G. *et al.* 1995. *In vivo* evidence of structural brain asymmetry in musicians. Science **267**(5198): 699–701.
2. PANTEV, C. *et al.* 1998. Increased auditory cortical representation in musicians. Nature **392**(6678): 811–814.
3. PASCUAL-LEONE, A., D. BARTRES-FAZ & J.P. KEENAN. 1999. Transcranial magnetic stimulation: studying the brain-behaviour relationship by induction of 'virtual lesions.' Philos. Trans. R. Soc. Lond. B. Biol. Sci. **354:** 1229–1238.
4. HEBB, D.O. 1949. The Organization of Behavior. Wiley. New York.
5. WIESEL, T.N. & D.H. HUBEL. 1963. Single-cell responses in striate cortex of kittens deprived of vision in one eye. J. Neurophysiol. **26:** 1003–1017.
6. BLAKEMORE, C. & G.F. COOPER. 1970. Development of the brain depends on the visual environment. Nature **228:** 477–478.

7. RAUSCHECKER, J.P. 1991. Mechanisms of visual plasticity: Hebb synapses, NMDA receptors and beyond. Physiol. Rev. **71**(2): 587–615.
8. RAUSCHECKER, J.P. 1995. Compensatory plasticity and sensory substitution in the cerebral cortex. Trends Neurosci. **18:** 36–43.
9. RAUSCHECKER, J.P. 1999. Auditory cortical plasticity: a comparison with other sensory systems. Trends Neurosci. **22:** 74–80.
10. LESSARD, N. *et al.* 1998. Early-blind human subjects localize sound sources better than sighted subjects. Nature **395**(6699): 278–280.
11. RÖDER, B. *et al.* 1999. Improved auditory spatial tuning in blind humans. Nature **400**(6740): 162–166.
12. RAUSCHECKER, J.P. & U. KNIEPERT. 1994. Auditory localization behaviour in visually deprived cats. Eur. J. Neurosci. **6:** 149–160.
13. KING, A.J. & C. PARSONS. 1999. Improved auditory spatial acuity in visually deprived ferrets. Eur. J. Neurosci. **11**(11): 3945–3956.
14. RAUSCHECKER, J.P. & M. KORTE. 1993. Auditory compensation for early blindness in cat cerebral cortex. J. Neurosci. **13**(10): 4538–4548.
15. KORTE, M. & J.P. RAUSCHECKER. 1993. Auditory spatial tuning of cortical neurons is sharpened in cats with early blindness. J. Neurophysiol. **70**(4): 1717–1721.
16. BUSHARA, K.O. *et al.* 1999. Modality-specific frontal and parietal areas for auditory and visual spatial localization in humans. Nat. Neurosci. **2**(8): 759–766.
17. WEEKS, R. *et al.* 2000. A positron emission tomographic study of auditory localization in the congenitally blind. J. Neurosci. **20**(7): 2664–2672.
18. ARNO, P., A.G. DE VOLDER, A. VANLIERDE, *et al.* 2001. Occipital activation by pattern recognition in the early blind using auditory substitution for vision. NeuroImage: 1–14. Published electronically February 16, 2001.
19. WESSINGER, C.M. *et al.* 2001. Hierarchical organization of human auditory cortex revealed by functional magnetic resonance imaging. J. Cognit. Neurosci. **13:** 1–7.
20. RAUSCHECKER, J.P. & B. TIAN. 2000. Mechanisms and streams for processing of "what" and "where" in auditory cortex. Proc. Natl. Acad. Sci. USA **97:** 11800–11806.
21. TIAN, B. *et al.* 2001. Functional specialization in rhesus monkey auditory cortex. Science. In press.
22. ROMANSKI, L.M. *et al.* 1999. Dual streams of auditory afferents target multiple domains in the primate prefrontal cortex. Nat. Neurosci. **2:** 1131–1136.
23. HALPERN, A.R. & R.J. ZATORRE. 1999. When that tune runs through your head: a PET investigation of auditory imagery for familiar melodies. Cereb. Cortex **9**(7): 697–704.
24. VAN LARE, J.E., B.A. ZIELINSKI & J.P. RAUSCHECKER. 1999. Anticipatory musical imagery: functional MRI studies of the human brain. Talk at International Society for Systematic and Comparative Musicology. Oslo, Norway.
25. TULVING, E. *et al.* 1994. Hemispheric encoding/retrieval asymmetry in episodic memory: positron emission tomography findings. Proc. Natl. Acad. Sci. USA **91:** 2016–2020.
26. MÜNTE, T.F. *et al.* 2001. Superior auditory spatial tuning in conductors. Nature **409**(6820): 580.
27. GRADY, C.L. *et al.* 1997. Attention-related modulation of activity in primary and secondary auditory cortex activation. Neuroreport **8:** 2511–2516.
28. NEWPORT, E.L. 1990. Cognit. Sci. **14:** 11–28.

# The Sounds of Poetry Viewed as Music

FRED LERDAHL

*Department of Music, Columbia University, New York, New York 10027, USA*

**Abstract: An extended parallel is developed between musical and prosodic structures, using the author's cognitively oriented music theory and recent work in generative phonology. For illustration, the sounds of a short poem by Robert Frost are treated entirely in musical terms. The poem is assigned a phonological stress grid and then musical grouping and meter. These structures enable a durational realization. Phonological stress also helps assign the poem's normative melodic contour. Finally, the similarities and differences in sound repetition are given hierarchical structure by means of musical prolongational theory. These formal parallels suggest a corresponding realization in brain localization and function. Evidence from the neuropsychological literature is cited in support of this view. The picture emerges that grouping, meter, duration, contour, and timbral similarity are mind/brain systems shared by music and language, whereas linguistic syntax and semantics and musical pitch relations are systems not shared by the two domains.**

**Keywords: Poetry and music; Music and poetry; Phonology, generative; Frost, Robert**

## INTRODUCTION

Comparisons between music and language have traditionally been couched in terms of syntax or rhetoric.[1] The more substantive parallels, however, are those between musical structures, on the one hand, and phonological and prosodic structures, on the other, and derive from the fact that both music and language consist of sounds organized in time. These parallels are explored by the development of a representational and partly computational account of the sounds of a poem treated entirely as music. This approach builds on work from previous collaborations and consultations.[2–5] As I hope to show, the analysis is suggestive for the neurobiology of music.

## A MUSICAL-POETIC ANALYSIS

Consider in some detail the first two lines of Robert Frost's short lyric, "Nothing Gold Can Stay," given in its entirety in Figure 1. Grouping structure, a fundamental component of music theory, segments a musical surface hierarchically into motives, phrases, and sections. Phonologists have developed the comparable concept of the prosodic hierarchy.[6] Hence, linguistics now offers two kinds of tree structure, a syntactic and a prosodic one. Even when the two coincide in their segmentations, their

Address for correspondence: Dr. Fred Lerdahl, Department of Music, Columbia University, New York, NY 10027. Voice: 212-854-1295.
fred@music.columbia.edu

Nature's first green is gold,
Her hardest hue to hold.
Her early leaf's a flower;
But only so an hour.
Then leaf subsides to leaf.
So Eden sank to grief.
So dawn goes down to day.
Nothing gold can stay.

**FIGURE 1.** "Nothing Gold Can Stay," by Robert Frost.

meanings differ. Syntactic trees represent parts of speech and syntactic phrases; phonological trees represent groupings of spoken sound.

At the lowest levels of the prosodic hierarchy are the syllable and the word. Words are categorized as either content words, which carry major semantic content, or function words, which mainly fill a syntactic role. For example, in the first line of the poem, "Nature's first green is gold," all the words are content words except "is." At the level above the word is the clitic group, which either coincides with a content word or combines content and function words that, together, sound as if they were one word. In the latter case, the stressed syllable of the content word is the clitic host, the sound to which the others are drawn. In "Nature's first green is gold," "is gold" sounds as one word, with "gold" as the clitic host. Likewise in the second line, "Her hardest hue to hold," the content words are "hardest," "hue," and "hold"; "Her hardest" and "to hold" are clitic groups, with "hard-" and "hold" as clitic hosts. Above the clitic group is a grouping of words, called the phonological phrase, for example, "first green." Above that is the intonational phrase, which conveys the melody of speech. The entire line, "Nature's first green is gold," is an intonational phrase. Finally, there is the utterance, which usually corresponds to a sentence. "Nature's first green is gold, / Her hardest hue to hold" is an utterance.

FIGURE 2 displays the grouping analysis of the first two lines of the poem. The syllable and word levels are omitted for convenience. The bracketing categories are indicated in boldface: **C** = clitic group, **P** = phonological phrase, **I** = intonational phrase, and **U** = utterance. A category can repeat from one level to the next. This representation abstractly resembles a parsing of a musical phrase into subgroups and

| [**C** Nature's] | [**C** first] | [**C** green] | [**C** is gold] | [**C** Her hardest] | [**C** hue] | [**C** to hold] |
|---|---|---|---|---|---|---|
| [**P** Nature's] | [**P** first | green] | [**P** is gold] | [**P** Her hardest | hue] | [**P** to hold] |
| [**I** Nature's | first | green | is gold] | [**I** Her hardest | hue | to hold] |
| [**U** Nature's | first | green | is gold | Her hardest | hue | to hold] |

**FIGURE 2.** Bracketed, leveled representation of the prosodic analysis of the first couplet.

motives. As in music, pauses are more likely to occur between, rather than within, these segments, with greater pauses between superordinate boundaries.

Now consider linguistic stress, whose musical equivalent is phenomenal accent. In both, the perception, whether of a syllable or a pitch event, is one of relative sonic prominence within its immediate context. Linguists have proposed a so-called metrical grid to represent linguistic stress in words and phrases.[7] From a musical perspective, this is a misnomer; it should be called a stress grid, for its purpose is to represent not hierarchical periodicities but hierarchical patterns of stress. Within a word, such a pattern is given lexically, as in the first word of the poem, "Náture's" (one would not say "Natúre's"). Within a phrase, the stress pattern obeys the nuclear stress rule, which, after the categorization of words into content and function words, assigns global stress to the strongest syllable of the last content word of a prosodic unit.[8] Thus, we say "first gréen," rather than "fírst green," because "green" is the last stressed syllable of its phonological phrase. This principle can be overridden by the rhythm rule, which ameliorates clashes in stress between adjacent and nearby syllables. According to nuclear stress, one says "hardest húe," but we might prefer to adjust it to "hárdest hue," so that the nearby words "hue" and "hold" are not comparably stressed. The rhythm rule implicitly invokes metrical periodicity: strong beats should be more or less evenly spaced. At this level, then, the choice is between "Her hardest húe to hóld," which obeys nuclear stress but challenges periodicity, and "Her hárdest hue to hóld," which violates nuclear stress in order to distribute the stresses. Finally, nuclear stress can be overridden by nonnormative focus on a particular word, to bring out a semantic nuance. Hence, if we say "Nature's fírst green is gold," the implication is that nature's second green is not gold. Generally, however, nuclear stress holds.

Setting the rhythm rule and focus to one side, a stress grid is constructed by cyclic generation of stresses from level to level in the prosodic hierarchy, observing nuclear stress starting at the phonological grouping level. As illustrated in FIGURE 3a, the procedure is first to assign an *x* to every syllable. Second, assign an *x* in the next row to a lexically relatively stressed syllable in a polysyllabic word or to the host of a clitic group. For the third row, shown in FIGURE 3b, add an *x* for the nuclear stress at the level of the phonological phrase. Finally, in FIGURE 3c, add an *x* for the nuclear stress at the level of the intonational phrase. There is little point in continuing to yet another level, for, as in music, distinctions in linguistic stress attenuate over long spans.

FIGURE 4 recasts the stress pattern in FIGURE 3 into an equivalent music-reductional format. Level *d* brackets the syllables and corresponds in content to the bottom row of the stress grid in FIGURE 3. Level *c* reduces out the syllables and monosyllabic words that in FIGURE 3 are assigned only one *x*. Similarly, levels *b* and *a* keep only those units that carry three and four *x*s, respectively. This layout is the same as that in musical time-span reduction, which selects relatively important pitch events at successively larger levels of rhythmic segmentation. These relationships could alternatively be displayed by a tree representation with right and left branching, again as in corresponding music theory.

Now turn to meter. In both music and poetry, metrical structure consists of hierarchically related periodicities inferred from the signal. It might be objected that periodicities in language, unlike those in the musical case, do not really exist because a)

```
   x          x       x        x          x         x        x
   x x        x       x      x x        x x x       x      x x
[C Nature's] [C first] [C green] [C is gold] [C Her hardest] [C hue] [C to hold]
```

b)

```
   x                  x        x                    x        x
   x          x       x        x          x         x        x
   x x        x       x      x x        x x x       x      x x
[P Nature's] [P first  green] [P is gold] [P Her hardest  hue] [P to hold]
```

c)

```
                               x                             x
   x                  x        x                    x        x
   x          x       x        x          x         x        x
   x x        x       x      x x        x x x       x      x x
[I Nature's  first  green]  is gold] [I Her hardest  hue  to hold]
```

**FIGURE 3.** Assignment of the stress grid in conjunction with the prosodic hierarchy applied to the couplet.

| | | | | | | | | |
|---|---|---|---|---|---|---|---|---|
| level *d:* | [Na][ture's] | [first] | [green] | [is][ gold] | [Her hard][est] | [hue] | [to] | [hold] |
| level *c:* | [Na- ] | [first] | [green] | [ gold] | [ hard- ] | [hue] | [ | hold] |
| level *b:* | [Na- ] | [ | green] | [ gold] | [ | hue] | [ | hold] |
| level *a:* | [ | | | gold] | [ | | | hold] |

**FIGURE 4.** The stress pattern in FIGURE 3 put in reductional format.

syllabic and phrasal durations are so variable. Yet, it would be misleading to say that musical durations are invariable. Expressive musical performance depends on deviations from isochrony. Like meter itself, temporal precision is a mental construct. While it is true that durations in verse are usually more variable than those in music, many poetic idioms, from nursery rhymes to sophisticated traditions, demonstrate considerable regularity.[9] As in music, these verbal idioms approach periodicity as a framework against which expressive deviations take place.

Once periodicity is understood as a relative matter, poetic and musical meter may appropriately be regarded as formally and cognitively equivalent. This has, in essence, long been recognized by music theorists in their use of prosodic foot notation for representing musical meter. Now, however, it is more usual in music theory to employ a grid notation, which represents the metrical hierarchy directly and does not confuse features of meter with those of grouping. A similar consensus has emerged in phonological theory.

```
a)                                     b)
x           x           x              x           x           x
x     x     x     x     x              x  .  x  x  x  x  x  x
x x x x x x x x x x x x x              x x x x x x x x x x x x x

c)
x             x             x
x   x   x     x   x   x     x
x x x x x x x x x x x x x x x
```

**FIGURE 5.** Three metrical grids: (**a**) 6/8 meter; (**b**) 3/4 meter; (**c**) 7/8 meter (2 + 2 + 3).

A poetic or musical meter exists when the perceiver infers conceptually regular levels of beats from the signal, so that a beat that is strong at one level is also a beat at the next larger level. The perceptually most prominent metrical level is called the tactus; it occurs at a moderate tempo, is usually at an intermediate level within a grid, and is often more restricted in regularity than are smaller or larger levels. Beats at a given level are two or three beats apart at the next smaller level. (Even lines in pentameter have intermediate strong beats, resulting in patterns of two and three.) Some poetic or musical idioms, while allowing a mixture of two or three beats apart across levels, discourage the mixture of two or three strong beats apart within a level. Thus, European classical tonal music uses compounds of two and three across levels, as in FIGURE 5a and b, but it regards as atypical the grid in FIGURE 5c, which alternates two and three beats within a single level. Other musical cultures, as those of the Balkan countries, would take FIGURE 5c as normative. The important point is not that different cultures prefer different metrical grids, but that the possible combinations of two and three, across or within levels, are very small. Cultural variation on poetic and musical meters is intrinsically limited.

In constructing a metrical grid, the perceiver seeks an optimal fit between the stresses provided by the acoustic signal and the culturally available repertory of metrical grids. To hear a musical phrase in 3/4 time or a poetic line in pentameter is to select that particular grid because, within the repertory, that grid gives the minimal number of conflicts between stress and beat. When the fit is imperfect, the effect is one of syncopation or, to use terms from traditional scansion, of inversion, substitution, or truncation. If the fit is seriously imperfect, the perceiver switches to another available grid or perhaps abandons metrical understanding altogether.

The interaction of grouping and meter causes intuitions of upbeat and afterbeat in relation to a downbeat. The meter is 3/4 in FIGURE 6a and b, but the two differ in that FIGURE 6a has afterbeats, while FIGURE 6b has an upbeat and an afterbeat. The terminology of traditional scansion recognizes this upbeat–afterbeat distinction. In current notation, FIGURE 6c and d represents the contrasting groupings of trochaic and iambic tetrameter, respectively. Traditional foot scansion becomes troublesome, however, in cases where the presumed foot contradicts the prosodic hierarchy. Supposing our lyric to be in iambic trimeter, in the line, "Her hardest hue to hold," the foot-grouping of "-est" is not with its own word stem, "hard-," but counterintuitively with "hue," as shown in FIGURE 6e. It is unclear in such cases what function the foot

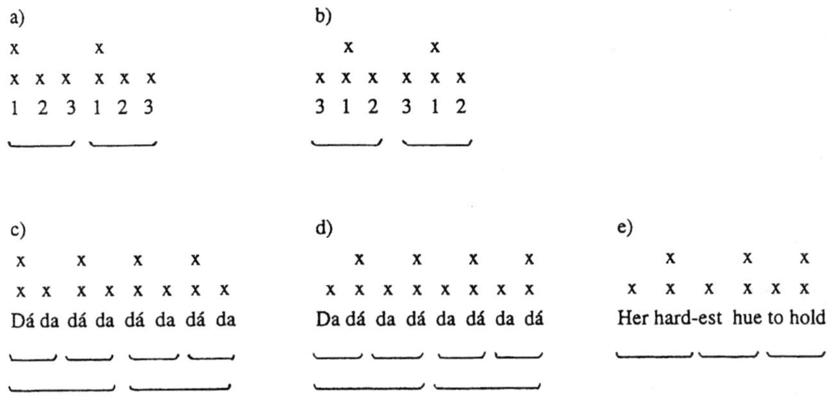


**FIGURE 6.** Interaction of grouping and meter: (**a**) 3/4 meter with afterbeats; (**b**) 3/4 meter with an upbeat and an afterbeat; (**c**) trochaic tetrameter; (**d**) iambic tetrameter; (**e**) conflict between word and foot boundaries at "hardest."

serves. In the present view, iambs, trochees, anapests, and so on arise not from a predetermined schema but directly from the prosodic groupings of beats. Moreover, the concept of the foot lacks a musical counterpart. While recognizing its historical practice, let us ignore the foot as a unit of rhythmic analysis.

Having set to one side the iambic dimension of our poem, I would now suggest that it is not in trimeter either. It is in truncated tetrameter, because of the punctuated pauses at the end of each line and, more elusively, by virtue of the historical style that it evokes. Many short lyrics, such as those of Emily Dickinson, employ so-called common meter, comprising couplets in a four-plus-three beat pattern with an implied silent final beat: da-dá, da-dá, da-dá, da-dá; da-dá, da-dá, da-dá, ---. Common meter is a variant of standard tetrameter, the final silent beat reinforcing the grouping structure. Our poem is, in turn, a variant of common meter. Frost uses it here because of the ineluctable decline that is the poem's theme. Every line falls into silence.

Given this paradigmatic meter, the first two lines of the poem receive the structural description in FIGURE 7. Above the lines appears the stress grid, taken from Figure 3c. Beneath, in boldface and inverted for visual convenience, is the metrical grid for the tactus and larger levels, showing four basic beats per line and their relative metrical strengths. The tactus is represented by uppercase *X*s. To clarify the grid notation, the beats are numbered as two 4/4 measures, grouped into two intonational phrases. Only stresses having three or four *x*s count for establishing the linchpin of the tactus. In accordance with the greatest stresses on "gold" and "hold," these syllables are assigned the strongest beats. The periodicity of the metrical grid requires that "Na-" and "hard-" receive the next strongest metrical position, with "green," "hue," and the silent beats getting the weaker beats. This treatment creates a slight syncopation at "hue." Observe that the strongest metrical beats take place at the end of each line, creating an out-of-phase relationship between the grouping boundaries and the spans between the strongest beats. This is a consequence of nuclear stress and is characteristic of poetry in English. In classical tonal music, by contrast, the strongest beat usually comes at or near the beginning of a phrase. For its ending

```
                                              x                       x
                       x          x           x                  x    x
Stress                 x     x    x           x           x      x    x
grid:                  x x   x    x    x x               x  x x  x x  x
                       Nature's first green is gold,   Her hardest hue to hold.
Metrical               X          X       X   X          X     X    X     X
grid:                  x                  x              x          x
                                          x                         x
Beats in 4/4:          3          4       1   2          3     4    1     2
Phrase boundaries:     └──────────────────────┘ └──────────────────────────┘
```

**FIGURE 7.** Stress and metrical analysis for the first two lines of the Frost poem.

rhythmic articulation, tonal music relies instead on the cadence, for which there is no linguistic counterpart. (In Balinese gamelan, by contrast, the strongest beat usually arrives at the end of a phrase.[10])

This analysis indicates conceptually equal durations between tactus beats. To derive subtactus levels, the two requisite principles are to seek the best possible fit (1) between stress and beat and (2) between duration and the prosodic hierarchy. Condition 1 has already been discussed; it is a matter of matching the metrical with the stress grid, starting with the tactus level. Condition 2 invokes the Gestalt principle that proximate objects tend to group together. It follows that grouping boundaries tend to exist between objects in a string that are relatively nonproximate. The prosodic hierarchy is meant to reflect perceptual groupings of sound; therefore, contextually longer durations—or, more precisely, distances between attack points—should occur between, rather than within, the constituents of the prosodic hierarchy. Once these subtactus levels are established, the metrical positions of all the syllables of the couplet can be translated into standard musical notation.

Let us apply these two conditions to the first three words of the poem "Nature's first green." FIGURE 8a fully satisfies condition 1: "Na-" and "green" provide the tactus frame, and in both grids "-ture's" receives a single *x* and "first" receives two *x*s. Translated into metrical durations, the result is a duple division with two sixteenths plus an eighth note, but this solution does not satisfy condition 2: the greater distance is between "first" and "green," weakening the perception of the phonological group "first green." The alternative in FIGURE 8b, an eighth note plus two sixteenths, is more problematic still, for it both violates condition 1 and breaks up the word "Nature's" with the longer duration. FIGURE 8c effects the preferable solution by smoothing out the durations in a triple-beat realization. Condition 1 is almost met, and there is no implication of an unwanted grouping boundary.

On the assumption that triple meter continues at the eighth-note level, FIGURE 9 completes the subtactus level for FIGURE 7 and realizes the rhythm for the two lines. The match between stress and beat is complete, and the durations reflect the constituents of the prosodic hierarchy: "is gold," "Her hardest," and "to hold" are all clitic groups, grouped musically by temporal proximity and surrounded on either side by longer distances between attack points.

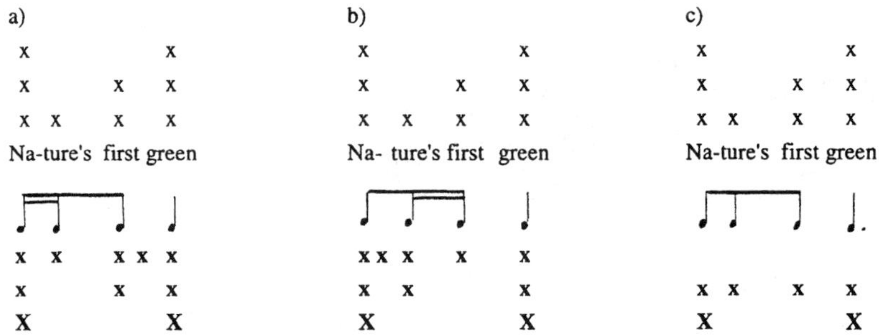


**FIGURE 8.** Duple and triple subtactus metrical realizations for the beginning of the first line.

**FIGURE 9.** Addition of the subtactus level to FIGURE 7, with translation into musical durations.

If this derivational procedure is carried out for the entire poem, the rhythmic derivation is as in FIGURE 10, cast in a 12/8 meter with two-beat anacruses for each line. Minor adjustments with duple divisions appear in the second, third, and fourth lines; these have to do with syllable length, a topic I pass over here (compare "hard-est" in FIGURE 9).

We turn now by way of stress to contour. Three acoustic variables contribute to stress: intensity (or loudness), pitch height, and duration.[11] While duration is an independent variable, intensity and pitch height covary, for the muscles that enable greater intensity also tend to raise pitch. Even when pitch height is artificially isolated, relative height induces the perception of relative stress. Thus, pitch height is a strong projector of stress. Since pitch height is a matter of contour, stress patterns can provide a basis for the derivation of the melody of speech. There are, however, many acceptable contours for an intonational phrase or utterance, each having its own nuance. If our concern were grammaticality, in keeping with much generative

**FIGURE 10.** Durational reading of the poem.

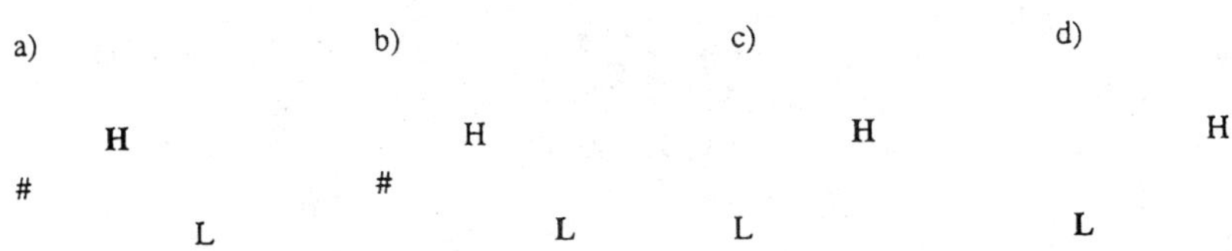


**FIGURE 11.** Four normative contour frameworks.

linguistics, well-formedness constraints would underdetermine the options for contour realizations. A different goal, and one congenial with a musical perspective, is to ask whether a grammatical contour is normative or not. We need then derive only normative contours from stress grids, letting the other grammatical contours be seen as nuanced variants.

To pursue this question, let us posit the restriction of contour features to a binary classification of high and low, designated by "H" and "L," and apply it to the two most stressed syllables in an intonational phrase.[12] The more stressed of the two is shown in boldface. This is a reductional system in a musical sense: dominating **H** or **L**, then secondary H or L, then all the rest of the syllables in the phrase. Suppose, further, that H does not occur at the left edge of the phrase, that is, that a normative intonational phrase does not begin with its most important peak. There is no symmetrical constraint for H at the right edge or for L at either edge. The four logical possibilities are given in FIGURE 11, with "#" standing for syllables to the left of the first H. The prototypical declarative contour is shown in FIGURE 11a and b: a rise and fall whose musical analogue is the tensing-relaxing pattern of a standard tonal phrase. The difference between the two is whether H or L dominates. In FIGURES 11c and d, H is to the right of L; this is the typical shape of an interrogative.

The claim that a higher pitch causes stress would seem to disqualify FIGURE 11b and d as normative, because in these cases **L** dominates. As a consequence of nuclear stress, however, **L** can indeed function in the position of the last clitic host of an intonational phrase. In other words, **L**'s location at the close of a major phrase boundary compensates for its lower pitch. Hence, FIGURE 11b is normative. By the same token, FIGURE 11d is not normative, because in this case **L** is not in a location to benefit from major nuclear stress. As there are no interrogatives in our poem, we can also ignore FIGURE 11c. This leaves FIGURE 11a and b as the remaining options. This very restricted repertory will provide the framework for mapping stress importance onto contour peaks and valleys.

The stress analysis in FIGURE 3 assigns three *x*s for the most stressed syllable in each phonological phrase and four *x*s for the most stressed syllable in each intonational phrase, the latter occurring at the end of the line. Hence, as shown in FIGURE 12a, the intonational contour falls in the pattern of FIGURE 11b. The two lines together, however, form an utterance, leading to the assignment in FIGURE 12b of H and L to four *x*s at this level, over "gold" and "hold," respectively. The assignment for "gold" is thereby reversed from L in FIGURE 12a to H in FIGURE 12b. This step does not yet determine whether H or L in FIGURE 12b is dominating; for reasons beyond the present purview, **H** is chosen. The next step is to combine the shapes at the utterance and intonational levels. In FIGURE 12a, "green" received an H, an assignment that is no longer preferred because it would contradict the **H** in FIGURE 12b. As shown in FIGURE 12c, the H on "green" becomes an L within its intonational phrase; the brackets signify its subordinate status. The H on "hue" remains an H, for it combines at the intonational level with an **L**, although it too is placed in brackets to indicate its lesser hierarchical position. This is the scaffolding within which more local contour shapes take place.

In moving from the abstract H-and-L representation to realized contours, it is useful to posit discrete levels of relative pitch height. Four levels appear to be optimal: three do not afford enough distinctions, and more than four would be fussy in view of the limited ability to categorize degrees of stress. The H-and-L representation is mapped onto these four levels. On the basis of the salience of registral extremes, **H** is placed in the highest slot and L is placed in the lowest slot. FIGURE 13a performs this step for the first couplet, using two lines and two spaces for the four pitch levels and representing degrees of hierarchical importance by relative durational values (that is, these durational values signify not durations but structural importance, as is often done in musical reduction theory). FIGURE 13b appends pitch heights for each [H] and [L], so that each [H] or [L] is placed one slot lower than its adjacent **H**. These locations preserve the rising–falling shape of the contour schema in FIGURE 11a and b. The H-and-L notation can be dispensed with at smaller levels. Except where **L** or L occurs at the right edge of an intonational phrase, the contour reflects the shape of

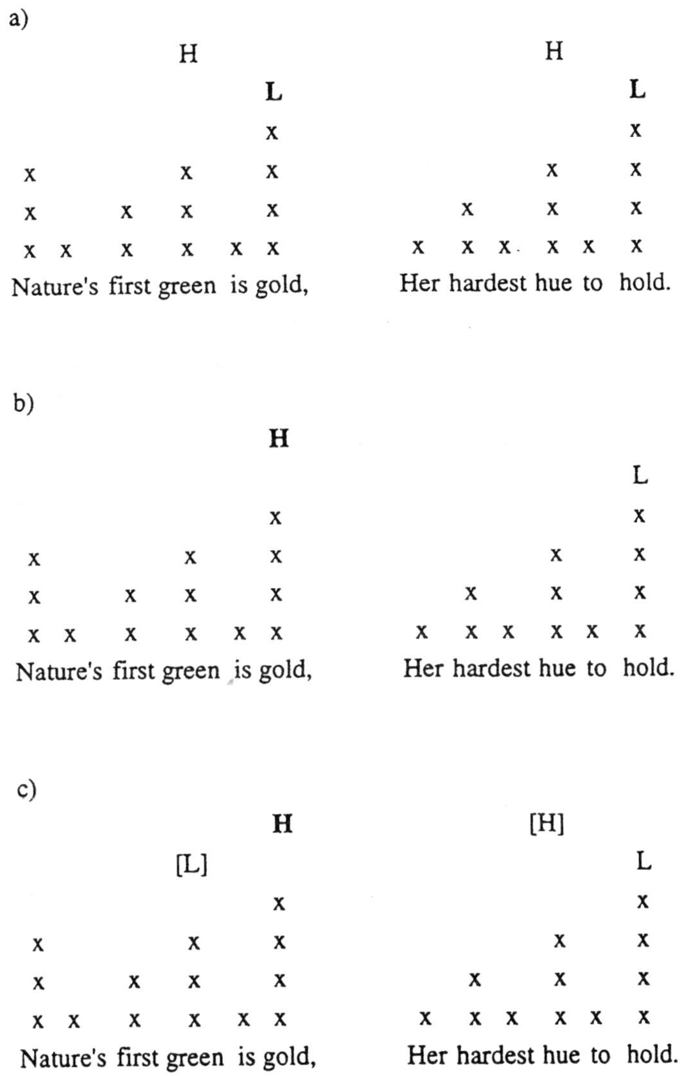


**FIGURE 12.** Global contour assignments for the first couplet.

the stress grid, with its peaks and valleys of *x*s. Thus, in FIGURE 13c the syllables with three and two *x*s, notated by sixteenth notes, appear one slot below the positions of the syllables that were already given three *x*s in FIGURE 13b. Then, in FIGURE 13d come the syllables with one *x*, represented by unstemmed noteheads. These syllables are cliticized and are typically spoken as schwas (the default neutral vowel for an unstressed syllable). They are placed one step below their clitic hosts, except when the latter are in the lowest slot, in which case they are located one step above the clitic

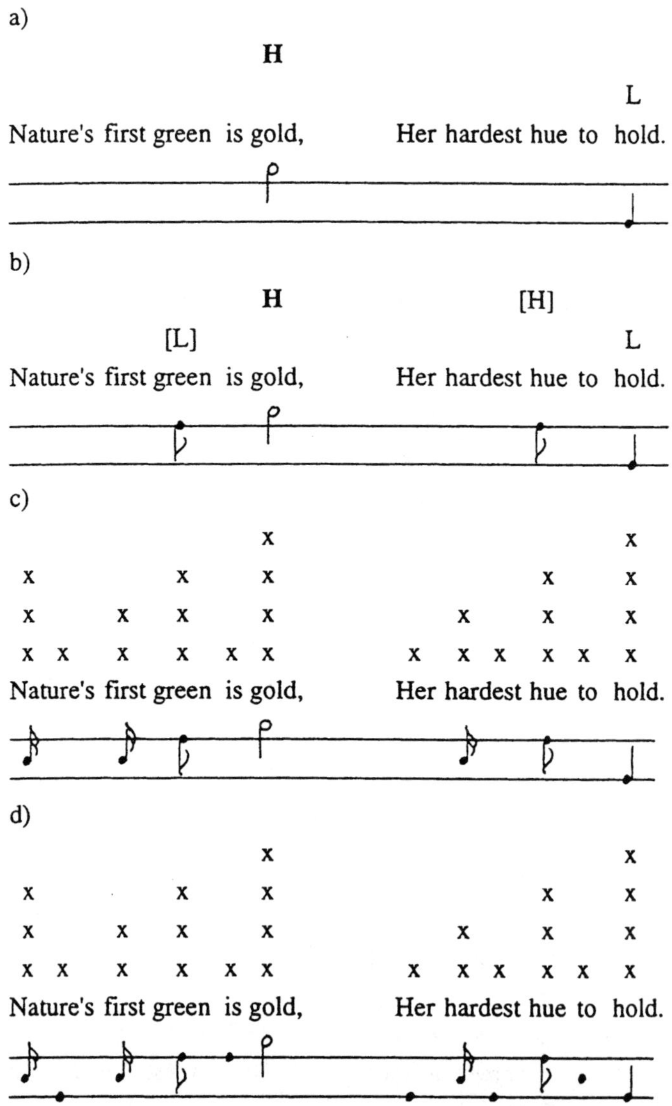


**FIGURE 13.** Assignment of pitch levels to the first couplet: (**a,b**) from the H-and-L analysis in FIGURE 12; (**c**) for unassigned syllables with three and two *x*s; (**d**) for syllables with one *x*.

host. Hence, "-ure's" is one step below "Na-" in "Nature's," and "is" is one step below "gold," but "to" is one step above "hold," because the latter is in the bottom slot.

This derivation of contour is musical not only in its attention to normative shapes but also in its hierarchical treatment. Rather than generate contours from left to right, as if without projection or memory, global highs and lows are established according

**FIGURE 14.** Meter and durations (from FIGURE 10) combined with a contour analysis of the entire poem.

to a few paradigmatic shapes, and the remaining syllables in the speech melody are filled in at successive levels of lesser stress. FIGURE 14 shows the result, combining the contour analysis of FIGURE 13 with the metrical/durational analysis of FIGURE 10, but now with these methods applied to the entire poem. Contained within this seemingly transparent musical notation are the structures of phonological stress, the prosodic hierarchy, the metrical grid, duration, and pitch height. The success of the derivation can be judged informally by its naturalness as a reading.

A usual prosodic analysis would have stopped by this point. From a musical perspective, however, there is yet another step to take, an analysis of the hierarchical patterns of recurrence of sounds. Traditional poetic analysis treats verbal recurrences such as rhyme, alliteration, and assonance as primitive sequential patterns: *aabb, abab,* and so forth. Music theory, by contrast, has a highly developed approach to recurrence in the form of prolongational structure. FIGURE 15 gives three kinds of prolongational connection: (1) strong prolongation or repetition, represented by a dashed slur; (2) weak prolongation or partial repetition, represented by a dotted slur;

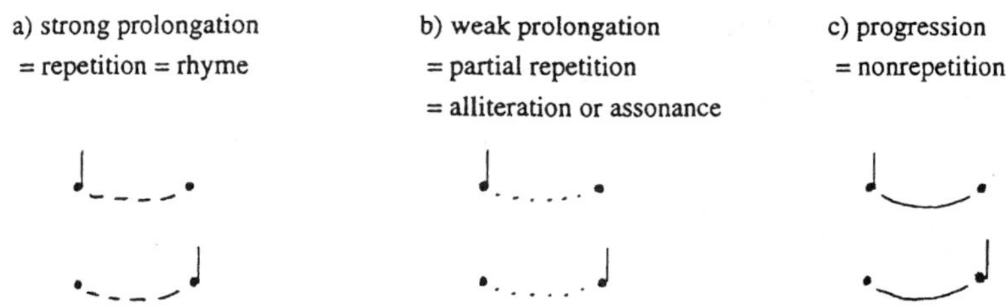


**FIGURE 15.** Repertory of prolongational connections.

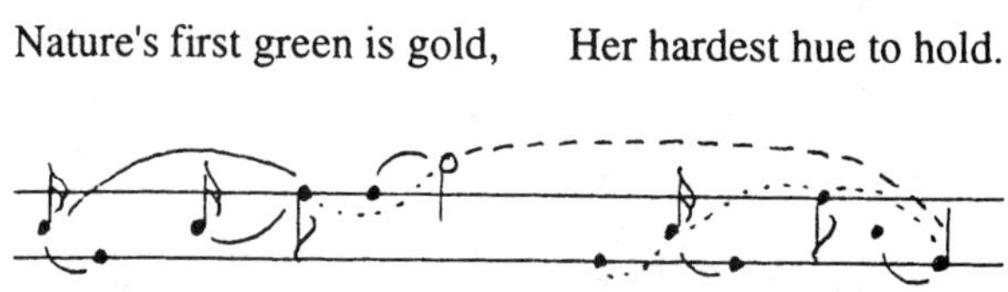


**FIGURE 16.** Prolongational structure for the first couplet. The contours are retained from FIGURE 14.

and (3) progression or nonrepetition, represented by a solid slur. In each case the superordinate element can precede or succeed the elaborating element.

Musical prolongational connections are derived through a combination of time-span segmentation and tonal pitch stability. Poetry incorporates the former through the prosodic hierarchy but lacks any counterpart to the tonal system. The better musical analogy, then, is to atonal music, for which event salience largely replaces event stability.[13] That is, when hearing atonal surfaces, and given the absence of acoustic or idiom-specific criteria to select stable from unstable events within hierarchical time spans, the listener tends to organize embellishing events in relation to events that have greater surface prominence within the framework of the time-span hierarchy. In phonological terms, relative surface prominence is equivalent to relative stress. Hence, prosodic prolongation can be viewed as a kind of atonal prolongation, in which the elements being prolonged are not pitch structures but degrees of timbral similarity mediated by stress within the prosodic hierarchy. (This approach extends beyond that of Lerdahl,[14] in which synthesized vowels are related prolongationally by similarity but without incorporating nested patterns of stress.) The three prolongational categories continue intact: the repetition of rhyme equals strong prolongation; the weaker repetition of alliteration or assonance equals weak prolongation; and syllabic nonrepetition equals progression.

The syllables connected in the prolongational analysis of the first couplet follow the parsing of the prosodic hierarchy in FIGURE 3 and the hierarchical contour in FIGURE 13d. Each connection forms a progression, weak prolongation, or strong prolongation, according to the degree of repetition of sound. FIGURE 16 illustrates, retaining the structural notation of FIGURE 13d. In the first line, "-ture's" attaches as a progression from "Na-," "first" to "green," and "is" to "gold." At the next level, "Na-" attaches to "green," because both have three *x*s. Next come the attachments of

[H]s and [L]s to Hs and Ls, and then of the latter to **H**s and **L**s. Thus, "green" attaches "gold" as a weak prolongation because of the alliteration between "green" and "gold." The second line creates a similar pattern, but with more internal alliteration. Next, at the level of the couplet, the two syllables with four *x*s, "gold" and "hold," form a rhyming strong prolongation. Note it is timbral similarity rather than pitch that is being connected prolongationally, for "gold" is in the highest pitch category while "hold" is in the lowest. The resulting graph shows not isolated instances of alliteration and rhyme, as in traditional prosodic analysis, but the richer relationship of partial repetitions nested within superordinate rhymes—"green" is to "gold" as "hue" is to "hold." Progressions between dissimilar syllables occur as relatively local connections. This approach can be extended to the poem as a whole.

This account of prosodic prolongational structure has skipped some derivational stages involving technicalities of the theory of musical reduction. The main point is that we can indeed transplant most of the mechanisms of the musical reduction theory to the poetic realm.

## CONNECTIONS

Let us step back from the derivation of the Frost analysis and assume that the music-analytic methods outlined here have validity for modeling the sonic organization of poetry in general. To what extent do these methods extend to the analysis of ordinary speech? Briefly, ordinary speech possesses phonological stress, the prosodic hierarchy, and contour; so the application of these components should continue intact. On the other hand, the phonological stresses of ordinary speech are irregular enough to inhibit the inference of a metrical grid. Similarly, ordinary speech lacks significant patterns of syllabic repetition, so that the sense of its timbral prolongational structure is weakened. These attenuated perceptions are reflected in the model by weakened metrical and prolongational derivations.

A complementary approach to studying the relationship between music and poetry would be to develop a rigorous theory of textsetting.[15] That listeners have strong intuitions about whether a poetic line is well- or ill-set to music is in itself an indication of the parallels between the two media. Generally, the stronger the correlation between the common formal structures of poetry and music, the more natural the textsetting. That is, a natural textsetting tends to match up the stresses, metrical positions, durations, phrasings, contours, and formal patterns of a poem and its musical realization. The idiom-specific traditions of *Sprechstimme* in early twentieth-century German musical expressionism and of hip-hop in current American popular music show how far music and language can merge along these dimensions.

The overall thrust of the Frost analysis is that the mental representation of the sounds of metrical, rhymed poetry and of music, whether texted or not, shares a good deal more organization than has usually been supposed. Perhaps this commonality has not been sufficiently recognized because in certain basic respects the two media are unlike: poetry has words and phrases with propositional meaning; music has a hierarchical, multidimensional pitch space.[16–18] The analysis of the Frost reflects these differences by not referring to the poem's imagery or meaning and by not invoking musical concepts such as specific pitches and intervals, scales, harmony, key, voice leading, counterpoint, or tonal tension and relaxation.

| *exclusively musical structures* | *common structures* | *exclusively linguistic structures* |
|---|---|---|
| pitches & intervals | durational patterns | syntactic categories & relations |
| scales | grouping (prosodic hierarchy) | word meaning (lexicon) |
| harmony | stress (contextual salience) | semantic structures (sense, reference, entailment) |
| counterpoint | metrical grids | phonological distinctive features |
| tonality | contour | (etc.) |
| pitch prolongations | timbral prolongations | |
| tonal tension & attraction | | |

**FIGURE 17.** Hypothesized brain organization of musical and linguistic structures.

It seems reasonable to suppose that well-founded theoretical parallels and distinctions have counterparts in brain structure and function. Appropriate theory can thus guide and interpret neurological findings. We may inquire, then, to what extent the commonalities and differences discussed here are instantiated in brain structure and function. The present analysis predicts, as one would expect, that there is not a single music area and a single language area, but that musical and linguistic functions are spread out in different structural and functional components. It further predicts that those brain modules that process rhythm, contour, and timbral relationships are the same in music and language, while those that process purely pitch structures and purely linguistic structures occupy different parts of the brain. FIGURE 17 illustrates.

Current neuropsychological evidence offers some support for this view. Studies of patients with brain lesions that affect musical and linguistic processing indicate that musical and phonological contours are processed in the same area of the right hemisphere but that musical pitch structures are processed in the left hemisphere, with a one-way channel from contour processing to intervallic pitch processing.[19–21] If so, individuals with deficits that inhibit their ability to process musical or linguistic contour are also unable to process specific pitch relations. This makes intuitive sense, for if a listener cannot tell up from down, how could he or she distinguish between a minor third and a perfect fourth? On the other hand, individuals unable to process musical pitch might still be able to process contour. Normal people who are musically tone-deaf do not speak in a monotone. (In this connection, tone languages such as Chinese depend not on fixed pitch categories and intervals but on relative pitch height and contour.[22] Thus, a person could remain competent in speaking a tone language despite an inability to distinguish musical pitch relationships.)

Neuropsychological evidence also suggests a bifurcation in musical processing between pitch and rhythm. There is partial justification for this from music theory, in which grouping and meter are distinct components, from both tonal reduction and tonal pitch space.[2,17] Support for this division comes not only from neuropsychology but also from experiments in music cognition.[23] In the details, however, the story is bound to be more complex. From a music-theoretic perspective, pitch is one of the inputs to the derivation of grouping and meter. For example, a tonal cadence signals a grouping boundary, and harmonic rhythm is a strong determinant of meter. Similarly, the derivation of pitch reduction depends on the position of pitches and chords within nested time spans based on grouping and meter. On the other hand, tonal space itself—the system (or systems) of pitches, intervals, scales, chords, tonal regions, tonal attractions, degrees of sensory and cognitive consonance, and degrees of mutual proximity and stability—remains, at a theoretical level, truly independent of duration, grouping, and meter. FIGURE 17 projects this overall picture: except for contour and timbral prolongations, all of the items listed under "common structures" belong to the domain of rhythm, which music and poetry share, and most of the items listed under "exclusively musical structures" belong to tonal space.

Note that the category "exclusively linguistic structures" in FIGURE 17 includes not just syntactic and semantic features but some phonological aspects as well, such as distinctive features.[8] Moreover, the perceptual organization of speech may not fully submit to a general, Gestalt-based auditory account.[24] These qualifications nevertheless do not undermine the convergence between theoretical and empirical treatments for the domains listed under "common structures."

Presumably these exclusive and common structures are a consequence of human evolution. In this view, the roots of music and language are the same, in the form of premusical and prelinguistic communicative and expressive auditory gestures involving shapes of duration, stress, contour, timbre, and grouping. We still communicate with infants and higher mammals in this manner. I believe that these elementary shapes lie at the basis of expressive utterance in language and of musical expression (an idea that goes back at least to Rousseau[25]). With evolution came specialization: music and language diverged in their most characteristic features, pitch organization in music and word and sentence meaning in language. Poetry straddles this evolutionary divergence by projecting, through the addition to ordinary speech of metrical and timbral patterning, its common heritage with music.

## ACKNOWLEDGMENT

I am grateful to the phonologist Bruce Hayes for advice on aspects of this study.

## REFERENCES

1. WINN, J. 1981. Unsuspected Eloquence: A History of the Relations between Poetry and Music. Yale University Press. New Haven.
2. LERDAHL, F. & R. JACKENDOFF. 1983. A Generative Theory of Tonal Music. MIT Press. Cambridge, MA.
3. LERDAHL, F. & J. HALLE. 1991. Some lines of poetry viewed as music. *In* Music, Language, Speech, and Brain. J. Sundberg, L. Nord & R. Carlson, Eds. Wenner-Gren International Symposium Series. Macmillan. London.
4. LERDAHL, F. & X. CHABOT. 1991. A Theory of Poetry as Music and Its Exploration through a Computer Aid to Composition. Proceedings of the International Computer Music Conference, 1991.
5. HAYES, B. Personal communication.
6. HAYES, B. 1989. The prosodic hierarchy in poetry. *In* Phonetics and Phonology: Rhythm and Meter. P. Kiparsky & G. Youmans, Eds. Academic Press. New York.
7. LIBERMAN, M. & A. PRINCE. 1977. On stress and linguistic rhythm. Linguistic Inquiry **8:** 249–336.

8. Chomsky, N. & M. Halle. 1968. The Sound Pattern of English, 2nd printing. MIT Press. Cambridge, MA.
9. Oehrle, R.T. 1989. Temporal structures in verse design. *In* Phonetics and Phonology: Rhythm and Meter. P. Kiparsky & G. Youmans, Eds. Academic Press. New York.
10. Becker, J. & A. Becker. 1979. A grammar of the musical genre *Srepegan.* J. Music Theory **23:** 1–44.
11. Handel, S. 1989. Listening. MIT Press. Cambridge, MA.
12. Pierrehumbert, J. 1980. The Phonology and Phonetics of English Intonation. MIT Ph.D. dissertation.
13. Lerdahl, F. 1989. Atonal prolongational structure. Contemp. Music Rev. **4:** 65–87.
14. Lerdahl, F. 1987. Timbral hierarchies. Contemp. Music Rev. **1:** 135–160.
15. Halle, J. & F. Lerdahl. 1994. A generative textsetting model. Curr. Musicol. **55.**
16. Krumhansl, C.L. 1990. Cognitive Foundations of Musical Pitch. Oxford University Press. New York.
17. Lerdahl, F. 1988. Tonal pitch space. Music Percept. **5:** 315–350.
18. Lerdahl, F. 1996. Calculating tonal tension. Music Percept. **13:** 319–363.
19. Peretz, I. 1993. Auditory agnosia: a functional analysis. *In* Thinking in Sound: The Cognitive Psychology of Human Audition. S. McAdams & E. Bigand, Eds. Oxford University Press. Oxford.
20. Peretz, I. 2001. Le cerveau musical. *In* Enciclopedia Einaudi della musica. J.J. Nattiez, Ed. In press.
21. Patel, A.D. & I. Peretz. 1997. Is music autonomous from language? A neuropsychological appraisal. *In* Perception and Cognition of Music. I. Deliége & J. Sloboda, Eds. Psychology Press. Hove, UK.
22. Yung, B. 1991. The relationship of text and tune in Chinese opera. *In* Music, Language, Speech, and Brain. J. Sundberg, L. Nord & R. Carlson, Eds. Wenner-Gren International Symposium Series. Macmillan. London.
23. Palmer, C. & C.L. Krumhansl. 1987. Independent temporal and pitch structures in determination of musical phrases. J. Exp. Psychol. Human Percept. Perform. **13:** 116–126.
24. Remez, R.E., P.E. Rubin, S.M. Berns, *et al.* 1994. On the perceptual organization of speech. Psychol. Rev. **101:** 129–156.
25. Rousseau, J.-J. 1760. Essai sur l-origine des langues. *In* On the Origin of Language. 1986. J. H. Moran & A. Gode, trans. University of Chicago Press. Chicago.

# Music and Nonmusical Abilities

E. GLENN SCHELLENBERG

*Department of Psychology, University of Toronto at Mississauga, Mississauga, Ontario, Canada L5L 1C6*

**ABSTRACT: Reports that exposure to music causes benefits in nonmusical domains have received widespread attention in the mainstream media. Such reports have also influenced public policy. The so-called "Mozart effect" actually refers to two relatively distinct phenomena. One concerns short-term increases in spatial abilities that are said to occur from listening to music composed by Mozart. The other refers to the possibility that formal training in music yields nonmusical benefits. A review of the relevant findings indicates that the short-term effect is small and unreliable. Moreover, when it is evident, it can be explained by between-condition differences in the listener's mood or levels of cognitive arousal. By contrast, the effect of music lessons on nonmusical aspects of cognitive development is still an open question. Several studies have reported positive associations between formal music lessons and abilities in nonmusical (e.g., linguistic, mathematical, and spatial) domains. Nonetheless, compelling evidence for a causal link remains elusive.**

**KEYWORDS: Musical ability; Music exposure; Mozart effect; Spatial ability; Nonmusical benefits of music training**

## INTRODUCTION

The present report evaluates claims that exposure to music produces benefits in nonmusical domains. These claims began to influence public policy as soon as they came to public notice.[1] For example, Zell Miller, the former Governor of Georgia, budgeted for the distribution of classical music recordings to each infant born in state. Moreover, Florida mandates daily doses of classical music in state-run preschools.

Researchers[2] and journalists[3–6] have generated confusion by failing to clarify the distinction between the short-term consequences of music listening and the long-term consequences of formal training in music. Indeed, results from both types of studies have been merged to yield the dictum, "music makes you smarter." However, passive listening to music, a ubiquitous activity, bears little resemblance to formal training, which involves lessons and systematic practice. Thus, separate evaluation of the short-term benefits of musical exposure and the long-term side effects of music lessons could help to clarify these issues.

Examination of the effects of previous experience on learning and behavior has a rich tradition in the history of psychology. "The *transfer* of training from old to new

Address for correspondence: E. Glenn Schellenberg, Ph.D., Department of Psychology, University of Toronto at Mississauga, Mississauga, Ontario, Canada L5L 1C6. Voice: 905-828-5367; fax: 905-569-4326.

g.schellenberg@utoronto.ca

situations is part and parcel of most, if not all, learning. In this sense the study of transfer is coextensive with the investigation of learning"[7] (p. 1019; emphasis added). In addition, hundreds of psychological examinations of *priming* have investigated how prior exposure to a stimulus affects subsequent processing of the same stimulus or a closely related stimulus.[8] If exposure to music causes welcome side effects, we would expect such effects to arise from transfer or priming. Moreover, some researchers who argue for the nonmusical benefits of exposure posit a specific neuropsychological basis for such benefits.[2,10–12] Presumably, this hypothesized cortical process would be compatible with other cortical processes that are demonstrably relevant to music.

## TRANSFER AND PRIMING

Transfer and priming occur in positive and negative forms. Positive transfer occurs when previous experience in problem solving makes it easier to solve a new problem,[7,13] typically by accelerating learning. As such, positive transfer describes successful generalization of a process or strategy. One example involves reasoning with analogies.[14] Previous exposure to analogies can lead to greater success at finding the missing piece in new analogies (e.g., *Lawyer* is to *client* as *doctor* is to *????*; the correct answer is *patient*). Similar research is available on metaphor and the transfer of skills.[13] A common theme across transfer effects is *similarity*;[13,15] positive transfer is more likely to occur when there are more similarities between the old and new problems.

Negative transfer is the opposite of its positive counterpart; previous experience interferes with solving a new problem.[7] Negative transfer, which is often called *interference*, can occur proactively or retroactively. Proactive interference is evident when previous learning makes subsequent learning relatively difficult. For example, a new problem is approached with an old mental set that is inefficient or inappropriate for the new context. By contrast, retroactive interference refers to difficulty accessing mental representations because of intervening experience between initial encoding and retrieval.

Roughly speaking, priming can be considered the "short-term" or "low-level" relative of transfer. Anderson[16] defines priming as "an enhancement of the processing of a stimulus as a function of prior exposure" (p. 459). In a classic experiment,[17] participants were asked to identify words presented briefly in the visual modality. Performance was superior for words that were seen prior to the word-identification test. The low-level nature of priming is evident in greater priming effects following open-ended instructions (e.g., study the word) compared to compulsory semantic processing (e.g., generate an antonym), the latter condition involving "deeper" levels of processing.[18] It is clear that priming does not require conscious awareness, as reflected in the priming effects observed in amnesics.[19]

Negative priming refers to situations in which the processing of a "target" stimulus is *inhibited* by prior exposure.[20] For example, when participants are presented with two words (a "target" and a "distractor") and required to name only one (the target), performance on subsequent trials is relatively slow when the target word was previously a distractor. Most priming studies examine *repetition* priming, or subsequent processing of an identical stimulus.[8] Nonetheless, cross-modal and cross-language priming effects are also observable. For people who are bilingual in Spanish–English, auditory presentation of a partial sentence in Spanish (the priming stimulus) can facilitate visual recognition of a target English word, provided that the target was implied by the sentential prime.[21] There are higher-level, *associative*[16] or *semantic*[9] priming effects, in which word processing (e.g., butter) is facilitated by previous presentation of an associated word (e.g., bread).[22]

This brief review of transfer and priming provides a context for evaluating specific claims about exposure to music. These claims posit remarkable positive side effects of exposure to certain types of music, side effects that, in principle, are closely related to transfer and priming. My intention is to situate claims that *music makes you smarter* in the context of cognitive psychology, which will permit a review and evaluation of such claims with reference to well-established cognitive phenomena. A secondary goal is to situate the neuronal mechanism advanced as the basis of associations between music and nonmusical abilities in the domain of cognitive neuropsychology.

## THE MOZART EFFECT

The current debate about musical exposure and its side effects was inspired, in part, by Rauscher *et al.*[23] who reported that brief exposure (10 minutes) to a Mozart sonata generates short-term increases in spatial-reasoning abilities (the *Mozart effect*). Each participant in their study was tested in three conditions. Participants in one condition listened to a Mozart sonata before completing one of three tests of spatial abilities. Participants in the other two conditions listened to a relaxation tape or sat in silence before completing a test. Performance was superior in the "Mozart" condition. This finding attracted considerable attention, because it appeared in a highly prestigious journal, *Nature*, and because the investigators translated their finding into an IQ-score improvement of approximately 8 points (i.e., half a standard deviation). Indeed, the popular conclusion that "music makes you smarter" followed directly from this IQ translation.

Closer examination of the method of that study raises questions about the validity of the findings. The choice of comparison conditions is particularly problematic. Sitting in silence or listening to a relaxation tape for 10 minutes is less arousing or interesting than is listening to Mozart. Moreover, mood states are known to influence performance on problem-solving tasks, with superior performance associated with positive affect.[24–26] Thus, the effect could have arisen from differences in mood or arousal rather than from exposure to Mozart.

Because the Mozart effect is at odds with the literature on priming and transfer, alternative explanations of the source of the effect (i.e., the mood/arousal hypothesis) seem all the more credible. Improved spatial skills following exposure to a Mozart sonata do not represent an instance of repetition priming (i.e., the priming stimulus was not repeated) nor are they an instance of associative priming. How is passive listening to a musical stimulus "associated" with performance on a visually presented test of spatial skills? Evidence for associative priming typically involves pairs of words with an obvious semantic association (nurse–doctor, bread–butter). How, then, could an auditory (musical) stimulus prime performance on a task with no obvious link to music? Transfer as an explanatory framework also raises more questions than it answers. Transfer typically involves applying a learned skill or a strategy to a new context. But what is learned by listening passively to a piece of music? Something about the music, no doubt, is learned, but it is difficult to rationalize how the transfer of such knowledge could yield improved performance on a spatial task.

In short, the Mozart effect is a radical claim about cognitive processes that is difficult to reconcile with known principles and findings in cognitive psychology. It comes as no surprise, then, that attempted replications have produced mixed results.[10,27] As of May 2000, there were about 20 published tests of the Mozart effect. Less than half replicated the effect. Because the Mozart-effect studies have been reviewed elsewhere, the present report focuses on the issues raised by selected studies.

Consider the replication reported by Rauscher *et al.*,[12] who "pretested" their participants with a paper-folding-and-cutting (PF&C) test (one of the spatial tests used in the original study). Participants were then divided into three groups of equivalent abilities. One group heard Mozart during three subsequent test sessions. A second group sat in silence during the three sessions. A third group heard a minimalist piece by Philip Glass during the first session, an audiotaped story in the second session, and a repetitive piece of dance music in the third session. After each session, the PF&C test was administered again. Although the Mozart group showed a significantly larger improvement in performance than did the other two groups after the first session, there was no difference between the Mozart and comparison groups after the next two sessions. The advantage of Mozart over silence and Glass conditions in the first test session did not extend or clarify the original finding. Participants may find repetitive, minimalist music as boring or unarousing as silence. The null findings in the second and third sessions also raise doubts about the reliability of the effect.

Rauscher and Shaw[10] suggest that the numerous replication failures can be explained primarily by differences in the spatial tasks that have been used as outcome measures. They claim that the effect can be obtained with "spatial-temporal" tasks (e.g., the PF&C task and other tasks involving mental transformation of visual images), but not with "spatial-recognition" tasks. This distinction is based on the idea that perceiving and remembering music involve identifying changes and systematic transformations in musical patterns (e.g., motives) that occur over time. Thus, transfer from music listening to the spatial domain should be limited to tasks involving mental manipulation of visual images, which also takes time. Indeed, the time required is linearly related to the amount of manipulation.[28] This distinction is curious in light of the original findings,[23] which indicated that the effect was identical across spatial tasks, temporal or otherwise. In a subsequent reanalysis of the original data,[10] however, the advantage of the Mozart effect proved to be significant on only one of the three spatial tests that were administered, the "temporal" PF&C task, and not on the two nontemporal tests. Nonetheless, mean scores were highest in the Mozart condition across tests, and the design precluded tests of the two-way interaction between the listening conditions and the spatial tests. In other words, despite their conclusion and interpretation, the data did not support Rauscher and Shaw's hypothesis, that is, that the influence of Mozart's music on spatial abilities depends on the temporal nature of the tasks. Moreover, the temporal/nontemporal distinction cannot explain why several attempts to replicate the original findings failed to do so, even though the outcome measure was a task that met Rauscher and Shaw's criteria for spatial-temporal status.[29,30] Finally, the distinction does not address the problem that the effect, when evident, may be a consequence of differences in mood or in arousal.

In *all* cases in which the Mozart effect has been evident, comparison conditions involved repetitive music, sitting in silence, or listening to relaxation tapes. As noted, these comparison conditions might seem boring to participants (compared to listening to music), promoting relatively negative mood states or low levels of cognitive arousal. As a first attempt to address this possibility, Nantais and Schellenberg[31] replicated and extended the original findings. In their first experiment, each participant was tested on the PF&C task twice, once after listening to 10 minutes of music and once after sitting in silence for 10 minutes. For some participants, the music was the same Mozart piece used by Rauscher and her colleagues. For others, a piece by Schubert (from the same compact disk performed by the same pianists) was used instead. This experiment was also the first to use a computer-controlled procedure administered to participants individually. Indeed, the potential impact of group dynamics on the results of earlier studies is unknown.[12,23,29,32,33] (Imagine a classroom of undergraduates being required to sit in silence for 10 minutes!)

As shown in FIGURE 1 (upper panel), performance on the PF&C test was better after listening to Mozart than sitting in silence. In other words, the Mozart effect was replicated. Nonetheless, an identical effect was evident when the Mozart composition was substituted with the piece by Schubert (FIG. 1, middle panel). One would predict such a "Schubert effect" if the comparison condition (silence) was depressing levels of performance. For both groups, performance also improved from the first to the second testing session, revealing a simple practice effect. In a second experiment, a Mozart condition was contrasted with a comparison condition that involved listening to a narrated short story (potentially as engaging as listening to music) instead of sitting in silence. The Mozart effect disappeared (FIG. 1, lower panel), as one would predict if the experimental (Mozart) and comparison (story) conditions were equally engaging and if the source of the Mozart effect stemmed from differences in mood or arousal. Perhaps even more important was the finding that performance interacted with listeners' preferences. Those who preferred Mozart over the story performed better on the PF&C test after listening to Mozart. Those who preferred the story performed better after listening to the story (FIG. 2). These findings provide support for the suggestion that short-term effects of music on tests of spatial abilities stem from differences in mood or arousal rather than from listening to Mozart. Although the FIGURE implies that participants who preferred Mozart performed better regardless of condition, the main effect of preference was marginal ($p = 0.09$).

Further support for the "mood or arousal" hypothesis comes from a metanalysis that provided an overview of 20 Mozart–silence comparisons conducted to date.[27] The overall advantage of listening to Mozart on subsequent tests of spatial skills was nonsignificant (i.e., equivalent to 1.4 IQ points). The effect size for studies that used a spatial-temporal outcome measure was also nonsignificant (2.1 IQ points) and smaller than the average test-retest variability in IQ for a single person. Moreover, successful replications of the Mozart effect were attributed to cognitive arousal, which is predominantly a right hemisphere function,[34–36] as are tests of complex

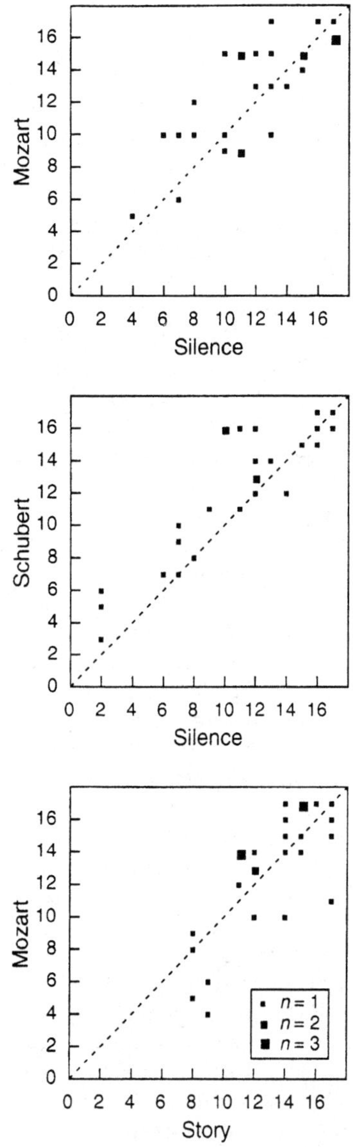


**FIGURE 1.** Scores on the paper-folding-and-cutting task for participants tested by Nantais and Schellenberg (1999). Each participant was tested twice. The *upper panel* illustrates scores after listening to Mozart or sitting in silence. The *middle panel* illustrates scores after listening to Schubert or sitting in silence. The *lower panel* illustrates scores after listening to Mozart or a narrated story. The line on the *diagonal* represents equivalent performance across conditions. The maximum score was 17.

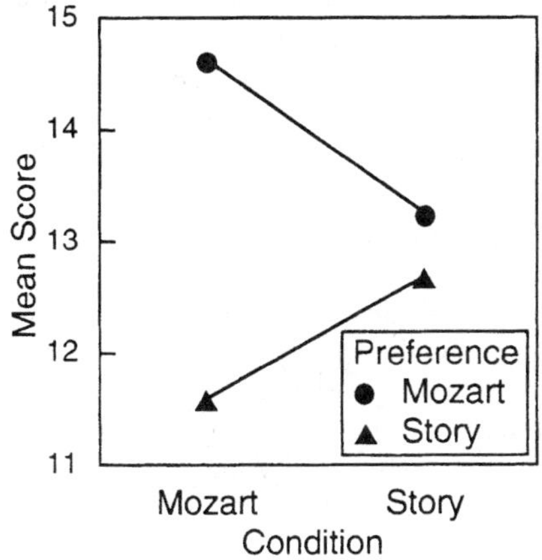


**FIGURE 2.** Scores from Nantais and Schellenberg's (1999) participants as a function of testing condition (Mozart or story) and preference (Mozart or story).

spatial abilities.[36,37] This view helps to explain why the Mozart effect tends to be slightly larger when the control condition consists of relaxation instructions, which are designed to reduce arousal, instead of sitting in silence.[27]

Another way to interpret the Mozart effect is provided by a new theory based on a large body of findings on the association between mood and cognition.[38] The theory proposes that positive mood states increase circulating levels of the neurotransmitter dopamine. During periods of positive affect, dopamine is released from the ventral tegmental area, which has projections to the prefrontal cortex. A variety of cognitive tasks that show improvement when positive affect is induced[25] may be influenced by the effects of dopamine on prefrontal function. It is possible, then, that the Mozart effect is another way in which positive affect influences performance in a problem-solving task. In short, although these seemingly mysterious effects of cross-modal priming (i.e., the Mozart effect) may indeed have a neuropsychological explanation, listening to music is just one of many ways to induce arousal or positive affect.

The metanalysis presented by Chabris[27] and the results of Nantais and Schellenberg[30] are consistent with the idea that differences in mood or arousal are the actual source of the Mozart effect, but neither report tested this hypothesis directly. Thompson *et al.*[39] attempted such a test using the PF&C task as their outcome measure. Each of their participants was tested once in a music condition and once in a silence condition (as in Nantais and Schellenberg,[30] Experiment 1). Participants' mood after listening to the music was measured using the Profile of Mood States.[40] For some participants, the music condition consisted of the same Mozart piece used in the original Mozart-effect study; for others, a piece by Albinoni was used instead. The Albinoni *Adagio* was selected, because it is considered to be a stereotypical example of slow, sad-sounding music.[41] By contrast, the Mozart sonata is pleasant and happy sounding. Hence, the prediction was that increases in performance on the PF&C task would be evident for music compared to silence in the Mozart group but not in the Albinoni group. This prediction was upheld by the data. Moreover, the performance advantage of the music over the silence condition in the Mozart group disappeared when differences in participants' mood scores were partialled out of the analysis. In short, the results were completely consistent with the notion that the Mozart effect is an epiphenomenon of mood or arousal.

The theoretical framework that Rauscher and Shaw use to explain the Mozart effect is called the "Trion model."[42,43] The model states that specific cortical firing patterns are present over large areas of the cortex when one performs, composes, or listens to music. Because these patterns are considered to be spatial-temporal in nature, they are said to be highly similar to patterns evident during spatial-temporal reasoning. Both processes involve perceiving and thinking about rule-governed transformations that occur over time. The model describes more than a simple associative or connectionist network, in which one group of neurons is connected to another group. Rather, it posits actual similarities in cortical firing patterns for (1) passive listening to music and (2) actively participating in a task requiring spatial-temporal reasoning.

If we examine the neuropsychological research on music processing, however, the basic tenets of the Trion model seem implausible. The research of Peretz and her colleagues is particularly relevant. Peretz has shown that much of music perception and cognition is relatively modular, and, moreover, that individual aspects of music cognition are relatively modularized and independent of other aspects.[44] For example, melody and rhythm are processed independently and in different parts of the brain,[45–47] lyrics are processed independently of tunes,[48] and perceiving musical emotion is independent of memory for music.[49,50] Most importantly, Peretz has studied brain-damaged patients with amusia, and none has exhibited accompanying deficits in spatial abilities. For example, one of her amusic patients could not discriminate tones that differed by gross differences in pitch, yet she continued to drive safely around Montreal. In short, there is substantial evidence for modularity of music processing and for independence of various aspects of music. Such evidence is inconsistent with the notion that cortical activity is similar across a variety of musical activities (performing, composing, and listening) and that such patterns of activation are identical during spatial-temporal reasoning.

## LONG-TERM SIDE EFFECTS OF MUSIC LESSONS

Although the short-term Mozart effect appears to be independent of Mozart in particular and of music in general, it is still possible that positive, relatively long-term cognitive side effects result from taking music lessons. Indeed, the two issues may be orthogonal. To anticipate the conclusion, the relevant findings reviewed below are consistent with the idea of an association between musical training and nonmusical benefits, but they fall far short of being conclusive. The studies are grouped according to design (correlational, quasiexperimental, or experimental).

### *Correlational Studies*

Several studies have examined whether musical ability (rather than musical training) is correlated with other kinds of abilities. Positive associations imply that improving one's musical ability through formal lessons would be accompanied by nonmusical benefits. In correlational designs, however, it is always impossible to make firm conclusions about the direction of causation when associations are discovered. It is also impossible to rule out the possibility that the association stems from a third, unidentified variable.

Gromko and Poorman[51] examined children between the ages of 4 and 13 who were enrolled in a private school. Their goal was to determine whether musical aptitude is related to children's ability to use symbols. In an initial testing session, children completed the tonal subtest of Gordon's[52,53] musical aptitude measures. During a second session, children were tested on two tasks, one that required them to match short melodies with graphic representations and another that required them to draw graphic representations of the contour of short melodies. Performance on all three measures improved with age, and each measure was significantly correlated with the other two. These findings confirm that children's musical aptitude is predictive of their ability to interpret and produce symbolic representations of music. Because each of the outcomes was associated with age, however, it is impossible to determine whether the associations would still be in evidence if differences in age were held constant (i.e., the authors did not report partial correlations).

In an examination of performance on musical and spatial tasks that required analogical reasoning, children from 6 to 12 years of age were tested on their ability to transfer a given relation between one pair of stimuli to a novel pair.[54] As the age of the children increased, performance on both tasks improved. Moreover, age-related improvements were virtually identical across tasks. As with the study by Gromko and Poorman,[51] however, the association between the music and spatial tasks could be a consequence of the fact that older children performed better on both tasks.

Lamb and Gregory[55] studied the association between reading and musical abilities in a sample of 5-year-old children. Reading abilities and phonemic awareness were positively associated with pitch-discrimination abilities but not with the ability to discriminate timbres. These associations remained in evidence when differences in age and nonverbal intelligence were held constant. Virtually identical associations between reading abilities and musical abilities (with differences in age and IQ held constant) were reported for a sample of 9-year-old children.[56] Although these findings do not address the issue of causation, they provide evidence of an association between reading and musical abilities that is independent of age or general intelligence.

Douglas and Willatts[57] tested a sample of 8 year olds to examine whether literacy and musical ability are associated. Pairs of tones were presented in a pitch-discrimination task that required children to identify whether the second tone was higher, lower, or the same as the first. A rhythm-discrimination test required children to respond "same" or "different" to pairs of sequences played on a wood block. Literacy was measured with tests of reading and spelling. All measures showed significant pairwise correlations. When differences in receptive vocabulary were held constant, however, reading and spelling measures were associated with rhythm-discrimination abilities but not with pitch-discrimination abilities. Whereas these findings suggest that rhythm-discrimination abilities are better than pitch-discrimination abilities at predicting literacy, the results of Lamb and Gregory[55] imply that pitch-discrimination abilities are a better predictor than timbre-discrimination abilities.

Finally, Lynn *et al.*[58] examined the association between musical aptitude and general intelligence (Spearman's *g*) in groups of children 10 years of age. Children were administered rhythm- and pitch-discrimination tasks as well as tests of general intelligence. Each of the music measures was positively associated with each of the measures of intelligence. These results suggest that musical aptitude is a function of general intelligence. Alternatively, musical aptitude may be a valid estimate of *g*. Although the association between musical aptitude and intelligence is provocative, it remains to be seen whether music lessons actually promote improvements in cognitive abilities.

### *Natural/Quasiexperiments*

Other researchers have tested for the possibility of differences between naturally occurring groups (e.g., those with and without musical training) in nonmusical abilities. Again, because we can never be sure that the groups are identical on other potentially relevant dimensions (e.g., socioeconomic status and overall IQ), unequivocal determinations of causation are impossible.

A classic example of a relevant quasiexperiment is a study by Chan *et al.*[59] of female college students in Hong Kong (mean age of 20 years). The authors compared the verbal and visual memory abilities of women with no musical training to those of women who had taken six years of music lessons before the age of 12. Although the groups did not differ on the visual-memory task, the musically trained group outperformed the untrained group on the verbal-memory task. Unfortunately, despite the authors' claim that the groups were matched according to years of education (with alpha = 0.01), closer inspection of the findings revealed that the musically trained group had significantly more education (with alpha set to a standard 0.05 value). In other words, it is impossible to determine whether the verbal advantage stemmed from music lessons rather than from additional years of education. Indeed, we would predict that better verbal skills would accompany higher levels of education.

Hassler *et al.*[60] examined verbal fluency and visual-spatial abilities in children 9 to 14 years of age. The children were classified into one of three groups: (1) musically talented and capable of composing or improvising, (2) musically talented but not capable of composing or improvising, or (3) nonmusicians. The groups did not differ on a test of spatial relations, but significant differences were found on tests of verbal fluency and visualization abilities, with the musically talented children outperforming the nonmusicians. At a follow-up test two years later, significant differences were found for each of the three outcome variables.[61] Nonetheless, students in the composing/improvising group had more music lessons than did the other musically talented group, yet no differences between these groups on the outcome measures were evident. As such, this study provides equivocal support for the idea that music lessons are accompanied by advantages in nonmusical domains.

Two studies compared the nonmusical abilities of children enrolled in a Kodály music program with those of a comparison group who were not taking music lessons.[62,63] The Kodály program is known for intensive training and for placing great emphasis on singing and on the development of sequential skills. The program also incorporates clapping, the use of hand signs, and simple musical notation. Hurwitz and his colleagues examined the sequencing and spatial skills of a group of seven year olds. Children in the Kodály group had taken music lessons for approximately seven months, with 40-minute lessons five days per week. The sequencing task involved tapping mechanical keys in a regular manner or in time with a metronome after the metronome was turned off or its rate had been changed. Children were also given tests of spatial abilities, plus a Stroop-like test of interference. The Kodály group outperformed the comparison children on the Stroop test and on some of the spatial tests. In a separate examination of children who had completed 1 year of Kodály instruction, the Kodály group performed better than a comparison group on a reading test even though the two groups had performed identically a year earlier. A subsequent study of four- and five-year-olds' understanding of prenumber concepts showed a benefit of Kodály training only for five-year-old girls.[63] These results suggest that training in music may lead to nonmusical improvements, yet it is impossible to ascertain whether nonmusical aspects of Kodály training or preexisting differences between groups may have influenced the results.

### *Experimental Studies*

The next group of studies had more-or-less random assignment of participants to experimental conditions. Thus, provided that comparison conditions were selected appropriately, we should be able to determine whether music lessons actually "cause" nonmusical cognitive advantages. As with most of the short-term (Mozart effect) studies, however, none of the studies in this group used comparison conditions that preclude the possibility of alternative explanations for the findings.

For example, six-year-old children who were taught music for seven months by means of the Kodály method showed improvements in mathematical and reading abilities that surpassed those of children without such training.[64] The researchers' goal was to examine possible by-products of a "test arts" (Kodály) program that was implemented in some first-grade classes but not in others. They examined two first-grade classes in each of two schools that were designated as "test arts" classrooms and another two from both schools that were "standard arts" classrooms. If we assume that the classrooms were assigned to the two arts programs at random, we can consider the design to approximate a "true" experiment. The reported advantage for the test-arts classes is remarkable when we consider that in the previous year children in the test-arts classes were actually *behind* the standard-arts children in terms of the proportion who had reached the national average grade level. Although these results are promising, children in the standard-arts classrooms did not participate in activities focusing on "sequenced skill development" as did children in the test-arts (Kodály) classrooms. Again, this confounding makes it impossible to attribute the remarkable recovery and achievements of the test-arts classrooms to training in music per se, rather than to other nonmusical aspects of the Kodály program.

In another study, four-year-old children who received individual 10-minute piano lessons once or twice a week for six to eight months performed better on a test of spatial skills than children assigned to comparison conditions.[11] Nonetheless, other aspects of the design question the reliability of the effect. For example, some of the children had 33% more lessons than other children, yet this additional training in music had no effect on performance. Moreover, the primary comparison condition involved playing with commercial software programs on a computer. Although a computer instructor provided one-on-one instruction about how to use the computer and open the programs, the software (not the instructor) was designed to teach the children basic skills in reading and arithmetic. As such, superior levels of performance in the piano group could be the consequence of additional instruction from an adult nonparent.

Standley and Hughes[65] found that children in prekindergarten classes (four to five years of age) who took 15 music lessons over a period of two months showed enhanced pre-reading and writing skills compared to other children. Children in the comparison condition were exposed to the regular prekindergarten curriculum but had no additional lessons of any kind. Again, it is impossible to determine whether the observed numerical and verbal benefits arose specifically from music instruction or from pedagogical differences that were independent of musical training. The investigators noted that "it was also apparent from the children's reaction that the music activities provided pleasure and excitement about academic participation, possibly generating long range motivation for reading and writing" (p. 83). Nonmusical activities that generate similar levels of pleasure and excitement could generate similar increases in motivation.

Gromko and Poorman's[66] study of three- and four-year-old children enrolled in a private Montessori school is similar to Standley and Hughes'[65] study just described. Children in the music group were provided with weekly group music lessons in addition to the regular curriculum, but the comparison group received no additional lessons of any sort. As such, the modest gains in nonverbal IQ witnessed for the music group relative to the comparison group can be attributed simply to additional educational instruction from an adult.

Three recent experimental studies suffer from similar methodological problems. Each compared young children enrolled in music-education programs with children in "control" groups who had no comparable extraschool activities.[67–69] One study provided three years of piano lessons free of charge to children in the fourth to sixth grades.[68] These "piano" children performed better than children in a control group on a comprehensive test of cognitive abilities after the first and second years, but the difference disappeared after the third year. Between-group differences during the first two years stemmed solely from differences in spatial abilities. In another study, kindergartners were provided with group keyboard lessons for eight months.[69] The keyboard children showed greater improvement than a control group on tests of spatial abilities, but there was no difference between groups on a test of recognition. A third study examined the influence of a 30-week structured music curriculum on cognitive development.[67] Treatment and control groups of six year olds were administered six subtests from the Stanford-Binet Intelligence Test before and after the curriculum. The treatment group showed relatively larger gains on a single subtest that measured capacity of short-term memory (Bead Memory).

Another recent study examined possible side effects of group keyboard lessons that were provided free of charge to children six to eight years of age.[70] A control group had computer lessons with a commercial software program designed to improve English language skills. Both groups were also given lessons intended to enhance spatial abilities by playing with a software program designed by the researchers. Unfortunately, the main outcome variable consisted of scores on a "testing" version of the same spatial software, which has unknown reliability and validity. Moreover, aggregate scores on the outcome tests did not differ between groups. The investigators reported a significant advantage for the keyboard group on a subtest of mathematical fractions and proportions, and they concluded that improved musical and spatial skills lead to improved mathematical abilities. These results would be more convincing if they had been obtained with standardized tests and if the piano group had performed better overall or at least on subtests for which clear predictions were made a priori.

The studies just reviewed provide consistent *suggestive* evidence that music lessons have positive nonmusical side effects. Nonetheless, specifics of the reported associations vary widely from study to study. If we suspend our disbelief, however, and assume that music education affects abilities in other areas, how could we account for this influence?

A number of neurological studies describe ways in which music lessons affect cortical development. Compared to nonmusicians, accomplished players of string instruments show increased representation in the cerebral cortex for the fingers of their left hand,[71] which implies that musical training can alter patterns of cortical organization. Indeed, cortical representations are especially large for those who begin music lessons at an early age when the brain is relatively plastic. Although the size of the corpus callosum is larger in musicians than in nonmusicians, this effect is particularly notable in musicians who began taking lessons before the age of 7.[72] Relatively large brain asymmetries are also evident among musicians who have absolute (perfect) pitch,[73] and this relatively rare ability to name and produce pitches in isolation is evident predominantly among musicians who begin lessons in early childhood.[74] Moreover, the representation of piano tones in the auditory cortex differs in musicians than in nonmusicians,[75] although genetic factors or simple exposure to music could also play a role.[76] Finally, specific cortical areas in the right hemisphere are activated when reading a musical score but not when reading one's primary or secondary language.[77]

Consequences of an enriched environment on other species (e.g., rats and mice) include denser patterns of dendritic branching and a greater number of hippocampal neurons.[78,79] If music education represents an enrichment of a child's environment, such enrichment could promote neurological development, which could, in turn, influence abilities in other domains. Music, however, is simply one of many ways to enrich a child's environment. Moreover, music education is a complex process that involves many different dimensions. As such, it may be more fruitful to examine the effects of music education at a behavioral level instead of attempting to map such effects directly onto cortical architecture.

We know that schooling improves a wide variety of cognitive skills and that this association is not simply a by-product of maturation.[80–83] For young children in particular, schooling is more effective in smaller classes.[84,85] Reviews of intervention programs for children who are at risk of academic failure suggest that extended one-on-one contact with a supportive adult is a common feature of successful interventions.[86–87] Thus, music lessons, which are typically taught individually or in small groups, may confer nonmusical benefits for children by providing close and extended contact with an adult other than a parent or teacher. If this is the case, then similar side effects should be evident with other types of lessons that provide similar levels of contact (e.g., chess, drawing).

Music lessons may be unique, however, because of their focus on a particular combination of factors, such as hours of individual practice, learning to read music, attention and concentration, timing, ear training, sight reading, constructive feedback from the instructor, and exposure to music.[88] Thus, positive transfer effects to nonmusical domains, such as language, mathematics, or spatial reasoning, could be similarly unique for individuals who take music lessons. On the other hand, music lessons are likely to improve many *general* skills, such as attending to rapidly changing temporal information, honing skills of auditory stream segregation, developing the ability to detect temporal groups, becoming attentive to signals of closure and other gestalt cues of form, developing emotional sensitivity and expressiveness, and developing fine motor skills. These general skills should be particularly likely to transfer to a variety of nonmusical domains.

As someone who took music lessons from the age of five and practiced regularly for the next 11 years, I feel changed—probably for the better—in ways that seem specific to my involvement with music. It remains to be seen, however, whether this personal observation will withstand the test of rigorous experimental investigation.

## ACKNOWLEDGMENTS

Preparation of this article was supported by a grant from the International Foundation for Music Research. Bill Thompson and Sandra Trehub provided helpful comments on an earlier draft.

## REFERENCES

1. Kolata, G. 1999. Muddling fact and fiction and policy. New York Times, WK5.
2. Shaw, G.L. 2000. Keeping Mozart in Mind. Academic Press. San Diego, CA.
3. Calamai, P. 1999. Mozart theory creates discord in the scientific community. The Toronto Star, A28, Aug. 26.
4. Ingram, J. 1997. More proof music lifts young IQs. The Toronto Star, F8, May 11.
5. Martin, S. 1994. Music lessons enhance spatial reasoning skills. APA Monitor, October: 5.
6. Motchuk, A. 1997. Can Mozart make maths add up? New Scientist, 17, March 15.
7. Postman, L. 1971. Transfer, interference and forgetting. *In* Woodworth & Schlosberg's Experimental Psychology, 3rd Edit. J.W. Kling & L.A. Riggs, Eds.: 1019–1132. Holt, Rinehart & Winston. New York.
8. Tulving, E. & D.L. Schachter. 1990. Priming and human memory systems. Science **247:** 301–306.
9. Neely, J.H. 1991. Semantic priming effects in visual word recognition: a selective review of current findings and theories. *In* Basic Processes in Reading: Visual Word Recognition. D. Besner & G.W. Humphreys, Eds.: 264–336. Erlbaum. Hillsdale, NJ.
10. Rauscher, F.H. & G.L. Shaw. 1998. Key components of the Mozart effect. Percept. Motor Skills **86:** 835–841.
11. Rauscher, F.H., G.L. Shaw, L.J. Levine, *et al.* 1997. Music training causes long-term enhancement of preschool children's spatial-temporal reasoning. Neurol. Res. **19:** 2–8.
12. Rauscher, F.H., G.L. Shaw & K.N. Ky. 1995. Listening to Mozart enhances spatial temporal reasoning: towards a neurophysiological basis. Neurosci. Lett. **185:** 44–47.
13. Robins, A. 1996. Transfer in cognition. Connec. Sci. **8:** 185–203.
14. Gick, M.L. & K.J. Holyoak. 1983. Schema induction and analogical transfer. Cognit. Psychol. **15:**1–38.
15. Gentner, D., M.J. Ratterman & K.D. Forbus. 1993. The roles of similarity in transfer: separating retrievability from inferential soundness. Cognit. Psychol. **25:** 524–575.
16. Anderson, J.R. 1995. Cognitive Psychology and Its Implications, 4th Edit. Freeman. New York.
17. Jacoby, L.L. 1983. Perceptual enhancement: persistent effects of experience. J. Exp. Psychol. Learn. Mem. Cognit. **9:** 21–38.

18. CRAIK, F.I.M. & R.S. LOCKHART. 1972. Levels of processing: a framework for memory research. J. Verb. Learn. Verb. Behav. **11:** 671–684.
19. GRAF, P., L.R. SQUIRE & G. MANDLER. 1984. The information that amnesic patients do not forget. J. Exp. Psychol. Learn. Mem. Cognit. **10:** 164–178.
20. MAY, C.P., M.J. KANE & L. HASHER. 1995. Determinants of negative priming. Psychol. Bull. **118:** 35–54.
21. HERNANDEZ, A.E., E.A. BATES & L.X. AVILA. 1996. Processing across the language boundary: a cross-modal priming study of Spanish-English bilinguals. J. Exp. Psychol. Learn. Mem. Cognit. **22:** 846–864.
22. MEYER, D.E. & R.W. SCHVANEVELDT. 1971. Facilitation in recognizing pairs of words: evidence of a dependence between retrieval operations. J. Exp. Psychol. **90:** 227–234.
23. RAUSCHER, F.H., G.L. SHAW & K.N. KY. 1993. Music and spatial task performance. Nature **365:** 611.
24. BENEDICT, R.H.B., M. DOBRASKI & M.Z. GOLDSTEIN. 1999. A preliminary study of the association between changes in mood and cognition in a mixed geriatric psychiatry sample. J. Gerontol. B Psychol. Sci. Soc. Sci. **54B:** 94–99.
25. ISEN, A.M. 1999. Positive affect. *In* The Handbook of Cognition and Emotion. T. Dalgleish & M. Power, Eds.: 521–539. Wiley. New York.
26. O'HANLON, J.F. 1981. Boredom: practical consequences and a theory. Acta Psychol. **49:** 53–82.
27. CHABRIS, C.F. 1999. Prelude or requiem for the 'Mozart effect'? Nature **400:** 826–827.
28. SHEPARD, R. & J. METZLER. 1971. Mental rotation of three-dimensional objects. Science **171:** 701–703.
29. STEELE, K.M., K.E. BASS & M.D. CROOK. 1999. The mystery of the Mozart effect: failure to replicate. Psycholog. Sci. **10:** 366–369.
30. STEELE, K.M., K.M. BROWN & J.A. STOECKER. 1999. Failure to confirm the Rauscher and Shaw description of recovery of the Mozart effect. Percept. Motor Skills **88:** 843–848.
31. NANTAIS, K.M. & E.G. SCHELLENBERG. 1999. The Mozart effect: an artifact of preference. Psycholog. Sci. **10:** 370–373.
32. STEELE, K.M., T.N. BALL & R. RUNK. 1997. Listening to Mozart does not enhance backwards digit span performance. Percept. Motor Skills **84:** 1179–1184.
33. STEELE, K.M., S. DALLA BELLA, I. PERETZ, *et al.* 1999. Prelude or requiem for the 'Mozart effect'? Nature **400:** 827.
34. HELLER, W. & J.B. NITSCHKE. 1997. Regional brain activity in emotion: a framework for understanding cognition in depression. Cognit. Emotion **11:** 637–661.
35. ROBERTSON, I.H., J.B. MATTINGLEY, C. RORDEN & J. DRIVER. 1998. Phasic alerting of neglect patients overcomes their spatial deficit in visual awareness. Nature **395:** 169–172.
36. TUCKER, D.M., A. HARTRY-SPEISER, L. MCDOUGAL, *et al.* 1999. Mood and spatial memory: emotion and right hemisphere contribution to spatial cognition. Biolog. Psychol. **50:** 103–125.
37. DITUNNO, P.L. & V.A. MANN. 1990. Right hemisphere specialization for mental rotation in normals and brain damaged subjects. Cortex **26:** 177–188.
38. ASHBY, F.G., A.M. ISEN & A.U. TURKEN. 1999. A neuropsychological theory of positive affect and its influence on cognition. Psychol. Rev. **106:** 529–550.
39. THOMPSON, W.F., E.G. SCHELLENBERG & G. HUSAIN. 2001. Arousal, mood, and the Mozart effect. Psycholog. Sci. **12:** 248–251.
40. MCNAIR, D.M., M. LORR & L.F. DROPPLEMAN. 1992. The Profile of Mood States, 3rd Edit. Educational and Industrial Testing Service. San Diego, CA.
41. KRUMHANSL, C.L. 1997. An exploratory study of musical emotions and psychophysiology. Can. J. Exp. Psychol. **51:** 336–352.
42. LENG, X. & G.L. SHAW. 1991. Toward a neural theory of higher brain function using music as a window. Concepts Neurosci. **2:** 229–258.
43. LENG, X., G.L. SHAW & E.L. WRIGHT. 1990. Coding of musical structure and the trion model of the cortex. Music Percept. **8:** 49–62.
44. PERETZ, I. 2001. Music perception and recognition. *In* The Handbook of Cognitive Neuropsychology. B. Rapp, Ed.: 521–540. Psychology Press. Hove, UK.

45. PERETZ, I. 1996. Can we lose memories for music? A case of music agnosia in a non-musician. J. Cognit. Neurosci. **8:** 481–496.
46. PERETZ, I. & R. KOLINSKY. 1993. Boundaries of separability between melody and rhythm in music discrimination: a neuropsychological perspective. Q. J. Exp. Psychol. **46A:** 301–325.
47. PERETZ, I. & J. MORAIS. 1989. Music and modularity. Contemp. Music Rev. **4:** 277–291.
48. BESSON, M., F. FAÏTA, I. PERETZ, *et al.* 1998. Singing in the brain: independence of lyrics and tunes. Psycholog. Sci. **9:** 494–498.
49. PERETZ, I. & L. GAGNON. 1999. Dissociation between recognition and emotional judgements for melodies. Neurocase **5:** 21–30.
50. PERETZ, I., L. GAGNON & B. BOUCHARD. 1998. Music and emotion: perceptual determinants, immediacy, and isolation after brain damage. Cognition **68:** 111–141.
51. GROMKO, J.E. & A.S. POORMAN. 1998. Developmental trends and relationships in children's aural perception and symbol use. J. Res. Music Ed. **46:** 16–23.
52. GORDON, E. 1979. Primary Measures of Music Audiation. G.I.A. Publications. Chicago, IL.
53. GORDON, E. 1982. Intermediate Measures of Music Audiation. G.I.A. Publications. Chicago, IL.
54. NELSON, D.J. & A.L. BARRESI. 1989. Children's age-related intellectual strategies for dealing with musical and spatial analogical tasks. J. Res. Music Ed. **37:** 93–103.
55. LAMB, S.J. & A.H. GREGORY. 1993. The relationship between music and reading in beginning readers. Ed. Psychol. **13:** 19–27.
56. BARWICK, J., E. VALENTINE, R. WEST & J. WILDING. 1989. Relations between reading and musical abilities. Br. J. Ed. Psychol. **59:** 253–257.
57. DOUGLAS, S. & P. WILLATTS. 1994. The relationship between musical ability and literacy skills. J. Res. Reading **17:** 99–107.
58. LYNN, R., R.G. WILSON & A. GAULT. 1989. Simple musical tests as measures of Spearman's *g*. Personality Individ. Diff. **10:** 25–28.
59. CHAN, A.S., Y.-C. HO & M.-C. CHEUNG. 1998. Music training improves verbal memory. Nature **396:** 128.
60. HASSLER, M., N. BIRBAUMER & A. FEIL. 1985. Musical talent and visual-spatial ability: a longitudinal study. Psychol. Music **13:** 99–113.
61. HASSLER, M., N. BIRBAUMER & A. FEIL. 1987. Musical talent and visual-spatial ability: onset of puberty. Psychol. Music **15:** 141–151.
62. HURWITZ, I., P.H. WOLFF, B.D. BORTNICK & K. KOKAS. 1975. Nonmusical effects of the Kodaly music curriculum in primary grade children. J. Learn. Disabil. **8:** 167–174.
63. NEUFELD, K.A. 1986. Understanding of selected pre-number concepts: relationships to a formal music program. Alberta J. Ed. Res. **32:** 134–139.
64. GARDINER, M.F., A. FOX, F. KNOWLES & D. JEFFREY. 1996. Learning improved by arts training. Nature **381:** 284.
65. STANDLEY, J.M. & J.E. HUGHES. 1997. Evaluation of an early intervention music curriculum for enhancing prereading/writing skills. Music Therapy Perspect. **15:** 79–85.
66. GROMKO, J.E. & A.S. POORMAN. 1998. The effect of music training on preschoolers' spatial-temporal task performance. J. Res. Music Ed. **46:** 173–181.
67. BILHARTZ, T.D., R.A. BRUHN & J.E. OLSON. 2000. The effect of early music training on child cognitive development. J. Applied Dev. Psychol. **20:** 615–636.
68. COSTA-GIOMI, E. 1999. The effects of three years of piano instruction on children's cognitive development. J. Res. Music Ed. **47:** 198–212.
69. RAUSCHER, F.H. & M.A. ZUPAN. 2000. Classroom keyboard instruction improves kindergarten children's spatial-temporal performance: a field experiment. Early Childhood Res. Q. **15:** 215–228.
70. GRAZIANO, A.B., M. PETERSON & G.L. SHAW. 1999. Enhanced learning of proportional math through music training and spatial-temporal reasoning. Neurol. Res. **21:** 139–152.
71. ELBERT, T., C. PANTEV, C. WIENBRUCH, *et al.* 1995. Increased cortical representation of the fingers of the left hand in string players. Science **270:** 305–307.
72. SCHLAUG, G., L. JÄNCKE, Y. HUANG, *et al.* 1995. Increased corpus callosum size in musicians. Neuropsychologia **33:** 1047–1055.

73. SCHLAUG, G., L. JÄNCKE, Y. HUANG & H. STEINMETZ. 1995. *In vivo* evidence of structural brain asymmetry in musicians. Science **267:** 699–700.
74. TAKEUCHI, A.H. & S.H. HULSE. 1993. Absolute pitch. Psychol. Bull. **113:** 345–361.
75. PANTEV, C., R. OOSTENVELD, A. ENGELIEN, *et al.* 1998. Increased auditory cortical representation in musicians. Nature **392:** 811–813.
76. MONAGHAN, P., N.B. METCALFE & G.D. RUXTON. 1998. Does practice shape the brain? Nature **394:** 434.
77. NAKADA, T., Y. FUJII, K. SUZUKI & I. KWEE. 1998. 'Musical brain' as revealed by high-field (3 tesla) functional MRI. Neuroreport **9:** 3853–3856.
78. GREENOUGH, W.T. & F.R. VOLKMAR. 1973. Pattern of dendritic branching in occipital cortex of rats reared in complex environments. Exp. Neurol. **40:** 491–504.
79. KEMPERMAN, G., H.G. KUHN & F.H. GAGE. 1997. More hippocampal neurons in adult mice living in an enriched environment. Nature **386:** 493–495.
80. CECI, S.J. 1991. How much does schooling influence general intelligence and its cognitive components? A reassessment of the evidence. Dev. Psychol. **27:** 703–722.
81. CECI, S.J. & W.M. WILLIAMS. 1997. Schooling, intelligence, and income. Am. Psychol. **52:** 1051–1058.
82. MORRISON, F.J., E.M. GRIFFITH & D.M. ALBERTS. 1997. Nature-nurture in the classroom: entrance age, school readiness, and learning in children. Dev. Psychol. **33:** 254–262.
83. MORRISON, F.J., L. SMITH & M. DOW-EHRENSBERGER. 1995. Education and cognitive development: a natural experiment. Dev. Psychol. **31:** 789–799.
84. BLATCHFORD, P. & P. MORTIMORE. 1994. The issue of class size for young children in schools: what can we learn from research? Oxford Rev. Ed. **20:** 411–428.
85. MOSTELLER, F. 1995. The Tennessee study of class size in the early school grades. Future Child. **5:** 113–127.
86. BARNETT, W.S. 1995. Long-term effects of early childhood programs on cognitive and school outcomes. Future Child. **5:** 25–50.
87. LAZAR, I. & R. DARLINGTON. 1982. Lasting effects of early education: a report from the Consortium for Longitudinal Studies. Monogr. Soc. Res. Child Dev. **47** (2–3, Serial No. 195).
88. LAMONT, A. 1998. Responses to Katie Overy's paper: respondent 3. Psychol. Music **26:** 201–204.

# Are Music and Language Homologues?

STEVEN BROWN

*Department of Clinical Neuroscience, Karolinska Institute,*
*Huddinge University Hospital M98, 141 86 Huddinge, Sweden*

**KEYWORDS: Music and language; Language and Music; "Musilanguage" system**

I have argued previously that music and language are homologous functions that evolved from a common ancestor that embodied their shared features, something that I have called the "musilanguage" system.[1] According to this model, the shared/parallel features of music and language evolved before their distinct, domain-specific features. These parallel features include use of a limited pool of discrete building blocks, arranged combinatorially, to generate structured phrases, modulated by expressive phrasing mechanisms.

In thinking about the evolutionary relationship between music and language, it is useful to make a distinction between three types of features and to consider models for their respective brain localizations (see FIG. 1): (1) shared features, (2) parallel features, and (3) distinct features. Shared features are those that are identical between music and language, and would include the general processes of vocalization as well as affective prosody, that is, the expression of emotional states in music or language. Parallel features are those that are analogous (but not identical) between music and language, and would include the features of discreteness, combinatoriality, phrase formation, and phrasing mentioned above. Finally, the distinct features are those that are specific to each domain and are therefore neither shared nor parallel, and would include music's use of isometric rhythms and pitch blends and language's use of words and propositional syntax.

I want to propose a model for how these three types of features could be instantiated in the modern brain starting from a bilaterally symmetric "musilanguage" system in the early hominid brain: (1) shared features are mediated by shared modules, (2) parallel features are mediated by duplicate modules, and (3) distinct features are mediated by diverse neural areas whose arrangements are not predictable a priori. The idea of shared modules implies that during the divergence of music and language from the "musilanguage" precursor, both functions came to adopt the same neural areas for the same functions. By contrast, the idea of duplicate modules suggests that during the divergence process, parallel functions developed as specializations in either the left or right hemisphere, where homologous functions came to occupy more-or-less corresponding positions in the two hemispheres. Finally, local-

Address for correspondence: Steven Brown, Department of Clinical Neuroscience, Karolinska Institute, Huddinge University Hospital M98, 14186 Huddinge, Sweden. Voice: +46-8-5858-6867; fax: +46-8-5858-6865.

steven.brown@neuro.ki.se

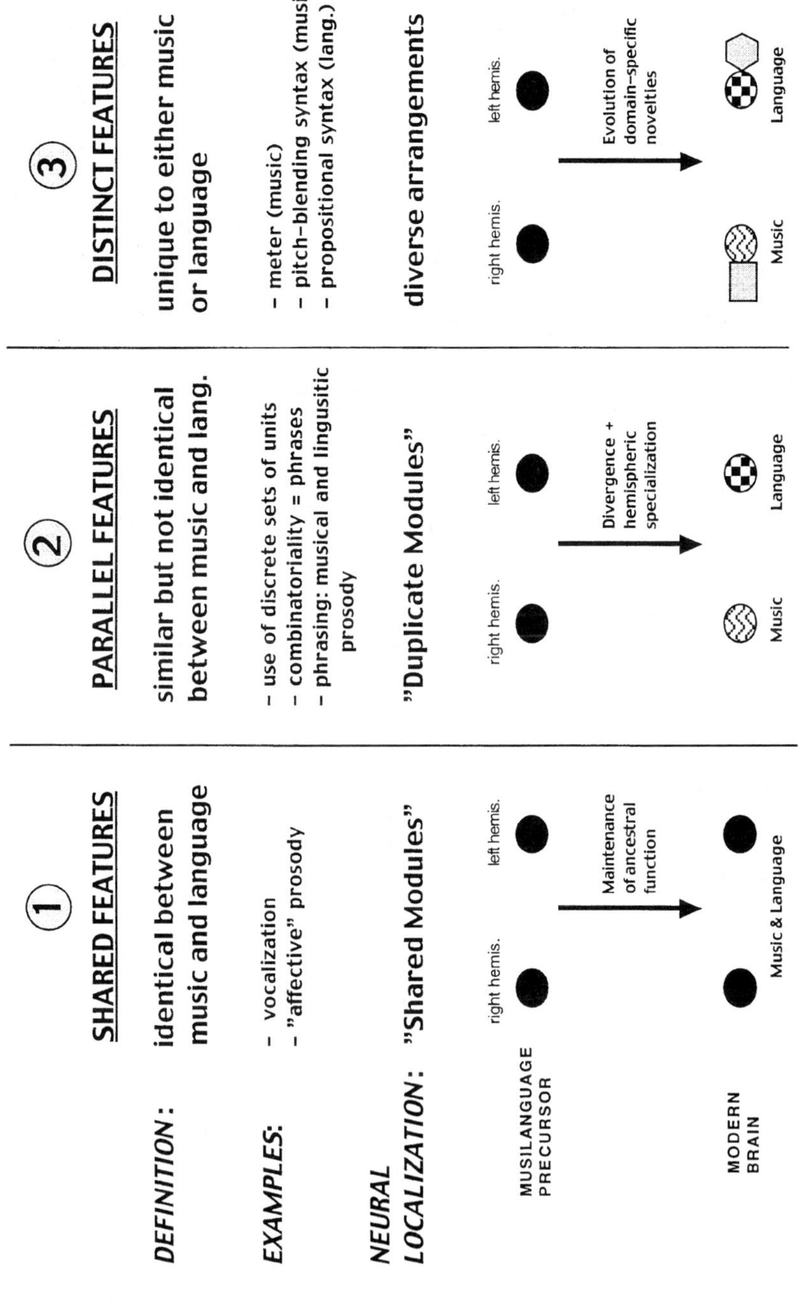

1
SHARED FEATURES
2
PARALLEL FEATURES
3
DISTINCT FEATURES
DEFINITION:
identical between music and language
similar but not identical between music and lang.
unique to either music or language
EXAMPLES:
- vocalization
- "affective" prosody
- use of discrete sets of units
- combinatoriality = phrases
- phrasing: musical and lingusitic prosody
- meter (music)
- pitch-blending syntax (music)
- propositional syntax (lang.)
NEURAL LOCALIZATION:
"Shared Modules"
"Duplicate Modules"
diverse arrangements
MUSILANGUAGE PRECURSOR
right hemis.
left hemis.
Maintenance of ancestral function
Divergence + hemispheric specialization
Evolution of domain-specific novelties
MODERN BRAIN
Music & Language
Music
Language


FIGURE 1.

ization of the distinct features does not depend on sharing or parallelism and therefore occurs in a diversity of arrangements.

From the neuroimaging literature, we can identify examples of each of these arrangements. Regarding sharing, the neural substrates for vocalizing[2] and reading[3] either music or language seem to involve a high degree of sharing or overlap. The evidence for duplication lies with the correspondence between the localizations of musical and language functions in the brain, notably with regard to the superior temporal cortex (area 22) and inferior frontal cortex (especially areas 44 and 45). For example, a study by Zatorre *et al.* of musical imagery for songs *with* words showed significant bilateral activation in these areas,[4] whereas Halpern and Zatorre's study of imagery for songs *without* words showed preferential activation of these same areas in the right hemisphere only.[5] Finally, the domain-specific features of music and language (such as meter, absolute pitch, word lexicons, propositional syntax) seem to show a diversity of arrangements that are distinct from those related to the shared and parallel features.

Such an overall trichotomy of cognitive features and neural localizations, while highly speculative, could be useful in designing neuroimaging experiments to disentangle music and language in the brain. The notion that music and language are homologues could explain much about the similarities and differences between these two human-specific forms of auditory communication.

## REFERENCES

1. BROWN, S. 2000. The "musilanguage" model of music evolution. *In* The Origins of Music. N.L. Wallin, B. Merker & S. Brown, Eds.: 271–300. MIT Press. Cambridge, MA.
2. PERRY, D.W., *et al.* 1999. Localization of cerebral activity during simple singing. Neuroreport **10:** 3979–3984.
3. NAKADA, T., *et al.* 1998. "Musical brain" revealed by high-field (3 Tesla) functional MRI. Neuroreport **9:** 3853–3856.
4. ZATORRE, R.J., *et al.* 1999. Hearing in the mind's ear: a PET investigation of musical imagery and perception. J. Cognit. Neurosci. **8:** 29–46.
5. HALPERN, A.R. & R.J. ZATORRE. 1999. When that tune runs through your head: a PET investigation of auditory imagery for familiar melodies. Cereb. Cortex **9:** 697–704.

# Ambiguous Musical Figures

## Sequential Grouping by Common Pitch and Sound-Source Location versus Simultaneous Grouping by Temporal Synchrony

MARTINE TURGEON AND ALBERT S. BREGMAN

*Auditory Laboratory, Psychology Department, McGill University, 1205 Dr. Penfield Avenue, Montreal H3A 1B1, Canada*

KEYWORDS: **Grouping, auditory; Fusion, auditory; Separation, auditory**

### INTRODUCTION

#### *Relative Contribution of Grouping Cues to the Perception of Concurrent Musical Events*

This research assumes that musical organization is governed by ecologically valid auditory grouping cues that exploit biologically relevant regularities in the acoustic signal.[1] Although the contribution of many specific cues to auditory grouping has been established empirically,[2] their interaction is poorly understood, especially in an open outdoor field. It is important to study grouping in the context of many interacting cues, because in real-world situations, including musical performances, no grouping cue acts in isolation. Furthermore, by creating competition among cues known to promote segregation or fusion, it becomes possible to determine which cues are most important in the perceptual organization of sounds produced by concurrently active sources, like the different instruments played simultaneously during a jazz performance. The present study was conducted in a open field with a semicircular array of speakers to look at how the spatial separation of sound sources interacts with two of the most robust cues for the grouping of concurrent tones: harmonicity[2] and temporal synchrony.[3]

#### *Rationale of the Rhythmic Masking Release (RMR) Paradigm*

We used the RMR paradigm[3,4] to study the relative contribution of onset asynchrony, deviations from simple harmonic ratios, and the spatial separation of sources ($\Delta\theta$) to the segregation of concurrent brief tones. In RMR, a rhythm is perceptually masked by embedding identical tones irregularly among the regular tones. The rhythm is camouflaged because no acoustic property distinguishes the regular subset

Address for correspondence and current affiliation: Martine Turgeon, Institut de Recherche en Coordination Acoustique/Musique (IRCAM), Perception et Cognition Musicales, 1 Place Igor-Stravinsky, 75004 Paris, France. Voice: +33-1-44-78-12-78; fax: +33-1-44-78-15-40.
martine.turgeon@ircam.fr

of tones from the irregular one. We refer to the irregular tones as "maskers"; though they do not mask the individual tones, they mask their rhythmic organization. "Captor" tones can be added in different critical bands simultaneously with the irregular maskers. These tones release the rhythm from masking when they fuse perceptually with the maskers.[3] The newly formed masker-captor units have emergent properties, such as a different timbre and a new pitch; this distinguishes perceptually the irregularly spaced tones from the regularly spaced ones. Consequently, the accurate perception of the rhythm is contingent upon the fusion of the irregular maskers and captors. Measuring the *listener's ability to identify the embedded rhythm thus provides an estimate of the degree of fusion* of the maskers and captors.

## METHODS

### *Subjects*

The listeners were 18 adults with normal hearing who were unaware of the purpose of the experiment.

### *Stimuli*

Signals were low-pass filtered at 5 kHz, using a flat amplitude (Butterworth) response with a roll-off of 48 dB/octave; they were presented at a rate of 20,000 samples per second. Listeners sat at the center of a semicircular array of 13 speakers, one meter away from each speaker. The speaker array was situated in the sound-attenuated chamber of Dr. R. Zatorre at the Montreal Neurological Institute. The head of the listener was fixed so as to point in the direction of the central speaker of the array. The root mean square intensity level was the same for all the four-partial tones; it was calibrated to be equal to that of a 1-kHz tone presented at 60 dB sound pressure level to the center of the listener's head. Each of the two to-be-discriminated rhythmic patterns repeated to form a sequence that had a total duration of 9.5 s, was composed of 15 tones, and had a tempo of 1.7 tones per second. The two rhythms were different temporal arrangements of a short 384-ms interstimulus interval (ISI) and a long 768-ms ISI. Rhythm 1 repeated an alternation of short, long, short, long ISIs three and a half times. This gave rise to perceptual grouping of tones by pairs. Rhythm 2 repeated a cycle of short, long, long, short ISIs three and a half times; this gave rise to perceptual grouping of tones in which triplets alternated with a single tone. To perceptually camouflage each rhythm, irregular maskers were interspersed among the rhythmic tones. The rhythms had a constant temporal density of one irregular masker for each 192-ms interval. The variability in the distribution of irregular intervals was the same in all conditions, including the no-captor controls. There was no overlap between the rhythmic and masking tones. In any condition, the same spectrum was used for the rhythmic and masker tones: the same four harmonics of 300 or 333 Hz of equal intensity. Together they formed the masked-rhythm sequence, which was presented in isolation for the no-captor control conditions. In all the other conditions, some captor tones were added; they were composed of four harmonics either of the same F0 as the maskers, or of a different F0. The four possible combinations of maskers and captors were odd and even harmonics of a 300-Hz F0, odd and even harmonics of a 333-Hz F0, odd harmonics of a 333-Hz F0 and even harmonics of a 300-Hz F0, and odd harmonics of a 300-Hz F0 and even harmonics of a 333-Hz F0. Each tone was 48-ms long, including a 8-ms onset and offset. The captors could be either simultaneous with the maskers or delayed from them by 12, 24, 36, or 48 ms. The maskers and captors were of the same duration; hence, for each onset asynchrony there was an offset asynchrony of the same duration. The masked-rhythm sequence and the captors were either both presented in the central speaker, or else in two different speakers, equally distant from the central speaker. The speakers could be off center by 30, 60, or 90 degrees; these relative positions of the sources of the maskers and captors yielded three angular separations ($\Delta\theta$): 60, 120, and 180 degrees.

### *Procedure*

Listeners had to judge *which one* of the two rhythms was embedded in the sequence and *how clearly* it was heard on a five-point scale. After each trial, feedback about the accuracy of rhythm identification was provided. During the training session, listeners were told that they would hear a warning tone followed by one of the two rhythms that they had previously heard in isolation. They were instructed to direct their attention to the location of the speaker that had sent the warning tone and to tell which of two rhythms was played. The two isolated rhythms were randomly played at each of the 13 possible speakers until the listeners reached the criterion of 10 correct identifications in a block. This was followed by a practice session, which randomly presented each combination of stimulus onset asynchrony (SOA), $\Delta\theta$, and harmonicity. This session allowed the listeners to become familiar with the task and to hear the variations across the conditions, so as to better use the full range of the rating scale. During the trials of the experiment proper, a 1-kHz warning tone was played in the speaker of the masked rhythm so that listeners could pay attention to the location of that rhythm; however, the listeners' heads remained fixed.

## RESULTS AND DISCUSSION

### *Computation of Scores*

Different accuracy measures were derived from listeners' responses: $d'$ scores, proportion-correct scores (PC), and weighted-accuracy (WA). WA weights the rated accuracy by the clarity of the identified rhythm. For parsimony purposes, this paper focuses on $d'$ scores, occasionally reporting PC scores. The $d'$ (sensitivity to rhythm 1) and $c$ scores (response bias) were evaluated according to standard procedures.[5,a] To obtain an estimate of the stimulus onset as SOA required for the perception of concurrent sounds as separate events, we determined the 75% SOA threshold from psychometric "Weibull" functions for the individual listeners.[6] Separate SOA thresholds were evaluated for each of the eight $\Delta\theta$-by-harmonicity conditions; the mean goodness of fit (as measured by $r$, the Pearson correlation coefficient) of the data was equal to, or larger than, 0.87 for each of these conditions.

### *No-captor Controls and Measures of Biases*

The no-captor controls yielded a mean PC of 0.54 (SE = 0.03) and a mean $d'$ of 0.26 (SE = 0.12); this is close to the chance level PC performance of 0.5 and $d'$ performance of 0. This verifies that the rhythm was perceptually masked in the absence of captors. The results also verified that response bias did not diminish the power of our statistical comparisons. For the conditions with captors, the mean response bias $c$ for the 18 listeners was −0.03 (SE = 0.05); for their no-captor counterparts, the mean $c$ across individuals was −0.009 (SE = 0.09). This is close to the no-bias $c$ value of zero.[a]

### *Effect of SOA, Harmonicity, and Δθ on Rhythm Sensitivity*

Temporal coincidence and deviations from it, as induced by onset and offset asynchrony, were by far the most important factor for the perception of short-duration tones as one or two sound(s). For each of the eight Δθ-by-harmonicity conditions with no temporal asynchrony, the rhythm-identification performance was at the ceiling value, namely, a PC of 0.99 and a $d'$ of 4.65 for each listener.[a] It thus seems that temporal coincidence caused frequency components to be perceptually fused, whether they were harmonically related or not, and whether they came from the same location or from spatially separated sources, 60, 120, or 180 degrees apart. There was a clear monotonic decrease of $d'$ with SOA ($p < 10^{-5}$). This powerful effect of SOA upon the fusion of the maskers and captors is consistent with past results found in the laboratory.[3,4]

There was a weak but consistent effect of harmonicity in promoting the fusion for asynchronous masker and captors, as measured by $d'$ scores for each listener ($p <$ 0.01). However, compared to SOA, the presence or absence of a common F0 does not seem to play an important role for the segregation of brief concurrent tones; this is shown by the small differences in PC and $d'$ obtained for harmonic and inharmonic maskers and captors. Furthermore, FIGURE 1 shows that it affected only weakly the temporal disparity needed for their segregation as separate events.

The spatial separation of sources (Δθ) did not affect at all rhythm sensitivity as estimated by $d'$ ($p > 0.1$). The $d'$ scores are compatible with the highly consistent mean SOA thresholds found across the different spatial and spectral relations, as shown on FIGURE 1. The mean thresholds all fell between 28.5 and 34.2 milliseconds. Note that a higher threshold indicates more fusion. From this FIGURE, it is clear that the effect of harmonicity was weak and that Δθ had no effect on fusion; still, har-

[a]In terms of Z (i.e., the inverse of the normal distribution function), $d'$ is defined as Z(H) − Z(F), where H is the proportion of Hits (i.e., rhythm 1 is reported when it is physically present) and F is the proportion of false alarms (i.e., rhythm 1 is reported when rhythm 2 is physically present). In Z-score units, $c$ is given by $0.5^{*}$ [Z(H) + Z(F)]. To avoid values of infinity in the computation of $d'$, proportions of 1 and 0 were converted into 0.999 and 0.001, respectively ($d'$ = 6.18 when H = 0.999 and F = 0.001). However, a lower value of $d'$, namely 4.65, is usually considered as the effective ceiling;[5] this is obtained when H = 0.99 and F = 0.01. As for response bias, a positive $c$ indicates a higher tendency to respond rhythm 1; a negative $c$ indicates a higher tendency to respond rhythm 2. Mean bias parameter $c$ close to the zero-bias point is considered as indicative of the absence of a systematic response bias.

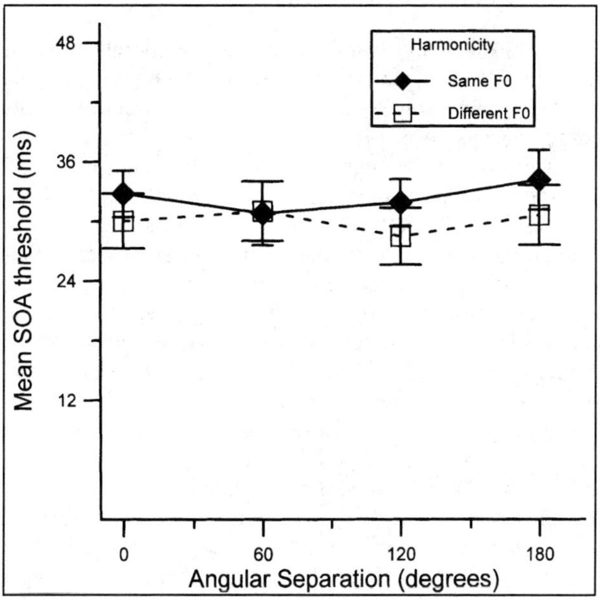


**FIGURE 1.** Mean SOA thresholds across individual listeners for different spectral and spatial relations between the masker and captor tones having the same F0 (harmonic) or different F0s (inharmonic), at four angular separations of their sources in a semicircular speaker array. Standard errors (SE) are indicated.

monicity slightly affected the temporal disparity needed for the perception of separate events—a mean SOA of 32.4 ms (SE = 3 ms) for harmonic stimuli, versus 30.1 ms (SE = 3.9 ms) for inharmonic ones. These results suggest that only spectrotemporal regularities matter for the segregation of concurrent brief tones in a open field, SOA making by far the greatest contribution.

The findings that the angular separation of sources ($\Delta\theta$) do not affect fusion (whether fusion is estimated from $d'$ or from SOA thresholds based on PC scores) appear to be inconsistent with past results obtained when the same sounds were presented over headphones.[6] It might be that dichotic separation is more efficient for sound segregation because it is an extreme case of interaural differences for sounds happening simultaneously, the stimulation of one sound being delivered to one ear only, while that of the other sound(s) is delivered to the other ear only. Open-field testing is more akin to real-world situations in which each of many individual sounds stimulates both ears, though at slightly different times and intensities, allowing for the computation of the location of each sound source. For this reason, the separation of sources in an open field is considered to be more representative of the true contribution of spatial separation to sound-source segregation. This contribution seems to be very weak when two sound sources are simultaneously active.

### *Estimates of Asynchrony Threshold for Perceiving Separate Events*

Another objective was to evaluate the minimum temporal deviation from perfect temporal synchrony that triggers the perception of concurrent tones as separate sound events. Onset asynchrony was expected to have a powerful effect on cross-spectral grouping, because it is a highly reliable cue for the segregation of sound-producing events. In a natural context, it is likely for sounds coming from different environmental sources to have some degree of temporal overlap; however, it is unlikely that they happen to be perfectly coincident in time. Given the adaptive value of detecting deviations from perfect coincidence, an empirical question of interest was to estimate the physical range of tolerance for the perceived simultaneity of sound events. We addressed this issue by estimating to what extent there could be a deviation from onset and offset synchrony before concurrent sounds were perceived as separate events 75% of the time. From the mean SOA thresholds estimated for the eight $\Delta\theta$-by-harmonicity conditions, an SOA between 26 and 37 seems to trigger the perception of concurrent brief tones as separate events. This is in good agreement with the estimated 23- to 42-ms range for the perception of the same tones as separate events when they were presented over headphones.[6] The 25- to 40-ms range of the mean SOA thresholds for the diotic, dichotic, and open-field segregation of brief tones agrees with the literature on auditory grouping, reviewed by Darwin and Carlyon,[7] showing that an SOA of 30 to 40 ms is required for removing a partial from contributing to the overall timbre, to the lateralization, and to the vowel identity of a complex sound.

### *Theoretical Considerations and Conclusions*

An important implication of this research is that when brief steady state complex tones happen at the same time, sound-source segregation (*how many* individual sources are perceived) is independent of sound-source separation (*where* individual sources are relative to each other). Further research is needed to establish firmly these results and their theoretical implications. It might be that temporal coincidence is a more robust cue than a common location in space for sound-source determination. Unlike reflected light, sounds go around and through rigid surfaces (as such, they are like transparent objects). As a consequence, in estimating the point of origin of a sound (i.e., the spatial location of the vibrating source), echoes may suggest more than one point of origin. We propose a working hypothesis: echoes and reverberation present the auditory system with a degraded signal, so that spatial information is often unreliable for the segregation of temporally overlapping sounds.

The finding that temporal coincidence fuses sounds perceptually, despite their frequency separation in different critical bands and the spatial separation of their sources, is not surprising when one considers that the perceptual fusion of the different frequency components produced by spatially separated sources is actually quite common in real orchestral performances. For instance, when struck at the same time, the sounds produced by different instruments, like those of a string section, typically blend together perceptually. If the brain were to always perceive sounds veridically, that is, to perceive distinct sound events when different sources are active (e.g., hearing the sounds of the different violins according to their spatially separated sources of origin, namely each individual violin), how could composers evoke the *perception of a single melodic line,* when, in fact, different violins are struck together at the same time? Many such musical chimeras that enchant our "ancient brain" would vanish with the disclosure of such sweet "illusions." It is a blessing for us, music listeners, that the brain can be "fooled" into believing that there is one source (the violin section) while there are actually many.

## ACKNOWLEDGMENTS

This research was presented as part of a Ph.D. thesis submitted to the Psychology Department of McGill University. Funding was provided in part by a grant from the National Sciences and Engineering Research Council of Canada (NSERC), and in part by a team grant from the FCAR to A.S. Bregman and R.J. Zatorre.

## REFERENCES

1. Bregman, A.S. 1990. Auditory Scene Analysis: The Perceptual Organization of Sound. MIT Press. Cambridge, MA.
2. Hartmann, W.M. 1988. Pitch perception and the organization and integration of auditory entities. *In* Auditory Function: Neurobiological Bases of Hearing. G.W. Edelman, W.E. Gall & W.M. Cowan, Eds.: 623–645. John Wiley and Sons. New York.
3. Turgeon, M. 1999. Cross-spectral grouping using the paradigm of rhythmic masking release. Ph.D. thesis. McGill University. Montreal.
4. Turgeon, M., A.S. Bregman & P.A. Ahad. Rhythmic masking release: contribution of cues for perceptual organization to the cross-spectral fusion of concurrent narrowband noises—asynchrony, frequency separation and spatial separation. J. Acoust. Soc. Am. In press.
5. Macmillan, N.A. & C.D. Creelman. 1991. Detection Theory: A User's Guide. MIT Press. Cambridge, MA.
6. Weibull, W.A. 1951. A statistical distribution function of wide applicability. J. Appl. Mech. **18:** 292–297.
7. Darwin, C.J. & R. Carlyon. 1995. Auditory grouping. *In* Hearing: The Handbook of Perception and Cognition, 2nd edit. B.C.J. Moore, Ed. Vol. 6: 387–424. Academic Press. London.
8. Kidd, G., C. Mason, T.L. Rohtla & P.S. Deliwala. 1998. Release from masking due to spatial separation of sources in the identification of nonspeech auditory patterns. J. Acoust. Soc. Am. **104:** 422–431.

# Explaining Harmony

## The Roles of Interval Dissonance and Chordal Tension

NORMAN D. COOK

*Informatics, Kansai University, Takatsuki, Osaka, Japan*

**KEYWORDS: Chord; Chordal tension; Dissonance, perception of; Harmoniousness, perception of**

The perception of the dissonance ("beating") of two tones whose frequencies are near to one another is thought to be due to physiological factors inherent in the human auditory system. By contrast, the degree of consonance of intervals that are not "physiologically dissonant" depends at least partly on acquired musical factors. The question arises, In the perception of the harmoniousness of chords consisting of three or more tones, what innate and acquired factors are involved? Although a chord that contains a dissonant interval is perceived as relatively "inharmonious," it is known that the harmoniousness of chords is *not* a simple summation of the consonance of the intervals contained therein. For example, the so-called unresolved diminished and augmented chords contain three consonant intervals but are perceived as less harmonious than the so-called resolved major and minor chords containing three similarly consonant intervals.

To study the factors contributing to the perception of the harmoniousness of chords, we presented 72 chords (consisting of three pure sine waves between G3 and G4) to 30 musically untrained undergraduates (mean age, 20 years), who evaluated their harmoniousness on a six-point scale. Not surprisingly, the results (FIG. 1) showed highly significant differences in the perception of resolved, unresolved, and dissonant chords ($p < 0.001$ in all combinations). More importantly, multiple regression analysis ($R = 0.7$) showed three significant factors influencing these judgments. The strongest factor was the presence of a dissonant (semitone or whole-tone) interval among the tones of the chord ($p < 0.0001$), and this factor alone accounted for roughly half ($R = 0.5$) of the variance in the data. In other words, when a dissonant interval was present, evaluations of the harmoniousness of a chord fell sharply.

The second factor was the total theoretical dissonance of the intervals contained within the chord—calculated using models based on the empirical data of Plomp and Levelt.[1] The total dissonance of the intervals was obtained in three ways: (1) using the model of Sethares[2] (in which only the dissonance among the three fundamental

Address for correspondence: Norman D. Cook, Informatics, Kansai University, 2-1-1 Ryozenji, Takatsuki, Osaka, 569, Japan.
cook@res.kutc.kansai-u.ac.jp

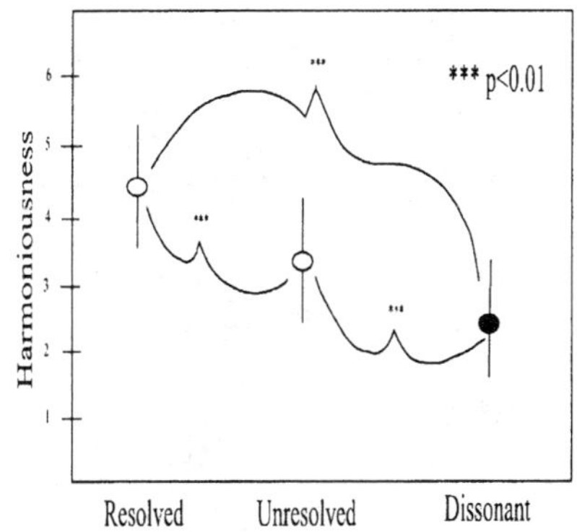


**FIGURE 1.** Mean (±SD) "harmoniousness" evaluations on a scale of 1–6 of the three types of chords used in the experiment: (1) the resolved (major/minor) chords, (2) the tension chords (including diminished/augmented chords, but excluding dissonant chords containing semitone or whole-tone intervals), and (3) the dissonant chords (any chord containing a semitone or whole-tone interval). All comparisons were highly significant ($p < 0.001$).

frequencies was included), (2) using the model of Kameoka and Kuriyagawa[3] (in which the dissonance among the three tones and their upper partials [F1–F5] was calculated), and (3) using the empirical dissonance-evaluation curve for intervals obtained from the same subjects in a separate experiment. In fact, the stimuli in the chord and interval experiments contained no upper partials whatsoever, but it was found that all three of the factors relating to the total dissonance of the intervals in the chords contributed significantly to the perception of the harmoniousness of the chords ($p < 0.001$). This finding indicates that, in addition to the effect of the presence of strongly dissonant intervals, the overall dissonance of the tone combinations had an influence on the perceived harmoniousness of the chords. It is surprising, however, that the dissonance of the intervals in the chords (with or without higher harmonics) was not a stronger predictor of the judgment of harmoniousness, because, from a reductionist perspective, every chord might be considered as simply the summation of its constituent intervals. On the contrary, it was found that either of the theoretical dissonance calculations alone or the empirical interval dissonance evaluations alone accounted for only a small part ($R = 0.1$) of the variance in the evaluation of the chords.

The third factor influencing the perception of harmoniousness was the presence of "chordal tension" ($p < 0.001$). Tension is defined as any three-tone chord containing two equivalent intervals. The concept of the tension chords is distinct from that of the unresolved chords, insofar as any chord containing a semitone or whole-tone interval is unresolved, but not necessarily "tense" (in the sense of having two equivalent intervals). In fact, any unresolved three-tone chord that does *not* contain a semitone or whole-tone dissonance can be transposed into a tension chord. In their root configurations, the tension chords have a clear physical definition and, perceptually, a commonality in being unresolved or "tense"—regardless of the presence or absence of dissonant intervals. There are a total of 12 types of tension chords

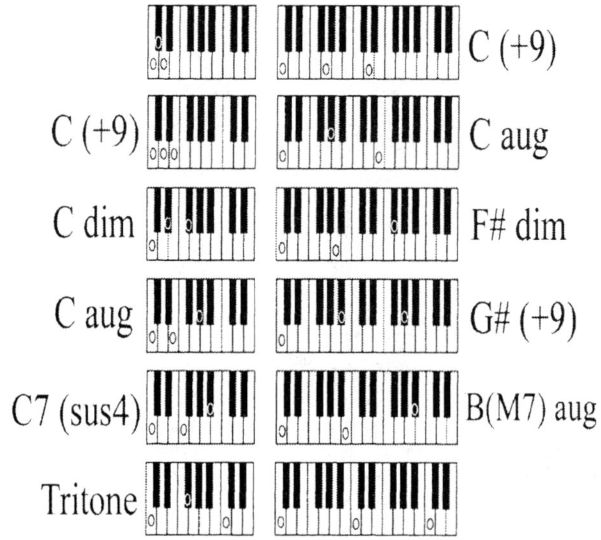


**FIGURE 2.** The full set of "tension chords" starting at C, among which are found augmented and diminished chords and the tritone. All of the tension chords contain two intervals with the same number of semitone steps, and all of them have an "unresolved" character. Although not recognized as a coherent set in traditional music theory, all of the tension chords have an unresolved character.

(FIG. 2). Although the perception of the dissonance of the intervals in the first two tension chords overpowers the perception of tension, the unresolved nature of any of the tension chords is evident when spread over several octaves. The inharmoniousness of these chords can be partly explained on the basis of dissonant intervals among upper partials, but the fact that the multiple regression analysis showed chordal tension to be a factor distinct from interval dissonance suggests that the perception of three-tone combinations is a gestalt that differs from the summation of the perception of intervals. This factor, unlike the presence of dissonant *two-tone* intervals or the total dissonance among the *two-tone* intervals of the upper partials, is related to the structure of the *three-tone* chord. In other words, it is a "three-body" problem, as distinct from the "two-body" problem of interval perception.

It is concluded that, in addition to an inherent sense of the consonance/dissonance of two-tone intervals, normal, musically untrained people have a sense of the tension inherent in three-tone chords. Specifically, any chord containing two equivalent intervals (or transposable into such a chord) is perceived as tense or unresolved, whereas chords that have neither such tension nor dissonant intervals are perceived as resolved. In the case of both intervals and chords, the central issue is the presence of the "unstable" dissonance or tension, not the location of the frequencies at which resolution is obtained. The nature of resolution is highly dependent on the musical culture and individual differences, but the detection of dissonance *and* the detection of tension may be universals inherent in the human auditory system. We suggest that, in the study of harmony, the three-body problem of chordal tension must be considered together with the two-body problem of interval dissonance.[4]

## ACKNOWLEDGMENTS

This research was supported by the Research for the Future Program from the Japan Society for the Promotion of Science (Project No. JSPS-RFTF99P01401).

## REFERENCES

1. PLOMP, R. & W.J.M. LEVELT. 1965. Total consonance and critical bandwidth. J. Acoust. Soc. Am. **38:** 548–560.
2. SETHARES, W.A. 1993. Local consonance and the relationship between timbre and scale. J. Acoust. Soc. Am. **94:** 1218–1228.
3. KAMEOKA, A. & M. KURIYAGAWA. 1969. Consonance theory. J. Acoust. Soc. Am. **45:** 1452–1459; 1460–1469.
4. COOK, N.D. 2001. Tone of Voice and Mind: The Connections among Intonation, Emotion, Cognition and Consciousness. John Benjamins Publ. Amsterdam. In press.

# Associations and Dissociations among Music Training, Tonal and Temporal Order Processing, and Cognitive Skills

CHRISTINE K. KOH,[a] LOLA L. CUDDY,[a] AND LORNA S. JAKOBSON[b]

[a]*Queen's University, Department of Psychology, Kingston, Ontario, K7L 3N6, Canada*

[b]*University of Manitoba, Winnipeg, Manitoba, R3T 2N2, Canada*

KEYWORDS: **Music training, effects of; Temporal order sensitivity; Tonality sensitivity**

## INTRODUCTION

Recent research has revealed a dissociation between performance on pitch (i.e., memory for, and abstraction of, relationships between tones) and nonmusical cognitive tasks across a wide range of music training and experience.[1] The relationships among auditory temporal order sensitivity, cognitive tasks, and levels of music training, however, have not yet been examined. The competence with which humans extract temporal regularities in the environment[2] and the absence of empirical investigation concerning the relationship between this and related competencies motivated the inclusion of auditory temporal order tasks in the present study.

The auditory system effectively positions sound events in time and in relation to other sound events.[2] Previous research has demonstrated that while temporal order discrimination for speech or music targets is possible at durations of 15–20 ms, discrimination is difficult when target items are unrelated (e.g., a sequence of hisses, tones, buzzes), even when target durations are increased to 200 ms (for a review, see Ref. 3). However, a subsequent study found that training improves sensitivity to changes in the temporal order of unrelated items.[3] The results of this research inform questions concerning the role of increased exposure (i.e., training) and target contexts on temporal order discrimination ability. They do not, however, specifically address how music training may affect temporal order processing skills.

## METHOD

Thirty-one university students participated in a probe tone study assessing recovery of the major and minor tonal hierarchy[4] for the keys of C major and C minor. The probe context was a four-note sequence (e.g., C-E-C-G), followed by a probe tone, one of the 12 steps of the chromatic scale beginning on C. Listeners rated the good-

Address for correspondence: Lola L. Cuddy, Queen's University, Department of Psychology, Kingston, Ontario, K7L 3N6, Canada. Voice: 613-533-6013; fax: 613-533-2499. cuddyl@psyc.queensu.ca ness of fit of the probe to the context. There were two blocks of randomized trials for each mode. The students also participated in a series of cognitive tests (Multdimensional Aptitude Battery-II: vocabulary, spatial ability; Wechsler Adult Intelligence Scale-III: forward digit span, backward digit span) and in a series of auditory perception tests (three subtests from the Test of Basic Auditory Capabilities, Modification III: pitch discrimination, temporal order, and syllable sequence). Each trial of each subtest had the same presentation structure: a standard sound followed by two test sounds. The participant was instructed to select the test sound that was different from the standard. Pitch discrimination assessed discrimination between pure tones differing in frequency. Temporal order assessed discrimination between four-tone sequences differing in the order of the second and third tones. Syllable sequence assessed discrimination between four-syllable speech sequences differing in the order of the second and third syllables. Cognitive and auditory tests were administered per standardized instructions. Years of music training ranged from 0 to 14 years.

## RESULTS

Data were subjected to factor analyses and model testing. The oblimin-rotated maximum-likelihood analysis revealed three factors of approximately equal weight that accounted for over half the variance in the data. Supporting previous findings,[1] recovery of the tonal hierarchy dissociated from cognitive skills. Both factors separated from a third, an independent factor implicating temporal order processing and years of music training.

## DISCUSSION

The separation of a tonal from a temporal order factor supports theoretical notions concerning functional auditory dissociations (e.g., Ref. 5). The absence of a strong relationship between music training and sensitivity to tonality was not unforeseen, given that research has shown that recovery of the tonal hierarchy fully occurs for both adults and elementary school–aged children[6] within the major triad context. The most intriguing new finding was that years of music training loaded on the temporal order factor, indicating a relationship between music training and sensitivity to temporal order of tones and speech.

The relationship between music training and sensitivity to temporal order suggests that temporal order discrimination may be an acquired skill. Research continues to question whether auditory temporal organization involves universal or acquired processes,[2] and here we have suggested that increased exposure to temporal orderings may improve discrimination. We acknowledge that the present findings are correlational and that an investigation of the direct influence of music training is required. Together, the results of both paradigms will inform theoretical questions concerning what processes are shared between musical and nonmusical skills and how shared processes influence performance on both.

## REFERENCES

1. STEINKE, W.R., L.L. CUDDY & R.R. HOLDEN. 1997. Dissociation of musical tonality and pitch memory from nonmusical cognitive abilities. Can. J. Exp. Psychol. **51:** 316–334.
2. DRAKE, C. 1998. Psychological processes involved in the temporal organization of complex auditory sequences: universal and acquired processes. Music Percept. **16:** 11–26.
3. WARREN, R.M. 1999. Auditory Perception: A New Analysis and Synthesis. Cambridge University Press. Cambridge, UK.
4. KRUMHANSL, C.L. 1990. Cognitive Foundations of Musical Pitch. Oxford University Press. New York.
5. PERETZ, I. 1993. Auditory agnosia: a functional analysis. *In* Thinking in Sound: The Cognitive Psychology of Human Audition. S. McAdams & E. Bigand, Eds.: 199–230. Oxford University Press. Oxford, UK.
6. CUDDY, L.L. & B. BADERTSCHER. 1987. Recovery of the tonal hierarchy: some comparisons across age and levels of musical experience. Percept. & Psychophys. **41:** 609–620.

# Playing the Piano by Heart

## From Behavior to Cognition

RITA AIELLO

*The Juilliard School, New York, New York 10023, USA*

*Aaron Copland School of Music, Queens College, City University of New York, Flushing, New York 11367, USA*

KEYWORDS: **Piano playing by heart; Memorization; Music memorizing**

## INTRODUCTION

Although research on memory[1,2] and piano performance has been extensive,[3–7] relatively few investigations consider how classically trained pianists memorize their repertoire.[8–10] This initial study addresses how classical pianists at two levels of expertise suggest using the information on the score to memorize a piano composition.

## METHOD

Semistructured interviews were conducted with four professional concert pianists and six undergraduate piano students enrolled in a leading music school in New York City. The subjects were asked to describe how they would recommend using the information written on the score to memorize the composition. Two well-known pieces from the piano literature were used: Chopin's Prelude in E Minor, Op. 28, No. 4, and J. S. Bach's Prelude in C Major from Book 1 of *The Well-Tempered Clavier.* Specifically, for each composition the subjects were asked to mark on the score the musical elements that they would suggest focusing on to memorize the piece, and to describe what they marked.

## RESULTS

The concert pianists wrote a total of 36 markings on the score of the Chopin prelude (FIG. 1). They indicated the two large phrases that form the overall structure of the composition; the repeated melodic line occurring in measures 1–4 and measures 13–15; the stepwise motion of the right hand in measures 1–9; the harmonic changes in the left hand that take place throughout the piece; the embellishment, the left hand

Address for correspondence: Rita Aiello, Aaron Copland School of Music, Queens College, City University of NewYork, Flushing, NY 11367.Voice: 212–316-3098; fax: 212–316-3099.
ritaaiello@aol.com

**FIGURE 1.** F. Chopin, Prelude E Minor, op. 28, no. 4.

**FIGURE 2.** J. S. Bach, Prelude in C Major, from Book 1 of *The Well-Tempered Clavier.*

**FIGURE 2.** *Continued.*

pattern, and the crescendo that occur on measure 16; the overall climax in measure 17 due to the pattern of the left hand, the right hand reaching its highest note, and the *forte* that should be reached there; and the very last chord of the piece. In marking the score of the Bach prelude (FIG. 2), they wrote a total of 43 markings. They indicated the arpeggiated chords and the chord progression upon which the entire piece is built, the importance of the bass line, the phrasing that they would create in performing the piece, and the relevance of the brief coda in measures 33–35. The concert pianists described in detail what they marked in both compositions. Their marking and verbal descriptions revealed that they would analyze the scores in detail to memorize the pieces.

Two of the six piano students, on the other hand, provided only one marking on each score. They reported that they memorize by rote. The remaining four piano students wrote a total of 21 markings on the score of the Chopin prelude (FIG. 1). On this score, they marked the two large phrases that form the overall structure of the piece; many of the harmonic chords in the left hand that occur throughout the piece; the embellishment and the crescendo in measure 16; the left hand pattern that occurs in measures 16 and 17; and the overall climax in measure 17. With the score of the Bach prelude (FIG. 2), the same four students wrote a total of 31 markings. They indicated many of the arpeggiated chords, the chord progression that occurs throughout the piece, and the important bass line. The students described the elements that they marked on both scores. In total, the students provided 56 score markings and the concert pianists provided 79.

## DISCUSSION

These data seem to indicate differences in the way the two groups of pianists use the information written on the score to memorize a piece. Whereas all the concert pianists provided detailed markings pertinent to the musical elements of the pieces and described what they marked, two of the six piano students were able to mark only one element on each score. These two students reported memorizing primarily by rote. They explained that they just repeated a piece enough times until they memorized it. The remaining four piano students were able to mark on the scores the musical elements that they considered valuable for memorizing the pieces and described what they marked. Although these four students seemed to provide fewer score markings than the concert pianists, overall they marked the same musical elements that the concert pianists did.

Inasmuch as all the subjects have the ability to perform the two pieces without memory lapses, these initial data seem to suggest that when the students memorize a piece, they elaborate less on the information written on the score than the professional pianists do. Data from the concert pianists seem to indicate that they have developed more precise cognitive plans for using the score to memorize the composition. As musical experience increases, there appears to be a shift in the way classically trained pianists use a score to memorize the music. Playing the piano by heart seems to progress from rote behavior to a more elaborate cognitive plan.

## REFERENCES

1. Anderson, J. 1983. The Architecture of Cognition. Harvard University Press. Cambridge, MA.
2. Bartlett, F. 1932. Remembering: A Study in Experimental and Social Psychology. Cambridge University Press. Cambridge.
3. Gabrielsson, A. 1999. Music performance. *In* The Psychology of Music, 2nd edit. D. Deutsch, Ed.: 501–602. Academic Press. San Diego, CA.
4. Palmer, C. 1997. Music performance. Annu. Rev. Psychol. **48:** 115–138.
5. Repp, B. 1990. Patterns of expressive timing in a performances of a Beethoven minuet by nineteen famous pianists. J. Acoust. Soc. Am. **88:** 622–641.
6. Shaffer, H. 1995. Musical performance as interpretation. Psychol. Music **23:** 17–38.
7. Williamon, A. 1999. Preparing for performance: an examination of musical practice as a function of expertise. Ph.D. thesis, University of London.
8. Aiello, R. & A. Williamon. 2002. Memorizing music for performance: perception, cognition and context. *In* Science and Psychology of Music Performance: Creative Strategies for Teaching and Learning. R. Parncutt & G. McPherson, Eds. Oxford University Press. New York. In press.
9. Chaffin, R. & G. Imreh. 1997. Pulling teeth and torture: musical memory and problem solving. Thinking Reasoning **3:** 315–336.
10. Clarke, E.F. 1988. Generative processes in performance. *In* Generative Processes in Music: The Psychology of Performance, Improvisation, and Composition. J.A. Sloboda, Ed.: 1–26. Clarendon Press. Oxford.

# Absolute Pitch, Early Musical Instruction, and Spatial Abilities

EUGENIA COSTA-GIOMI, RAMONA GILMOUR, JEANNE SIDDELL, AND ELIZABETH LEFEBVRE

*Faculty of Music, McGill University, Montreal, QC H3A 1E3, Canada*

KEYWORDS: **Absolute pitch; Musical instruction, early; Relative pitch; Pitch; Spatial abilities**

## INTRODUCTION

Those who possess absolute pitch differ from nonmusicians in certain structural and functional characteristics of the brain.[1] It is known that most absolute pitch possessors start music instruction by age seven and that absolute pitch is unlikely to develop if music instruction is initiated after this age.[2] Interestingly, early musical instruction has been associated with the development of selected spatial abilities.[3,4] Studies that provided six- to nine-year olds with music instruction did not find improvement in these abilities or found them to be only temporary.[5,6] The present study focused on the relationship between absolute pitch, early musical training, and spatial abilities.

## MATERIAL AND METHODS

A total of 12 musicians with absolute pitch (AP), 11 musicians with relative pitch (RP), and 12 nonmusicians (NMs) with no formal musical training participated in the study. On average, musicians had received 15.5 years of music instruction. Participants completed a questionnaire, the Hidden Figure Test, and two subtests of the Performance part of the Multidimensional Aptitude Battery, Form L: The Spatial and Object Assembly. Subjects who reported that they were generally unable to identify pitch in isolation were considered to have relative pitch. Those who reported having absolute pitch completed an absolute pitch test individually. The test, an adaptation of that by Zatorre *et al.*, consisted of 100 tones between F♯3 and C♯5, played in random order with a synthesized timbre unfamiliar to the subjects.[1]

## RESULTS

Test scores of AP possessors, RP possessors, and NMs were analyzed through three analyses of variance. The results showed that the subject's group affected

Address for correspondence: Dr. Eugenia Costa-Giomi, McGill University, 555 Sherbrooke West, Montreal, QC H3A 1E3, Canada. Voice: 514-398-4535 ext. 0693; fax: 514-398-8061. costa@music.mcgill.ca scores in the Hidden Figures Test, $F$ (2,34) = 8.33, $p = 0.001$, but did not affect subjects' performance in the Object Assembly and Spatial Subtests. Scheffe comparisons indicated that AP possessors obtained significantly higher scores, $p < 0.05$, than did RP possessors and NMs in the Hidden Figure Test. There were no differences between the Hidden Figure Test scores of NMs and musicians with RP.

The correlation between age at which formal music instruction began and percentage of correct identifications in the AP test was only moderate but significant, $r = 0.56$, $p = 0.01$. The average age at which musicians began formal music instruction was 7.36 years for the RP possessors and 4.5 years for the AP possessors.

Data were also analyzed to study the effects of age at which music instruction was initiated (i.e., starting age) on the performance in the three spatial tests through three ANOVAs. Subjects were classified into three groups: musicians who began music instruction by age 5, musicians who began music instruction at age six or later, and NMs who never participated in music instruction. Results indicated that starting age affected the scores in the Hidden Figure Test, $F$ (2,40) = 43.99, $p < 0.01$, the Spatial Subtest, $F$ (2,40) = 69.18, $p = 0.02$, and the Object Assembly Subtest, $F$ (2,40), = 75.04, $p = 0.03$. Scheffe comparisons showed that musicians who began music instruction when they were 3, 4, or 5 years of age obtained significantly higher scores in the Hidden Figure Test and in the Object Assembly Subtest than did those who began later or never received formal music training, $p = 0.05$. The Spatial Subtest scores of musicians who began lessons later were slightly (and not significantly) lower than those of NMs and significantly lower than those of the other musicians. To study whether the differences between AP possessors and RP possessors could be attributed to the early musical training of the former group, data from musicians was analyzed further. ANOVAs with group (AP or RP possessors) and starting age (3, 4, 5, or 6+ years) as the independent variables indicated that the subject's group affected the Hidden Figure Test scores $F$ (1,19) = 4.53, $p = 0.05$. No other significant main effects or interactions were found.

## SUMMARY

The results of the study show that AP possessors performed better in the Hidden Figure Test than did NMs and musicians with RP and that age at which music instruction is initiated affects performance in specific spatial tasks. Individuals who begin extensive musical instruction at a very early age may have an advantage when completing selected spatial tests over those with no formal musical training. However, characteristics other than starting age of musical instruction are associated with AP and seem to affect musicians' performance in the Hidden Figure Test. The differences found between the performance of AP possessors and RP possessors in this nonmusical task suggest further research.

## REFERENCES

1. ZATORRE, R.J., D.W. PERRY, C.A. BECKETT, *et al.* 1998. Functional anatomy of musical processing in listeners with absolute pitch and relative pitch. Proc. Natl. Acad. Sci. USA **95:** 3172–3177.

2. SERGEANT, D.C. 1969. Experimental investigation of absolute pitch J. Res. Music Educ. **17:** 135–143.
3. GROMKO, J.E. & A. SMITH POORMAN. 1998. The effect of music training on preschoolers' spatial-temporal task performance. J. Res. Music Educ. **46:** 173-181.
4. RAUSCHER, F., G. SHAW, L. LEVINE, *et al.* 1997. Neurol. Res. **19:** 2–8.
5. COSTA-GIOMI, E. 1999. J. Res. Music Educ. **47:** 198–212.
6. HURWITZ, I., P. WOLFF, B. BORTNICK & K. KOKAS. 1975. J. Learn. Disabil. **8:** 45–52.

# Infant Long-Term Memory for Music

JENNY R. SAFFRAN, MICHELLE M. LOMAN, AND RACHEL R. W. ROBERTSON

*Waisman Center and Department of Psychology, University of Wisconsin-Madison, Madison, Wisconsin 53706, USA*

KEYWORDS: **Infants, long-term music memory in; Music, infants' long-term memory for**

## INTRODUCTION

To what extent do young infants maintain specific musical experiences in long-term memory? Recent findings in the language domain suggest that 8-month-old infants can remember words from stories over time delays of two weeks.[1] Because music, along with language, is among the most complex of auditory learning tasks facing infants, it is possible that infant memory for musical stimuli is equally powerful. As yet, however, it is unknown if infants remember particular musical pieces heard during passive listening experiences.

## METHODS AND RESULTS

To determine if infants show long-term memory for music, we exposed a group of seven-month-old infants to a 10-minute recording of two Mozart piano sonata movements, KV 281 in B-flat major, Andante, and KV 282 in E-flat major, Adagio. Each infant listened to the music once a day, at home, for 14 consecutive days. Following a two-week retention interval, during which the infants did not hear these pieces, we assessed the infants' long-term memory for the Mozart sonatas in the laboratory. To do so, we measured listening preferences using the head-turn preference procedure, a standard computer-automated method used in studies of infant auditory perception.

We hypothesized that if infants remembered the music with which they were familiarized at home, they would show recognition by listening longer to passages of familiar music than to novel passages (as observed in research on infant linguistic long-term memory). The test consisted of two familiar and two novel passages, all 20 seconds in duration. The familiar test passages were drawn from the middle of each of the two familiarized sonata movements. The novel test stimuli were drawn from two other Mozart sonatas (taken from the same CD recording and also composed during late 1774–early 1775: KV 280 in F major, Adagio, and KV 283 in G major, Andante). To ensure that arbitrary properties of the test stimuli could not ac-

Address for correspondence: Jenny R. Saffran, Department of Psychology, University of Wisconsin-Madison, Madison, WI 53706. Voice: 608-262-9942; fax: 608-262-4029.
jsaffran@facstaff.wisc.edu

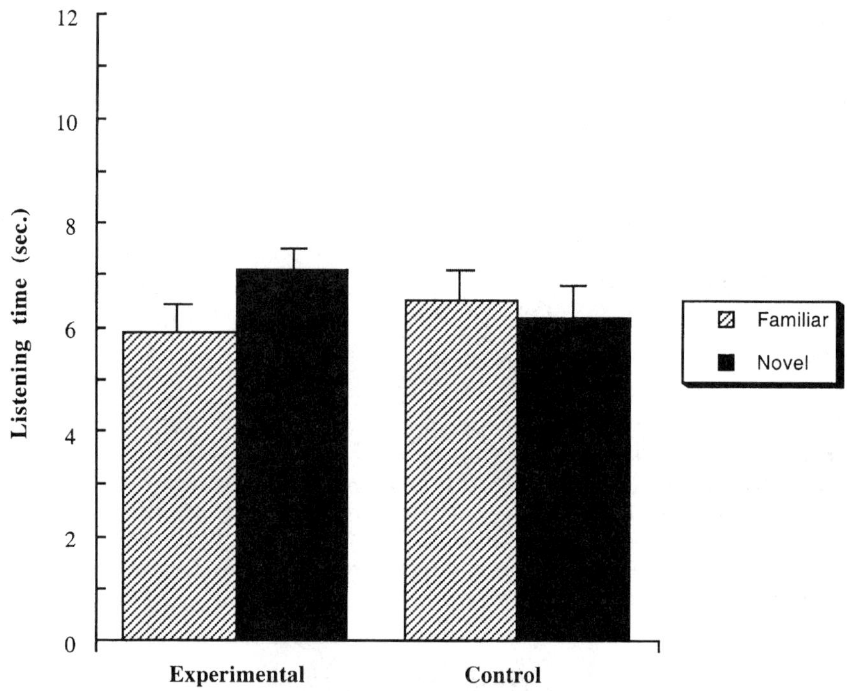


**FIGURE 1.** Experiment 1: Listening times for familiar and novel test passages for infants from the experimental (prior exposure) and control (no prior exposure) groups.

count for the infants' performance, we also tested a control group of eight-month-old infants who received no prior exposure to any of these pieces.

Mean listening times for the familiar and novel musical passages were calculated for each infant (FIG. 1). Consistent with the hypothesis that infants maintain musical experiences in long-term memory, there was a significant difference in listening times for infants in the experimental group: $t(10) = 2.52$, $p < 0.05$. The direction of the preference, however, was contrary to our hypothesis; the infants preferred the novel passages to the familiar passages. To ensure that this pattern of results was not due to biased test materials, we also examined the performance of the infants in the control group. These infants did not show a significant preference for either type of test passage, suggesting that the preferences that emerged for the experimental group reflected long-term memory for the familiar passages: $t(13) = 0.55$, n.s.

Why did the infants prefer the novel passages in the first experiment? This direction of preference is the opposite of the results from Jusczyk and Hohne's linguistic study,[1] in which infants preferred familiar stimuli after a two-week delay. There are (at least) two possible explanations: (1) Mastery: if infants mastered the familiarized music, they might now prefer to hear something new, despite the two-week retention period. (2) Musical context: the familiar test passages were taken from the middle of the familiar sonata movements. Thus, when they were played during testing, they were removed from the musical context within which they were originally learned.

Because infants typically prefer to listen to stimuli that are maximally natural, passages removed from context may not have been preferred.

The second experiment was designed to test these two hypotheses. A second group of seven-month-old infants was familiarized with the same two sonata movements, following procedures identical to those in experiment 1. Following a two-week retention interval, the infants were tested in the laboratory on two different types of *familiar* passages: passages taken from the middle of the sonatas (which were the familiar passages used in experiment 1) and the beginning passages from the two familiarized sonatas. If the results of experiment 1 reflected mastery, as reflected in relative disinterest in the familiar music, then the infants should now prefer the *middle* passages, as the beginnings are presumably more memorable than the middle passages. However, on the musical context hypothesis, infants should prefer the familiar passages taken from the *beginning* passages of the sonatas. Even in the context of the full sonata movements, the beginning passages were always preceded by silence. Thus, unlike the passages taken from the middle of the sonatas, the beginning test passages were not removed from their musical context when played in isolation.

As in experiment 1, there was a significant difference in listening times: $t(11) = 2.86$, $p < 0.05$ (FIG. 2). The direction of the preference was consistent with the mu-

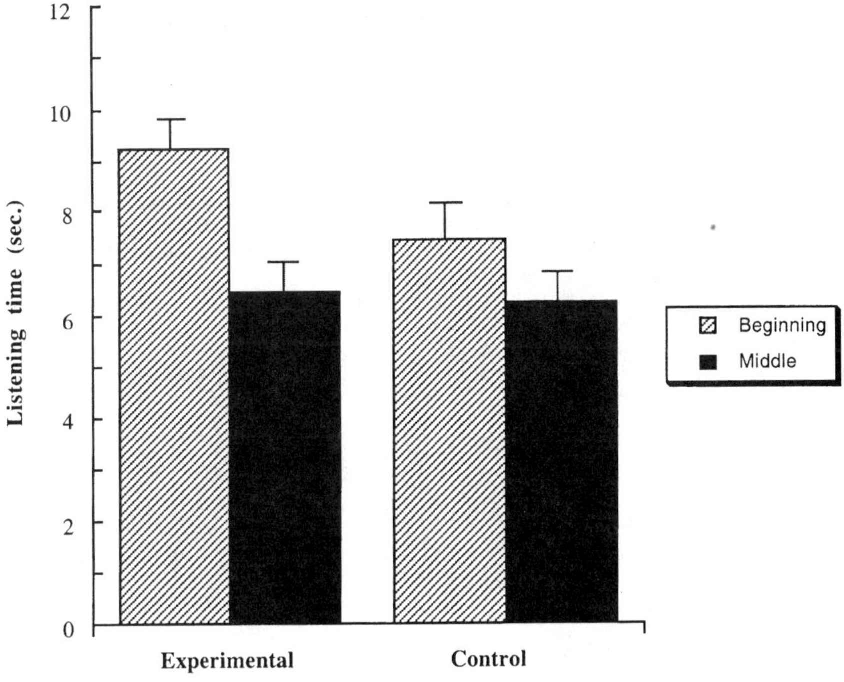


**FIGURE 2.** Experiment 2: Listening times for passages from the beginning and middle of the familiar pieces for infants from the experimental (prior exposure) and control (no prior exposure) groups.

sical context hypothesis: infants preferred passages from the beginning rather than the middle of the familiarized sonatas. Again, we tested a separate control group to ensure that biased test passages could not account for the experimental group's performance; these infants did not show a significant preference for either type of test passage: $t(13) = 0.98$, n.s. Along with experiment 1, the results of experiment 2 strongly suggest that infants do remember the music that they hear; the existence of a preference for passages drawn from the beginning, rather than the middle, of the familiar passages indicates that the infants remembered the pieces last heard at home two weeks before. Moreover, the infants preferred music that was not taken out of context. The control infants showed no such bias; even out of context, these passages sounded coherent to listeners who were not familiar with the pieces from which they were drawn.

## CONCLUSIONS

These results, along with other findings pertaining to infant short-term musical memory, discrimination abilities, and musical preferences, suggest that infants are sophisticated musical listeners.[2–4] Their representations of music in long-term memory are not undifferentiated sequences of notes. Instead, passages are linked together to form coherent musical events. The current results lend further support to an emerging picture of infants as remarkably adept at implicitly learning and remembering the structured information that characterizes the environment in which they develop.

## REFERENCES


1. JUSCZYK, P.W. & E.A. HOHNE. 1997. Infants' memory for spoken words. Science **277:** 1984–1986.
2. SCHELLENBERG, E.G. & S.E. TREHUB. 1996. Natural musical intervals: evidence from infant listeners. Psychol. Sci. **7:** 272–277.
3. ZENTNER, M. & J. KAGAN. 1998. Infants' perception of consonance and dissonance in music. Infant Behav. Dev. **21:** 483–492.
4. TREHUB, S., E.G. SCHELLENBERG & D.S. HILL. 1997. The origins of music perception and cognition: a developmental perspective. *In* Perception and Cognition of Music. I. Delige & J. Sloboda, Eds.: 103–128. Psychology Press. East Sussex, UK.

# Phases in Children's Rhythmic Development

SILVIA MALBRÁN
*Calle 57 N° 1404, (1900) La Plata, Argentina*

KEYWORDS: **Synchronization; Rhythmic development in children; Phase model**

## INTRODUCTION

Tapping or clapping according to the most salient pulse from a musical fragment can be considered a form of processing information.[1,2] The subject (1) processes aural information input according to the more salient traits; (2) changes the information into an inner temporal scale according to the stimulus temporal scale; and (3) translates the information into a motor chain. According to Shaffer[3] and Ivry and Hazeltine,[4] these kinds of performance demand an inner mechanism of synchronization.

The successive intertap intervals have been considered an important index of regularity.[5] Steady and synchronic adjustment is an essential component of rhythmic behavior. Previous studies have considered performance with a pulse fragment as a predictor of adjustment and rhythmic development.[6]

## MATERIAL

A computerized test was designed to explore synchronic tapping in children. Responses were analyzed according to the four dimensions to be described. Data analysis allowed us to distinguish the kind of performance, which was assigned to three phases or modalities from asynchrony to synchrony.

### *Dimensions*

(1) *Correspondence*: matching onset to beat (one to one). Levels: number of tactus from the total beats in the piece. (2) *Sustainability*: action persistence during the entire performance. Levels: number of tactus in each chunk (a minimum of four successive tactus forms a chunk). (3) *Adjustment*: performance in time; distance to the strict zero point. Levels: each onset was measured on the tick scale to estimate the degree of coincident tap beats (soft setting, 480 ticks per quarter note). (4) *Regularity*: stability of performance in terms of levels of adjustment. Levels: relationship between the number of tactus and the level of adjustment.

Address for correspondence: Silvia Malbrán, Calle 57 #1404, (B1900ENM) La Plata, Argentina. Voice/fax: +54-221-453-9201.
smalbran@netverk.com.ar

### *Phases*

(1) *Global (diffuse)*: low tap beat correspondence. Many onsets were nonvalid. Children apparently cannot build a chunk. (2) *Fluctuating (oscillating)*. Some accurate chunks were followed by inaccurate ones. (3) *Precise (accurate)*. There was high adjustment and regularity.

## METHODOLOGY

The sample consisted of 314 children, aged 3–12 years.

*Hypothesis.* Accuracy in synchronic tapping to musical stimuli can be measured in terms of correspondence, sustainability, adjustment, and regularity.

*Independent Variable.* First 10 measures (40 beats) of a musical piece in moderate and stable tempo and in its original timbre (Prokofiev's "March," from *Music for Children*, op. 65, for piano).

*Dependent Variable.* Degree of synchronic adjustment to the most salient pulse in a piece of music.

*Testing Procedure.* The test consisted of tapping the pulse (tactus) on a digital drum simultaneously with a musical fragment.

## RESULTS

The results showed a turning point between the ages of 8 and 9. Before that age, children generally showed great difficulty in building the correspondence between one clap and one pulse in an entire fragment and in sustaining the action throughout an entire musical piece.

### *Performance Tendencies*

#### *Diffuse Phase (Fig. 1)*

The most inaccurate children's performance showed: low correspondence and sustainability; most invalid onsets, with no tactus; absence of chunks, but some isolated tactus; some successive tactus (not more than three); start of the performance immediately with the stimulus (responses showed that the children did not pay attention to the salient traits at the beginning); lack of internal consistency in the whole fragment; and very low adjustment and regularity throughout the entire fragment.

#### *Fluctuating Phase (Fig. 1)*

Oscillating performance (inaccurate–accurate) showed: intermediate correspondence and sustainability; oscillating regularity: peaks towards low adjustment levels (leaps generally covered more than one level); progress in adjustment and regularity towards the end of the chunks; low regularity intrachunk; improved adjustment intrachunk: responses beginning at a low level of adjustment towards the end; similar design interchunks with similar tendency towards greater adjustment, beginning at the lowest levels and improving towards the end; interchunk adjustment changes: each chunk was developed at a different level of adjustment; unstable adjustment and

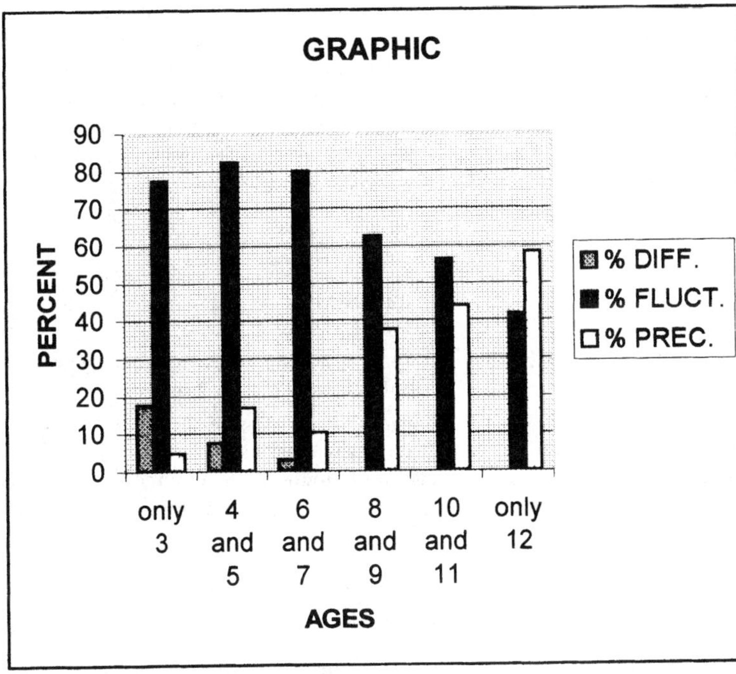


FIGURE 1.

regularity interchunks; intrachunk stability (changes in adjustment/regularity levels were noted between chunks); high but brief adjustment and regularity alternating with alternative inaccurate moments; and discontinuity in adjustment and regularity during the performance.

*Precise Phase (Fig. 1)*

Stable performance showed: high adjustment and regularity with sustainability in more than 75% of the musical fragment: short lack of correspondence and sustainability during the action (less than 25%); start with security for two or three tactus later (waiting for stimuli to internalize the temporal scale); chunks with high regularity; and internal consistency between isolated tactus and chunks.

## DISCUSSION

Delay is the tendency observed most commonly. Anticipated responses are less frequent; the most frequent ones appear in highly synchronic subjects. Strict onset responses are infrequent and are generally produced by the most accurate performers. Correspondence and sustainability can be considered critical variables in children between the ages of 3 and 7; adjustment and regularity are important for those of all ages. Basically, performance requires correspondence and sustainability, but the critical variables are adjustment and regularity. Children in the early years show adjustment and regularity only in a few cases and generally cannot obtain correspondence and sustainability. Eight-year-old children and older show correspondence and sustainability, but not always adjustment and regularity. Adjustment responses showed important differences at all ages, but the most frequent responses were the delayed ones. At ages 3–7, most children are in the fluctuating phase. A slight relationship between age and phase was observed; however, age was not a reliable criterion in predicting rhythmic phase.

## CONCLUSIONS

Previous studies reported that anticipation was the most frequent response.[7] Our studies show strong discrepancies with this principle. The high precision in assessment, obtained through electronic means, can be considered a suitable way of assessing this kind of performance. Synchrony is usually considered in terms of all-or-nothing ability. However, as a dimension of rhythmic development, distinguishing phases on a continuum seems to be a more appropriate way of conceptualizing this trait.

## REFERENCES

1. LERDAHL, F. & R. JACKENDOFF. 1981. On the theory of grouping and meter. Musical Q. **25:** 45-90.
2. PARNCUTT, R. 1994. Pulse salience and metrical accent. Music Percept. **11:** 409-464.
3. SHAFFER, L. 1982. Rhythm and Timing in Skill. Psychol. Rev. **89:** 109-121.
4. IVRY, R. & E. HAZELTINE. 1995. Perception and production of temporal intervals across a range of durations: evidence for a common timing mechanism. J. Exp. Psychol. **21:** 3-18.
5. WING, A. & A. KRISTOFFERSON. 1973. Response delays and the timing of discrete motor responses. Percept. Psychophysics **14:** 5-12.
6. SHAFFER, L., E. CLARK & N. TODD. 1985. Metre and rhythm in piano playing. Cognition **20:** 61-67.
7. POUTHAS, V. 1995. Développement de la perception du temps et des régulations temporelles de l'action chez le nourrisson et l'enfant. *In* Naissance et développement du sens musical. I. Deliege & J. Sloboda, Eds.: 133–166. Presses Universitaires de France. Paris.

# Variance Decomposition of Tempo Drift in Isochronous Rhythmic Tapping

GEOFFREY L. COLLIER[a] AND R. TODD OGDEN[b]

[a]*Department of Psychology and Sociology, South Carolina State University, Orangeburg, South Carolina 29212, USA*

[b]*Department of Statistics, University of South Carolina, Columbia, South Carolina 29208, USA*

KEYWORDS: **Rhythm; Tempo drift**

## INTRODUCTION

Timing has come to play an increasing role in our understanding of motor control, with rhythmic timing playing a central role. It provides a bridge between the psychology of music and issues of general interest in human cognition and motor control. For over 25 years, the Wing-Kristofferson model of timing in rhythmic tapping has played a key role.[1] This model decomposes variability in isochronous rhythmic tapping into two components, a central or clock component and a motoric component. This model has been fruitfully applied in a variety of domains, including special populations such as children or persons with Parkinson's disease.

A problem with the model that has been repeatedly noted is that its mathematics requires the clock process and the motor delay process to be independent of each other and, more critically, serially independent. However, this assumption of serial independence in the clock process is abrogated whenever there is a drift in the tempo. Such a drift in tempo has been widely observed, even among trained musicians whose tempo is generally considered to be stable. Tempo drift among the untrained and among special populations can be even more egregious.

## MATERIAL AND METHODS

We have developed a statistical approach to this problem by explicitly assuming, in addition to clock and motor variance, a slow drift process as a third source of variance. To examine this hypothesis, we developed a test statistic for the presence of drift and provided statistical estimators for the three drift components: motor variance, drift variance, and (drift-free) clock variance. With these we hope to provide the field with new approaches to examining the effects of various independent variables and individual differences.

Address for correspondence: Dr. Geoffrey L. Collier, South Carolina State University, 343 Hunters Blind Drive, Columbia, South Carolina 29212. Voice/fax: 803-732-2019.
rhythmpsyc@aol.com

The derivation of the statistics is provided by Ogden and Collier (submitted). Here we provide a brief informal description. The test statistic, which we call Q, is nonparametric. We believe that this nonparametric approach is critical, because it is evident that drift can take many forms, both deterministic and stochastic; therefore, a broad net must initially be cast. Q estimates variance components based on the "pseudoresiduals" (related to the Von Neumann test; see, e.g., Ref. 2) Heuristically, this approach is based on the sum of squared first and second differences, where differencing removes the effect of drift.

Turning first to the statistical properties of the Q statistic for the presence of drift, formal derivations show it to display the proper behavior under the null model and that the components of variance are unbiased. This was verified with a simulation study. In 10,000 simulations of the null (nondrift) model, with clock SD = 25 ms, motor SD = 10 ms, and drift SD = 0 (null), ~3–5% of the trials were rejected at alpha 0.05. Mean variance estimators were almost exactly equal to 625, 100, and 0 ms, respectively. However, the standard error of these statistics was large, resulting in many negative variances. This is to be expected, given that the model is nonparametric. Power analyses obtained from non-null simulations also show the model to have reasonable power, given large enough drift standard deviations. It is conservative in the sense of failing to reject the null except when the effect size is fairly large.

Turning next to an empirical application, the model was applied to a large data set. Forty-six subjects tapped sequences of 72 taps with no metronome, using the standard entrainment-continuation paradigm. They did this separately for each of four limbs, at each of eight tempos, with three replications per condition. (Some data were lost due to experimenter or equipment error.)

Turning first to the *p* values associated with the Q statistic, across all conditions 37% were rejected at the 0.05 level. Given that the statistic is conservative (as already stated), we can infer that drift is as ubiquitous as we postulated.

As for the effect of various variables on drift, we examined the estimates of variance due to drift, motor, and (nondrift) clock processes. To do so, negative variance estimates were set to zero. (This has the effect of introducing bias into the estimators, while reducing their standard errors; the latter effect is sufficiently dramatic to justify the trade-off with the former, in our view). Statistics were bases on standard deviations rather than variances (which were more skewed).

## RESULTS AND DISCUSSION

FIGURE 1 demonstrates that drift increases dramatically with slower tempos, but it shows little or no effect as to which limb is used. Neither do the two factors interact. A similar picture is seen for clock variance. According to the classical Wing and Kristofferson results,[1] motor variance does not depend on tempo, whereas our results show some dependence on tempo at the extremes (FIG. 2), indicating that either the classical results are wrong at the extremes or our motor variance estimator is contaminated by other sources. On the other hand, the motor standard deviations are the only ones that differentiate the limbs, with feet more variable than hands. An unexplained exception is the right foot at the extremes, where it is the worst performer. Tests of the effects of subjects were run, collapsing across both limbs and tempos.

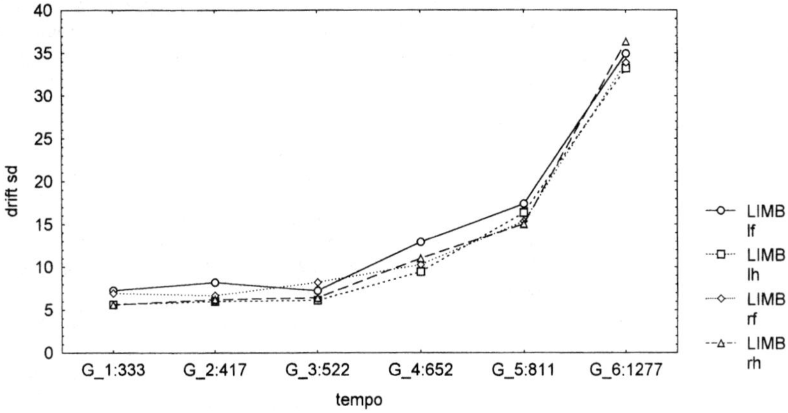


**FIGURE 1.** Drift SD as a function of tempo and limb. Tempo: $F(5, 2828) = 242.89$, $p < 0.000$; limb: $F(3, 2828) = 1.91$, $p < .1264$; 2-way interaction: $F(15, 2828) = .44$, $p < .9690$.

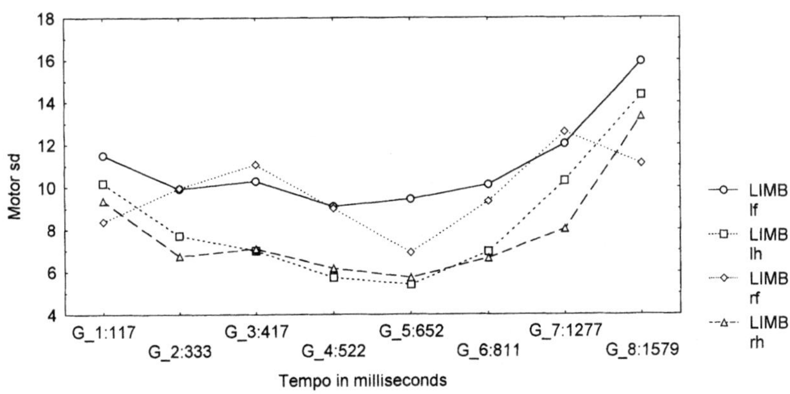


**FIGURE 2.** Motor SD as a function of tempo and limb. Tempo: $F(7, 3776) = 12.19$, $p < .000$; limbs: $F(3, 3776) = 10.68$, $p < .000$; 2-way interaction: $F(21, 3776) = 1.04$, $p < .4118$. Statistical significances don't change when two extreme tempos are omitted.

Subjects differed significantly in all three factors, indicating the importance of individual differences.

## SUMMARY

The statistical techniques developed herein show strong promise as a nonparametric extension of the classical Wing-Kristofferson decomposition of variance in isochronous tapping to include tempo drift. It is now important to establish the relative role of the three sources of variance in discriminating among unselected individuals as well as selected populations differentiated by such factors as age, gender, and motor control impairment. We are currently making the estimation techniques available through the web, so that they can be applied to extant and new data. We can be contacted by e-mail for further information.

## REFERENCE

1. WING, A.M & A.B. KRISTOFFERSON. 1973. Response delays and the timing of discrete motor responses. Percept. Psychophysics **14:** 5–12.
2. GASSER, T., L. SROKA & C. JENNEN-STEINMETZ. 1986. Residual variance and residual pattern in nonlinear regression. Biometrika **73:** 625–633.

# Effects of Music Perception and Imagery on Sensorimotor Synchronization with Complex Timing Patterns

BRUNO H. REPP

*Haskins Laboratories, 270 Crown Street, New Haven, Connecticut 06511-6695, USA*

**KEYWORDS: Expressive timing; Imagined music; Movement and music; Music and movement; Musical imagery; Timing**

There is a close relationship between music perception and action. Music unfolds in time, and listeners often move along in synchrony. Western classical music may give rise to experiences of inner movement while overt movement is suppressed. The rate at which this music unfolds often varies continuously; this is expressive timing, which is especially associated with motion.[1] A particular musical structure tends to favor a particular, typical expressive timing pattern (ETP). However, ETPs of individual artists sometimes deviate considerably from this norm.[2]

The present study investigated whether overt movement (finger tapping) in synchrony with an ETP is facilitated when the ETP is carried by music rather than just by a sequence of clicks, and whether this facilitation is greater when the ETP is typical of the music rather than atypical. The study further examined whether such facilitation occurs when the music is merely imagined. Musically inappropriate timing patterns (TPs), for which no facilitation was expected, served as comparison.

Two experiments both used the opening of Chopin's Etude in E Major, op. 10, no. 3, as the musical excerpt. Each experiment included three conditions requiring synchronized finger tapping: (1) with a sequence of clicks, (2) with clicks accompanied by music, and (3) with clicks while only imagining the music, in that order. The ETPs (T0, T1, T2, T4) were derived from a principal components analysis of 115 expert performances.[2]

In experiment 1, there were 5 TPs: isochronous, T1 (fairly typical), T2 (atypical), T4 (atypical), and R1 (random). The design was blocked by condition, and each TP was presented 10 times in succession in each condition. In experiment 2, there were 9 TPs: isochronous, T0 (highly typical), T1, T2, T4, as well as T0′, T1′, T2′, T4′ (phase-shifted versions misaligned with the musical structure). The design was blocked by TP. Each TP was presented 20 times in succession, two times for practice (clicks only) and then six times per condition, without breaks. In both experiments, the isochronous TP was presented first as practice; the other TPs followed in a coun-

Address for correspondence: Bruno H. Repp, Haskins Laboratories, 270 Crown Street, New Haven, Connecticut 06511-6695. Voice: 203-865-6163 ext. 236; fax: 203-865-8963.
repp@haskins.yale.edu

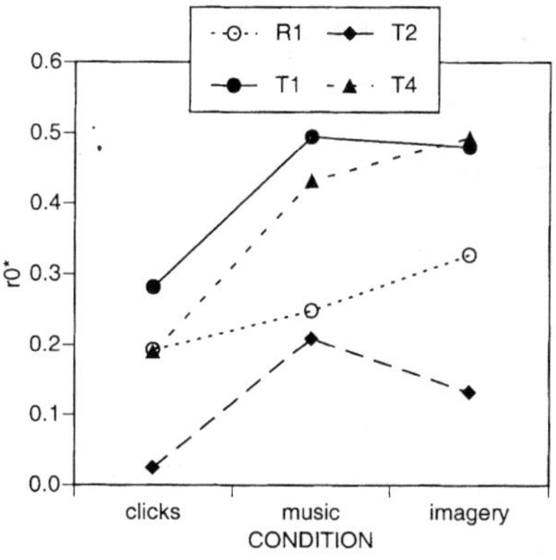


**FIGURE 1.** Results of experiment 1.

terbalanced order. Twelve musically trained undergraduate students participated in each experiment.

The measure of synchronization accuracy for the anisochronous TPs was the *prediction index,* $r0^* = (r0 - ac1)/(1 - ac1)$, where r0 was the correlation between the TP's interonset intervals and the intertap intervals, and ac1 was the TP's lag-1 autocorrelation. The ac1 set a lower limit to r0 because lag-1 tracking typically occurs when synchronization with an unpredictable TP is attempted.[3]

FIGURE 1 shows the average $r0^*$ indices for each TP in each condition of experiment 1. Since the conditions were in a fixed order, there was a gradual improvement due to practice for all TPs. However, performance improved more for the ETPs than for R1 between the click and music conditions, $F(3,33) = 3.2$, $p < 0.04$, regardless of ETP typicality. Conversely, comparing the music and imagery conditions, ETPs suffered more from the absence of audible music than did R1, $F(3,33) = 4.9$, $p < .007$. Between the click and imagery conditions, there was likewise a significant interaction with TP, $F(3,33) = 4.1$, $p < 0.02$, suggesting a beneficial effect of imagery for at least one ETP (T4) relative to R1.

FIGURE 2 shows the results of experiment 2, in which each ETP could be compared to its phase-shifted version. In each case, synchronization accuracy improved more for the ETP than for its phase-shifted version between the click and music conditions, $F(1,11) = 24.0$, $p < 0.0006$. The opposite was true between the music and imagery conditions, $F(1,11) = 36.6$, $p < 0.0002$. However, the Version × Condition interaction was nonsignificant between the click and imagery conditions. There were no triple interactions with TP.

These results provide convincing evidence that synchronization of a simple motor activity with an ETP is facilitated when the ETP is carried by an appropriate musical structure than when it is carried by an unstructured click sequence. Moreover, this facilitation occurred regardless of the typicality of the ETP, which supports the notion that radically different ETPs, observed in expert performances,[2] can be compatible with the same musical structure. However, there was little evidence that musical imagery was an effective substitute for audible music.

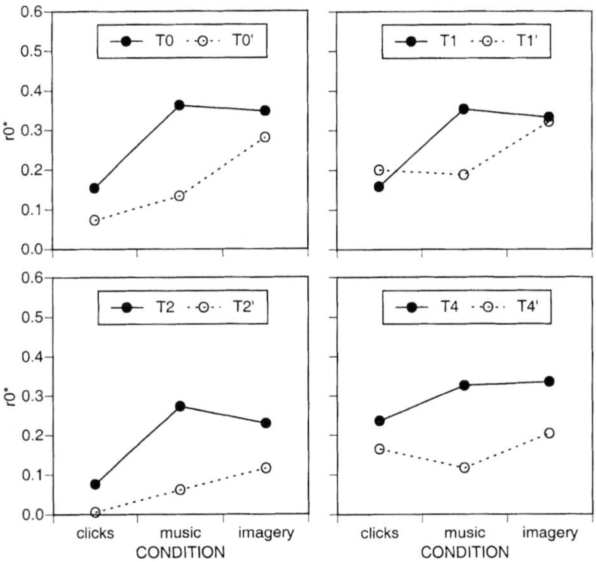


**FIGURE 2.** Results of experiment 2.

Nevertheless, imagery did have an interesting effect on synchronization with *isochronous* sequences in both experiments. The intertap intervals in the music condition showed small but systematic deviations from regularity, as observed previously.[4] Thus, the musical structure influenced the timing of the synchronized action, without participants' awareness. The very same pattern of deviations, albeit somewhat reduced in magnitude, emerged in the imagery condition. This proves that participants did imagine the music and that imagined music can have involuntary effects on the timing of accompanying motor behavior. These effects, however, may be unrelated to the ability to predict ETPs.

## ACKNOWLEDGMENTS

This research was supported by NIH Grant No. MH-51230.

## REFERENCES

1. TODD, N.P. MCA. 1995. The kinematics of musical expression. J. Acoust. Soc. Am. **97:** 1940–1949.
2. REPP, B.H. 1998. A microcosm of musical expression: I. Quantitative analysis of pianists' timing in the initial measures of Chopin's Etude in E major. J. Acoust. Soc. Am. **104:** 1085–1100.
3. MICHON, J. 1967. Timing in Temporal Tracking. Van Gorcum. Assen, the Netherlands.
4. REPP, B.H. 1999. Detecting deviations from metronomic timing in music: effects of perceptual structure on the mental timekeeper. Percept. & Psychophys. **61:** 529–548.

# Estimates of Sequence Acceleration and Deceleration Support the Synchronization of Internal Rhythms

AMANDINE PENEL,[a] MARIE RIVENEZ,[b] AND CAROLYN DRAKE[b]

[a]*Haskins Laboratories, 270 Crown Street, New Haven, Connecticut 06511-6695, USA*

[b]*Laboratoire de Psychologie Expérimentale, CNRS UMR 8581 and Université René Descartes, Boulogne, France*

**KEYWORDS: Expressive timing; Rhythm, internal; Tempo change; Synchronization; Timing**

Work on expressive timing in music performance has placed greater emphasis on explaining lengthenings compared to shortenings, and indeed the measure of temporal deviations from the musical score suggests that expressive timing consists mainly of lengthenings away from a maximum tempo rather than of lengthenings and shortenings around a mean tempo.[1] We wondered whether this bias in time production may originate from a bias in the perception of accelerations compared to decelerations. Compared to accelerations, are decelerations more or less salient? Some work seems to indicate the second possibility,[2,3] and we investigated that further.

In a series of experiments, we used a method of direct estimation of the rate of tempo change, asking participants to indicate on a continuous scale how much a sequence of nine identical sounds accelerated or decelerated. Results did not reveal any general under- or overestimation of accelerations compared to decelerations. However, interesting asymmetries were found depending on the tempo within the presented range. Generally, at the fastest tempi within a range of tempi presented in an experimental session, accelerations were judged as being greater than decelerations of the same physical magnitude. The opposite was observed for the slowest tempi within a range. FIGURE 1 gives an example of the results obtained for one experiment. All sequences had a tempo change of 20% ($IOI_{max}$–$IOI_{min}$ = 20% of the mean IOI of the sequence [IOI, interonset interval]), and they could have a mean IOI of 500, 600, 700, 800, or 900 milliseconds. Twenty-two participants took part in this experiment, and the average of their normalized ratings is plotted for accelerations and decelerations as a function of the tempo zone (mean IOI of the sequence). It can be seen that for sequences in the 500 ms IOI tempo zone, accelerations were judged as greater than decelerations. The tempo changes of accelerating and decelerating

Address for correspondence: Amandine Penel, Psychology Department, The Ohio State University, 200C Lazenby Hall, 1827 Neil Ave., Columbus, OH 43210. Voice: 614-292-8667; fax: 614-292-5601.

penel.1@osu.edu

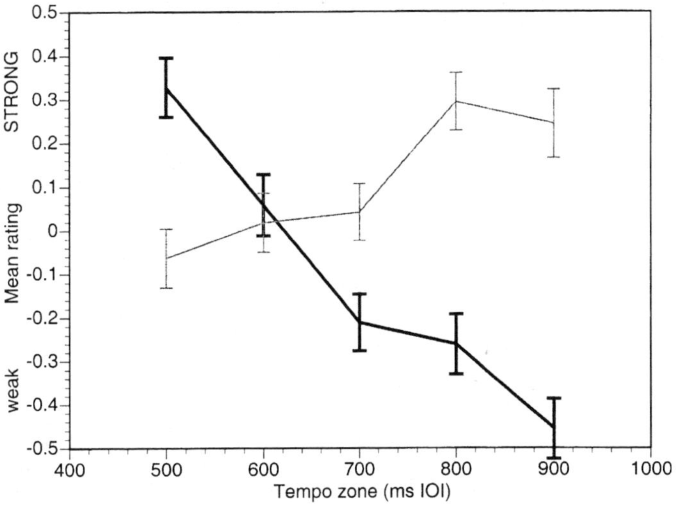


**FIGURE 1.** Mean tempo change rating, from weak to strong, for accelerating (*dark*) and decelerating (*light*) sequences around 500, 600, 700, 800, and 900 ms IOI (in all cases, the tempo change was 20%).

sequences were considered as equivalent at 600 ms IOI. Finally, at 700, 800, and 900 ms IOI, decelerations were judged as greater than accelerations (tempo zone/direction of tempo change interaction: $F(4, 84) = 15.5$; $p < 0.0001$).

This finding is consistent with the hypothesis of the synchronization of internal rhythms with events in sequences, even if their IOIs are irregular. At fast tempi, the internal rhythm synchonizing with events is also, in the same time, attracted toward an intermediate slower tempo, making an acceleration more salient than a deceleration. At slow tempi, the synchronized internal rhythm is also attracted toward an intermediate faster tempo, making a deceleration more salient than an acceleration.

This attractor tempo varied with the presented range, but 600 ms IOI seems to be a fixed one because it was observed in various experimental contexts (with large or narrow ranges, whether it was in the middle of the range or not, as in FIGURE 1). This is coherent with Fraisse's[4] results suggesting that 600 ms IOI is a preferred internal rhythm.

## REFERENCES

1. PENEL, A. 2000. Variations temporelles dans l'interprétation musicale: processus perceptifs et cognitifs. Ph.D. thesis, Université Paris VI. Paris.
2. KUHN, T.L. 1974. Discrimination of modulated beat tempo by professional musicians. J. Res. Music Educ. **22:** 270–277.
3. MADSEN, C.K. 1979. Modulated beat discrimination among musicians and nonmusicians. J. Res. Music Educ. **27:** 57–67.
4. FRAISSE, P. 1984. Perception and estimation of time. Annu. Rev. Psychol. **35:** 1–36.

# Frontal Lobe Contributions to Perception of Rhythmic Group Structure

## An EEG Investigation

PHILIP G. HARRIS, RICHARD B. SILBERSTEIN, GEOFF E. NIELD, AND ANDREW PIPINGAS

*Brain Sciences Institute, Swinburne University of Technology, 400 Burwood Road, Hawthorn, 3122, Victoria, Australia*

**KEYWORDS: Frontal lobe; Group structure; Interonset interval; Perception of group structure; Rhythmic group structure**

Substantial research supports the salience of relatively longer interonset intervals (IOIs) in determining rhythmic group structure,[1–4] yet little is understood about the neural underpinnings of the perceptual grouping process. Gestalt-based interpretations of perceptual processes underlying rhythmic groupings have been proposed[5] and converge with orienting response theory, which proposes enhanced neural sensitivity following a change in IOI.[6] In support of this, enhancement of the N1-P2 component of event-related potential (ERP) responses to the first sound in short rhythmic groups has been demonstrated.[7] The current study investigated cortical responses to perception of rhythmic group structure by examining changes in EEG responses associated with participants' active and passive perception of pitch/rhythm patterns. The study used the steady state visually evoked potential (SSVEP) to examine ongoing changes in 13-Hz EEG responses while participants listened to pitch/rhythm patterns.

Forty-three females, aged 18–42 years (mean = 23.0, SD = 5.6) with a wide range of musical training (mean = 4.0, SD = 7.0, range = 0–32) performed two tasks while their EEG was recorded. Pitch memory was tested with a probe recognition task, which required participants to actively memorize 10-note pitch/rhythm patterns for subsequent recognition. A passive task required participants to detect the rare playing of two-note pitch pairs among a single-note pitch/rhythm series. Stimuli in both tasks consisted of diatonic pitch series at 600 ms IOI with a semirandom ($p = 0.15$) doubling of IOI to 1200 ms. Neither task placed explicit demands on perception of rhythmic structure of the pitch series. Brain electrical activity was recorded from 64 scalp electrodes, including all International 10–20 positions, referenced to linked ear lobes with nose as ground. The SSVEP was elicited by a 13-Hz flicker delivered to

Address for correspondence: Philip G. Harris, Brain Sciences Institute, Swinburne University of Technology, 400 Burwood Road, Hawthorn, 3122, Victoria, Australia. Voice: +61-3-92148832; fax: +61-3-92145525.

pharris@bsi.swin.edu.au

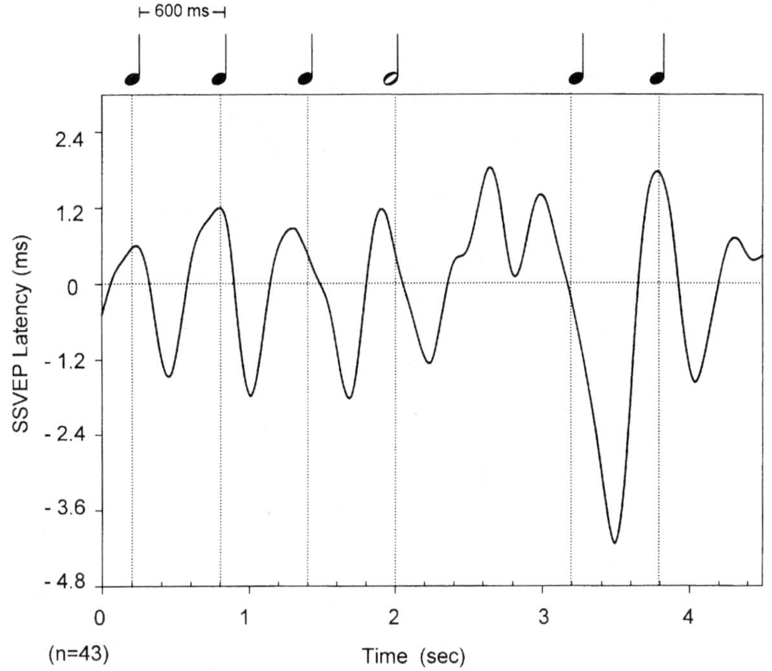


**FIGURE 1.** Averaged SSVEP latency series of 43 participants at a left frontal site during probe recognition task listening phase. A note onset following a lengthened IOI elicits a strong reduction in latency response.

participants via LED arrays mounted on a set of goggles. An across-subject SSVEP latency time series for a 5-s period, centered on the occurrence of lengthened IOIs in each task, was calculated at each electrode site at a temporal resolution of 190 milliseconds.

Longer IOIs significantly influenced participants' performance in the probe recognition task. Pitch memory performance was significantly better when memory test note groups matched rhythmic group structure established by longer IOIs (66.0% correct), than when note groups straddled over longer IOIs (59.9% correct), ANOVA $F$ (1,42) = 12.3, $p$ = 0.001. EEG responses to the pitch/rhythm patterns showed rhythmic 13-Hz SSVEP latency reductions over bilateral frontal areas in both tasks following approximately 200 ms after each note onset. A strong enhancement of SSVEP latency reduction was apparent at left posterior frontal sites in response to a note following a longer IOI (see FIG. 1). A repeated measures analysis of variance that compared latency responses in a 100–300 ms window following each note onset revealed a significant enhancement of SSVEP latency reduction at a left posterior frontal site in response to a note following a longer IOI, compared to latency reductions elicited by notes following standard IOIs in the active task, $F(2.9, 42) = 2.9$, $p < 0.05$), but not the passive task, $F(2.9, 42) = 1.3$, $p > 0.05$. Post hoc $t$ tests comparing peak latency values appear in FIGURE 2.

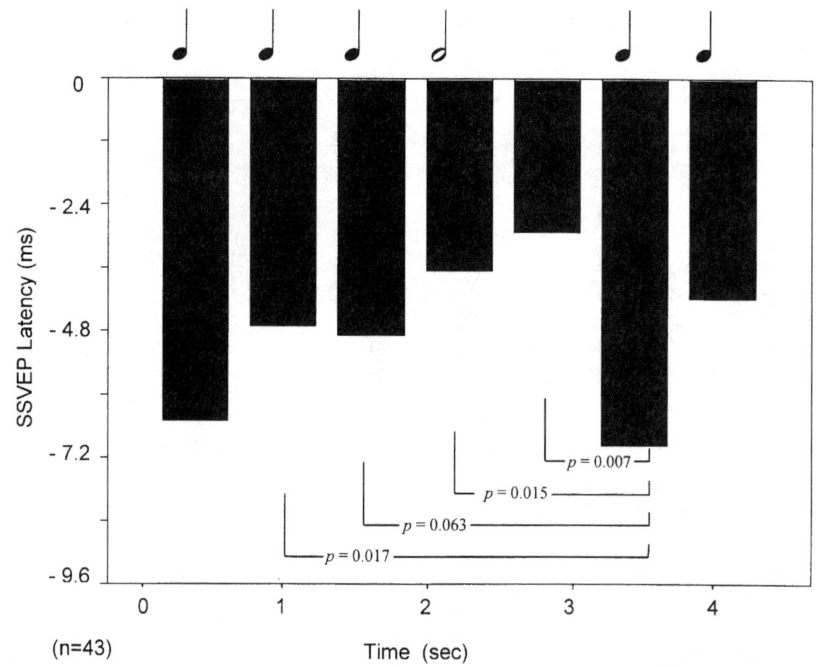


**FIGURE 2.** Peak analysis of 43 subjects during listening phase of probe recognition task. Each *bar* represents the average of individual participants' peak latency reduction in a 100–300 ms window following note onsets. *p* Values represent paired *t* test comparisons between lengthened IOI peak latency compared to peak latencies following standard IOIs.

Steady state latency reductions in the frontal lobes have previously been linked with increases in frontal neural processing speed,[8,9] inferred to represent frontal excitatory processes. In view of behavioral results indicating that longer IOIs significantly influenced internal representations of the patterns, the left frontal latency reduction following a longer IOI may index a frontal contribution to the coding of the internal representation of the pattern temporal structure. Longer IOIs elicited transient left frontal excitation, but only in a task requiring the holding of an internal representation of the pattern. In summary, the current study finds support for gestalt and orienting response theory proposals of a special perceptual and neural status for an event following a relatively longer-time interval. Left frontal excitation in response to longer IOIs, while not a primary perceptual response, appears to be related to the coding of an internal representation of the rhythmic structure of a pitch/rhythm pattern.

## REFERENCES

1. DOWLING, W.J. 1973. Rhythmic groups and subjective chunks in memory for melodies. Percept. & Psychophys. **14:** 37–40.

2. DELIEGE, I. 1987. Grouping conditions in listening to music: an approach to Lerdahl and Jackendoff's grouping preference rules. Music Percept. **4:** 325–360.
3. PERETZ, I. 1989. Clustering in music: an appraisal of task factors. Int. J. Psychol. **24:** 157–178.
4. HARRIS, P.G., R.B. SILBERSTEIN, A. PIPINGAS, *et al.* 1999. Perceptual grouping of pitch sequences in the steady-state visually evoked potential responses of musically trained subjects. *In* Music, Mind and Science. S.W. Yi, Ed.: 149–170. Western Music Research Institute. Seoul.
5. LERDAHL, F. & R. JACKENDOFF. 1983. A generative theory of tonal music. MIT Press. Cambridge, MA.
6. SOKOLOV, E.N. 1975. The neuronal mechanisms of the orienting reflex. E.N. Sokolov & O.S. Vinogradova, Eds. Lawrence Erlbaum Associates. Hillsdale, NJ.
7. FRAISSE, P. & A. LAVIT. 1986. Auditory evoked potentials and the perception of rhythms. Int. J. Psychophysiol. **4:** 85–89.
8. SILBERSTEIN, R.B., P.J. CADUSCH, G. NIELD, *et al.* 1996. Steady-state visually evoked potential topography dynamics and cognition. *In* Recent Advances in Event-related Brain Potential Research: Proceedings of the 11th International Conference on Event-related Potentials (EPIC), Okinawa, Japan. C. Ogura, Y. Koga & M. Shimokochi, Eds.: 379–385. Elsevier. Amsterdam.
9. SILBERSTEIN, R.B., M. FARROW, F.L. LEVY, *et al.* 1998. Functional brain electrical activity mapping in boys with attention-deficit/hyperactivity disorder. Arch. Gen. Psychiatry **55:** 1105–1112.

# Cortical Representations of Temporal Structures in Music

MICHAEL GROSSBACH, HELEN KUCK, MARC BANGERT, AND ECKART ALTENMÜLLER

*Institute for Music Physiology and Performing Arts Medicine, State Academy for Music and Drama, Hanover, Germany*

**KEYWORDS: Cortical activation and musical temporal structure; Temporal musical structure, cortical representations of; Rhythm processing**

## INTRODUCTION

In the visual modality local and global features are processed in different areas of the brain, whereas in the auditory domain the existence of a similar distinction is subject to discussion. Several studies have addressed this question, with different results.[1–3] Our aim was to investigate differences in cortical activation while subjects were listening to simple musical sequences that could vary in local (i.e., rhythm) or global (i.e., meter) temporal structure.

## METHODS

Subjects seated in a dimly lit room were subjected to auditory stimuli via two stereo speakers at comfortable sound levels (~73 dB SPL), and instructions were displayed on a 14-inch monitor approximately 1.5 m away. Each acoustic stimulus consisted of two monophonic sequences with both metric structure and rhythmic variation. The second sequence was either identical to or different from the first one. Differences were either metric or rhythmic in nature. The subject's task was to judge whether or not the sequences were identical.

Slow cortical potentials were monitored by 32-channel (DC) electroencephalography. After artifact reduction and baseline correction, DC potentials were epoched for the tasks, and trials of one task were averaged for each subject and over all subjects (grand average, $n$ = 18, eight females).

## RESULTS

When cortical activation patterns during rhythm and meter processing were compared, focusing on differences between the first and the second sequence, both tasks

Address for correspondence: Dr. Michael Grossbach, Plathnerstr. 35, 30175 Hanover, Germany. Voice: +49-511-3100-562; fax: +49-511-3100-557.
michael.grossbach@gmx.de

yielded bilaterofrontal activation during the first sequence, whereas the second sequence evoked frontotemporal activation in the right hemisphere.

Trigger-locked grand averages of all trials in which changes occurred in the second sequence revealed different temporal dynamics while the subjects were processing rhythm and meter. Activation patterns established an earlier right frontolateral peak during meter tasks than during rhythm tasks. Difference pictures underline this result.

## DISCUSSION AND CONCLUSIONS

The first sequence seems to be encoded in frontocentral neuronal networks regardless of the task, supporting the idea that the prefrontal lobe is involved in working memory function.[4] Additionally, no significant differences in activation amplitudes were noted between rhythm and meter tasks in accordance with our aim of designing all sequences with the same algorithm in the first place. Only after a change had been introduced into the second sequence were subjects potentially able to judge for rhythm or meter tasks (although they were not supposed to do so). The right frontolateral activation during the second sequence might reflect the neurophysiologic correlate of comparing the actual with the encoded (first) sequence.

Because no significant differences were found in the activated regions during rhythm and meter processing, a more thorough examination using higher electrode numbers and additional imaging techniques (functional magnetic resonance imaging) will follow to clarify the contradictory results of lesion studies[2,3] in which patients with left hemispheric stroke were found to have impaired rhythm processing.

An early established activation pattern during meter processing possibly reflects the order in which temporal information in music is analyzed; a global survey precedes local analysis.

## ACKNOWLEDGMENT

This work was supported by Grant DFG 1046 from the Deutsche Forschungsgemeinschaft.

## REFERENCES

1. PERETZ, I. 1990. Processing of local and global musical information by unilateral brain-damaged patients. Brain **113:** 1185–1205.
2. LIÉGEOIS-CHAUVEL, C., I. PERETZ, M. BABAI, *et al.* 1998. Contribution of different cortical areas in the temporal lobes to music processing. Brain **121:** 1853–1867.
3. SCHUPPERT, M., T.F. MÜNTE, B.M. WIERINGA & E. ALTENMÜLLER. 2000. Receptive amusia: evidence for cross-hemispheric neural networks underlying music processing strategies. Brain **123:** 546–559.
4. BADDELEY, A. 1992. Working memory. Science **255:** 556–559.

# Neapolitan Chords Activate the Area of Broca

## A Magnetoencephalographic Study

STEFAN KOELSCH, BURKHARD MAESS, THOMAS C. GUNTER, AND ANGELA D. FRIEDERICI

*Max-Planck-Institute of Cognitive Neuroscience, Leipzig, Germany*

**KEYWORDS: Neopolitan chords; Broca's area; Harmony, perception of; Music; Syntax; MEG**

## INTRODUCTION

In recent electroencelphalographic studies,[1,2] we found that in a major–minor tonal environment, harmonically unrelated chords (infrequently presented among in-key chords) may elicit an early right anterior negativity (ERAN) in the event-related brain potential (ERP). The present experiment was aimed at localizing the neural substrates of the ERAN using magnetoencephalography (MEG).

Six nonmusicians listened to chord sequences, each sequence consisting of five chords. Some sequences consisted of in-key chords only and built up a musical context towards the end of the sequence. Randomly, a chord at the third or fifth position of the sequence was a Neapolitan chord. Neapolitan chords are consonant major chords; but according to music theory, Neapolitan chords contain out-of-key notes with respect to the preceding harmonic context. From a music psychological view, chords with out-of-key notes are perceived as unexpected.[3,4] Moreover, in some sequences, a chord was played not on a piano, but on a deviant instrument (e.g., marimba, trumpet); participants were asked to detect the deviant instrument by pressing a response button. Event-related magnetic fields were averaged from the continuous MEG for each subject and each condition (in-key versus Neapolitan chords). Dipole solutions were then obtained for each subject, transformed into a standard brain, and grand averaged across subjects.

In-key chords elicited the magnetic equivalent of the P2 (P2m, maximal around 200 ms). Neapolitan chords elicited a particularly early magnetic field effect, which was greatest around 200 ms, the magnetic equivalent of the ERAN (mERAN). The field maps of the mERAN revealed a dipolar pattern over each hemisphere; in all subjects, the fields of the mERAN had virtually an inverse "polarity" compared to the fields of the P2m. Moreover, the steepest field gradients of the mERAN are located more anteriorly than those of the P2m, indicating that the neural generators of the mERAN are located anteriorly to those of the P2m. The grand-average dipole solution for the mERAN indicated that in each hemisphere one dipole was located

Address for correspondence: Dr. Stefan Koelsch, Stephanstr. 1a, D-04103 Leipzig, Germany. Voice: +49-3 41/99-40-121; fax: +49-3-41/99-40-113.
koelsch@cns.mpg.de

within the lower part of the pars opercularis (in the inferior frontal gyrus, part of BA 44, in the left hemisphere, classically called Broca's area).

The mERAN, like the ERAN, presumably reflects the brain's response to a sound-expectancy violation. Chords preceding the Neapolitan chord strongly establish a tonal key. During such a sequence, listeners build up a "hierarchy of harmonic stability,"[3,4] which specifies expectancies for certain chords to follow. At the fifth position of the chord sequences, a tonic chord was most expected. Instead of a tonic, a Neapolitan chord occurred, which was an inappropriate chord function and contained out-of-key notes. Neapolitan chords therefore sounded unexpected in the established tonal environment. The ability to perceive distances between chords (and keys, respectively) and to expect certain harmonies to a higher or lower degree can rely only on a representation of the principles of harmonic relatedness described by music theory. These principles, or rules, were reflected in the harmonic expectancies of listeners and may be taken as musical syntax. Surprisingly, the generators of the mERAN were *not* located within the temporal lobe, but in the lower BA 44. Notably, during language perception these areas are thought to be responsible for the syntactic analysis of incoming language input (in the sense of determining grammatical relationships in a sentence) and are specialized for fast and automatic parsing processes.[5]

The present results demonstrate that Broca's area (and its right-hemisphere homologue) is also involved in processing musical syntax in the sense of determining harmonic relationships within a musical phrase. This finding indicates that these areas process information with considerably less domain specificity than previously believed, suggesting that the brain processes syntax also in domains other than language. From a functional-neuroanatomical view, results reveal a strong relationship between language and music. This relationship might, at least partly, account for influences of musical training on verbal abilities.[6] Notably, effects were elicited in nonmusicians who listened to the stimuli under instructions to ignore the harmonies, supporting the hypothesis of an implicit musical ability of the human brain[1,2] and enabling broad generalization of the present findings.

## REFERENCES


1. KOELSCH, S., T.C. GUNTER, A.D. FRIEDERICI & E. SCHROGER. 2000. Brain indices of music processing: "non-musicians" are musical. J. Cognit. Neurosci. **12.3:** 520–541.
2. KOELSCH, S., E. SCHROGER, T.C. GUNTER & A.D. FRIEDERICI. 2000. Pre-attentive musicality of the human brain. Submitted.
3. KRUMHANSL, C.L., J. BHARUCHA & M.A. CASTELLANO. 1982. Key distance effects on perceived harmonic structure in music. Percept. & Psychophys. **32:** 96–108.
4. BHARUCHA, J. & C. KRUMHANSL. 1983. The representation of harmonic structure in music: hierarchies of stability as a function of context. Cognition **13:** 63–102.
5. FRIEDERICI, A.D. 1998. Language Comprehension: A Biological Perspective: 281–289. Springer. New York.
6. CHAN, A.S., Y.C. HO & M.C. CHEUNG. 1998. Music training improves verbal memory. Nature **396:** 128.

# Cortical Dynamics and the Perception of Tone Sequence Structure

ANIRUDDH D. PATEL AND EVAN BALABAN

*The Neurosciences Institute, San Diego, California 92121, USA*

**KEYWORDS: Brain activity; Melody processing; Perception of tone sequence**

## INTRODUCTION

Melodies are a special subset of auditory sequences. Their acoustic raw materials may be extremely simple (e.g., a few dozen pure tones), yet the arrangement of these materials in time can create structures that engage a host of interacting mental processes, including chunking, melodic expectancy, and the perception of meter. Studying these processes is a principal goal for the cognitive science of melody and is being actively pursued by many research groups.[1]

Discovering the neural correlates of melodic processing is a challenge for cognitive neuroscience. Progress in this area has focused on average neural responses to individual events in sequences via event-related potentials[2] or on the brain's average response to entire sequences via positron emission tomography.[3] These techniques continue to provide valuable information, yet it is evident that to "tap into the moment-to-moment history of mental involvement with the music,"[4] techniques that measure patterns of neural activity are needed as perception unfolds within individual sequences.

With these goals in mind, we set out to determine if aspects of tone sequence structure are reflected in dynamic neural responses and if melody-like sequences have a distinct neural signature. Full details of this study were presented by Patel and Balaban.[5] This abstract presents a brief overview of our methods and results.

## METHODS

Subjects listened to one-minute-long, pure-tone sequences while neural signals were recorded using 148-channel whole-head magnetoencephalography. Sequences were random, deterministic (scalar), or one of two categories of "fractal" sequences differing in their balance of predictability and unpredictability. One of these fractal categories had melody-like statistics.[6] All sequences were confined to a two-octave range (220 and 880 Hz) and used pitches from Western musical scales. Amplitude modulation of the tone sequences was used to generate ongoing, identifiable cortical activity[7] whose amplitude and timing (phase) could be studied as a function of tone

Address for correspondemce: Dr. Aniruddh D. Patel, The Neurosciences Institute, 10640 John Jay Hopkins Drive, San Diego, CA 92121. Voice: 858-626-2085; fax: 858-626-2099.
apatel@nsi.edu

sequence structure. Amplitude modulation gave the tone sequences a slightly "warbly" quality, without disrupting their perceived pitch pattern; listeners heard them as sequences of pitches at the underlying pure tone frequencies. (Examples of these sequences can be heard at www.nsi.edu/users/patel/tone_sequences.)

## RESULTS

We found that timing patterns of stimulus-related brain activity showed a strong dependency on the structure of the tone sequence. At certain sensor locations, the phase of the cortical signal tracked the pitch contour of the tone sequences with increasingly accurate tracking as sequences became more predictable. This suggests that processes of expectancy may play a role in this brain response. By contrast, interactions between brain regions (as measured by phase coherence) were greatest for the tone sequences with melody-like statistics. Spatial analysis revealed that this pattern was driven by interactions between the left posterior hemisphere and the rest of the brain, including anterior and posterior right hemispheric regions. This is of interest because previous neuropsychological research on melody suggests that left superior temporal regions are involved in the discrimination of precise interval sizes, whereas right frontotemporal circuits are involved in the perception of more global contour patterns.[8,9] Thus, the observed pattern of coherence may reflect the dynamic integration of local and global pitch perception and suggests that this integration is greatest when tone sequences resemble musical melodies.

## CONCLUSION

Dynamic neural responses reveal a neural correlate of pitch contour in the human brain and show that interactions between brain regions are greatest when tone sequences have melody-like statistical structure. Notably, these results are based on temporal aspects of neural activity. This suggests that dynamic imaging can provide an important complement to techniques such as functional magnetic resonance imaging and positron emission tomography, which are sensitive to the amount of neural activity but insensitive to the temporal structure of this activity. The convergence of multiple techniques is likely to provide the best approach for mapping brain activity associated with music perception.

## ACKNOWLEDGMENTS

This work was supported by the Neurosciences Research Foundation as part of its research program on Music and the Brain at The Neurosciences Institute.

### REFERENCES

1. Krumhansl, C., J. Louhivuori, P. Toiviainen, *et al.* 2000. Melodic expectation in Finnish spiritual folk hymns: convergence of statistical, behavioral, and computational approaches. Music Percept. **17:** 151–196.

2. BESSON, M. & F. FAÏTA. 1995. An event-related potential (ERP) study of musical expectancy: comparison of musicians with nonmusicians. J. Exp. Psychol. Hum. Percept. Perform. **21:** 1278–1296.
3. ZATORRE, R.J., A.C. EVANS & E. MEYER. 1994. Neural mechanisms underlying melodic perception and memory for pitch. J. Neurosci. **14:** 1908–1919.
4. SLOBODA, J. 1985. The Musical Mind: The Cognitive Psychology of Music. Clarendon Press. Oxford.
5. PATEL, A.D. & E. BALABAN. 2000. Temporal patterns of human cortical activity reflect tone sequence structure. Nature **404:** 80–84.
6. NETTHEIM, N. 1992. On the spectral analysis of melody. Interface **21:** 135–148.
7. GALAMBOS, R., S. MAKEIG & P.J. TALMACHOFF. 1981. A 40-Hz auditory potential recorded from the human scalp. Proc. Natl. Acad. Sci. USA **78:** 2463–2647.
8. LIÉGEOIS-CHAUVEL, C., I. PERETZ, M. BABAÏ, *et al.* 1998. Contribution of different cortical areas in the temporal lobes to music processing. Brain **121:** 1853–1867.
9. PATEL, A.D., I. PERETZ, M. TRAMO, *et al.*. 1998. Processing prosodic and musical patterns: a neuropsychological investigation. Brain Lang. **61:** 123–144.

# On Practice: How the Brain Connects Piano Keys and Piano Sounds

MARC BANGERT, UDO HAEUSLER,[a] AND ECKART ALTENMÜLLER

*Institute for Music Physiology and Performing Arts Medicine, State Academy of Music and Drama, Hanover, Germany*

[a]*German Primate Center, Goettingen, Germany*

**KEYWORDS: Piano practice; Brain, joint activation of auditory and sensorimotor areas of; Brain, representaiton of ear and hand in**

## INTRODUCTION

This review of our unpublished data demonstrates how practice promotes a joint mental representation of ear and hand, which is a phenomenon familiar to most musicians: silent dexterity drills produce "audible tones inside the head," and, vice versa, sounding music "flashes right into the fingers." By dissociating auditory and motor features of piano playing in the experimental paradigm, we tested the effects of short-term (20 minutes), long-term (5 weeks), and lifetime (>15 years, in professional musicians) piano practice.

## MATERIAL AND METHODS

Twenty-four right-handed subjects, two groups of nine novices each ("map group" and "no-map group," see below) and an expert performer group of six professional pianists, were examined. A single session consisted of (1) an EEG recording during a set of 60 purely auditory tasks (passive listening to piano melodies) and 60 purely motor tasks (pressing keys on a silent piano keyboard with the right hand) in randomized order, (2) an adaptive and interactive practicing procedure (duration of 20 minutes) in which the novices had to replay acoustically presented melodies (the complexity level of subsequent tasks was adjusted according to the online analysis of the player's performance), and (3) another EEG recording identical to the first one, again on the silent keyboard.

The map group then worked with the training system twice a week over a period of five weeks with EEG monitoring during the first, sixth, and eleventh (last) session. The no-map group attended the same program, but the key-to-note assignment was

Address for correspondence: Dr. Marc Bangert, Institute for Music Physiology and Performing Arts Medicine, State Academy of Music and Drama, Plathnerstrasse 35, 30175 Hanover, Germany. Voice: +49-511-3100-552; fax: +49-511-3100-557.
marc.bangert@hmt-hannover.de

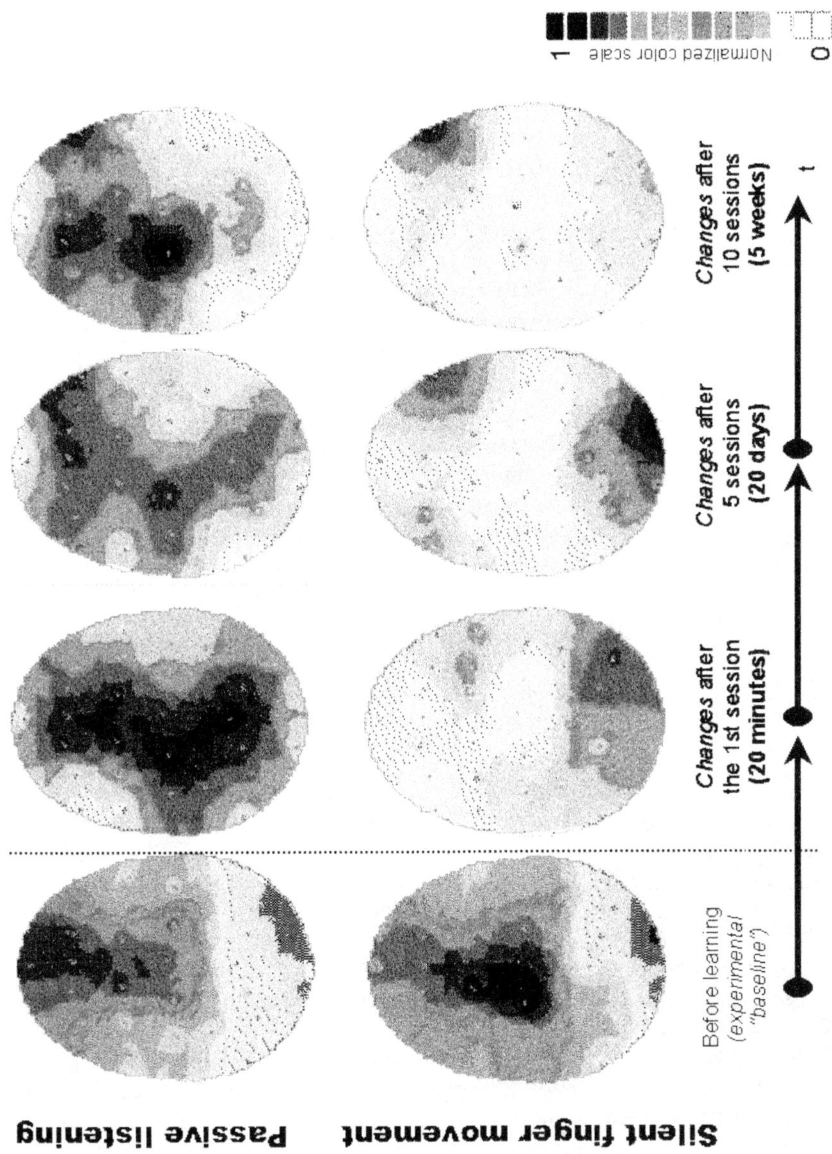


**FIGURE 1.** *See following page for caption.*

shuffled after each single trial. In other words, they had the same chance to acquire timing and dynamics skills as the map learners, but no chance to learn any connection between fingers and pitches.

Analysis of the 32-channel EEG[1] included the topographical distribution of task-related DC-EEG changes, task-related interelectrode coherence.[2] Based on DC-EEG data, the following regions of interest (ROI) were predefined for statistical analysis: frontotemporal areas (left and right) and motor/premotor areas of the left hemisphere. The investigated frequency bands comprised alpha (8–13 Hz), low beta (13–21 Hz), gamma (35–45 Hz), and delta (0.1–4 Hz) frequencies. A vectorial description of the potential distributions allowed the dot product to be used as a measure of pattern similarity. Additionally, the subjects' performance based on MIDI data (note onset times and key velocities) was examined.

In a subsequent study, classical conditioning of the eyeblink reflex by means of an airpuff to one of five notes was used to determine if an intermodal transfer of the stimulus (silent piano keys instead of audible pitches) is possible in highly trained pianists. During the procedure the subjects (15 nonpianists, 16 pianists) were distracted by verbal semantic and working memory tasks mentally.

## RESULTS

The five-week training schedule as well as a single training session induced common activation of auditory and motor areas in the DC-EEG (FIG. 1). In the auditory task the additional activation of premotor and motor areas as well as parietal cortex and frontolateral areas was observed, whereas in the silent motor task the additional activation of the temporal lobes, active inhibition of the sensorimotor field of the unused left hand, and again activity in the right lateral frontal lobe were established with increasing practice time. Vector analysis of cortical pattern similarities indicated that in the novices the activation patterns for the physically different tasks in "passive listening" and "silent motion" were different in the first place but became more similar in the course of practice.

A comparison between the two training groups (map vs. no map) produced a right frontotemporal region ($p < 0.01$), possibly reflecting an additional skill that only the map group had been able to acquire.

The task-related coherence, in the predefined ROIs serving as a measure of functional coupling of mental ears and fingers, yielded (specific to the map group) increased coupling for the passive auditory and the silent motor task, after one session constrained to specific bands and after five weeks broadband coupling of all ROIs.

---

**FIGURE 1.** Differences in cortical task-related DC-potential patterns of the map group compared to the normal prestudy activation (e.g., motor cortex when moving), which is displayed on the *left*. The three panels on the *right* show exclusively the additional activity due to training. Each pattern was acquired 2–3 seconds after stimulus/movement onset and was averaged across 1,000 ms and 60 trials in 9 subjects. Dark = negative potential (depolarization/activation) or increased activity; light = decreased activity or inhibition.

### *Behavioral Data*

Under ordinary practicing conditions (i.e., with auditory feedback), the no-map group performed much better after five weeks with respect to timing and dosage of fine finger forces. A possible reason for this effect may be the additional interference for the map group originating from processing the pitch (melody) feedback. Under conditions of missing feedback, the two groups performed similarly. Five weeks of practice yielded improvements of finger velocity, velocity variances, and timing.

In the blink reflex experiment, after successful conditioning, the subjects were asked to voluntarily press keys on a silent keyboard. As expected, nonmusicians showed no specific reaction. Professional pianists, however, exhibited a markedly intensified eyeblink when pressing the corresponding key.

## CONCLUSIONS

Piano playing provides a model for a motor system under auditory control with high demands on quick and precise coordination. After practice, cortical auditory and sensorimotor hand areas are jointly activated for purely auditory as well as for mute motor tasks. The topographical distributions induced by each of the involved modalities are similar. The degree of similarity correlates to the accumulated practice time. In addition, a right dorsolateral prefrontal area engages in this corepresentation in beginners and experts, but not in beginners who could not establish a mental map of the keyboard. This map area is connected to motor fields functionally, as indicated by increasing coherences during practice. Interestingly, the strong corepresentation is established during the first minutes of training,[3] is consolidated within weeks, and may provide the basis for further skill achievement. After years of practice, the described sensorimotor corepresentation is automatized to a high degree and can even be activated preattentively.[4]

## REFERENCES

1. Jasper, H.H. 1958. The ten-twenty electrode system of the international federation. Electroencephal. Clin. Neurophysiol. **10:** 371–375.
2. Andres, F.G. *et al.* 1999. Functional coupling of human cortical sensorimotor areas during bimanual skill acquisition. Brain **122:** 855–870.
3. Classen, J. *et al.* 1998. Rapid plasticity of human cortical movement representation induced by practice. J. Neurophysiol. **79:** 1117–1123.
4. Koelsch, S., E. Schroger & M. Tervaniemi. 1999. Superior pre-attentive auditory processing in musicians. Neuroreport **10:** 1309–1313.

# Electrical Brain Activity Associated with Automatic and Controlled Processing of Melodic Contour and Interval

L. J. TRAINOR,[a,b] K. L. McDONALD,[b] AND C. ALAIN [b,c]

[a]*McMaster University,* [b]*Rotman Research Insititute, Baycrest Centre for Geriatric Care, and* [c]*University of Toronto, Toronto, Canada*

**KEYWORDS: Melodic contour processing; Melodic interval processing; Listening, active; Listening, passive; Pitch processing, automatic**

## INTRODUCTION

Melodic information is thought to be encoded in the brain in two different forms, a contour code, consisting of the up/down pattern of pitch changes, and an interval code, consisting of the exact pitch distances.[1] In both cases, the important melodic information lies more in the relative relation between tones than in the absolute pitch of the tones. Because the organization of auditory cortex appears to be tonotopic, it is interesting to ask at what stage of processing relational pitch information is extracted. A related question concerns the extent to which relational melodic contour and interval are automatically processed in the absence of focused attention. These two questions were investigated using event-related brain potentials (ERPs).

## MATERIAL AND METHODS

Eleven musically untrained adults listened to sequences of five-note melodies. On 80% of the trials, a melody from a standard set was played; on the remaining 20% of the trials, a melody from a deviant set was played. In the contour condition, the standard melodies all had a rising pitch contour (intervals varied across melodies); on deviant melodies, the final interval descended. In the interval condition, standard melodies consisted of the first five notes of a major scale (starting note varied); deviant melodies were the same as standards except that the final interval (but not contour) was changed. Thus, detection of the deviant melodies in both contour and interval conditions relied on the abstraction of relational pitch information, because the absolute pitch information changed from trial to trial. Melody note onset-to-

Address for correspondence: Dr. L.J. Trainor, Department of Psychology, McMaster University, Hamilton, Ontario L8S 4K1, Canada. Voice: 905-525-9140 ext. 23007; fax: 905-529-6225. ljt@mcmaster.ca onsets were 300 ms; trials were separated by 2,000 milliseconds. Each individual listened to each of the contour and interval conditions twice, once while reading a book (passive listening) and once while indicating by button pressing whether a standard or oddball trial was heard (active listening).

The EEG was digitized continuously (250-Hz sampling rate per channel; band-pass 0.05–50 Hz) from 47 electrodes. ERP epochs were extracted off-line, re-referenced to an averaged reference, and included 200-ms prestimulus activity and 1,200-ms poststimulus activity, starting from the onset of the final melody note in each trial. Eye movements and blinks were corrected for each individual average, using an ocular source components approach.

## RESULTS AND DISCUSSION

During the passive listening condition, both interval and contour deviant trials elicited a significant mismatch negativity (MMN) wave (FIG. 1, *left*). MMN is thought to reflect the updating of traces in auditory sensory memory, as it is elicited for infrequent deviant sounds in a set of standard frequent sounds, even in the absence of attention.[2] Thus, relative contour and interval information are both extracted automatically and encoded in sensory memory. The MMN peak was earlier in the contour (150 ms) condition than in the interval (180) condition, but the amplitude was similar in both cases.

When individuals were required to listen actively to the sequence, the deviant trials generated an additional frontal positive wave that was maximum over the fronto-central region (FIG. 1, *right*). This frontal positivity peaked earlier for the contour (317 ms) than for the interval (379 ms) condition, but was significantly greater for the interval than for the contour condition. Frontal positivities, named P3as, were reported previously for salient nontarget stimuli in either attended or unattended channels[3,4] and were linked to the inadvertent capture and/or orienting of attention. Our frontal positivity for contour and interval deviants was found for target stimuli only in the attended channel, so it is not a typical P3a. It is possible that the frontal positivity reflects the ongoing processing of a patterned sequence of sounds. Whether this frontal positivity is specfic to musical processing remains a question for future research.

In the active condition, the frontal positivity was followed by a P3b wave maximum over the parietal regions. The P3b is associated with analysis of information in conscious working memory.[5] The P3b did not differ significantly between the contour and interval conditions in either latency (580 ms) or amplitude at Pz (FIG. 1, *right*).

**FIGURE 1.** *Left*: Difference waves (oddball–standard) for the contour and interval passive conditions showing mismatch negativity (MMN). *Right*: Difference waves (oddball–standard) for the contour and interval active conditions showing frontal positivity (FP) and P3b components.

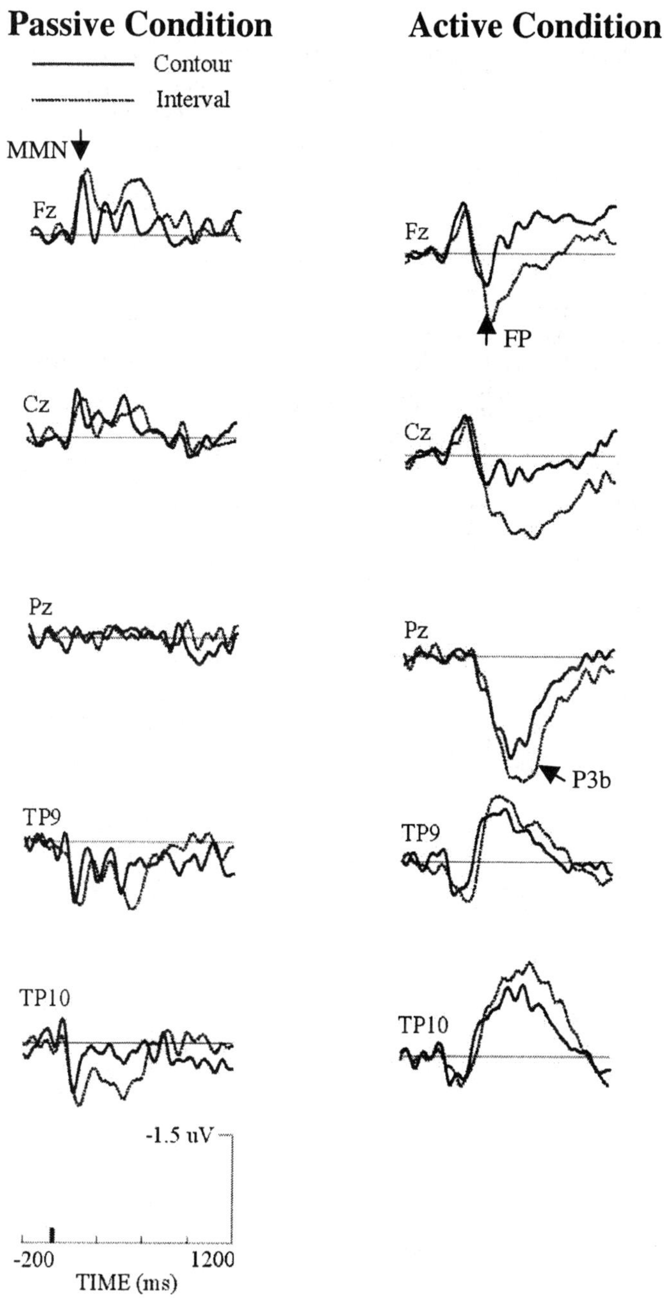


**FIGURE 1.** *See previous page for caption.*

## CONCLUSIONS

These findings indicate that both relational contour and relational interval information are extracted automatically in relatively early stages of processing in auditory cortex. They further indicate that processing melodic contour and interval information involves a widely distributed neural network that can be dissociated into automatic and controlled processes.

## REFERENCES


1. DOWLING, W.J. 1978. Scale and contour: two components of a theory of memory for melodies. Psychol. Rev. **85:** 341–354.
2. NÄÄTÄNEN, R. 1992. Attention and Brain Function. Erlbaum. Hillsdale, NJ.
3. ESCERA, C., K. ALHO, L. WINKLER & R. NÄÄTÄNEN. 1998. Neural mechanisms of involuntary attention to acoustic novelty and change. J. Cognit. Neurosci. **10:** 590–604.
4. KATAYAMA, J. & R. POLICH. 1998. Stimulus context determines P3a and P3b. Psychophysiology **35:** 23–33.
5. PICTON, T.W. 1992. The P33 wave of the human event-related potential. J. Clin. Neurophysiol. **9:** 456–479.

# Transcranial Magnetic Stimulation Produces Speech Arrest but Not Song Arrest

LAUREN STEWART,[a] VINCENT WALSH,[b] UTA FRITH,[a] AND JOHN ROTHWELL[c]

[a] *Institute of Cognitive Neuroscience, 17 Queen Square, London WC1N 3AR, United Kingdom*

[b] *Department of Experimental Psychology, South Parks Road, Oxford OX1 3UD, United Kingdom*

[c] *MRC Human Movement and Balance Unit, Institute of Neurology and National Hospital for Neurology and Neurosurgery, Queen Square, London WC1N 3BG, United Kingdom*

KEYWORDS: **Transcranial magnetic stimulation; Speech arrest**

Transcranial magnetic stimulation (TMS) is a tool that can be used to disrupt cortical processing for a few tens of milliseconds and, when combined with cognitive paradigms, can be used to look at the role of specific brain regions. TMS can be described as a way of creating virtual neuropsychological patients, but can also extend these findings. It can be delivered focally in time and therefore has the advantage of being able to provide information about the time course of cortical events. In addition, because "virtual lesions" are transient, the interpretation of behavioral effects are not complicated by the functional recovery that results when a damaged brain reorganizes.

One of the more famous neuropsychological patients is Paul Broca's patient from 1861, who came to be known as "Tan" because this was the only sound he could produce. Postmortem examination of his brain revealed a lesion to the left inferior frontal region, which subsequently became known as Broca's area. Using direct electrical stimulation in subjects undergoing surgery for epilepsy, Penfield and Rasmussen[1] demonstrated that speech arrest could be elicited when stimulation was applied over the inferior frontal zone of the dominant hemisphere. This can be distinguished from the motor type of speech arrest elicited by stimulation over rolandic cortex of either hemisphere, which is associated with contralateral activation of the tongue and lip muscles. TMS should, in principle, be able to produce the same results as those seen with direct electrical stimulation; after all, TMS is merely an alternative way of inducing electrical current in the brain; it has the advantage of being painless and can be performed on normal subjects.

Address for correspondence: Lauren Stewart, Institute of Cognitive Neuroscience, 17 Queen Square, London WC1N 3AR, UK. Voice: +44-171-391-1168; fax: +44-20-7813-2835.
l.stewart@ucl.ac.uk

Recently, we attempted to reproduce Penfield's results using TMS. Subjects were seated comfortably, and TMS was applied at 10 Hz for 1 s at 140% motor threshold while the subject counted briskly upward, from 1 to 10. TMS was applied over an area of left frontal cortex at an arbitrary point in their counting. In 6 out of 9 subjects, speech was disturbed. Occasionally, speech was stopped entirely; in other subjects it was slowed or the quality of speech was altered. EMG measurements from the contralateral orbicularis oris muscle confirmed that the speech arrest was nonmotor in origin. Such effects could not be produced by stimulation over the homologous area of the right hemisphere, in keeping with the findings from neuropsychological patients.

The relationship between speech deficits and melodic production deficits is not clear-cut. Some neuropsychological evidence suggests that the two are independent; for instance, Tan was reported to be able to sing "La Marseillaise."[2] Similar findings have been reported in aphasic patients with left frontal lesions.[3–5] However, other reports that patients with left frontal lesions may present with impaired speech and singing[6,7] support the idea that these two functions arise from the same brain area. Gordon and Bogen[8] provide evidence that melodic production is a right hemisphere function; injection of sodium amylobarbitone into the right carotid artery produced deficient singing. In addition, Alexander *et al.*[9] observed impaired singing following damage to the internal capsule of the nondominant hemisphere.

In the second part of our study we attempted to produce "song arrest" using TMS. First, TMS was applied over the left frontal region until speech arrest was elicited. The subject was then asked to sing or hum a familiar tune such as "Happy Birthday," twice with TMS and twice without, over the same area at which speech arrest had been produced. The song produced was compared with and without TMS for quality of melody and clarity of words. The coil was systematically moved across many different locations of the left frontal lobe. The procedure was repeated over the right frontal lobe. None of the five subjects tested showed any flattening or loss of melody with stimulation applied anywhere over the left or right frontal lobe. Furthermore, despite all subjects' showing disturbed speech when stimulation was given over a specific region of the left frontal lobe speech, when they sang with words, the clarity of the words was completely unaffected.

The finding that stimulation over a region of the left frontal lobe affects speech but not melody supports the idea that these two functions are subserved by different brain areas. The fact that production of words while singing was unaffected even when speech alone was disturbed is counterintuitive and suggests that there may be something special about word production while singing—perhaps the melodic context facilitates word production. Yamadori *et al.*[3] found that 12 out of 21 of Broca's aphasic patients produced good text words while singing. Stimulation over the right frontal lobe did not impair melodic production. Perhaps it is the case that while the speech generation site can be well localized, the circuitry underlying melodic production is more diffuse. Indeed, even though the neuropsychological literature points to the right hemisphere's involvement, there is little consensus as to where in the right hemisphere it may reside. Another possibility is that melodic production may just be more robust than speech production and hence less susceptible to interference.

## REFERENCES

1. PENFIELD, W. & T. RASMUSSEN. 1949. Vocalizaton and arrest of speech. Arch. Neurol. Psychiatry **61:** 21–27.
2. BEHIR, M. 1836. Cited by Edgren. 1895. (See below.)
3. YAMADORI, A., S. OSUMI & M. OKUBO. 1977. Preservation of singing in Broca's aphasia. J. Neurol. Neurosurg. Psychiatry **40:** 221–224.
4. MORGAN, O. & R. TILLUCKDHARRY. 1982. Preservation of singing function in severe aphasia. West Indian Med. J. **31**(3)**:** 159–161.
5. CADALBERT, A., T. LANDIS, M. REGARD & R. GRAVES. 1994. Singing with and without words: hemispheric asymmetries in motor control. J. Clin. Exp. Neuropsychol. **16**(5)**:** 664–670.
6. EDGREN, J. 1895. Amusie (musikalische Aphasie). Deutsche Z. Nervenheilkunde. **6:** 1–64.
7. HENSCHEN, S. 1925. Clinical and anatomical contributions on brain pathology. Arch. Neurol. Psychiatry **13:** 226–249.
8. GORDON, H. & J. BOGEN. 1974. Hemispheric lateralization of singing after intracarotid sodium amylobarbitone. J. Neurol. Neurosurg. Psychiatry **37**(6)**:** 727–738.
9. ALEXANDER, M.P., D.F. BENSON & D.T. STUSS. 1989. The frontal lobes and language. Brain Lang. **37**(4)**:** 656–691.

# Development of the Happy–Sad Distinction in Music Appreciation

## Does Tempo Emerge Earlier Than Mode?

S. DALLA BELLA,[a] I. PERETZ,[a] L. ROUSSEAU,[b] N. GOSSELIN,[a]
J. AYOTTE,[a] AND A. LAVOIE[a]

[a]*Department of Psychology, University of Montreal, Montreal, Quebec, Canada*

[b]*Department of Psychology, Laurentian University, Sudbury, Ontario, Canada*

**Keywords: Tempo; Mode; Affect in music**

It is widely held that happiness and sadness in music are associated with two structural properties: tempo (i.e., the number of beats per minute) and mode (i.e., the specific arrangement of intervals among the subset of pitches used to write a given musical excerpt). More specifically, fast tempi tend to evoke a happy tone, whereas slow tempi tend to evoke a sad tone.[1–3] Similarly, the major mode is associated with happiness, and the minor mode is associated with sadness.[4]

There is compelling evidence that the manipulation of tempo and mode affect happy–sad judgments in adults.[5] Yet, it is less clear at what stage of development listeners become sensitive to these manipulations. The capacity to detect changes in tempo seems to be present very early in life, before one year of age;[6,7] nevertheless, sensitivity to tempo has never been considered separately from mode in developmental studies of emotions. By contrast, the ability to emotionally respond to mode has been studied independently from tempo and has been found to emerge later, around the age of seven to eight years.[8,9] However, there are theoretical and empirical grounds for thinking that mode, as well as tempo, may be present very early in development. The major mode is viewed as more consonant than the minor mode.[10] Since sensitivity to consonance emerges within the first year of life,[11] it is likely that the capacity to analyze mode will appear earlier than seven years of age. One study supporting this prediction shows that three-year-old children recognize the emotions associated with modes.[12]

In order to examine the respective roles of tempo and mode in development, the material and procedure employed effectively with adults in a previous study by Peretz and collaborators was adopted here as well. Thirty-two excerpts were taken from the classical repertoire. They were selected so that half ($n$ = 16) evoked a sense of happiness and the other half ($n$ = 16) a sense of sadness. Happy selections were

Address for correspondence: Isabelle Peretz, Department of Psychology, University of Montreal, C.P. 6128, succ. Centre-ville, Montreal, Quebec H3C 3J7, Canada. Voice: 514-343-5840; fax: 514-343-5787.

isabelle.peretz@umontreal.ca

**TABLE 1. Mean percentages of correct responses and standard errors provided by adults, and six- to eight-year-old and five-year-old children, in the different conditions defined by the type of modification applied to the musical excerpts**

| | Group | | | | | |
|---|---|---|---|---|---|---|
| | Adults | | Six to Eight Year Olds | | Five Year Olds | |
| Condition | *M* | (SE) | *M* | (SE) | *M* | (SE) |
| Original | 86.8 | (3.9) | 88.2 | (4.0) | 84.2 | (4.9) |
| Tempo change | 81.8 | (3.2) | 74.7 | (3.3) | 69.2 | (4.0) |
| Mode change | 70.7 | (3.2) | 75.6 | (3.4) | 79.4 | (4.1) |
| Tempo + mode change | 56.4 | (2.9) | 58.5 | (3.0) | 59.6 | (3.6) |

played at a fast tempo and were written in major mode; sad selections were played at a slow tempo and were written in minor mode. Tempo and mode were manipulated in three different conditions. In the "tempo change" condition, all tempi were set to the median value of the original tempi. In the "mode change" condition, the excerpts were changed to the opposite mode (from major to minor or vice versa). Finally, in the "tempo + mode change" condition, both tempo and mode manipulations were employed. A further condition, referred to as "original," was employed in which excerpts did not undergo any modification.

Adults ($n$ = 24; $M$ = 20.4 years of age) and three groups of children (six to eight year olds, $n$ = 22; five year olds, $n$ = 15; three to four year olds, $n$ = 30) were presented with the full set of 32 excerpts in each of the four conditions.[c] Adults were required to judge on a 10-point scale whether the excerpts were happy or sad (1 = sad; 10 = happy). Children were asked to judge whether the excerpts were happy or sad by pointing to one of two face drawings designed to express happiness and sadness.

The mean percentages of correct responses obtained in each condition and provided by adults, six- to eight-year-old, and five-year-old children are presented in TABLE 1. A response was considered correct when it was the same as the intended emotion. For the sake of comparison, adults' responses were dichotomized, considering scores above and below five as happy and sad responses, respectively. Three- to four-year-old children were performing at chance in all conditions. Therefore, they were not considered in the following analyses. The analysis of variance showed that tempo and mode had a different impact on the emotional judgment for the three age groups, as attested by a significant group × condition interaction [$F(6, 174) = 3.08$, $p < 0.05$]. Post hoc comparisons (by Tukey HSD, with $p < 0.05$) revealed an effect of both tempo and mode manipulations in adults and in six- to eight-year-old children. By contrast, five-year-olds' performance was affected exclusively by tempo manipulation.

The present findings clearly show that, from five years of age, children are able to discriminate between happy and sad excerpts. To do so, five-year-old children use

[c]Three- to four-year-old children were unable to focus their attention on the excerpts for the entire duration of the experiment. Therefore, for this group only 16 excerpts per condition (8 happy and 8 sad) were used instead of 32. The full set of 32 excerpts, however, was used across participants.

information about tempo exclusively, whereas six to eight year olds use both tempo and mode as adults do. Altogether, these results substantiate the hypothesis that sensitivity to tempo (i.e., emerging around 5 years of age) appears earlier than sensitivity to mode (i.e., emerging around 6–8 years of age). This is consistent with the idea that sensitivity to tempo requires more fundamental processes than sensitivity to mode. The failure to document sensitivity to the happy–sad distinction in three- to four-year-old children is probably related to the inadequacy of the musical material (i.e., classical music up to twentieth-century composers) for this age group.

## REFERENCES

1. HEVNER, K. 1937. The affective value of pitch and tempo in music. Am. J. Psychol. **49:** 621–630.
2. GABRIELSSON, A. & P. JUSLIN. 1996. Emotional expression in music performance: between the performer's intention and the listener's experience. Psychol. Music **24:** 68–91.
3. BALKWILL, L-L. & W.F. THOMPSON. 1999. A cross-cultural investigation of the perception of emotion in music: psychophysical and cultural cues. Music Percept. **17**(1): 43–64.
4. HEVNER, K. 1935. The affective character of the major and minor modes in music. Am. J. Psychol. **47:** 103–118.
5. PERETZ, I., L. GAGNON & B. BOUCHARD. 1998. Music and emotions: perceptual determinants, immediacy, and isolation after brain damage. Cognition **68**: 111–141.
6. BARUCH, C. & C. DRAKE. 1997. Tempo discrimination in infants. Infant Behav. Dev. **20**(4): 573–577.
7. PICKENS, J. & L.E. BAHRICK. 1995. Infants' discrimination of bimodal events on the basis of rhythm and tempo. Br. J. Dev. Psychol. **13:** 223–236.
8. GERARDI, G.M. & L. GERKEN. 1995. The development of affective response to modality and melodic contour. Music Percept. **12**(3): 279–290.
9. GREGORY, A., L. WORRAL & A. SARGE. 1996. The development of emotional responses to music in young children. Motiv. Emotion **20:** 341–349.
10. HELMHOLTZ, H. VON. 1954. On the sensations of tone. Dover. New York. (Original work published 1877.)
11. SCHELLENBERG, E.G. & S.E. TREHUB. 1996. Natural musical intervals. Psychol. Sci. **7**(5): 272–277.
12. KASTNER, M.P. & R.G. CROWDER. 1990. Perception of the major/minor distinction: IV. Emotional connotations in young children. Music Percept. **8:** 189–202.

# Frontal EEG Responses as a Function of Affective Musical Features

C. D. TSANG, L. J. TRAINOR, D. L. SANTESSO, S. L. TASKER, AND L. A. SCHMIDT

*Department of Psychology, McMaster University, Hamilton, Ontario L8S 4K1, Canada*

**KEYWORDS: Affect in music; Mode; Tempo; Brain**

Recent models of emotional processing have suggested that the pattern of frontal EEG activity changes with the emotion experienced.[1,2] Greater left frontal activity has been found to be associated with positive affect, while greater right frontal activity has been associated with negative affect. In a previous study, we showed that this general approach applied to emotions induced by music as well.[3] Specifically, we found greater relative left frontal activation to musical excerpts expressing joy and happiness and greater relative right frontal activation to those expressing fear and sadness.

For the most part, music is abstract in that it does not refer directly to specific events in the world. This implies that emotion in music must be conveyed through the structure and rendering of the music itself. Two of the major determinants of musical emotion are mode (major mode tends to sound more positive than minor mode) and tempo (increased speed tends to be associated with positive emotions). In the present study, we tested whether frontal asymmetric activation effects would be elicited by changes in mode or tempo. Six musical excerpts ["Adagio" by Albinoni; "Solvejg's Leid," from *Peer Gynt Suite No. 2* by Grieg; *Utrecht Te Deum* by Handel; Piano Concerto No. 23 (3d movement) by Mozart; "Finale," from the *Carnival of the Animals* by Saint-Saëns; and "Brindisi," from *La Traviata* by Verdi] were collected and modified such that we had a piano-timbre slow-minor, slow-major, fast-minor, and fast-major version of each (based on stimuli of Peretz, Gagnon, and Bouchard[4]). Adults listened to one of the six musical excerpts in its different variations of mode and tempo. Every 60 s, the mode of the excerpt or the tempo, or both, changed.

EEGs were recorded in 55 young right-handed undergraduate students while they listened to the 60-s excerpts from left and right frontal (F3, F4) and left and right parietal (P3, P4) sites, referenced to Cz, a mid-central site. We were interested in power in the alpha band at these sites because it is thought to vary negatively with neural activation.[5] EEGs were digitized at 512 Hz and bandpass filtered between 1 and 100 Hz. Segments of EEG with artifacts due to eye blinks, eye movements, and other motor movements were eliminated. All artifact-free EEG data were subjected

Address for correspondence: Christine D. Tsang, Department of Psychology, McMaster University, Hamilton, Ontario L8S 4K1, Canada. Voice: 905-525-9140 ext. 27114; fax: 905-529-6225.

tsangcd@mcmaster.ca

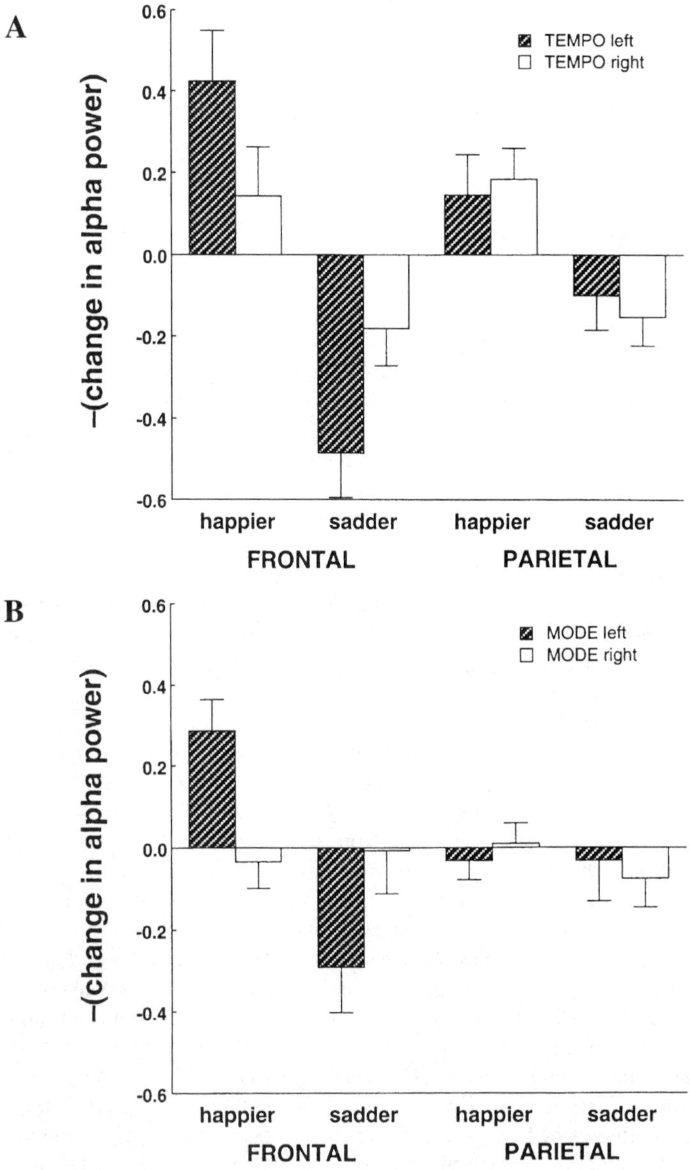


**FIGURE 1A and B.** Alpha power is plotted as a function of happier mode and sadder mode conditions at left frontal (F3), right frontal (F4), left parietal (P3), and right parietal (P4) sites. The negative of alpha power is shown because alpha power is negatively related to activation.

to a discrete Fourier transform (DFT) with a Hanning window of 1 s, and 50% overlap between windows and power in the alpha band (8 to 13 Hz) was determined.

Because we were interested in the effects of changes in tempo and changes in mode, for each pair of successive 60-s excerpts (across which either tempo or mode, or both changed), we subtracted the power in the alpha band for the first excerpt from that for the second excerpt at each site (F3, F4, P3, P4). For each subject at each site, we then derived four average alpha-power difference scores: one for increases in tempo (happier tempo), one for decreases in tempo (sadder tempo), one for changes from minor to major (happier mode), and one for changes from major to minor (sadder mode).

To examine the effects of tempo, we performed an analysis of variance (ANOVA) with condition (happier tempo, sadder tempo), hemisphere (left, right), and region (frontal, parietal) as within-subject factors (FIG. 1A). There were no main effects, but significant interactions were found: tempo × region, $F(1, 54) = 9.40$, $p < 0.003$; tempo × hemisphere, $F(1, 54) = 10.91$, $p < 0.002$; and tempo × region × hemisphere, $F(1, 54) = 32.75$, $p < 0.0001$. To examine these interactions, separate ANOVAs were performed for frontal and parietal regions, with condition and hemisphere as factors in each case. No effects were significant in the parietal analysis. However, the frontal analysis revealed a significant condition × hemisphere interaction, $F(1, 54) = 43.44$, $p < 0.0001$. Specifically, an increase in tempo (happier tempo) induced increased frontal activation in the left hemisphere, whereas a decrease in tempo (sadder tempo) induced decreased frontal activation in the left hemisphere (FIG. 1A). Because alpha power is thought to be negatively related to activation, these results are consistent with emotion models of frontal asymmetry in which it is postulated that positive emotions are associated with relatively more left than right frontal activation and negative emotions with relatively more right than left frontal activation.

To examine the effects of mode, we performed a similar analysis of variance (ANOVA) with condition (happier mode, sadder mode), hemisphere (left, right), and region (frontal, parietal) as within-subject factors (FIG. 1B). There was a significant main effect of mode, $F(1, 54) = 7.39$, $p < 0.009$, and a number of significant interactions: Mode × Region, $F(1, 54) = 9.29$, $p < 0.004$; Mode × Hemisphere; $F(1, 54) = 9.14$, $p < 0.004$; and mode × region × hemisphere, $F(1, 54) = 18.55$, $p < 0.0001$. To examine these interactions, separate ANOVAs were performed for frontal and parietal regions, with condition and hemisphere as factors in each case. The only significant effect for the parietal analysis was a main effect of mode, $F(1, 54) = 4.38$, $p < 0.04$. However, the frontal analysis revealed a significant effect of mode, $F(1, 54) = 9.03$, $p < 0.004$ and a significant condition × hemisphere interaction, $F(1, 54) = 36.69$, $p < 0.0001$. Specifically, a change from minor to major (happier mode) induced increased frontal activation, especially in the left hemisphere, whereas a change from major to minor (sadder mode) induced decreased frontal activation, especially in the left hemisphere (FIG. 1B). These results parallel those of the tempo analyses and are again consistent with emotion models of frontal asymmetry.

In summary, changes in both tempo and mode in the happier direction resulted in greater relative left frontal activation, whereas changes in both tempo and mode in the sadder direction resulted in greater relative right frontal activation. These results extend the findings of our previous study by showing that structural aspects of musical composition affect emotional processing in the brain.

## REFERENCES

1. DAVIDSON, R.J. 1998. Anterior electrophysiological asymmetries, emotion, and depression: conceptual and methodological conundrums. Psychophysiology **35:** 607–614.
2. FOX, N.A. 1994. Dynamic, cerebral processes underlying emotion regulation. *In* The Development of Emotion Regulation: Behavioral and Biological Considerations. N.A. Fox, Ed.: 152–166. Monographs of the Society for Research in Child Development, Vol. 59. University of Chicago Press. Chicago, IL.
3. SCHMIDT, L.A. & L.J. TRAINOR. Frontal brain electrical activity (EEG) distinguishes *valence* and *intensity* of musical emotions. Cognition Emotion. In press.
4. PERETZ, I., L. GAGNON & B. BOUCHARD. 1998. Music and emotion: perceptual determinants, immediacy, and isolation after brain damage. Cognition **68:** 111–114.
5. HELLER, W. & J. LEVY. 1981. Perception and expression of emotion in right-handers and left-handers. Neuropsychologia **19:** 263–272.

# Musical Processing in High Functioning Children with Autism

P. HEATON, L. PRING, AND B. HERMELIN

*School of Social Sciences, University of Greenwich, Eltham, London SE9 2BR, United Kingdom*

KEYWORDS: **Autism; Savant, musical; Absolute pitch**

The weak central coherence (WCC) theory of autism[1] proposes that autism is characterized by a cognitive style that biases processing towards local features at the expense of global, context-dependent meaning or gestalt.[2] However, some autistic individuals, known as savants, show exceptional abilities within specific domains, one of which is music. Musical savants all have absolute pitch ability and in addition demonstrate detailed knowledge about key relationships in their improvisations and memory reproductions.[3] They do not appear to demonstrate deficits in global processing, at least within their talent domain. In order to explore the link between autistic cognition and savant musical ability, a series of experiments with musically naive autistic children and intelligence- and age-matched normal controls were carried out.

The first investigation replicated previous findings[4] that showed enhanced pitch memory in musically naive autistic children. In that study, scores for pitch discrimination and memory correlated with scores on the Block Design test, a marker for WCC.[5] This then raised the possibility that one developmental outcome of such a cognitive style might be a predisposition toward absolute pitch ability.

In the study, subjects were exposed to musical tones that were presented in conjunction with animal pictures. After completion of these familiarization trials, the preexposed tones were presented in randomized order, and subjects were asked to point to the animals with which they had previously been paired. The findings from the experiment again showed that the children with autism were able to pair animals and tones to a significantly higher level than controls.

In a second experiment children were presented with 48 melodic pitch intervals and were asked to judge whether they moved up or down. The results showed that children with autism demonstrated equally high levels of performance with small (1–4 semitones), medium (5–8 semitones), and large (9–12 semitones) intervals, whereas controls could only achieve such good performance with medium and large intervals.

As previously suggested, WCC theory predicts that individuals with autism show reduced susceptibility to gestalt forces; this has been demonstrated most recently in

Address for correspondence: Dr. Pamela Heaton, Psychology Department, Goldsmiths College, University of London, New Cross, London SE14 6NW, United Kingdom.
p.heaton@gold.ac.uk

a task using visual illusions.[6] Musical chords are powerfully gestalt, and in two experiments the ability to disembed individual tones from within chords was tested. In the first phase, the children were again preexposed to four tone/animal pairs. In a second phase, three of these component tones were played together and the children were asked to point to the animal that was not included in the chord. The results showed that half of the children with autism were able to identify 75% of the missing animal/tone pairs. In a second chord experiment, the children were not preexposed to any individual tones and no paired associate stimuli were provided. Subjects heard major and minor tonic triads followed by an individual tone and were asked to judge whether the target tone belonged to the chord that preceded it. In this study, no significant difference between the performance of the autistic children and their controls was found, and the findings suggested that normal holistic processing was in evidence. Similar findings were obtained in a study where subjects were asked to make same/different judgments about pairs of musical contours that were (1) unchanged, (2) maintained, or (3) violated. It had been hypothesized that autistic subjects would detect changes in melody pairs even when they preserved their global characteristics. However, no group differences emerged, and the findings suggested normal patterns of holistic processing. Overall, the findings from these studies suggest that the information processing style characteristic of those with autism conveys some advantage and no disadvantage when processing musical stimuli. This might then help explain why music is one domain within which savant ability flourishes.

## REFERENCES

1. FRITH, U. 1989. Autism: Explaining the Enigma. Blackwell. Oxford.
2. HAPPE, F.G.E. 1999. Autism: cognitive deficit or cognitive style? Trends Cognit. Sci. **3:** 216–222.
3. MILLER, L. 1989. Musical Savants: Exceptional Kills in the Mentally Retarded. Erlbaum. Hillsdale, NJ.
4. HEATON, P., B. HERMELIN & L. PRING. 1998. Autism and pitch processing: a precursor for savant musical ability? Music Percept. **154:** 291–305.
5. SHAH, A. & U. FRITH. 1993. Why do autistic individuals show superior performance on the block design task? J. Child Psychol. Psychiatry **34:** 1351–1364.
6. HAPPE, F.G.E. 1996. Studying weak central coherence at low levels: children with autism do not succumb to visual illusions. A research note. J. Child Psychol. Psychiatry **37:** 873–877.

# Impact of Music Therapy on the Communication Skills of Toddlers with Pervasive Developmental Disorder

YEOU-CHENG M. MA, JOSEPH NAGLER, MATHEW H. M. LEE, AND IZUMI N. CABRERA

*Department of Pediatrics, Albert Einstein College of Medicine, Bronx, New York 10461, USA*

**KEYWORDS: Music therapy; Pervasive developmental disorders; Communication problems**

## INTRODUCTION

Children with pervasive developmental disorder (PDD) often exhibit pronounced delays in the development of communication skills.[1] Our intent was to explore the efficacy of music therapy as a form of intervention in children with PDD. Over the last three decades, the study of infant/toddler behavior and development has advanced at a phenomenal rate. Yet despite this rapid expansion of our knowledge, there is a paucity of research on the effect of music on the developmentally delayed child. Infants as young as four months of age perceive music as a communicative style distinct from speech. Research by Rose *et al.* demonstrates a relationship of affect to attention and learning. Bergeson and Trehub,[2] in the last decade, studied the influence of such elements as pitch, tempo, intensity, and tone of voice on young children.

Literature on music therapy and autism suggests that music can play a role in enhancing communication skills in individuals with delayed or impaired social interaction and communication.[3–5,8] We sought to evaluate the effect of music therapy sessions in a group setting on the communication skills in developmentally delayed children.

## METHODOLOGY

Nineteen preschool-aged children (18–48 months) enrolled in the Early Intervention Program at the Mt. Sinai-NYU/Rusk Institute Preschool and Infant Development Program were invited to participate in this study. Of those 19 children, 6 met the following criteria: (1) primary diagnosis of PDD, (2) significant reported speech and communication delays, and (3) ability to participate in at least four of the six sessions of the study.

Address for correspondence: Yeou-Cheng M. Ma, MD., Albert Einstein College of Medicine of Yeshiva University, 1410 Pelham Parkway South, Bronx, NY 10461. Voice: 718-430-8600 ext. 8322; fax: 718-460-6465.

einzma@aol.com

Data were collected through use of the Rosetti speech and language scale, a music therapy rating scale designed for this study, and panel analysis of videotapes of each session. The six children included in our study were allowed to interact with each other and the music therapist while participating in interactive musical play activities and songs. Six music therapy playgroup sessions were performed over a three-week period and were incorporated into a preexisting school schedule.

At the time of the study, the average chronological age of the children was 35.3 months, with a range of 31–40 months. Rating scales of the Southern California Ordinal Scales of Development indicated developmental presentations averaging 17.9 months, with a range of 16.4–19 months. Children were also evaluated using the Rosetti, PLS3, Bayley, and Muller Scales.

Six children attended at least four of the six sessions of a 30-minute interactive music therapy group. Each session included a series of the following eight interactive songs: "Greeting Song," "I'm in the Mood for," "Fingers in the Air," "Beat the Drum," "Bells," "This Is the Way," "Five Little Ducks," and "Good-Bye Song." Each song had a specific purpose and intent. The "Greeting Song" was used to provide a general assessment of the gestalt of the group. The intent was to establish eye contact and to set the tone for the opening of each session. "I'm in the Mood for" was an opportunity to stimulate gross motor activity while identifying receptive language abilities. The intent was to gauge the interactive stimulus of activity when combined with the instruments. "Fingers in the Air" was used to stimulate gestural language. The intent was to determine how the children would communicate through gestural prompting. "Beat the Drum" allowed the opportunity to monitor the presence of number concepts. The intent was to foster greater impulse control and tactile interaction. "Bells" allowed each child the opportunity to display his or her fine motor abilities. The intent of this activity was to monitor each child's responses to and communication of directions, impulse control, and tactile interactions. "This Is the Way" was used to promote activities of daily living. The intent was to monitor each child's imitation abilities. "Five Little Ducks" was used to stimulate the use of American sign language. The intent was to provide modeling of mood states and number concepts. Finally, the "Good-Bye Song" was used to end each session. The intent of this song was to serve as a transitional piece for the children.

Each song was performed using at least one of the following instruments: drums of different sizes and vibrations; acoustic guitar, using basic chords and strumming; autoharp strum with stick or fingers; and bells tuned to a C major pentatonic scale.

Evaluations were made using a rating scale derived from the Rosetti scale system. The following four categories containing a total of 24 points were used to evaluate the subjects' progress over the length of the intervention: motoric/gestural language, abilities interaction, expressive language, and receptive language. The data assessment was performed as a panel through videotape, creating uniformity in the recorded observations of all research assistant observers.

## RESULTS

Our findings, while preliminary, indicate increased communication skills through child-centered therapeutic musical activities.[6] The six children included in our study were allowed the opportunity to interact with each other and the music therapist, while participating in interactive musical play activities and songs. Our findings indicate that children exposed to early intervention through music therapy display enhanced communication skills. Most remarkably, these increased communication skills were found to allow the opportunity for individuation from care providers. These responses include both musical and verbal responses to different musical play songs and activities.

The musical activities that provided both verbal prompting with tactile musical stimulation facilitated the greatest level of response from all subjects. The song activities that relied solely on verbal cueing garnered the poorest responses. Music therapy sessions in a group setting provided an opportunity for individual interaction between child and care provider.

Interactive music activities, with the music selection tailored to the unique abilities of the children (with respect to harmony, melody, and timbre of instruments), produced better results than did singing alone.

## CONCLUSIONS

In examining musical activities with different instruments, those involving vibration and tactile senses elicited the strongest response from children with pervasive developmental disorders and developmental delays. Mastery Rates of goals in the four rated categories ranged from 13–38%, suggesting the need for continued intervention. Other aspects such as song order, population selection (size, biological variables), length of study, and variables of therapy environment warrant further investigation.[7,9]

## REFERENCES

1. ROSE, S.A., L.R. FUTTERWEIT & J.J. JANKOWSKI. 1999. The relation of affect to attention and learning in infancy. Child Dev. **70:** 549–559.
2. BERGESON, T.R. & S.E. TREHUB. 1999. Mothers' singing to infants and preschool children. Infant Behav. Dev. **22:** 51–64.
3. BAKER, B.S. 1982. The use of music with autistic children. J. Psychosoc. Nurs. Ment. Health Serv. **20:** 31–34.
4. EVERS, S. 1992. Music therapy in the treatment of autistic children. Medico-sociological data from the Federal Republic of Germany. Acta Paedopsychiatry **55:** 157–158.
5. CHARMAN, T. 1999. Autism and the pervasive developmental disorders. Curr. Opin. Neurol. **12:** 155–159.
6. LEWKOWICZ, D.J. 1998. Infants' response to the audible and visible properties of the human face: discrimination of differences between singing and adult-directed speech. Dev. Psychobiol. **32:** 261–274.
7. MOGFORD-BEVAN, K. 2000. Developmental language impairments with complex origins: learning from twins and multiple birth children. Folia Phoniatr. Logop. **52:** 74–82.
8. SWETTENHAM, J., S. BARON-COHEN, T. CHARMAN, *et al.* 1998. The frequency and distribution of spontaneous attention shifts between social and nonsocial stimuli in autistic, typically developing, and nonautistic developmentally delayed infants. Child Psychol. Psychiatry **39:** 747–753.
9. TOOLAN, P.G. & S.Y. COLEMAN. 1994. Music therapy, a description of process: engagement and avoidance in five people with learning disabilities. J. Intellect. Disabil. Res. **38:** 433–444.

# Music and Testosterone

## A New Hypothesis for the Origin and Function of Music

HAJIME FUKUI

*Nara University of Education, Nara, Takabatake, 630-8528, Japan*

**KEYWORDS: Music; Testosterone**

### INTRODUCTION

The origin and function of music have been discussed extensively,[1] but they remain a mystery. There have been numerous discussions about the close relationship between spatial ability and testosterone,[2] spatial ability and musical ability,[3] and testosterone and musical ability.[4] The reason that a sex hormone such as testosterone is closely related to music is still unknown despite the many reports that music affects the human endocrine system.[5] I reported findings that music suppressed testosterone in the male.[6] The current study investigates whether listening to music affects the testosterone level in females and whether any sex differences exist in the effect of listening to music on testosterone level. Finally, I propose a new hypothesis for the evolutionary function and origin of music.

### METHODS

Seventy college students, all in good health and not taking any medication, were recruited and asked to sign a consent form. There were 35 males and 35 females, ranging in age from 19–25 years, the average age being 21. Forty were music majors and 30 were nonmusic majors. Only women claiming to have regular menstrual cycles were selected. Samples taken in the luteal and follicular phases were used for analysis.

In the experiment, subjects had six choices. They were given 30 minutes to listen to (1) their favorite music ($n$ = 10: five men and five women each), (2) Gregorian chant ($n$ = 10), (3) Mozart ($n$ = 10), (4) jazz ($n$ = 10), (5) popular music ($n$ = 10), or (6) no music at all ($n$ = 20). Males and females were separated, and a same-sex research assistant conducted each group. The experiment was carried out in a quiet laboratory room at a specific time (between 1400 and 1700 hours) because of the diurnal variation in testosterone secretion. Stimuli were presented by audio speakers. Subjects rested for 10 minutes and were given 30 minutes of music stimuli or si-

Address for correspondence: Hajime Fukui, Nara University of Education, Nara, Takabatake, 630-8528, Japan. Voice: +81-742-48-1302; fax: +81-742-48-1261.
fukuih@nara-edu.ac.jp

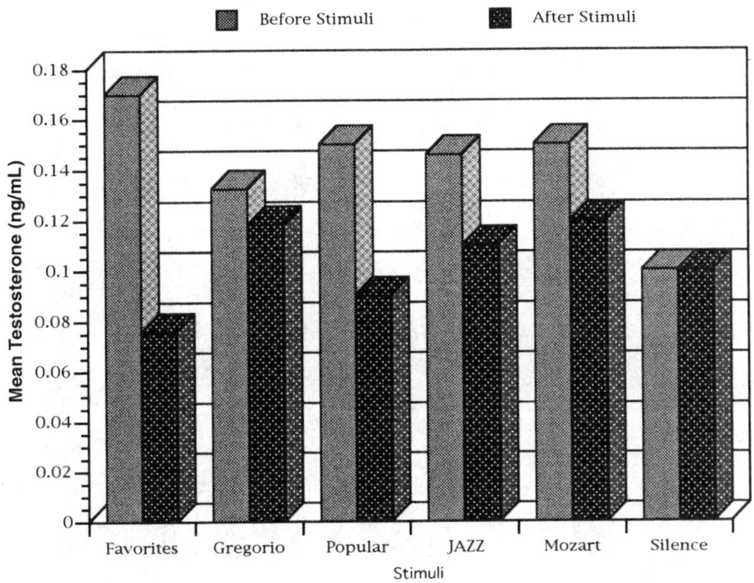


**FIGURE 1.** Testosterone levels of 35 male subjects sampled at 30-minute intervals. ANOVA revealed that the main effect of the stimuli was significant ($p < 0.0001$).

lence. They submitted 7-mL saliva samples before and after the stimuli. Because of individual differences in testosterone pulse, their order was randomized. Only free testosterone levels were assessed by radioimmunoassay, measured as described elsewhere.[7] Salivary testosterone concentrations are highly correlated with serum concentrations[8] and represent the free and biologically active steroid fraction. Taking a saliva sample is less invasive and less stressful than is drawing blood.

Music stimuli used for the experiments were (1) the subject's favorite music (chosen from Japanese pop music composed in the American-European pop style, American pop music, and jazz), (2) music chosen for the experiment, for example, Gregorian chant (Canto Gregoriano, EMI, TOCE-8374), Mozart (Sonata in D Major, K. 448, for Two Pianos, CBS/SONY, 32DC627), jazz (pieces from "100 Gold Fingers," ALLART, KICJ137), and (3) popular music (pieces from "The Swinging Star," EPIC/SONY, ESCB1350).

## RESULTS

FIGURES 1 and 2 illustrate the saliva testosterone pulse profiles. Testosterone decreased under music conditions compared with silence, but not significantly. However, significant sex differences were noted in the effects of music. Testosterone *decreased* in males under all conditions ($p < 0.0001$), whereas testosterone *increased* in females under all conditions ($p = 0.0007$). ANOVA revealed that in both sexes the main effects of stimuli were significant ($F = 23.224$, $p < 0.0001$ and $F = 8.429$,

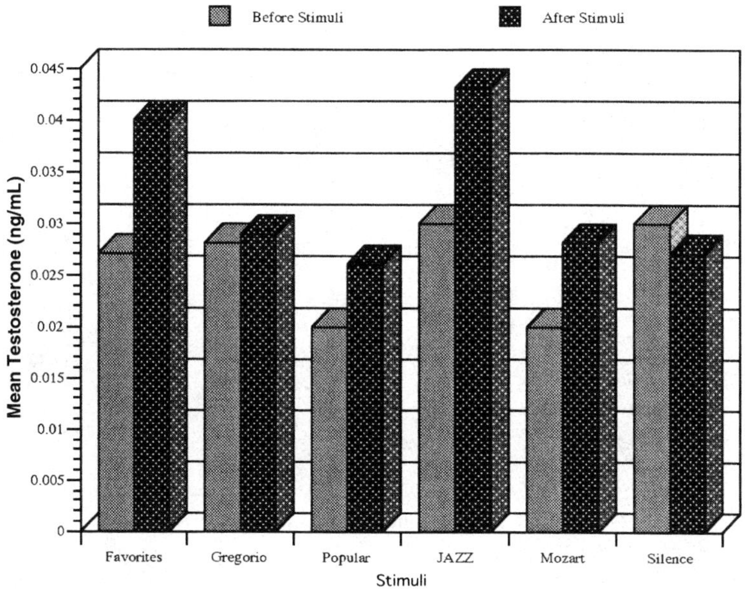


**FIGURE 2.** Testosterone levels of 35 female subjects sampled at 30-minute invervals. ANOVA shows that the main effect for stimuli was significant ($p < 0.0001$).

$p < 0.0001$, respectively). Post-hoc testing revealed that favorite-music conditions significantly differed from other stimuli in males and from Gregorian chant and silence in females. There were no significant effects from testosterone due to subject major and menstrual cycle, and neither was there any noticeable difference between music liked and that disliked by the subjects and between music offered and music of the subject's own choice.

## DISCUSSION

This study revealed that music affects the testosterone level in both sexes. It is known that the interaction of testosterone with behavior is bidirectional: testosterone can influence behavior and behavior can alter testosterone levels. Yet, the evolutionary function of testosterone changes is unknown. Testosterone has been related to libido, activity level, sensation seeking, and dominance in animals and human beings,[9] and it also plays a part in aggressiveness.[10] Taken together with this evidence, I propose that music has a biological and evolutionary function. However, it not only developed as a signal of courtship, as Darwin remarked,[11] but it also evolved to control human love in a complicated way.

During evolution, the early humans adopted such new reproductive strategies as monogamy, mother–father families, and group living. However, group living intensified all kinds of competition for the resources of reproduction, including food and mates. Human beings needed a system to control these states, which stemmed from sexual, aggressive, and dominant behavior due to excessive testosterone. I suggest that music originated from that need, with its function to control sexual and aggressive behavior. Because testosterone responses to libido differed according to sex, men with lower testosterone levels had low libido,[12] whereas women with higher testosterone levels had higher libidinal capacity, yet they had lower rates of intercourse.[13] Therefore, decreased testosterone levels in males would avoid confrontation, and increased testosterone levels in females suppress sexual behavior.

This theory is convincing if we recall that in music throughout the world, and even in animal sounds (monogamous birds), music and certain sounds function to ease tension or strain, strengthen social bonds, bring pleasure or ecstasy, and bring cohesiveness among people. Music is used empirically and universally in various social situations where stress, tension, and anxiety exist, such as ceremonies, funerals, war, and even banquets.

## REFERENCES


1. WALLIN, N.L. *et al.*, Eds. 2000. The Origins of Music. MIT Press. Cambridge.
2. GOUCHIE, C. & D. KIMURA. 1991. The relationship between testosterone levels and cognitive ability patterns. Psychoneuroendocrinology **16:** 323–334.
3. HASSLER, M., N. BIRBAUMER & A. FEIL. 1985. Musical talent and visual-spatial abilities. Psychol. Music **13:** 99–113.
4. HASSLER, M. 1991. Testosterone and musical talent. Exp. Clin. Endocrinol. **98:** 89–98.
5. MOCKEL, M., L. ROCKER, T. STORK, *et al.* 1994. Immediate physiological responses of healthy volunteers to different types of music: cardiocascular, hormonal and mental changes. Eur. J. Appl. Physiol. **68:** 451–459.
6. FUKUI, H. 1998. The effect of music listening on testosterone secretion. J. Music Percept. Cognit. **4:** 10–18.
7. FERGUSON, D.B. 1984. Steroid Hormones in Saliva. Karger. Basel.
8. KHAN-DAWOOD, F.S., J.K. CHOE & M.Y. DAWOOD. 1984. Salivary and plasma bound and free testosterone in men and women. Am. J. Obstet. Gynecol. **148:** 441–445.
9. WALLEN, K. & J. LOVEJOY. 1993. Sexual behavior. *In* Hormonally Induced Changes in Mind and Brain. J. Schulkin, Eds.: 71–97. Academic Press. San Diego.
10. CARTER, S.C. 1992. Oxytocin and sexual behavior. Neurosci. Behav. Rev. **16:** 131–144.
11. DARWIN, C. 1874. The Descent of Man. John Murray. London.
12. ALEXANDER, G.M. *et al.* 1997. Androgen-behavior correlations in hypogonadal men and eugonadal men. Horm. Behav. **31:** 110–119.
13. KEMPER, T. 1990. Social Structure and Testosterone. Rutgers University Press. New Brunswick, NJ.

# A Dynamic Systems Model of Musically Induced Emotions

## Physiological and Self-Report Evidence

JULIAN F. THAYER AND MEREDITH L. FAITH

*National Institute on Aging, National Institutes of Health, Baltimore, Maryland 21224, USA*

KEYWORDS: **Musically induced emotions; Emotions, musically induced**

## INTRODUCTION

Music has long been considered the language of emotions. However, few studies have examined musically induced emotions in the context of a comprehensive model of emotion. We recently proposed such a comprehensive model of emotion based on a dynamic systems theory and here present an overview of data on musically induced emotions.[1] Much literature supports the idea that emotions can be described by their location in a state space defined by the dimensions of valence (pleasantness/unpleasantness spectrum) and arousal (degree of activation). These dimensions can be conceptualized as motivational control parameters of affective experience, and the discrete emotions as attractors in state space.

## MATERIAL AND METHODS

In experiment 1,[2] we show that these dimensions of valence and arousal can be recovered in musically induced emotions. Thirty-three subjects without musical training listened to 17 musical excerpts.[3,4] They scaled their subjective responses (e.g., happiness, sadness, excitement, disgust, anger) and rated each selection on eight cognitive–perceptual responses (e.g., fast/slow, high-pitched/low-pitched). Observations were pooled and factor analyzed. Two factors were retained by scree test and accounted for 72% of the variance. Factor 1 represented a bipolar factor: all positive emotion terms loaded positively; all negative emotion terms loaded negatively. Independent of valence, states of low activation loaded negatively on factor 2, while states of high activation loaded positively. Perceived tempo loaded predominately on factor 2 (activation), whereas perceived pitch level loaded more highly on factor 1 (pleasant/unpleasant).

Address for correspondence: Julian F. Thayer, National Institute on Aging, 5600 Nathan Shock Drive, National Institutes of Health, Baltimore, MD 21224. Voice: 410-558-8612; fax: 410-558-8108.

thayer@lpc.grc.nia.nih.gov

**TABLE 1. Partial listing of musical selections used in studies**

| | |
|---|---|
| Sad | |
| Tschaikovsky | Fourth Movement from the Sixth Symphony |
| Rodrigo | "Adagio" from the Concerto for Guitar and Orchestra |
| Beethoven | "Marche Funèbre" from the Third Symphony |
| Grieg | "The Death of Ase" from Peer Gynt Suite No. 1 |
| Chopin | Sonata for Piano No. 2 in B-flat Minor (*Funeral March*) |
| Mozart | "Lacrymosa" from the Requiem |
| Verdi | "Lacrymosa" from the Requiem |
| Serene | |
| Debussy | *Prelude to the Afternoon of a Faun* |
| Traditional | Gregorian Chant |
| Offenbach | "Barcarole" from *The Tales of Hoffmann* |
| Vivaldi | "Largo" from *Spring* from *The Four Seasons* |
| Bizet | "Intermezzo" from Carmen Suite No. 1 |
| Dvorak | "Largo" from the Ninth Symphony |
| Gerschwin | "Lullabye" |
| Traditional Bulgarian | "Tulip or Hyacinth" |
| Happy | |
| Traditional | "Rockinover River" |
| Rossini | *William Tell* Overture |
| Wagner | Prelude to Act III of *Lohengrin* |
| Beethoven | Fourth Movement from the Ninth Symphony |
| Dvorak | Slavonic Dances, Opus 46, no. 1 |
| Bizet | "Les Toreadors" from Carmen Suite No. 1 |
| Strauss, Johann | "Unter Donner und Blitz" |
| Traditional Romanian | "E Nikita" |
| Stravinsky | *Pastorale for Woodwinds* |
| Vivaldi | "Allegro" from *Spring* from *The Four Seasons* |
| Davis, Miles | *Ascenseur pour L'Echafaud* |
| Agitated | |
| Holst | "Mars" from *The Planets* |
| Prokofiev | "Death of Tybalt" from Romeo and Juliet Suite No. 1 |
| Bartok | *Concerto for Two Pianos, Percussion, and Orchestra,* "Assai Lento"/"Allegro Molto" |
| Stravinsky | *The Rite of Spring* |
| Shostakovich | "Adagio" from the Eighth Symphony |

In experiment 2,[2] using facial electromyography (EMG), we show that music can elicit the discrete emotions of happiness, sadness, agitation, and serenity. Facial EMG allows the detection of covert muscle actions and provides a precise, linear index of isometric muscle contraction. Voluntary and involuntary manipulation of fa-

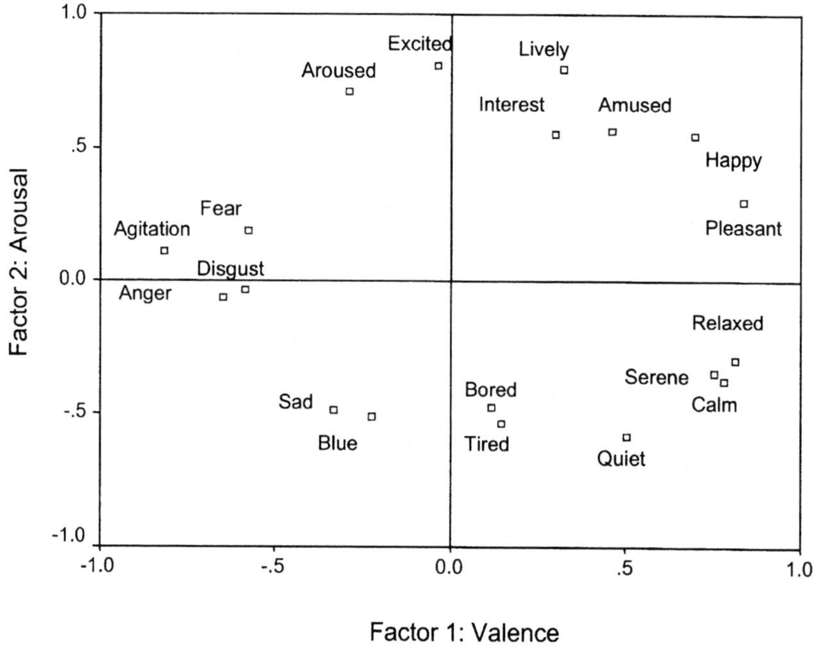


**FIGURE 1.** Representative factor structure of emotional experience.

cial musculature has been shown to produce emotive states. Twelve subjects listened to eight music selections; each of four emotion categories was represented by two selections. Electrodes were placed over the brows (corrugator, "frown" muscle) and on the cheeks (zygomatic, "smile" muscle). Subjects rated the intensity of each of the four emotions of joy, serenity, agitation, and sadness. Main effects among music selections for each rating scale were significant ($F$s > 10; $p$s < 0.001). MANOVA analysis found significant main effects for valence ($F$ (3, 82) = 3.01, $p < 0.05$). Univariate $F$ was significant for cheeks ($F$ (1, 84) = 6.07, $p < 0.05$). Ratings of joy correlated positively with left cheek activity ($r = 0.29, p < 0.05$); sadness ratings correlated positively with left brow activity ($r = 0.23, p < 0.05$).

In experiment 3,[4] we show that these same emotions can be recovered from cardiorespiratory responses. Twenty-six subjects listened to 12 music fragments and one neutral "white noise" condition, and then rated their responses. Electrocardiograms (ECGs) and respiratory signals were simultaneously obtained. Factor analysis revealed two factors, interpretable as valence and arousal, that explained 64% of the variance. The classification of cardiorespiratory variables into the four emotion categories of happy, serene, sad, and agitated, by discriminant analysis, resulted in 46.5% "hits" ($z = 8.1, p < 0.0001$, P1H1 = 86% above chance). The successful discrimination was due primarily to two autonomic components: a respiratory component, which was related to the arousal dimension of self-reported emotions, and a chronotropic component, which was related to the valence dimension.

In experiment 4,[2] we show that a standard musical composition can be altered to produce different discrete emotions. Specifically, by transposing the composition up or down an octave and doubling or halving the tempo, the emotions of happiness, sadness, agitation, and serenity can be elicited. One hundred twenty-five subjects listened to a computer-altered Schubert composition and rated their subjective state and perceived qualities of the music (e.g., rhythm, tempo, pitch, harmony). Subjects were divided according to five conditions, each assigned an emotional tone a priori, based on previous research: (1) music as written, *neutral*; (2) pitch raised one octave, tempo doubled, *happy*; (3) pitch raised one octave, tempo halved, *serene*; (4) pitch lowered one octave, tempo doubled, *agitated*; (5) pitch lowered one octave, tempo halved, *sad*. Response patterns were submitted to discriminant analysis and classification results compared to chance. Correct classifications were as follows: happy (65%), neutral (36%), serene (60%), agitated (75%), and sad (52%) ($z = 10.4$, $p < 0.0001$, P1H1 = 185% above chance).

In experiment 5,[2] we show that these altered musical compositions can produce changes in brain electrical activity consistent with the self-reported emotions. Thirteen subjects listened to seven musical excerpts, three of which were three versions of a piano piece altered in tempo and pitch. Subjects provided EEGs (sites F3, F4, P3, P4, C3, C4, T3, and T4) and completed a subjective response scale. Faster tempo was associated with less alpha activity, indicating greater cortical activation ("natural" music selections: $F\ (1,12) = 5.88$, $p < 0.03$, $r^2 = 0.33$; "altered" selections: $F\ (2,22) = 2.86$, $p < 0.07$, $r^2 = 0.20$). This relationship was more pronounced in the temporal region. The valence × hemisphere interaction was marginally significant for temporal alpha ($F\ (1,12) = 3.61$; $p < 0.08$, $r^2 = 0.23$); this finding is consistent with recent research positing hemispheric specialization for valence in that the left hemisphere is preferentially involved in positive emotions and the right in negative emotions.

In experiment 6,[5] we show that music can produce cardiovascular responses that represent discrete emotion attractors in a state space defined by valence and arousal. Thirty-two pieces of music were selected, designated a priori as eliciting four emotional states: happy, sad, agitated, and serene. Six subjects listened to 32 music selections and provided ratings. Electrocardiograms and respiratory signals were obtained. ECGs were decomposed into oscillatory components indicative of sympathetic and parasympathetic input to the heart. Results were analyzed within subjects, and pooled analyses were completed. Cardiac variables and reports of degree of pleasantness and arousal were analyzed by discriminant analysis. Pooled classification accuracy for emotions were sad (54%), happy (71%), agitated (74%), and serene (72%) ($z = 12.0$, $p < 0.0001$, P1H1 = 149% over chance). Individual and pooled factor analyses yielded the predicted dimensions of pleasure and arousal.

## RESULTS AND DISCUSSION

The foregoing studies clearly showed that music elicits a broad range of emotion. These emotional states can be measured reliably by assessing self-descriptive information as well as a range of physiological responses including cardiovascular, respiratory, electromyographic, and electroencephalographic activity. Consistent with a body of other research, the foregoing studies show that substantial variance in emotional responses can be captured by the two dimensions of valence and arousal. These dimensions are captured in self-report and physiological data reduction. Our model proposes that the specific, discrete emotional states are well represented as basins of attraction, or preferred configurations, within a state space defined by the parameters of valence and arousal.[1] The differentiation of lower-order discrete emotions takes place along a limited number of the higher-order dimensions of valence and arousal.

Emotions are self-regulatory responses that allow the efficient coordination of the organism for goal-directed behavior. Theoretically, valent tone and arousal may reflect primitive motivational subsystems. Valence represents the evaluative outcome necessary to initiate an approach or withdrawal response, and arousal reflects the resource investment in the directional tendency. The neurovisceral underpinnings of this model include the link between the auditory nerve and the amygdala and the interaction of the medial prefrontal cortex and brain stem inputs to the cardiovascular system via the vagus nerve.[6] This comprehensive model of emotion can accommodate the extant literature and produce testable hypotheses for future research.

## REFERENCES

1. FAITH, M. & J.F. THAYER. 2001. A dynamical systems interpretation of a dimensional model of emotion. Scand. J. Psychol. **42** (2): 121–133.
2. THAYER, J. 1986. Multiple indicators of affective response to music. Dissertation. Abstr. Int. **47:** 12.
3. CAPUSO, A., V.R. FISICHELLI, L. GILMAN, *et al.* 1952. Music and Your Emotions. A Practical Guide to Music Selections Associated with Desired Emotional Responses. Liverright Publishing Corp. New York, NY.
4. NYKLICEK, I., J.F. THAYER & L.J.P. VAN DOORNEN. 1997. Cardiorespiratory differentiation of musically-induced emotions. J. Psychophysiol. **11:** 304–321.
5. FAITH, M. & J.F. THAYER. 2000. Autonomic pattern classification of autonomic and self-report indices of emotion. Manuscript in preparation.
6. THAYER, J.F. & R.D. LANE. 2000. A model of neurovisceral integration in emotion regulation and dysregulation. J. Affective Dis. **61:** 201–216.

# Subject Index

# Index of Contributors